CISM COURSES AND LECTURES

Series Editors:

The Rectors of CISM
Sandor Kaliszky - Budapest
Mahir Sayir - Zurich
Wilhelm Schneider - Wien

The Secretary General of CISM
Giovanni Bianchi - Milan

Executive Editor
Carlo Tasso - Udine

The series presents lecture notes, monographs, edited works and proceedings in the field of Mechanics, Engineering, Computer Science and Applied Mathematics.
Purpose of the series is to make known in the international scientific and technical community results obtained in some of the activities organized by CISM, the International Centre for Mechanical Sciences.

INTERNATIONAL CENTRE FOR MECHANICAL SCIENCES

COURSES AND LECTURES - No. 386

LOCALIZATION AND FRACTURE PHENOMENA IN INELASTIC SOLIDS

EDITED BY

PIOTR PERZYNA
POLISH ACADEMY OF SCIENCES

SpringerWienNewYork

Le spese di stampa di questo volume sono in parte coperte da
contributi del Consiglio Nazionale delle Ricerche.

This volume contains 233 illustrations

This work is subject to copyright.
All rights are reserved,
whether the whole or part of the material is concerned
specifically those of translation, reprinting, re-use of illustrations,
broadcasting, reproduction by photocopying machine
or similar means, and storage in data banks.
© 1998 by CISM, Udine
Printed in Italy
SPIN 10679631

In order to make this volume available as economically and as
rapidly as possible the authors' typescripts have been
reproduced in their original forms. This method unfortunately
has its typographical limitations but it is hoped that they in no
way distract the reader.

ISBN 3-211-82918-0 Springer-Verlag Wien New York

PREFACE

The main purpose of the CISM ADVANCED SCHOOL ON LOCALIZATION AND FRACTURE PHENOMENA IN INELASTIC SOLIDS, Udine, September 29 - October 3, 1997 was the discussion of some important aspects of localization and fracture phenomena in inelastic solids (single crystals, polycrystalline solids and geological materials). Emphasis was laid on experimental and physical foundations as well as on mathematical constitutive modelling and numerical solutions of initial boundary value problems.

The present volume offers the lecture notes of all contributions to the above course. They reflect the state-of-the-art in this field of mechanics and present new results of recent research work in this challenging domain. The lecture notes have an original character and contain the research results which have been mainly developed by the authors.

In the discussion of experimental aspects strain localization phenomena in tensile and torsional tests are considered. The explanation of different cooperative mechanisms responsible for the initiation of localization phenomenon are presented.

In the physical aspects the discussion are concentrated on the mechanical characteristics of single crystal and polycrystalline materials. Particular attention is focused on the evolution of the dislocation substructure and its importance for the proper interpretation of instability phenomena. Micro-damage mechanisms are considered and the localized fracture phenomenon in inelastic solids during dynamic loading processes is discussed. This kind of fracture can occur as a result of an adiabatic shear band localization generally attributed to a plastic instability implied by micro-damage and thermal softening during dynamic plastic flow processes.

In the mathematical constitutive modelling aspects it is intended to formulate the theory of elastoviscoplasticity within a framework of the rate type covariance material structure with finite set of the internal state variables. This constitutive theory is used to describe monocrystal bodies, polycrystalline solids and broad class of geological materials. Analytical criteria for adiabatic shear band localization of plastic deformation are formulated by assuming that some of eigenvalue of the acoustic tensor for rate independent response is equal to zero. Various effects are discussed and examined. Cooperative phenomena are considered and synergetic effects are investigated.

Particular attention is focused on the viscoplastic regularization procedure for the solution of the dynamic initial boundary value problems by means of finite element method with localization of plastic deformation. Many numerical solutions of the initial-boundary value problems (evolution problems) are presented. In these solutions the ductile fracture is treated as a final stage of the entire plastic flow process. The criterion of fracture does depend on the evolution of the constitutive structure of inelastic solids.

It is a pleasure of the authors and the editor to express our gratitude to the Scientific Council of the CISM for granting and supporting the course as well as permitting this publications. We also thank all participants for the fruitful discussions.

We do hope the present volume will constitute a useful source of information on some present-day problems of localization and fracture phenomena.

Piotr Perzyna

CONTENTS

Page

Preface

Examples of Strain Localisation
by H.P. Stüwe 1

Structural and Mechanical Aspects of Homogeneous and
non-Homogeneous Deformation in Solids
by A. Korbel 21

Constitutive Modelling of Dissipative Solids for Localization and Fracture
by P. Perzyna 99

Computational Modelling of Localisation and Fracture
by L.J. Sluys 243

Numerical Solutions of Initial-Boundary-Value Problems
with Shear Strain Localization
by R.C. Batra 301

Numerical Solutions of Initial-Boundary-Value Problems
for Metals and Soils
by T. Lodygowski 391

EXAMPLES OF STRAIN LOCALISATION

H.P. Stüwe
Erich-Schmidt Institute for the Physics of Rigid Bodies, Leoben, Austria

ABSTRACT

On the atomic scale plastic strain is always localised in the form of discrete dislocations. The concept of homogeneous plastic strain has meaning only on a macroscopic or perhaps on a mesoscopic level. This chapter treats strain localisation on a macroscopic level for specimens deformed in tension and in torsion, on a mesoscopic level for ductile fracture and on an atomistic level for fracture in fatigue.

1 STRAIN LOCALISATION IN BULK DEFORMATION [1]

1.1 The tensile test

Fig. 1 shows schematically a specimen of cross section A and length ℓ, tested in tension.

It will yield plastically when the load P reaches a critical value $P_y = A\sigma_y$ where σ_y is the flow stress which, in turn, may be a function of plastic strain φ, strain rate $\dot\varphi$, temperature T, etc. Assume that a partial volume of length ℓ_1 experiences a local fluctuation of strain $\Delta\varphi$. With $A = A_0 \exp(-\varphi)$ this will lead to a local fluctuation in

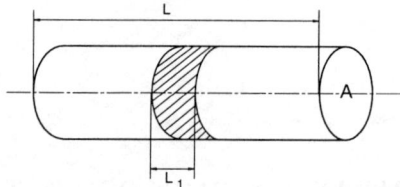

Figure 1: Specimen deformed in tension or torsion, schematic

specimen strength

$$\frac{\Delta P_y}{P_y} = \frac{\Delta \sigma_y}{\sigma_y} - \Delta \varphi \qquad (1)$$

If $\Delta P_y > 0$ then further strain will smooth out the fluctuation and the test is called "stable". If $\Delta P_y < 0$ then further strain will be localised in ℓ_1 and the test becomes "unstable".

1.1.1 First Example: $\sigma_y = \sigma_y(\varphi)$

If the flow stress is a function of strain alone then Eq. (1) transforms to

$$\frac{\Delta P_y}{P_y} = \frac{d\sigma_y}{d\varphi} \frac{\Delta \varphi}{\sigma_y} - \Delta \varphi \qquad (2)$$

With $\Theta = \frac{d\sigma}{d\varphi}$ the test is unstable for

$$\Theta < \sigma \qquad (3)$$

This is the "Considère criterion" [2] which is usually assumed to indicate the maximum load and the beginning of necking.

Strain is localised immediately when $\Theta < 0$ as, e.g., in the transition from the upper to the lower yield point like in many structural steels. This will lead to a finite localised strain (called "Lüders Strain") as indicated schematically in Fig. 2a).

Homogeneous deformation will be resumed after the "Lüders Band" has swept the entire specimen.

This instability can be avoided by an enforced homogeneous deformation e.g. by rolling ("skin pass"). Fig. 2b) shows that deformation much smaller than the Lüders strain is sufficient which should be obvious from Fig. 2a).

1.1.2 Second example: $\sigma_y = \sigma_y(\tau_y(\varphi), M_s(\varphi))$

The treatment of the preceding paragraph is classical, but oversimplified. This becomes evident when the specimen is a single crystal oriented for single slip. Here σ_y is not a

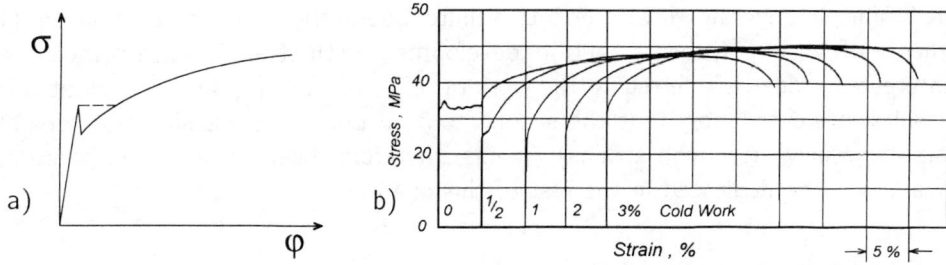

Figure 2: a) Stress-strain curve with upper and lower yield point, Lüders strain (schematic); b) Elimination of Lüders strain by a "skinpass" ($\sigma = P/A_o$ and $\epsilon = \Delta l/l_0$ are the engineering units for stress and strain)

material parameter but depends on crystal orientation according to Schmid's law [3]

$$\sigma_y = M_s \tau_y \qquad (4)$$

where τ_y is the critical resolved shear stress in the active slip system – the "real" material parameter related to the motion of dislocations. The "Schmid-factor"[1]

$$M_s = \frac{1}{\cos \chi \cos \lambda} \qquad (5)$$

depends on crystal orientation which may also change with strain. χ is the angle between the tensile axis and the normal on the slip plane and λ is the angle between the tensile axis and the slip direction. The shear in the slip system, a, is related to strain by

$$da = M_s d\varphi \qquad (6)$$

In these terms Eq. (2) must be replaced by

$$\frac{\Delta P_y}{P_y} = \left(\frac{1}{\tau_y} \frac{d\tau_y}{d\varphi} + \frac{1}{M_s} \frac{dM_s}{d\varphi} - 1 \right) \Delta\varphi \qquad (7)$$

Depending on orientation the additional second term on the right hand side may be quite large. It may also be negative which can produce spectacular instabilities before the condition (3) is fulfilled. Therefore it is imperative to use Eq. (7) instead of (2) when discussing the plasticity of single crystals.

In polycrystals, on the other hand, several slip systems are operating simultaneously in each crystal. The orientation factor M_s is then a weighted average over those of all operating systems. It will therefore depend on the texture of the material i. e. on the distribution of individual orientations occurring in the polycrystalline matrix. During strain this distribution usually changes in the direction towards a final "deformation

---

[1] In this paper the index s is used to distinguish M_s from M which shall be used later to mean "momentum".

texture". This, in general, will also lead to a finite "geometrical" second term in Eq. (7). The theory of M_s in polycrystals and its development with strain is quite complicated. The interested reader is referred to the work of, e.g., L. Toth [4] and P. v. Houtte [5]. The results, however, show that the second term in Eq. (7) is usually small for polycrystals (especially after high strains). It has, therefore, been neglected in paragraph 1.1.1. and shall be neglected in the rest of this chapter.

1.1.3 Third example: $\sigma_y = \sigma_y(\dot{\varphi})$

If σ_y is a function of strain rate alone then Eq. (1) transforms to

$$\frac{\Delta P}{P} = \frac{\Delta \dot{\varphi}}{\sigma} \frac{d\sigma}{d\dot{\varphi}} - \Delta \varphi \tag{8}$$

If the overall strain rate was $\dot{\varphi}_0$ before localisation and we assume that a fluctuation limits all further deformation to ℓ_1 then $\dot{\varphi}_1 = \dot{\varphi}_0 \ell / \ell_1 = \Delta \dot{\varphi}$. Using the definition of strain rate sensitivity

$$m \doteq \frac{d \log \sigma}{d \log \dot{\varphi}} \tag{9}$$

Eq. (8) transforms to

$$\frac{\Delta P}{P} = m - \Delta \varphi = m - \dot{\varphi}_1 \Delta t \tag{10}$$

This is a remarkable result. It means that local variations of strain rate will not lead to strain localisation unless they lead to a **finite** difference in strain $\Delta \varphi$. This is not caused by the assumption of a "sudden" localisation. Assuming a gradual development of the disparity of strain rates inside and outside the inverval ℓ_1 does not change the result. Setting, e.g.,

$$\left. \begin{array}{rl} \dot{\varphi}_2 &= \dot{\varphi}_0 - \alpha \Delta t \\ \dot{\varphi}_1 &= \dot{\varphi}_0 + \alpha \frac{\ell_2}{\ell_1} \Delta t \\ \Delta \dot{\varphi} &= \dot{\varphi}_1 - \dot{\varphi}_2 \\ &= \alpha \frac{\ell}{\ell_1} \Delta t \end{array} \right\} \begin{array}{rl} [\alpha] &= s^{-2} \\ \alpha \Delta t &\leq \dot{\varphi}_0 \\ \Delta \varphi &= \int \Delta \dot{\varphi} dt \\ &= \frac{\alpha}{2} \frac{\ell}{\ell_1} (\Delta t)^2 \end{array} \tag{11}$$

then

$$\frac{\Delta P}{P} = \alpha \Delta t \left[\frac{d\sigma}{d\dot{\varphi}} \frac{1}{\sigma} - \frac{\Delta t}{2} \right] \tag{12}$$

which leads to instability for

$$m < \frac{\Delta \varphi_{\max}}{2} = \overline{\Delta \dot{\varphi}} \Delta t \tag{13}$$

which corresponds to Eq. (10). The necessity for finite fluctuations in certain cases is the reason why in this chapter finite values Δ are used instead of a differential description (see Appendix). The stabilising effect of strain rate sensitivity is well known for fluids of high viscosity ($m = 1$) which can be drawn out to long filaments without necking. (For fluids of low viscosity m is also equal to 1 but here capillary forces – which have been neglected in this chapter – become more important.)

1.1.4 Fourth example: $\sigma_y = \sigma_y(\varphi, \dot\varphi)$

If σ_y is a function both of strain and strain rate, then Eq. (1) transforms to

$$\frac{\Delta P}{P} = m - \Delta\varphi\left(1 - \frac{\Theta}{\sigma}\right) \qquad (14)$$

This equation shows that instability begins only after a finite fluctuation beyond the Considère condition Eq. (3). Four cases can be distinguished:

For $m > 0$ and $\sigma < \Theta$ the test is stable. This applies to typical metals at moderate strains.

For $m > 0$ and $\sigma > \Theta$ the test is unstable, if

$$\Delta\varphi > \frac{m}{1 - \Theta/\sigma} \qquad (15)$$

This case applies to superplastic alloys where Θ is small and m is large (usually $m > 0.3$). They permit large uniform strains far beyond the Considére condition (3).

For $m < 0$ and $\sigma < \Theta$ the test is unstable if

$$\Delta\varphi < \frac{m}{1 - \Theta/\sigma} \qquad (16)$$

This applies to alloys showing the Portevin-leChatelier effect (serrated yielding) before condition (3) is fulfilled. Fig. 3 shows that $\Delta\varphi$ (and, hence, $\Delta\sigma$) is increasing as the specimen approaches the Considère condition.

For $m < 0$ and $\sigma > \Theta$ the test is unstable.

1.1.5 Fifth Example: $\sigma_y = \sigma_y(\varphi, \dot\varphi, T)$

If the test is not isothermal, then Eq. (14) expands to

$$\frac{\Delta P}{P} = m - \Delta\varphi\left[1 - \frac{\Theta}{\sigma} - \frac{1}{\sigma}\frac{dT}{d\varphi}\frac{d\sigma}{dT}\right] \qquad (17)$$

In metals usually $\frac{d\sigma}{dT} < 0$. This tends to destabilise the test. If $\frac{d\sigma}{dT} > 0$ then the test is stabilised.

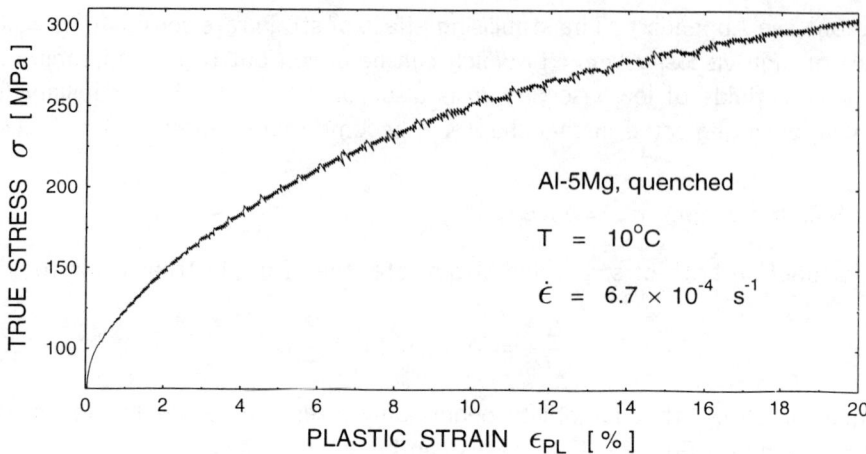

Figure 3: Example for serrated yielding (Portevin-LeChatelier effect)

1.2 The torsion test

1.2.1 General considerations

Fig. 1 can also be seen as a cylindrical specimen loaded in torsion. If its radius is a the momentum is given by

$$M = 2\pi \int_0^a \tau(r) r^2 dr \qquad (18)$$

If no longitudinal stresses are imposed the specimen will usually change its length during plastic torsion so a is generally not constant because of constancy of volume.

If the material is ideally plastic ($\tau = \tau_0$) then

$$M = \frac{2\pi a^3}{3} \tau_0 \qquad (19)$$

If the deformation is purely elastic then

$$M = 2\pi \int_0^a \frac{\gamma_a}{a} G r^3 dr = \gamma_a \frac{\pi}{2} G a^3 \qquad (20)$$

where γ_a is the shear strain on the surface of the specimen.

The work hardening of metals is sometimes described by linear work hardening as

$$\tau = \tau_0 + \gamma \vartheta \qquad (21)$$

Fig. 4 shows that this is a good approximation in a wide range of large strains. It is not good for small strains which are usually not investigated by torsion tests. τ_0, of

Examples of Strain Localisation

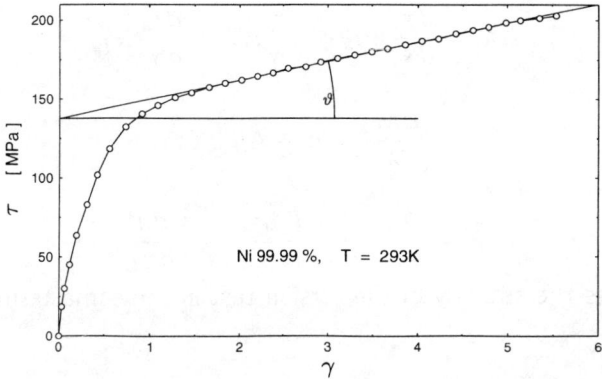

Figure 4: Stress-strain curve of nickel up to high strains

course, is a meaningless constant that should not be confounded with the end of the elastic regime which in the scale of Fig. 4 is somewhere in the lower left hand corner.

Inserting Eq. (20) in Eq. (18) gives

$$M = \frac{2\pi a^3}{3}(\tau_0 + \frac{3}{4}\gamma_a \vartheta) \qquad (22)$$

or

$$M = \frac{2\pi a^3}{3}(\tau_a - \frac{\gamma_a \vartheta}{4}) \qquad (23)$$

Eq. (22) yields

$$\frac{d(M/a^3)}{d\gamma_a} = \frac{\pi}{2}\vartheta \qquad (24)$$

With Eq. (23) this gives

$$\tau_a = \frac{3M}{2\pi a^3} + \frac{\gamma_a}{2\pi}\frac{d(M/a^3)}{d\gamma_a} \qquad (25)$$

This equation is generally used to evaluate torsion tests [7]. Here it has been proved for linear work hardening but it works reliably for other forms of strain hardening as well. The second term on the right side is frequently small enough to be neglected. This is the approximation we shall use in the following equations.

1.2.2 Strain localisation in torsion

In analogy to Eq. (18) we obtain

$$\frac{\Delta M_y}{M_y} = \frac{\Delta \tau_{ya}}{\tau_{ya}} + 3\frac{\Delta a}{a} \qquad (26)$$

Using
$$\Delta \tau_a = \frac{d\tau_a}{d\gamma_a}\Delta\gamma_a + \frac{d\tau_a}{d\gamma}\Delta T + \frac{d\tau_a}{d\dot\gamma_a}\Delta\dot\gamma_a \qquad (27)$$
this gives
$$\frac{\Delta M_y}{M_y} = m + \Delta\gamma\, aX \qquad (28)$$
where
$$X = \frac{\vartheta}{\tau_a} + \frac{1}{\tau_a}\frac{dT}{d\gamma_a}\frac{d\tau_y}{dT} + \frac{3}{a}\frac{da}{d\gamma_a} \qquad (29)$$

We can now discuss the stability of the torsion test in the same terms as in paragraph 1.1.4.:

If $m > 0$ and $X > 0$ the test is stable

If $m > 0$ and $X < 0$ the test is unstable for $\Delta\gamma > -\frac{m}{X}$.

If $m < 0$ and $X > 0$ the test is unstable[2] for $\Delta\gamma < -\frac{m}{X}$.

If $m < 0$ and $X < 0$ the test is unstable.

It is, however, not so easy to discuss the value of X in general terms.

The first term is identical to the first term in Eq. (18). It is usually positive in metals. Exemptions, like the transition from upper to lower yield point have been treated under section 1.1.1.

The second term consists of two factors. The first factor describes the heating of the specimen due to the plastic work spent. This is illustrated in Fig. 5.

It shows that adiabatic heating with

$$\frac{1}{\tau_a}\frac{dT}{d\gamma_a} = \frac{1}{\rho c_p} \qquad (30)$$

is a good approximation for fast running machines, at least in the first stages of the test. The second factor is typically negative in metals.

The third term, although easily measured in experiment, is difficult to discuss in general. For many metals it is negative at low temperatures and positive at high temperatures. Since it reflects the anisotropy of the material it may reverse sign when the direction of twist is reversed [7]. At least one can say that this term is usually small compared to the other two.

The importance of the second term can be illustrated very nicely by the experiment shown in Fig. 6.

---

[2] This case is illustrated by a film [8]

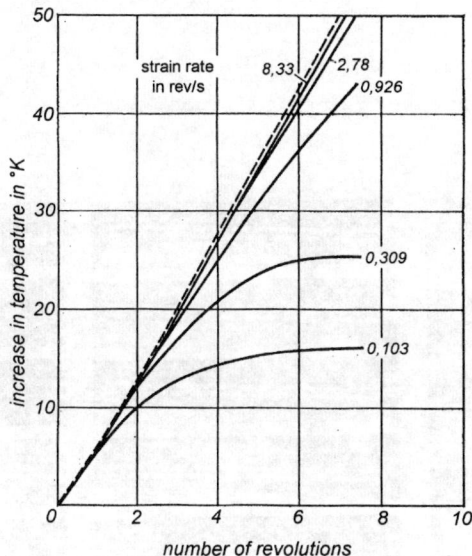

Figure 5: Temperature increase during torsion tests on aluminium. The dashed line corresponds to adiabatic heating

Fig. 6 shows a torsion test on a specimen of structural steel. Two longitudinal lines (one thick and one thin) that have been drawn on the surface are winding up into screws. After 2.5 revolutions the deformation localises to form a band that subsequently expands over the length of the specimen. After completion at 6 revolutions a second deformation band starts at the right side of the specimen and spreads again. Short before completion, at 12 revolutions, the specimen breaks. In Fig. 7 the experiment is repeated under running water, i. e., under isothermal conditions. No strain localisation is observed.

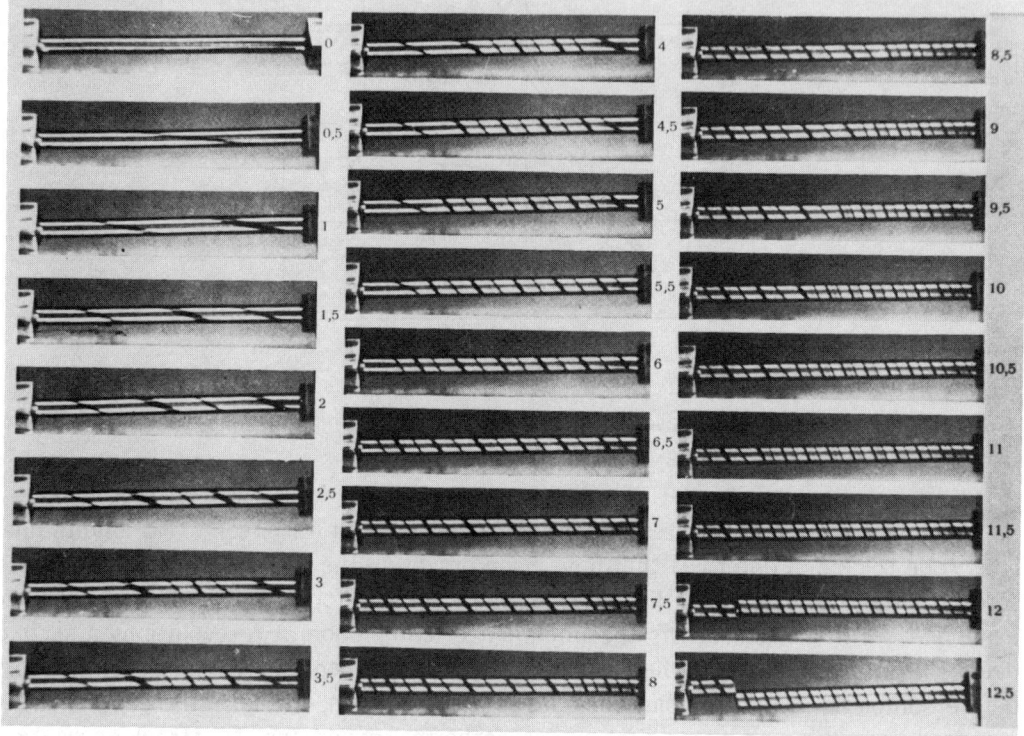

Figure 6: Strain localisation in the torsion of a carbon steel

Examples of Strain Localisation

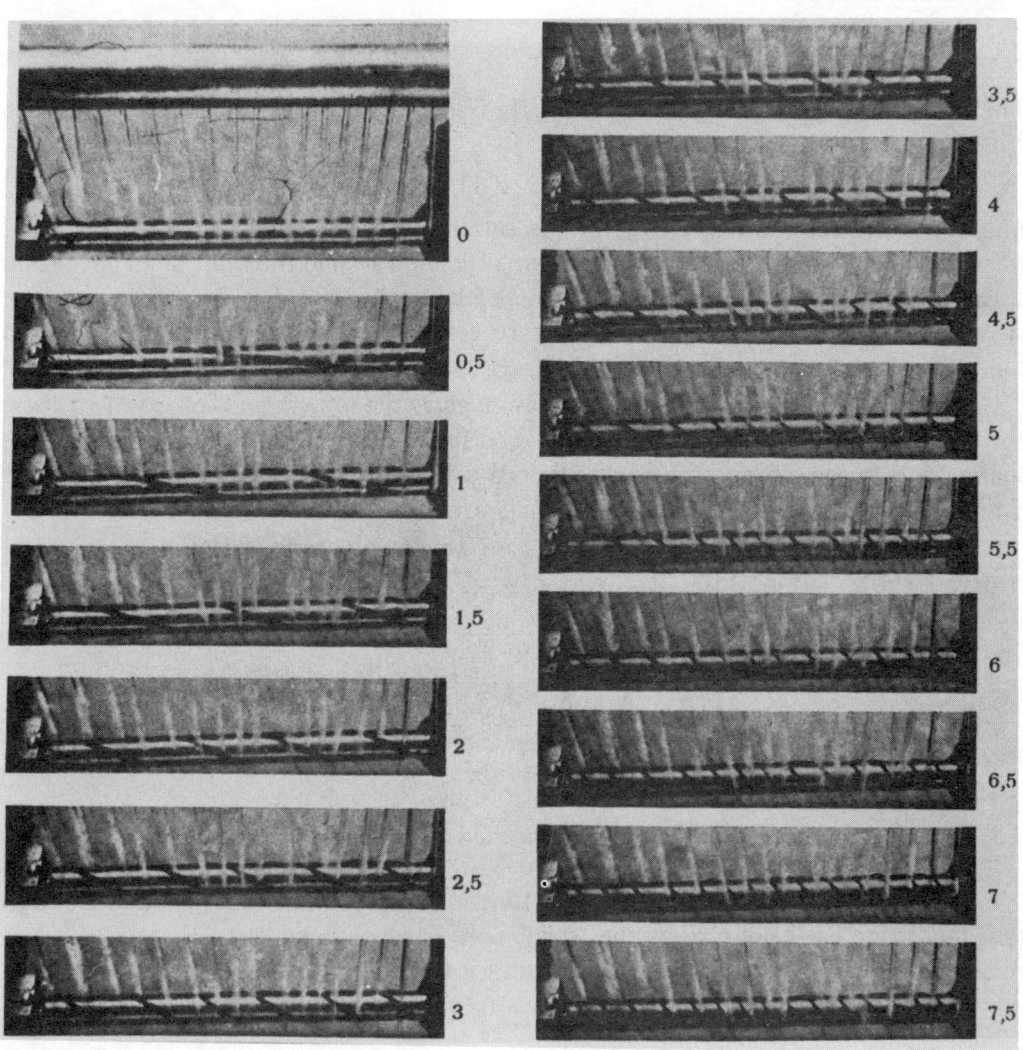

Figure 7: Experiment like in Fig. 6, but under water cooling

2 STRAIN LOCALISATION IN FRACTURE

2.1 Ductile fracture

2.1.1 Plastic work to form a ductile fracture surface

The stress intensity leading to ideally brittle fracture can be described by the Griffith criterion

$$K_{IC} = \sqrt{2\gamma E} \qquad (31)$$

where γ is the specific free energy of the surface of the material. In cases of failure where plastic deformation is localised in a small volume around the front of an advancing crack linear elastic fracture mechanics can also be applied but γ in Eq. (31) must then be replaced by γ_B, the plastic work spent to create the fracture surface. K_{IC} is then called "fracture toughness".

The growth of a crack is frequently described as the growth and linkage of cavities; but for the purpose of this book it would be better to describe it as the shrinking of the ligaments between existing (or newly formed) cavities. Either way the fracture surface is finally characterised by the typical "dimple structure".

The work necessary to form this structure can be easily estimated for the simplified structure shown in Fig. 8 [9].

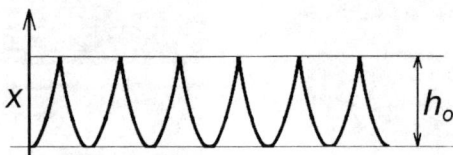

Figure 8: Schematic simplification of the dimple structure

The surface consists of identical features of height h_0. A cross section through this "landscape" at height x has the "solid" part F which is equal to F_0 at $x = 0$ and zero for $x = h_0$. According to the elementary theory of plasticity the local strain is $\varphi = \ln F/F_0$. With a suitable average flow stress $\bar{\sigma}$ the plastic work spent to form one volume element is $\bar{\sigma}\varphi$ and γ_B is given by

$$\gamma_B = \frac{1}{F_0} \oint_V \bar{\sigma}\varphi dV \qquad (32)$$

$$= \bar{\sigma} \int_0^{h_0} \frac{F}{F_0} \ln \frac{F_0}{F} dx = \bar{\sigma} h_0 \int_0^1 z \ln \frac{1}{z} dx = \bar{\sigma} h_0 S \qquad (33)$$

The details of the dimple contour assumed in Fig. 8 are contained in the function $z(x)$ and, hence, in S. A more detailed study [10] shows that the value of S does not depend very sensitively on the specific function assumed for $z(x)$. For all reasonable dimple contours it is always near $S \approx 1/4$. This gives an estimate for fracture toughness as

$$K_{IC} = \sqrt{\frac{\bar{\sigma} h_0 E}{2}} \qquad (34)$$

The same paper [10] gives a an estimate for $\bar{\sigma}$ from values gained in a tensile test of the bulk material. Comparison to experiment [9, 11] shows that this estimate is quite good in some cases. In other cases the measured K_{IC} values are higher than indicated by Eq. (33) which means that the strain is not entirely localised in the dimple structure but that some plastic strain extends into the bulk of the material [10].

2.1.2 Determination of the deformed volume

In practical cases the determination of h_0 as a measure for the deformed volume is not as easy as might be suggested by Fig. 8. This is because real fracture surfaces do not consist of <u>identical</u> features. Fig. 9 illustrates that h_0 cannot be determined – not even in principle – from a real fracture surface alone without looking at the opposing surface.

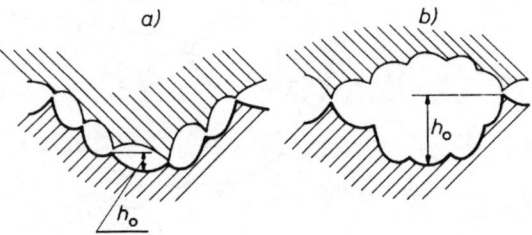

Figure 9: Two possibilities for the mismatch between two opposing fracture surfaces (schematic)

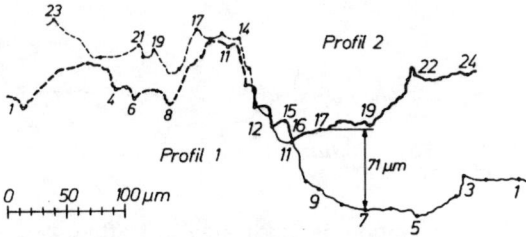

Figure 10: Critical crack opening displacement taken from the corresponding profiles of opposite fracture surfaces (from [13])

This has been the motive for developing quantitative stereographic photogrammetry of fracture surfaces. It works like the evaluation of satellite photographs for geographical topography. The pictures are taken in the scanning electron microscope from a specimen tilted under two angles. Details have been published elsewhere [12]. Evaluation of such pictures permits the point by point determination of profiles along certain lines on the fracture surface. Fig. 10 shows that two such profiles permit an unambiguous measurement of the true crack opening displacement.

Fig. 11 shows that by shifting two such profiles against each other one gets a fairly realistic insight into the growth of a crack by the growth and linkage of cavities or better, by the shrinkage of the ligaments between them.

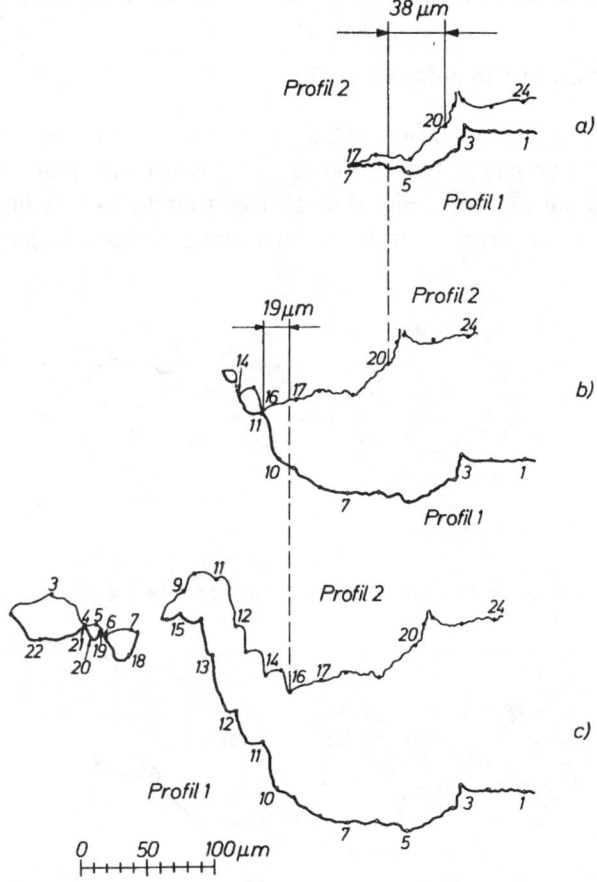

Figure 11: Ductile crack propagation demonstrated by shifting two measured crack profiles (from [13])

The method also permits to make two-dimensional maps of the two fracture surfaces but the labour involved in the manual determination point by point is almost prohibitive. The procedure was therefore automatised. It is now possible to get realistic maps on the basis of, say, 10^4 measured points in a preselected area within a few minutes [14, 15]. It is still left to the experimenter's skill to find the two corresponding areas on both flanks of the fracture. The task is made more difficult by the fact that even in an experiment that is macroscopically purely mode I the local fracture shows other components as well. A small mode II component, for instance, can be seen in Fig. 11 by comparing the relative position of the points 20 and 3. This problem is solved in our computer program by an appropriate distorsion of the two measuring grids on the basis of a few salient points on the two fracture surfaces. The resulting maps can be used in a variety of ways; one of them is to superimpose them in analogy to Fig. 10 to gain more realistic values for the deformed volume in Eq. (32) [15].

2.2 Growth of fatigue cracks

In classical fatigue (as opposed to low cycle fatigue) the bulk of the loaded specimen is deformed only elastically and plastic strain is localised in a narrow zone around the front of the advancing crack. Fig. 12 shows a typical crack growth curve. It has two features which are not easily explained by classical continuum mechanics:

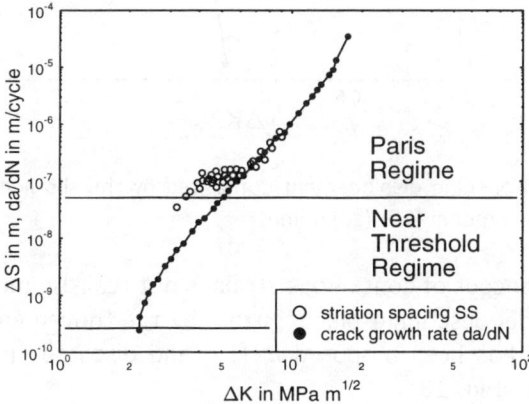

Figure 12: Typical crack growth curve da/dN vs. ΔK and striation spacing (from [16])

1. The Paris relation

$$\frac{da}{dN} \sim \Delta K^n \tag{35}$$

does not hold for very small amplitudes. Instead, there is a threshold value ΔK_{th} where the growth rate drops to zero.

2. While at higher amplitudes the spacing of striations (also entered in Fig. 12) corresponds to the crack growth rate this is not so for lower amplitudes. Instead, striation spacing remains practically constant at around 10^{-7} m. A similar lowest value has been reported for many materials [17].

2.2.1 The threshold value ΔK_{th}

It is generally assumed that the crack growth rate is linked to the plastic crack opening displacement CTOD. Fig. 12 shows CTOD as a function of ΔK according to classical continuum mechanics [18] using the material constants of iron. In the double logarithmic plot the result is a straight line extending down to values for CTOD of a few Burgers vectors only and even smaller.

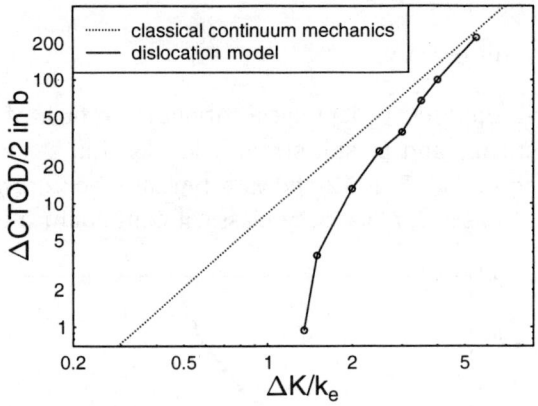

Figure 13: Crack tip opening displacement computed by the slip line model [18] (dotted line) and a dislocation model [19] (solid line)

Obviously the concept of continuous strain is not realistic on this scale and should be replaced by a model describing plastic strain by the movement of discrete dislocations. Such a model has been proposed in [19] and discussed in more detail in [20]. One result is shown in Fig. 13.

One sees that both models lead to the same result for CTOD-values of more than about 100 Burgers vectors but that only the discrete model leads to a threshold value determined by the condition that the stress should be sufficient to emit at least one dislocation from the crack front.

2.2.2 Striation spacing

If one extends the calculations to higher stress levels then many dislocations are emitted from the crack tip upon loading. Some of them will return to the crack tip to be annihilated during unloading while the rest remain in the lattice forming a pile-up and leaving behind a surface step on the crack. In subsequent cycles, again many dislocations are emitted but all or most of them will return to be annihilated because of the repulsive forces of the existing pile-up. Therefore they will not leave a surface step

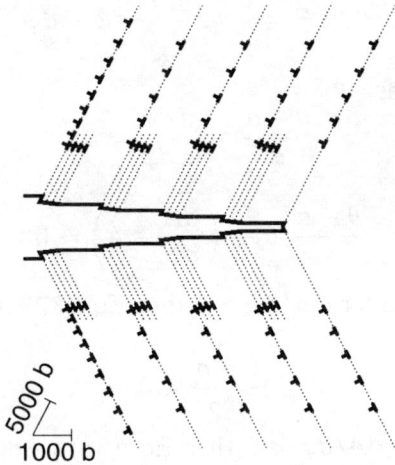

Figure 14: Calculated dislocation arrangement in the material after many cycles (from [19])

while the crack front advances a distance given by CTOD. After a certain number of cycles the emitted dislocations will be far enough from the first pile-up and can now remain in the lattice leaving behind a new step on both fracture surfaces. They will also form a new pile-up that will influence dislocations produced in subsequent loading cycles. The results of the computation are shown in Fig. 14.

The distance between the steps visible on the fracture surface is about $10^3\,b$. This result is fairly independent on the material constants assumed and corresponds reasonably well with the minimum striation spacing of 10^{-7} m shown in Fig. 12. For higher amplitudes of ΔK, CTOD and da/dN are larger than this minimum spacing and then, of course, each loading cycle will leave their own surface steps with a spacing equal to da/dN.

The minimum spacing of about 10^{-7} m between parallel dislocation pile-ups given in [10] does not seem to be a consequence of this particular model. It has been predicted in much more general terms by Orowan [2] for other forms of plastic deformation. It is therefore probably not a coincidence that a similar spacing is observed for the thickness of micro shear bands as described in this book in the chapter by Korbel.

APPENDIX

In differential form and for $\sigma = \sigma(\varphi, \dot{\varphi})$ Eq. (1) can be written as

$$d \ln P = \frac{\theta}{\sigma} d\varphi + \frac{m}{\dot{\varphi}} d\dot{\varphi} - d\varphi \tag{36}$$

For constant load this yields

$$\frac{\theta - \sigma}{\sigma} + \frac{m}{\dot{\varphi}} \frac{d\dot{\varphi}}{d\varphi} = 0 \tag{37}$$

or

$$\frac{\theta - \sigma}{\sigma} + \frac{m}{\dot{A}} \left(\frac{d\dot{A}}{d\varphi} + \dot{A} \right) = 0 \tag{38}$$

If instability is assumed for $d\dot{\varphi}/d\varphi > 0$ then Eq. (37) yields the criterion proposed by Estrin and Kubin [22]:

$$\frac{\theta - \sigma}{m\sigma} < 0 \tag{39}$$

If instability is assumed for $d\dot{A}/d\varphi < 0$ then Eq. (38) yields the criterion proposed by Hart [23]:

$$\frac{\theta}{\sigma} + m < 1 \tag{40}$$

ACKNOWLEDGEMENT

Work on this paper was in part supported by the Austrian National Bank under Project Nr. 6635.

Thanks are also due to Dr. P. Les, Prof. Dr.E. Pink and Dr. A.W. Zhu for interesting discussions and to Mrs. M. Rattner for her help with the manuscript.

REFERENCES

1. Stüwe, H.P., H.O. Asbeck: Instabilitäten im Zug- und Verdrehversuch, Arch. Eisenhüttenw. 40 (1969) 125–130

2. Considère, M.: Die Anwendung von Eisen und Stahl bei Konstruktionen, Gerold-Verlag, Wien (1888)

3. Schmid, E., W. Boas: Kristallplastizität, Springer, Berlin (1935)

4. Toth, L.S., P. Gilormini, J.J. Jonas: Effect of rate sensitivity on the stability of torsion textures, Acta Met., 36 (1988) 3077–3091

5. van Houtte, P., E. Aernoudt: Solution of the generalised Taylor theory of plastic flow, Z. Metallk., 66 (1975) 202–209

6. Stüwe, H.P., O. Kolednik: Shape instability of thin cylinders, Acta Met. 36 (1988) 1705–1708

7. Stüwe, H.P., H. Turck: Zur Messung von Fließkurven im Torsionsversuch, Z. Metallk. 55 (1964) 699–703

8. Witzel, W.: Torsionsverformung von Metallen – Bewegung von Verformungsfronten bei den Aluminiumlegierungen AlCuMgPb und AlCu3, Inst. f. d. wissenschaftl. Film, Göttingen, Film Nr. E 1899

9. Stüwe, H.P.: The work necessary to form a ductile fracture surface, Engng Fract. Mech. 13 (1980) 231–236

10. Stüwe, H.P.: The plastic work spent in ductile fracture, threedimensional constitutive equations and ductile fracture, Ed. S. Nemat-Nasser, North Holland Publishing Comp. (1981) 213–221

11. Kolednik, O., H.P. Stüwe: Abschätzung der Rißzähigkeit eines duktilen Werkstoffes aus der Gestalt der Bruchfläche, Z. Metallk. 73 (1982) 219–223

12. Kolednik O.: Ein Beitrag zur Stereophotogrammetrie am Rasterelektronenmikroskop, Prakt. Metallographie 18 (1981) 562–573

13. Kolednik, O.: Stereogrammetrische Untersuchungen des Rißwachstums bei duktilen Materialien, Gefüge und Bruch, Eds. K.L. Maurer und M. Pohl, Gebr. Borntraeger, Berlin–Stuttgart (1990) 193–198

14. Stampfl, J. S. Scherer, M. Gruber, O. Kolednik: Reconstruction of surface topographies by scanning electron microscopy for application in fracture research, Appl. Physics A63 (1996) 341–346

15. Stampfl, J., S. Scherer, M. Berchthaler, M. Gruber, O. Kolednik: Determination of the fracture toughness by automatic image processing, International J. of Fracture 78 (1996) 35–44

16. Serdyuk, V.A., N.M. Grinberg: The plastic zone and growth of fatigue crack in Magnesium MA12 alloy at room and low temperatures, Int. J. Fatigue 5 (1983) 79–85

17. Davidson, D.L., J. Lankford: Fatigue crack growth in metals and alloys: mechanisms and micromechanisms, International material reviews 37 (1992) 45–76

18. Rice, J.R.: Mechanics of crack tip deformation and extension by fatigue, Fatigue Crack Propagation, ASTM STP 415, Am. Soc. Testing Mats. (1967) 247–311

19. Riemelmoser, F.O., R. Pippan: Investigation of a growing fatigue crack by means of a discrete dislocation model, Materials Science and Engineering A234–236 (1997) 135–137

20. Riemelmoser, F.O., R. Pippan, H.P. Stüwe: An argument for a cycle by cycle propagation of fatigue cracks at small stress intensity ranges, Acta Mat., submitted 1997

21. see Nabarro, F.R.N.: Theory of crystal dislocations, Eds. M.F. Mott, E.C. Bullard, D.H. Wilkinson, Oxford University Press, London (1967)

22. Estrin, Y. L.P. Kubin: Plastic instabilities: phenomenology and theory, Materials Science and Engineering, A 137 (1991) 125–134

23. Hart, E.W.: Theory of the tensile test, Acta Met. 15 (1967) 351–355

STRUCTURAL AND MECHANICAL ASPECTS OF HOMOGENEOUS AND NON-HOMOGENEOUS DEFORMATION IN SOLIDS

A. Korbel
Academy of Mining and Metallurgy, Cracow, Poland

ABSTRACT

The aim of this work is to provide a basic experimental information about the physical nature of plastic deformation of crystalline bodies and to show the analytical workshop within which discreet micro-structural events of the plastic flow and the accompanying effects may be accounted for. The attention is focused upon the slip which is a dominating micro-structural mechanism of deformation. The criterion for slip in the slip system is shown and discussed in terms of the effect of geometrical constraints upon the stress state and the choice of the operating system. The experimental patterns of slip during homogeneous and localised deformation are analysed in terms of the evolution of slip intrinsic features and the feedback between the mechanical, geometrical and structural aspects of slip in crystal. The analysis of the evolution of slip in crystals is supplemented by an essential information about the mechanism of slip and properties of dislocations. Interactions between dislocations of different slip systems are analysed from the point of view of the mechanisms of the strain hardening (formation of the obstacles network) and the softening mechanisms. The correlation between slip intrinsic features and global mechanical performance of crystals is made. It is shown that the change from a stable into an unstable mode of plastic flow is caused by the change of slip from a „fine slip" into a „coarse slip" in single crystals and into shear bands in polycrystals. The latter is shown to take the origin in the mechanical instability of the obstacles network. The factors controlling the evolution of slip and responsible mechanisms are discussed in terms of the slip geometry and interactions between dislocations.

1. INTRODUCTION

If a body is subjected to the action of balanced external forces one can say that it is under load. The behaviour (reaction) of a body during loading and under the load depends upon the material of which the body is made and upon the loading conditions (loading rate, temperature). It may concern the change of of a body shape as well as its internal structure.

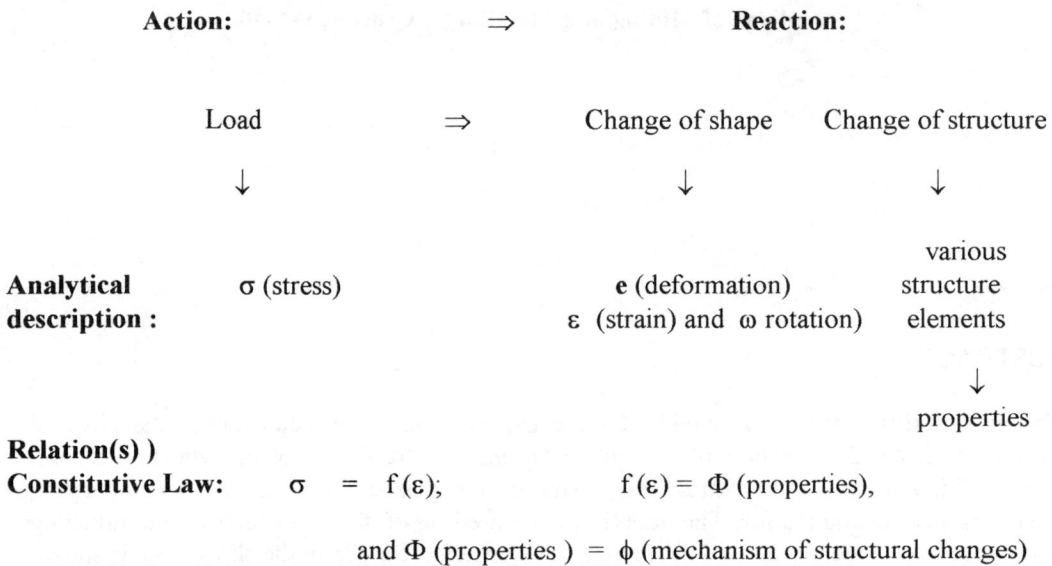

Regardless, however, of the kind of material and loading conditions the response of the material to the applied load shows some characteristic features. They become revealed with increasing stress, so that three physically different forms of the response (reaction) of the body to the load (or more precisely to the scheme of loading) can be identified. They are classified and analysed in terms of the material properties known as elasticity, plasticity and decohesion (fracture). Our attention will be focused upon elasto-plastic behaviour of crystalline materials. Hence, a distinction between them has to be made first.

For an analytical description of the relationship between the action (load) and reaction (change of the body shape) we used the quantities which are precisely defined in the framework of the mechanics of continuum. Hence the state of loading is expressed by the stress tensor while, the geometrical changes of the body by the tensor of the deformation gradients. The physical links of these two tensors (constitutive laws) which decide about the form of the response are governed by the discreet nature of the material. It is important, therefore, to expose these features which may be used as discriminants of the particular form of the material response. This may be done basing upon a typical experimental tensile curve which relates the value of the load with the corresponding elongation of a test piece, as it is schematically shown in Fig.1.1.

Elastic deformation is fully reversible with stress and varies in tact with the stress variations. Hence, during unloading the deformation reverses as shown by arrows in the figure. It is an atomic-scale uniform deformation of the body which means that it results in a change of the distance between the neighbouring atoms causing the change of the proper volume of the material. In crystals it causes (except of purely hydrostatic stressing) the change of the crystal symmetry.

Plastic deformation is a permanent change of the body shape. If during loading the material is brought to such a state, then the following unloading removes only the elastic deformation and leaves some residuum (plastic strain). Such a response (reaction) has a different physical pattern in a continuum medium, like a viscous liquid whose internal structure can be approximated by a hypothetical continuum, and a different pattern in crystals, possessing well defined internal structure. The global plasticity of crystalline materials results from several, micro-scale events typical for a discreet structure materials. Hence, the plastic deformation of crystalline materials, except of very special cases (twinning and martensitic transformation) is non-homogeneous in the atomic scale.

FEATURES

Elastic deformation: **Plastic deformation:**

continuum:

Elastic deformation	Plastic deformation
Varies in tact of stress	Initiates at some (critical) stress) **yield criterion**
Fully reversible	Leads to permanent change of the body shape
Causes of the change of the body volume (change of the material density)	No- volume change - **mechanism(s)**

Body with internal structure (crystals):

Homogeneous in the atomic scale	Non-homogeneous in the atomic scale
Anisotropic (crystal symmetry dependent)	Anisotropic (deformation mechanism dependent)
Causes the **change of crystal symmetry** No other structural changes	Substructural changes: generation of crystal lattice defects change of size and form of crystallites change of texture

The load (stress) at which these events of deformation start to play the dominating role in the behaviour of polycrystalline metals is referred to as the yield point or a proof stress (Fig.1.1). The local events of deformation show, in fact, the features of atomic scale processes. They lead to structural changes of the material during plastic deformation. These changes are responsible for the evolution of the material properties. The rise of the load (stress) with elongation shown in Fig.1.1 is the evidence of the increasing material resistance to plastic deformation. This effect of the deformation is called strain hardening. It is temperature and strain rate dependent.
Figure 1.2 gives the examples of the effect of temperature and strain rate on the strain hardening of metals [1].

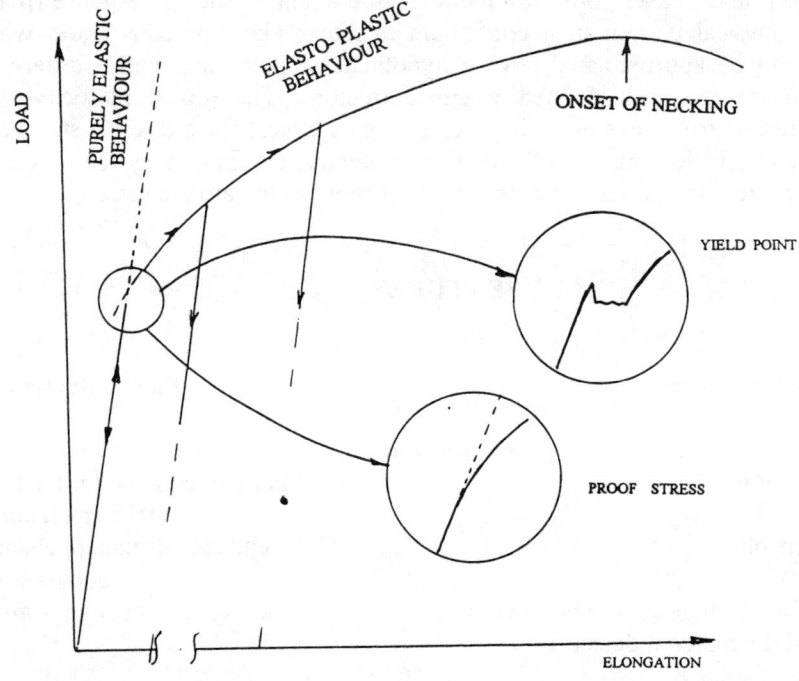

Fig.1.1 A schematic plot of a typical load versus elongation curve for metallic materials.

Focusing the attention upon the microstructural aspects of plastic deformation we should to identify the deformation processes. These, in general, may be divided into two groups: those which are associated with the mass transport (diffusion), and others which are responsible for slip. This is because only due to diffusion and due to slip the two basic features of plastic deformation, namely the permanent change of shape and constancy of the material volume can be ensured. This may be easily seen (Fig.1.3) if one notices that atoms from one face of the body can be brought by diffusion onto another (say orthogonal) one, and that due to slip the parallel atomic layers (planes) may be displaced in such a way that the interspacing between the layers does not change. The latter may be accomplished if the direction of the relative atomic planes displacement and the slip plane combine into a „slip system", which means that the slip direction belongs to the plane of the slip. In both cases the interatomic distances are sustained and so is the crystal density. The role of diffusion controlled processes becomes important at high temperature and/or at very low rates of plastic straining (creep). They may play a significant role also in deformation accompanying relaxation processes within the material microstructure. In most cases, however, the slip is the deformation bearing process. Hence, the homogeneous deformation and conditions

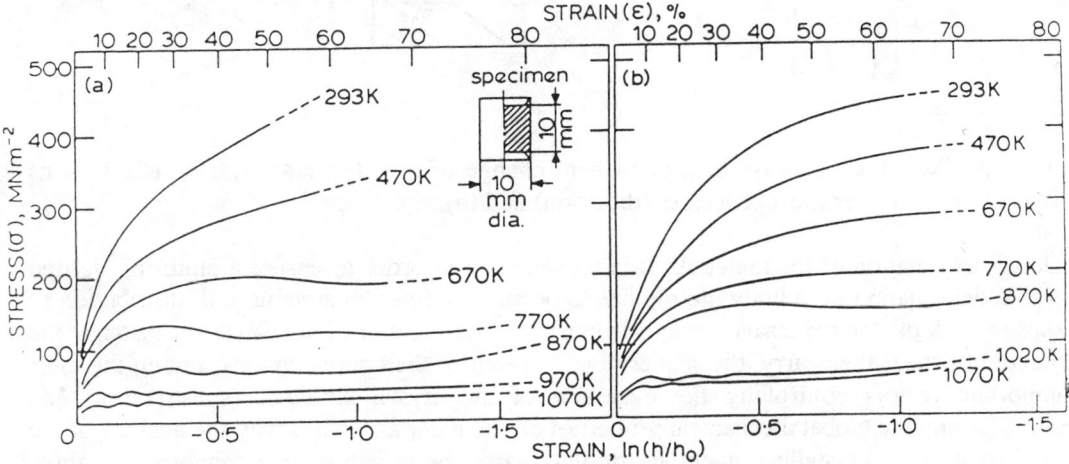

Fig. 1.2 The effect of temperature on stress-strain curves for compression tests on copper at initial strain rates of $8 \cdot 10^{-4} s^{-1}$ (left) and $8 \cdot 10^{-2} s^{-1}$ [1].

under which it changes into a non-homogeneous process will be analysed in terms of the slip in crystals. Particular attention will be paid to some intrinsic features of slip and the associated microstructure effects which are responsible for the evolution of the mechanical properties of crystals. In order to initiate a slip process some „critical value of the stress" has to be applied to the crystal. This critical value reflects the resistance of the crystal lattice to slip and it is a highly anisotropic property of crystals. This implies that slip is not only a local event within the crystal volume, but also that it is an anisotropic feature of the crystal deformation. If one specifies the slip system and the amount of shear in such a system, then the analytical account for the contribution of slip in a global deformation of a crystal may be found. The answer is provided by the slip geometry under the condition that slips are evenly disposed within the crystal volume. One may conclude, therefore, that factors which control the distribution of slips in crystals are the controlling factors of the homogeneity of the

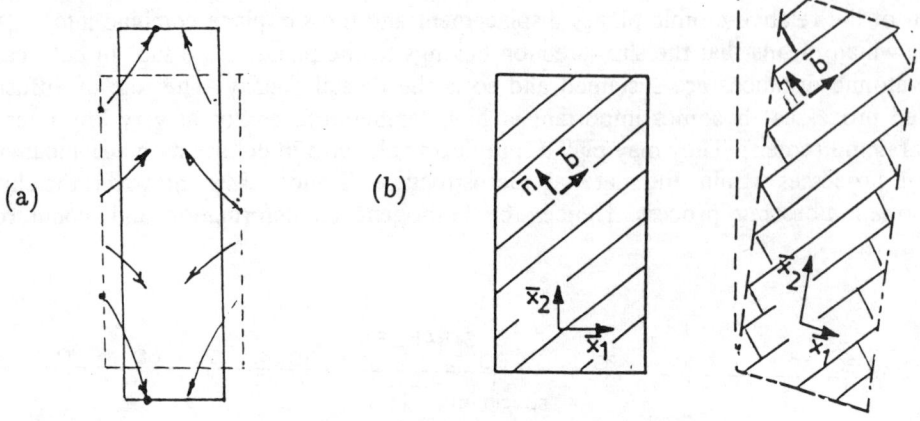

Fig.1.3 Two possible ways of a permanent change of a body shape with no effect on its density: (a) mass transport processes (diffusion) and (b) slip.

global deformation of the material. It is obvious that in order to ensure a uniform, desired global deformation of a body the slip has to occur in a few slip systems and distribution of slips in each of the necessary systems is uniform. The number of activated slip systems, the amount of shear they carry, the sequence and instant of their activation are among the most important factors controlling the evolution of the crystal structure (substructure) and consequently the global mechanical properties of the material. That is why the analysis of the performance of crystalline materials during plastic deformation must incorporate three interrelated aspects of plastic deformation:

mechanistic, **geometrical,** **structural.**

The mechanical aspect concerns the criterion of slip in a system(s) under the applied scheme of loading and the evolution of the loading scheme in the course of deformation. Such an evolution is closely related to geometrical changes of the crystal form (crystallite in a polycrystalline aggregate) which are caused by slip (shear) and to the rotation of the crystal lattice. The relation between geometrical changes and the change in the scheme of loading of a crystal is a natural consequence of the fact that some geometrical changes of the crystal shape are not allowed under a given scheme of straining. This is because of the external and internal constraints, which are often referred to as the „reaction stress". And, finally, the „structural aspect" of slip concerns the mechanism of slip in the crystal lattice, the nature of the slip opposing obstacles, formation of the obstacles network (substructure) and its mechanical stability. Formation of the obstacles network resulting from the mechanism of slip (glide of dislocations and dislocation accumulation) is the primary cause of the strain hardening of crystals and a limiting factor for the slip extent. Therefore, the hardening rate, anisotropy of the hardening (known as the latent hardening effect) and stability of the

VARIOUS ASPECTS OF SLIP IN CRYSTALS

Mechanical	Geometrical	Structural
Slip criterion (applied stress only) $\tau = b\sigma n$	Single system slip: b, n, γ - simple shear	Mechanism of slip and Slip intrinsic features: $\gamma = b\lambda\rho$
	Distortion of the body:	Storage of dislocations Formation of the obstacles network Latent hardening
	Change of the body shape and „position" (rotation)	
Evolution of the stress state (geometrical constraints)	Lattice rotation	
	Slip in secondary system	Storage of dislocations in secondary system(s)
	Development of texture	Monotonic strain hardening Evolution of the obstacles network : development - instability Global behaviour : hardening - softening Deformation: homogeneous - non-homogeneous

obstacles network are related to the activity of the slip systems engaged into deformation and the slip intrinsic features. There exists, therefore, a clear physical feedback between slip behaviour and the evolution of the global mechanical properties of the material. The existence of such a feedback provides the opportunity to consider the plastic flow in crystalline materials in terms of the slip properties and evolution of slip in the course of straining.. Such an approach to the crystal plasticity is the subject of the present paper in which „structural and mechanical aspects of homogeneous and non-homogeneous deformation in crystals" are discussed in terms of the micro-mechanisms of deformation and the governing physical and geometrical rules. Obviously, the arguments which stand behind such an approach come from the experimental measurements and observations. Hence, the most important observations about the nature of plastic deformation in single and polycrystalline metals have to be shown first.

2. EXPERIMENTAL PATTERN OF PLASTIC DEFORMATION.

2.1 Macro and micro patterns of plastic deformation

The behaviour of a polycrystalline material is definitely much more complex than the behaviour of a single crystal under similar plastic deformation conditions. The differences in the plastic flow of single crystals and polycrystals must be, therefore, clearly identified in order to extract those features of single crystals which are the most important for the mechanical performance of polycrystals.

Polycrystals are composed of thousands of very fine (from a few to several micrometers in size) crystallites (grains). Because of the multitude of grains in a massive aggregate and their approximately random orientation, the material as a whole is considered isotropic. Then, according to the continuum mechanics, the change of the material response from elastic to elasto-plastic requires that some critical (elastic) energy is provided to the body or, equivalently, a critical state of stress is attained [2-4]. The requirement of the constancy of material volume makes it necessary that the determinant of the deviatoric part of the applied stress tensor is used as the measure an „equivalent shear stress" which may lead to plastic deformation. The „critical" value of such a macroscopic measure of the stress at the onset of plastic yielding (yield criterion) depends upon the kind of material and its internal structure. A very well documented effect of the grain size upon the critical value of a macroscopic flow stress provides a good basis for the analysis of the mechanical performance of polycrystals. In the case of a pure, well annealed, metal the size of grains is a factor differentiating the material structure. The macroscopic flow stress (σ_y) follows then the Hall-Petch relationship:

$$\sigma_y = \sigma_o + k_y d^{-0.5}, \qquad (2.1)$$

where, d is an average grain size in the material and k_y is a material constant. σ_o has to be also considered as a constant because, in general, it does not correspond to the critical value of the shear stress in a single crystal. Hence, the most important task of the „structural plasticity" is to establish physical foundations to the role the material structure in slip behaviour within an individual crystal and further in the behaviour of a polycrystalline aggregate. The observations of the deformation pattern appear helpful in such attempts. Examples are shown in Fig.2.1 and 2.2. Figure 2.1 shows macro picture of a metal sample (copper in this case) composed of 5 up to a few centimetres large crystals after tensile

Structural and Mechanical Aspects

deformation. Elongation of the sample was accompanied by change of its external faces. A surface, very flat before deformation, received, as may be seen in the figure, a mountain like profile. If the material has a fine grain structure, then a surface flat and shining prior to

Fig 2.1 The surface pattern of a copper sample composed of 5 very big crystallites after tensile deformation

50 μm

Fig. 2.2 Micrograph of the surface of the polycrystalline metal which has undergone small (5%) elongation in tension.

stretching becomes rough already after a few percent elongation. To reveal the surface pattern, the microscope observations are then necessary. The picture in Fig.2.2 is the optical microscope image of the surface of a polycrystalline metal after a very small plastic deformation in tension. Prior to stretching the surface was well polished to make it very flat

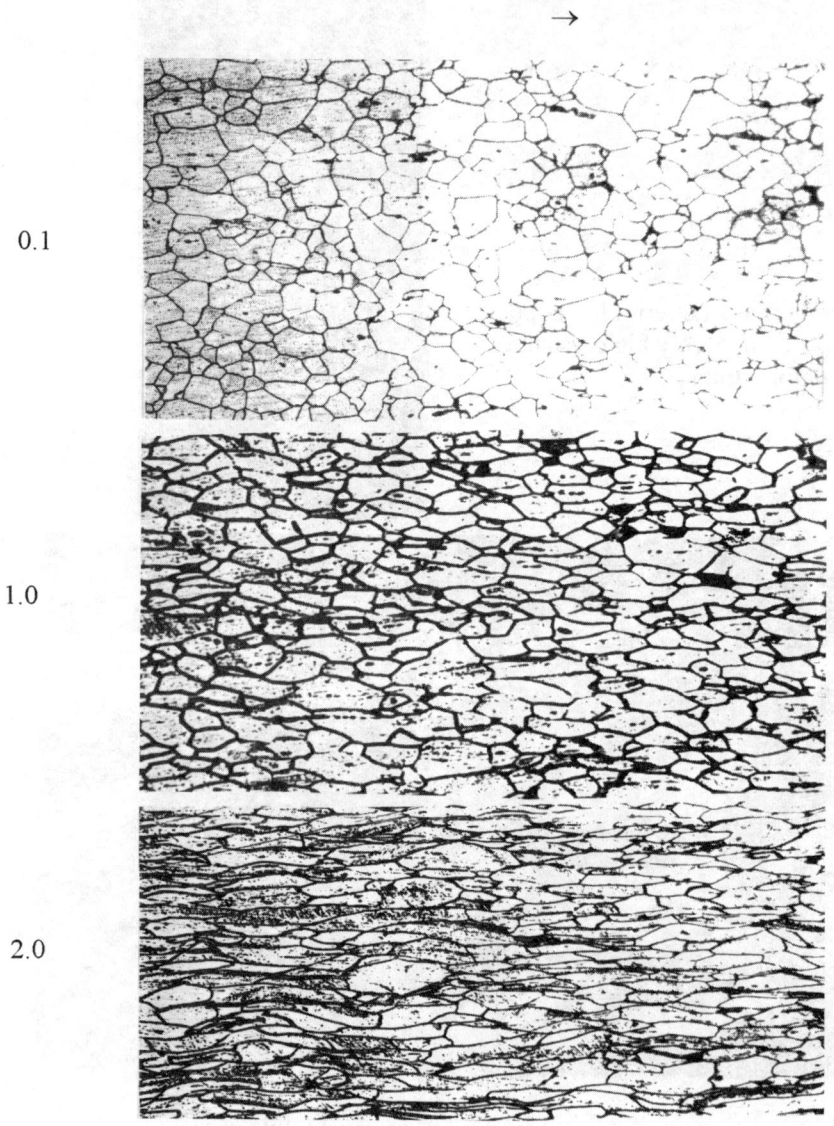

Fig 2.3. Micrographs of showing the change of the grain shape with deformation during rolling. Pictures are taken in section parallel to rolling direction

and shining. Therefore the patterns seen in Fig.1.4 and 1.5 reflect the effect of deformation on the shape of grains at the free surface of the sample which is the same regardless of the grain size. One may notice the slip lines within individual grains which in turn may be distinguished from each other due to the surface relief. Such a relief is caused by „rotation" of the external surface of the grain and may occur only if the surface is not constrained by the tool or the neighbouring grains.

Different positions of slip lines in different grains and the surface relief give the best evidence that deformation results in fact from slip and that slip is a highly anisotropic property of crystals. It is obvious that within the material the grains are short of such a freedom as they have at the surface. Internal faces of grains must match each other or, in other words, the deformation of the neighbouring grains must be compatible. Figure 2.3 shows the arrangements of grains in the sections through a piece of metal after different plastic deformation by rolling. Grain boundaries were revealed by chemical etching of the flat (polished) surface parallel to the rolling direction. It is seen that grains elongate along the rolling direction with a clear tendency to maintain the same orientation of the faces in the test piece. One may conclude that the faces of individual crystals parallel to the external surfaces of the material remain parallel to these surfaces also after deformation, or that the shape of a crystallite follows the change of the shape of the material. The neighbourhood of adjacent grains sets, therefore, severe constraints upon the behaviour of individual crystals within the body. One may say, that there is an interaction between grains from the very beginning of plastic yielding. Such an interaction is due to the demand of the material continuity (grains have to match each other, while slips within the particular grains lead to their different forms). This demand is accomplished by local elastic deformation - elastic accommodation. Therefore, it would be hard to accept that the experimental yield stress (or proof stress) value reflects the resistance of the crystal lattice to slip. In fact, the onset of macroscopic yielding (yield point) is preceded by „micro- yielding". This term is used to denote very small plastic strains ($< 10^{-3}$) in the transient between elastic and plastic behaviour of the material. The real value of the critical shear stress for slip can be obtained only while testing single crystals. The surrounding grains influence the slip in crystallites not only at the onset of plastic flow but throughout the entire deformation of a body. In a rigorous description one has to express the role of the crystal surroundings in terms of the internal stresses and to link these stresses with geometrical changes of the crystal. However, the evaluation of the internal stresses in polycrystals is probably the most difficult task of theoretical considerations. It is also the reason why it is so difficult to analyse the macroscopic instability of plastic flow and strain localisation in polycrystalline materials in terms of micro-scale slip events of a single grain. In order to get some insight into this problem let us first look at the experimental pattern of unstable plastic flow in single and polycrystalline metals.

2.2 Instability of plastic flow - monotonic deformation

By instability of the plastic flow we understand a state in which plastic deformation has a choice to follow a different „path" in its performance under the same scheme of straining. (monotonic straining). In a continuum mechanics of solids, which sees the deformation in terms of a constitutive law (equation), such a state is associated with reaching the conditions of more than one solution of the constitutive equation (bifurcation) or a change in the type of the governing constitutive law. In such an approach the instability of plastic flow is made equivalent to the change from a homogenous into a heterogeneous deformation of the material, on the basis that different parts (volumes) of the material may then follow different deformation paths. The basic problem which arises then concerns the distinction of those volumes in the material in which the deformation processes (slips) lead to different geometrical changes (state of strains) and/or to different physical performance (strain rate). From Figs.2.1 and 2.2 it is already known that plastic deformation employs different slip systems in each grain of a polycrystal leading to their different shapes. Despite of this, the deformation is „stable" in the sense that load increases in a monotone way with deformation. In a coarse grain material (Fig.2.1), non-homogeneities of the deformation appearing in a macro-scale are seen with the naked eye. Refining of grains causes that the deformation becomes homogeneous in macro-scale, still being non-homogeneous in the micro-scale (Fig.2.2). Hence, it seems impossible to give an exact definition of non-homogeneity of deformation, simply because it is a scale dependent feature. It is possible, however, to identify the instability of plastic flow either in terms of the instability of the applied load in the course of deformation or in terms of macroscopic strain localisation. The term „strain localisation" defines more adequately the instability of plastic flow because it refers to a macroscopic strain gradient and consequently to strain rate gradient in the material. Therefore, the instability of the applied load in the course of deformation and the occurrence of macroscopic localisation in the material are the two patterns of plastic flow instability. They may reflect different physical processes, and therefore they may reveal themselves with different amplitude (load) and in different scale (localisation) as it is illustrated in Fig. 2.4. Figure2.4 a,b, delivers the examples of the load - elongation curves of Zn single crystals recorded in tension at different temperatures [5]. They show also the effect of the crystal orientation on the plastic flow. Figs 2.5 and 2.6 are photographs of these crystals after some tensile deformation. Discontinuities (jerkies) on a load curve, while the macroscopic extension rate is constant, prove that plastic flow is unstable and accelerates in some periodic manner in the course of crystal stretching. The photographs, in turn, show that deformation is highly non-uniform. Some of these heterogeneities are closely related to jerkies of the load, but some develop while the curve is smooth and continuously rising. (compare the tensile curve for 473K- Fig.2.4 with the corresponding photograph of a diffuse neck in the crystal- Fig.2.5). It is worth noticing two important facts. The first is that jerkies develop on the background of a rising envelope of the load curve and that amplitude of the load drops tends to increase with the load. This means that they reflect an instant, but not a global instability of plastic flow of crystals, and that such an instant instability tends to turn into a global instability.

The second fact is that jerkies are associated with the presence of very distinct deformation markings occurring locally, but across the entire crystal. (Fig.2.5). Regardless of the

Structural and Mechanical Aspects

particular mechanism of such localisation (coarse slip, twinning, kink banding) deformation markings result from large shear. The failure of the crystal, the onset of which coincides with global load instability, is caused either by highly concentrated shear or it has its origin at the zone of localised shear.

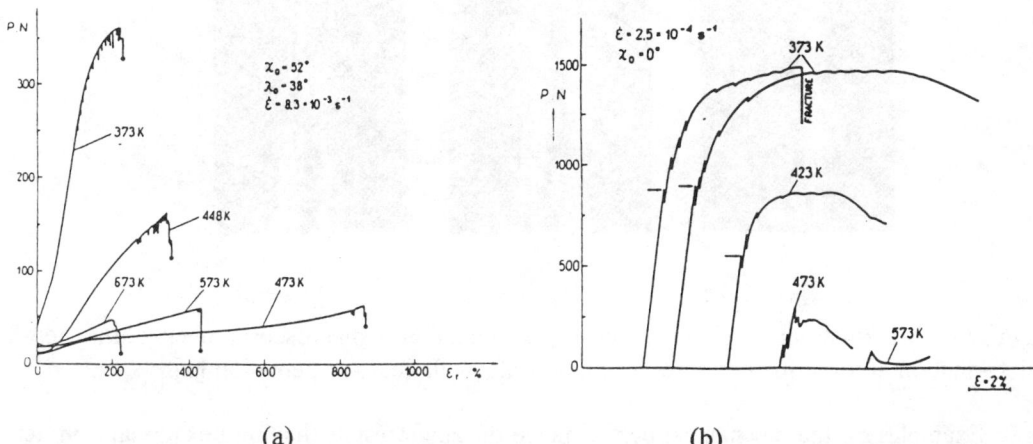

(a)   (b)

Fig.2.4 Tensile curves of zinc single crystal with „soft" orientation (a) and „hard" orientation [5].

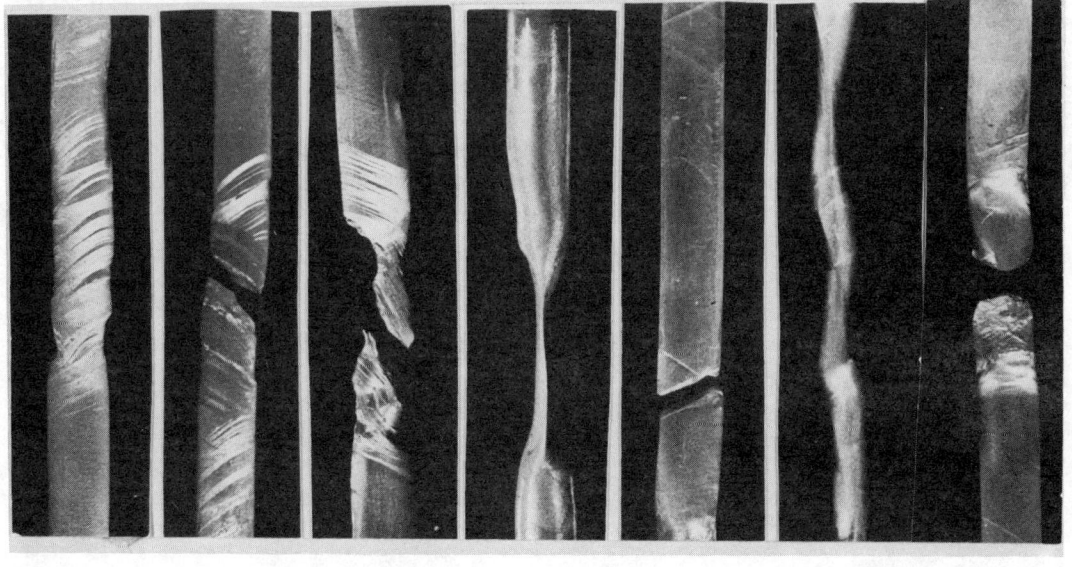

a) T=373K   b) T=373K   c) T=448K   d) T=473K   e) T=473K   f) T=573K   g) T=573K
ε=150% ε=220% ε=350% ε=150% ε=800% ε=300% ε=420%

Fig.2.5 Macro-structure of zinc crystals tensile tested as in Fig. 2.4 (a) [5].

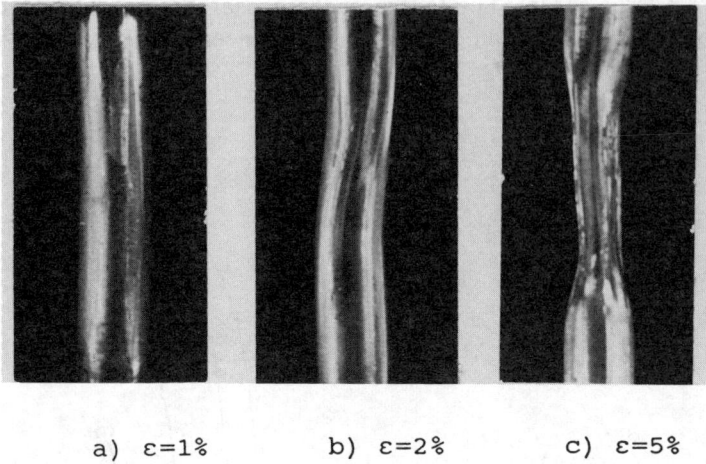

a) ε=1%   b) ε=2%   c) ε=5%

Fig. 2.6 Macro-structure showing the development of a macroscopic neck during tensile deformation at room temperature in zinc crystals with „hard" orientation [5].

Examples of the mechanical performance of zinc crystals (having hexagonal symmetry of the crystal lattice) were chosen because, due to the „deficit" of slip systems in hexagonal crystals, different (although not all) patterns of instability and localisation of slip in single unconstrained crystals may be observed in the same metal. Similar examples might be shown (Fig.2.5 a-c) for the others single crystals of metals and alloys. Owing to this it was possible to consider the phenomenon from the point of view of the mechanism of slip in a crystal lattice and its stability. At the same time, however, one may ask the question, whether localisation of slip within a grain shows the same features and follows the same mechanism(s) as macroscopic localisation in a polycrystalline material. The answer to this question is a primary demand of the microstructural approach to global instability of metallic materials.

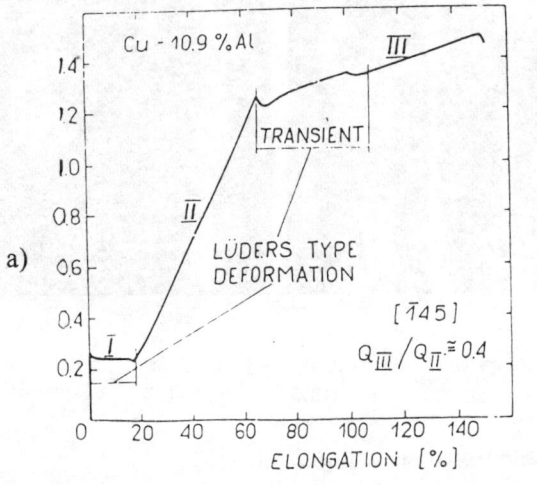
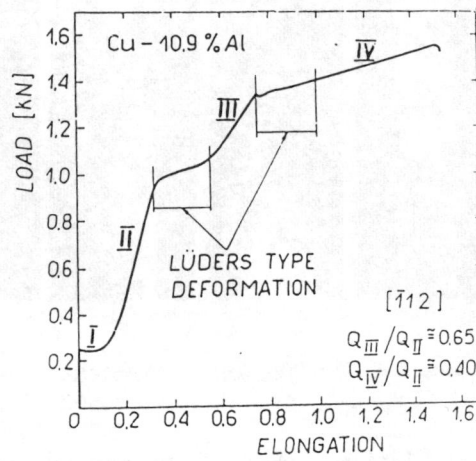

Structural and Mechanical Aspects

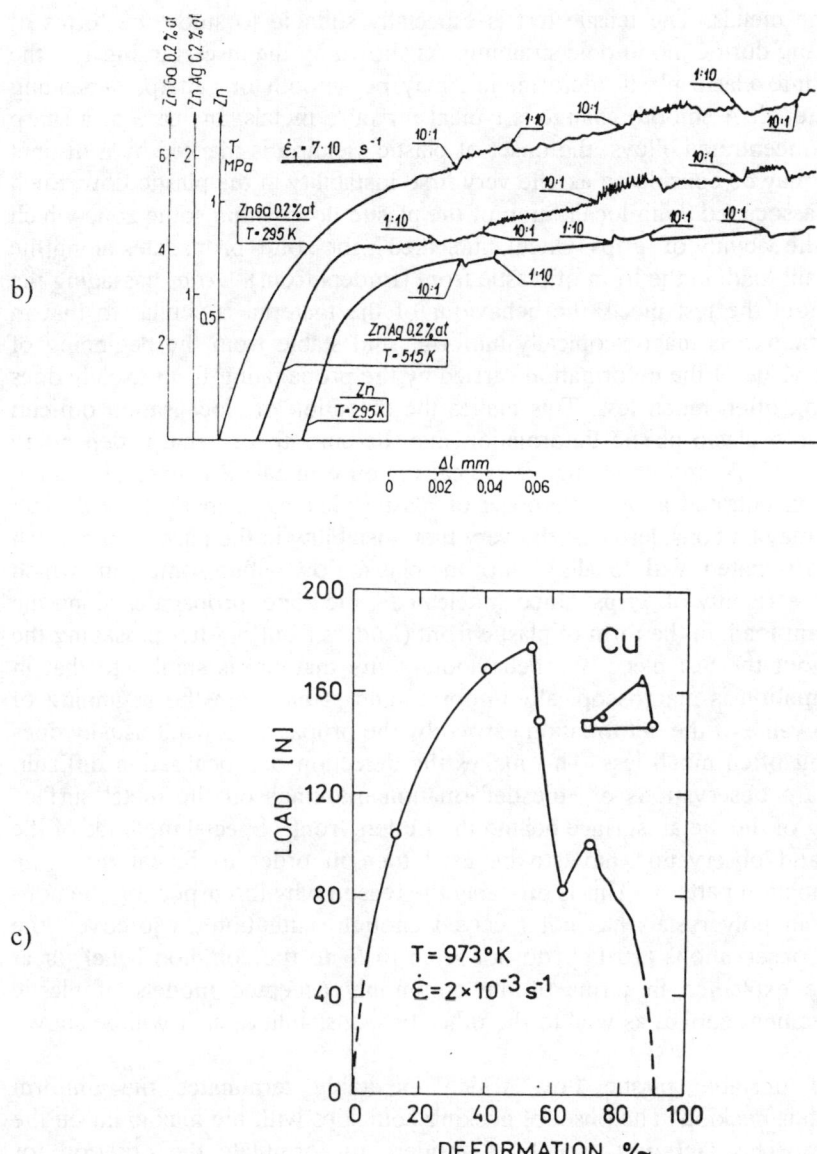

Fig. 2.7 Instabilities of the load recorded on tensile curves: (a) Cu-Al single crystals -tests at room temperature (effect of crystal orientation) [6], (b) Zn, Zn-Ga and Zn-Ag single crystals at room temperature [7], and (c) copper single crystal deformed at high temperature [8].

Let us therefore look at typical experimental patterns of unstable, localised plastic flow in polycrystalline metals. The tensile test is especially suitable to study the forms of unstable flow occurring during monotonic straining. As shown by the inserts in Fig.1.1 the change from elastic into elasto-plastic deformation may be smooth or sharp, depending upon the kind of material. A smooth change is typical for pure metals. In case of a sharp change, typical for concentrated alloys, the onset of plastic yielding is marked by a distinct „yield point" which may be considered as the very first instability in the plastic flow. Such an instability is often associated with localisation of the plastic flow within some zone which usually originates in the vicinity of grips. Once „nucleated", the zone propagates along the sample under a constant load in the form of plastic front (Luders front). After passaging the Luders front throughout the test piece, the behaviour of the material is similar to that in Fig.1.1, where deformation is macroscopically uniform and stable from the beginning of plastic straining. The value of the deformation carried by the propagating front usually does not exceed 2%, being often much less. This makes the detection of localisation difficult change from elastic into elasto-plastic deformation may be smooth or sharp, depending upon the kind of material. A smooth change is typical for pure metals. In case of a sharp change, typical for concentrated alloys, the onset of plastic yielding is marked by distinct „yield point" which may be considered as the very first instability in the plastic flow. Such an instability is often associated with localisation of the plastic flow within some zone which usually originates in the vicinity of grips. Once „nucleated", the zone propagates along the sample under a constant load in the form of plastic front (Luders front). After passaging the Luders front throughout the test piece, the behaviour of the material is similar to that in Fig.1.1, where deformation is macroscopically uniform and stable from the beginning of plastic straining. The value of the deformation carried by the propagating front usually does not exceed 2%, being often much less. This makes the detection of localisation difficult. Even more difficult are observations of the deformation markings on the metal surface except the roughening of the metal surface behind the Luders front. Special methods of the surface preparation and observation have to be used then in order to reveal the fine structure of the deformation pattern. This is probably the reason why this aspect of the non-uniform deformation in polycrystals has not received enough attention. Moreover, the existing experimental observations [10,11] do not seem to fit to the common belief, or at least, they cannot be explained in terms of the commonly accepted models of plastic deformation. This comment applies as well to the other flow instabilities as it will be shown later.

The form of unstable plastic flow which inevitably terminates the uniform deformation in tension is necking. The onset of necking coincides with the maximum on the load-elongation curve. This fact was used by Considere to formulate the criterion for necking [9]. In strain rate independent materials (a low temperature case) this criterion is given by the argument that:

$$\sigma = d\sigma/d\varepsilon \qquad (2.2)$$

where σ is the true stress and $d\sigma/d\varepsilon$ is the true strain rate, which simply expresses that localisation begins when the increase of hardening cannot balance the decrease of the cross section of the elongating sample.

Depending upon the material composition and deformation conditions (temperature and strain rate) the tensile curve may show the other instant flow instabilities in the course of a monotonic deformation. The very common for alloys is the periodically unstable, localised deformation known as the Portevin-LeChatelier Deformation (P-LChD) [12] or „dynamic strain ageing effect". The load-elongation curves typical for this effect are shown in Fig. 2.8 (a,b). Their characteristic feature is the presence of several load instabilities (load drops). The „shape" of the curve at an instability point was used to distinguish the „serrated yielding" (Fig.1.11a) from „jerky flow" (Fig. 2.8b). The essential difference between these two types of Portevin-LeChatelier deformation is the mode the deformation develops in the material. Measurements of the elongation on a very small, when compared with the sample size, gauge length show that deformation proceeds by formation and propagation of the bands of concentrated deformation (also shown in Fig1.11). Load drops are associated with formation of bands. During serrated yielding a single band nucleates and similarly as a Luders front, more or less smoothly propagates along the test piece. Then a new band nucleates at a higher stress level. Before the nucleation of a band the load rises slightly giving a typical „load tooth" on the tensile curve. During jerky flow very dynamic bands form in a discontinues manner, and they sequentially consume undeformed material.

During high temperature deformation the load-elongation (contraction in compression) curve of a polycrystalline material may show still another load instability which appears in the form of round load maximum or maxima on the load-deformation characteristic. The character of the load change (oscillating with some wavelength - multi-peak behaviour or systematically decreasing after reaching the maximum - single peak behaviour) in the course of deformation, along with the dependence of the behaviour on material composition, temperature and strain rate have turned the attention to the role of the structure restoration processes (recrystallization) as the physical reason of the instability [13] which is now commonly known as „dynamic recrystallization". Figure 2.9 gives examples of high temperature instability in polycrystalline copper (a) and nickel (b).

There is not definite answer as yet whether the load instability is then accompanied by a macroscopic localisation of strains or not. The available experimental observations [14,15] suggest that also in this case the global instability is associated with a kind of macro-localisation of deformation which differentiates the strain rate in the material giving a chance to soften (due to high temperature) the material in these volumes in which the strain rate is instantly low.

Among different forms of unstable, localised deformation the formation of shear bands is particularly important and, hence, a very attractive object of research and theoretical considerations. Like the formation of a neck (if not preceded by fracture) in tension, shear banding is also an inevitable form of plastic flow instability at large deformation. In a monotonic deformation the range of strains, in which shear bands can be relatively easily revealed, often coincides with the limit of uniform deformation in tension. Fig.2.10 shows shear bands within the zone of a neck in the deformed tension sample. They are seen on the micrograph as dark lines extending across grains which are also seen in the figure.

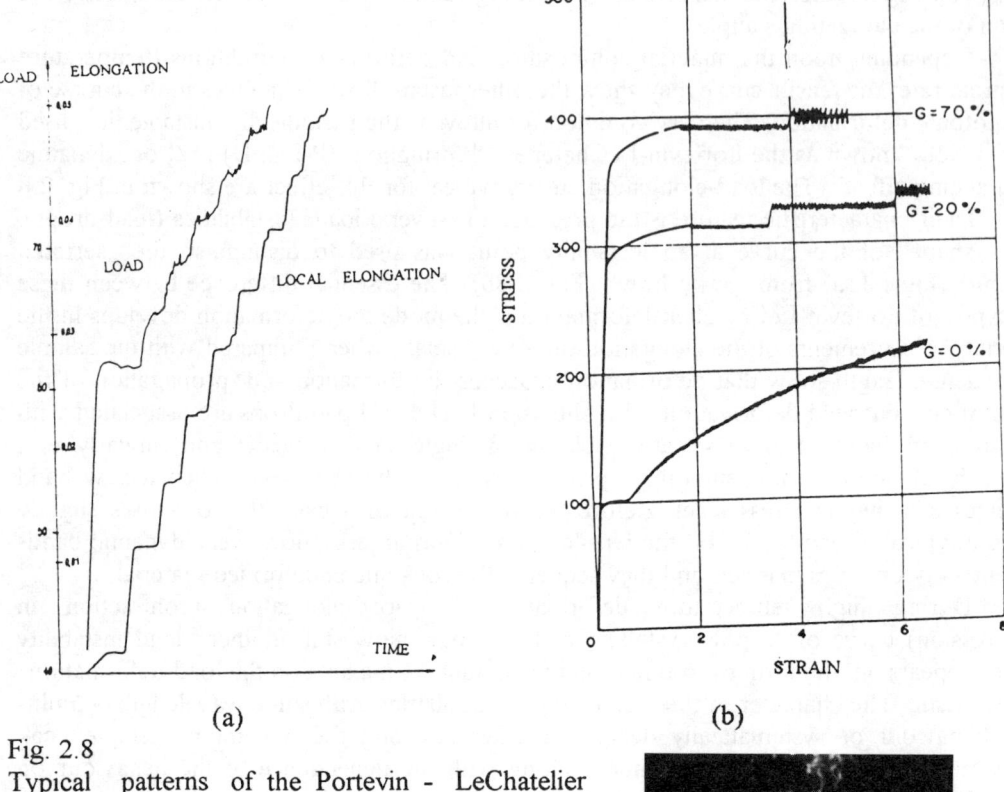

Fig. 2.8
Typical patterns of the Portevin - LeChatelier Deformation: (a) serrated yielding in polycrystalline Cu - Zn brass at 400 K, (b) jerky flow at room temperature in as annealed Al-Mg alloy and after some prestrain in rolling, c) bands of localised deformation formed during jerky flow in prestrained (20%) Al-Mg alloy [16].

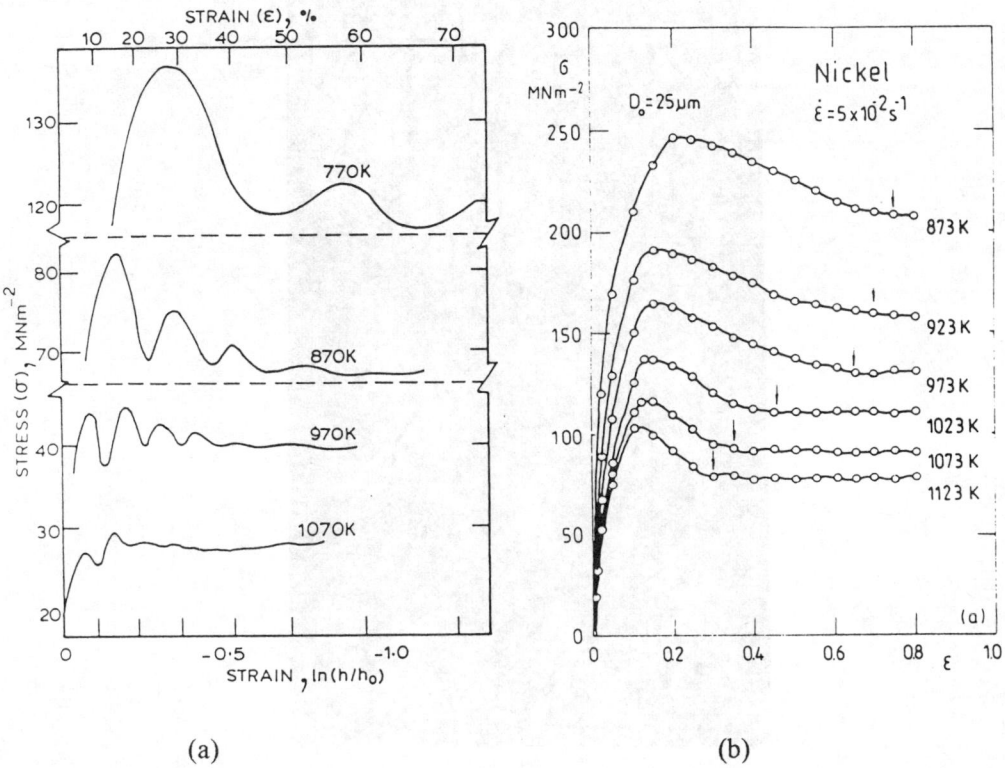

(a) (b)

Fig.2.9 High temperature compression curves for polycrystalline copper (a) [1] and nickel [14].

Associated with the change of the sample cross section the loss of the global stability at necking makes uncertain the judgement whether necking is initiated by shear banding or shear bands developing because of large deformation in the neck. The experimental observations of the development of the neck during tensile deformation under high hydrostatic pressure [17] show that shear banding gives rise to the formation a geometric neck being the only mode of deformation from the onset of necking to the ultimate shearing off the sample (Fig.2.11). This is possible, because high hydrostatic pressure prevents the development of dilatation damage and thus the sample fracture. Nevertheless, the geometrical defect in the form of a local change of the sample cross section helps to concentrate deformation within shear bands, so they receive a macroscopic size. Otherwise, like in the case of a large deformation imposed in metal forming operations, shear bands may not be revealed, until the metallographic observations are performed.

Fig. 2.10 Formation of the neck in a polycrystalline sample of Aluminium alloy associated with shear-banding.

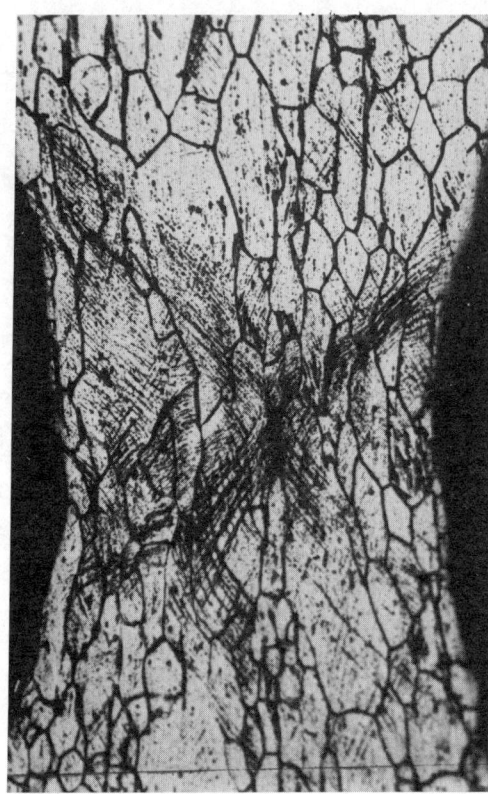

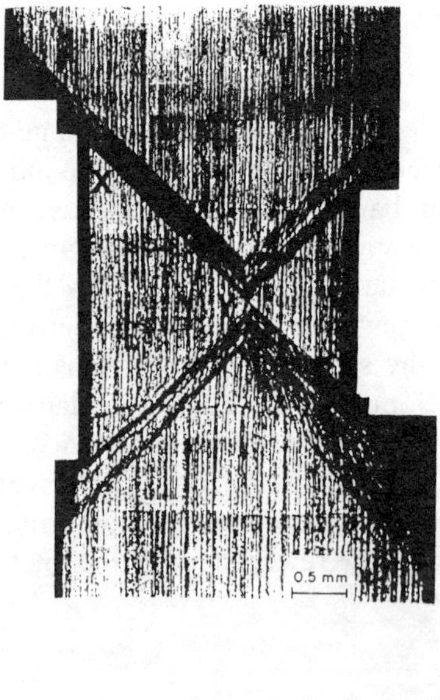

Fig. 2.11. The neck formed by conjugate shear bands during tension under hydrostatic pressure [17].

From this very brief overlook of the forms of plastic flow instability in single and polycrystalline materials it appears that some of them lead to a macroscopic localisation of deformation only in single crystals (twinning , kink banding) and the others do this as well in single as in polycrystalline metals (Luders deformation, P-LChD, necking). Our attention, will be, of course, concentrated only upon these forms of localised deformation which occur in polycrystals. A parallel occurrence of similar form of localisation in a single and polycrystalline material provides, however, a particular opportunity to approach the problem from the point of view of mechanism and mechanics of plastic deformation.

2.3 Instability of plastic flow - non monotonic deformation

As it has been already shown, there is a spectrum of apparently different forms of localised deformation in metallic materials. They may be revealed in simple tests, like tension, compression or torsion, in which the scheme of straining is kept constant (monotomic deformation). This spectrum must be supplemented by examples of localised deformation which occurs when the deformation conditions vary in the course of straining. The term non-monotonic deformation will be used is such a case. It refers to the change of such parameters of deformation as temperature, strain rate and scheme of macroscopic straining. The latter is often referred to as the change of the deformation „path". As a simple as well as very instructive example of non-monotonic deformation we may consider the tension of a crystal accompanied by a change of the temperature. Practically, after some deformation, the test is interrupted for a while, the temperature is then changed for the test to be restored again when a new temperature stabilises. Upon such an experiment the very important rule of the crystals plasticity was established by Cottrell and Stokes [18]. Testing single crystals of aluminium they have found that the ratio of the low temperature flow stress (prior to the temperature change) to a high temperature flow stress (at the onset of deformation at higher temperature) is constant throughout the deformation (Cottrell-Stokes Law). It depends, however, upon the initial temperature and the temperature difference. They have found also, that an increase of the temperature leads not only to decrease of an instant plastic flow on restoring the deformation, but it may lead to the global mechanical instability of the crystal. The effect was called „strain softening", because of the decreasing flow stress during high temperature straining, although the associated effect was localisation of deformation either into a Luders like front (small deformation before the temperature increase) or into the permanent neck.

Similar experiments were performed on single crystals of copper and copper alloy and led to similar results as it is shown in Fig 2.12. [19]. The experiments have shown also, that if the temperature is decreased, then the flow stress rises and the deformation is stable. If one takes into account that in very pure metals like aluminium or copper the critical shear stress of virgin single crystals is almost temperature independent, then the reason for the decrease of the flow stress after the temperature rise must be sought in the influence of temperature on the recovery processes in the metal substructure (softening). Hence, the experiments on single crystals show that an increase of the temperature in the course of deformation appears as an important factor destabilising the process. Such experiments show also, that by increase of temperature one may induce a Luders like deformation in the material which does not show the tendency toward strain localisation until necking. In this way it is possible as well to cause necking far before the onset of necking in a monotonic straining. These results point to the role of „the stability of metal substructure" in the mechanical performance of crystals during plastic deformation and provide a link between the metallurgical and the mechanical factors influencing plastic flow. This may be more clear if one notices that a rearrangement or collapse of the substructure elements (elements of the obstacles network) due to e.g. thermal processes results in a local softening of the material which, in turn, must lead to local acceleration of strain within a softer volume.

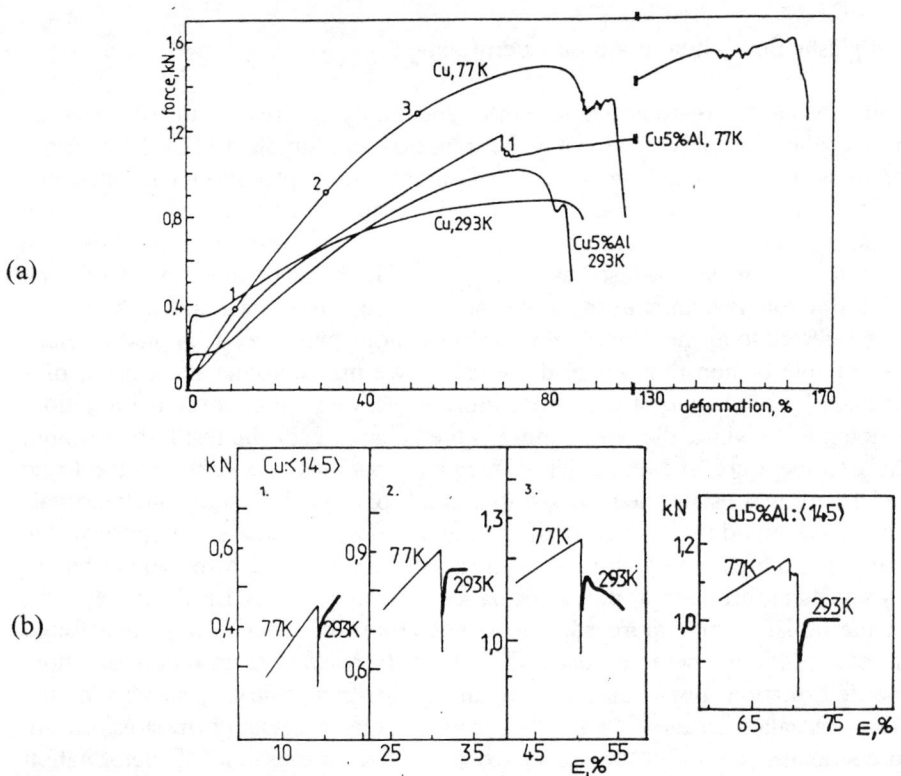

Fig.2.12 Tensile (monotonic) curves of copper and Cu-5%Al single crystals strained along <145> crystal direction at two temperatures (a) and the effect of the temperature increase on the plastic flow of these crystals (b) [9]. The strains at which the temperature was changed are denoted by numbers, and corresponding effects on the tensile curve are shown in (b) respectively.

The effect of decrease of the strain rate is similar to that caused by the rise of temperature except that it is much weaker. Therefore, only at higher temperatures the decrease of strain rate may lead to the loss of global stability of deformation.

Much more profound is the effect of change of the deformation path. In a single crystal a change of the scheme of straining means the change in the deformation bearing slip system and may be accomplished by reorientation of the crystal with respect to the applied load. The results of studies of this effect in different metals, summarised in [20], show the increase of the flow stress at the onset of a secondary straining when compered to that at the end of the primary deformation path . They show also that the secondary deformation is unstable and localised, receiving in tension the form of a Luders front if the primary deformation is small or a permanent neck for larger primary strains. The experiments on single crystals proved that the change of the deformation path is, besides of the temperature, the second controlling factor of the plastic flow instability. In a single crystal , where due to slip anisotropy it is possible to apply the load to the crystal in such a way (which is in fact the most frequent case) that only one system is the deformation carrying one, the instability of slip in such a system leads to global mechanical instability of the crystal. This is the important conclusion which, however, must be augmented by the information what physically means the instability of slip in a crystal. This will be done later, because it must be preceded by some basic information about the nature (mechanisms) of slip in crystals. Now it seems more important to see if a change of the scheme of straining leads to similar effects in polycrystals. Fig. 2.13 shows the tensile curves of polycrystalline iron which has undergone the deformation by rolling prior to tension . For comparison the tensile curve of a virgin material is also shown ($\varepsilon=0$).

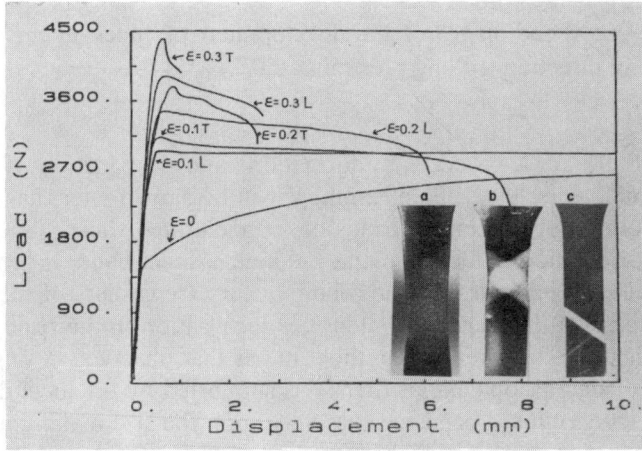

Fig.2.13. Typical tensile curves of samples prestrained (ε) by rolling. Tensile directions are parallel (L) and perpendicular (T) to the rolling direction [21].

The change from rolling to tension results in a very unstable flow. The effect increases with the increase of the rolling deformation being more profound in tension along direction perpendicular to the rolling direction. One may notice that already after a small ($\varepsilon=0.1$) rolling deformation the plastic flow in tension is unstable. The load instability is accompanied by localisation which at small rolling strains spreads out like a Luders front forming a diffuse neck. In Fig.2.14 both these aspects of plastic flow instability are shown. The effect of the strain path change is much more drastic in tension along the sample tested along the perpendicular direction as it is shown in Fig.2.15. Then, due to much more intense localisation of deformation one can distinguish two clear macroscopic shear bands. Similar bands, although less intense, may be noticed in Fig.2.14. Where they intersect a diffuse neck forms (Fig.2.14).

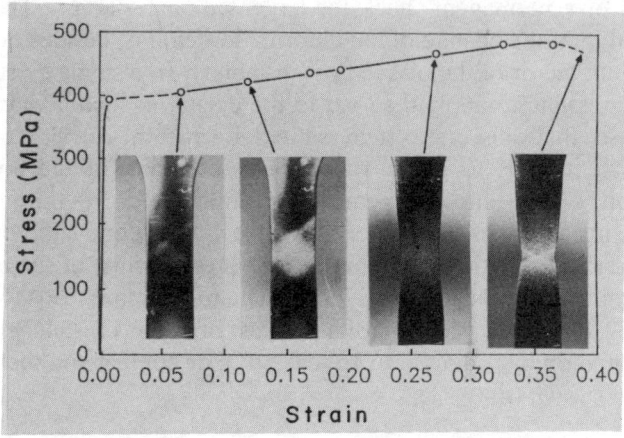

Fig.2.14 The tensile curve and stages of the development of neck in the sample during tension along the rolling direction (rolling prestrain $\varepsilon = 0.2$) [21].

Hence, the growth of the neck proceeds by formation and propagation of the fronts of macroscopic shear bands along the sample length. It will be shown later that „propagation" of the band results from the formation of bands of concentrated shear. Such micro-shear bands keep the same orientation in the sample as the macro-shear band. In tension a band is 54° off the tensile direction regardless of the rolling history (regardless of the anisotropy of the sample). The increase of the prestrain intensifies localisation in the band, as shown in Fig. 2.16. The effect is, therefore, very much the same as that observed in single crystals. In tension the growth of macroscopic bands (neck) is supported by the local decrease of the sample cross section. Due to this, a band receives very soon the size making it visible with a naked eye. The cause of the loss of global stability of the deformation in tension is the same. Most of the other schemes of straining (compression, rolling, ect), do not favour necking (clustering of micro-shears), hence the identification of the onset of flow instability is difficult and conclusion about the instability is based upon the observation of the strain distribution in the material. In general, metallographic techniques are used. The resolution of such techniques limits the reliability of the identification of strain localisation in the material. especially if the microstructural features of deformation are revealed by chemical etching. Therefore, where it is possible, the surface observations of deformation markings are preferred.

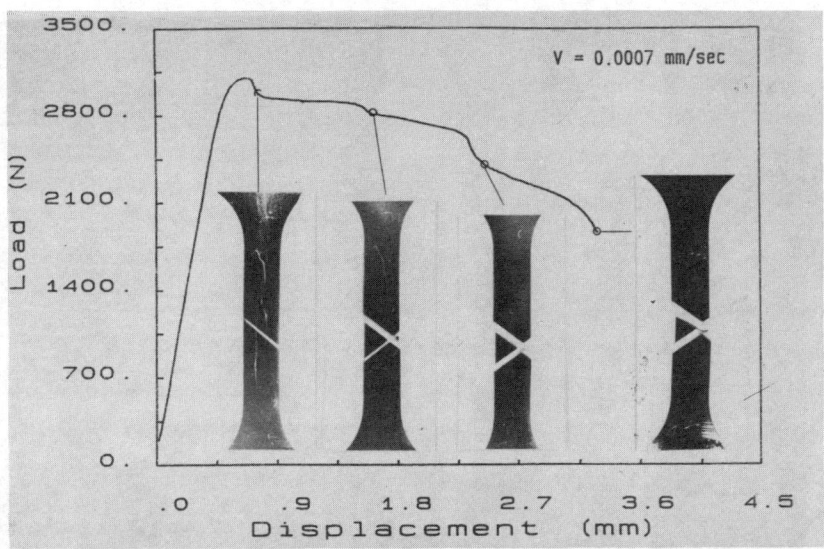

Fig.2.15. Development of strain localisation in a sample prestrained by rolling (ε=0.2) during tension along the transverse to rolling direction [21].

Fig.2.16 is an example of very distinct deformation markings on the initially flat and shining lateral face of a test piece of aluminium alloy. These are intense shear bands, as may be seen from the offsets the band of one family leaves on the other. They develop during the rolling + rolling scheme of deformation [22]. This example is one of several similar experimental observations of the effect of the change of the deformation path in polycrystals [23], although only under special conditions shear bands assume the form of a network with such a large mesh-length as in Fig.2.16. A general conclusion which results from such observations is that the change of the scheme of deformation, like in e.g. rolling-tension or rolling-transverse rolling experiments, always leads to almost instant development of shear bands during secondary deformation while, during equivalent monotonic deformation no shear bands are observed. In other words, the change of the scheme of straining (or scheme of loading) accelerates the onset of strain localisation in metals in comparison to monotonic straining. Localisation itself reveals the feature of macroscopic shear(s) concentrated within more or less evenly distributed bands. Shear bands maintain an approximately constant position in the test piece, 35° to the most recent rolling direction, regardless of the scheme of primary deformation.

The experimental studies provide, therefore, very important information not only about the factors influencing the onset of unstable flow, but also about the morphological features of localised deformation. This aspect of localised deformation

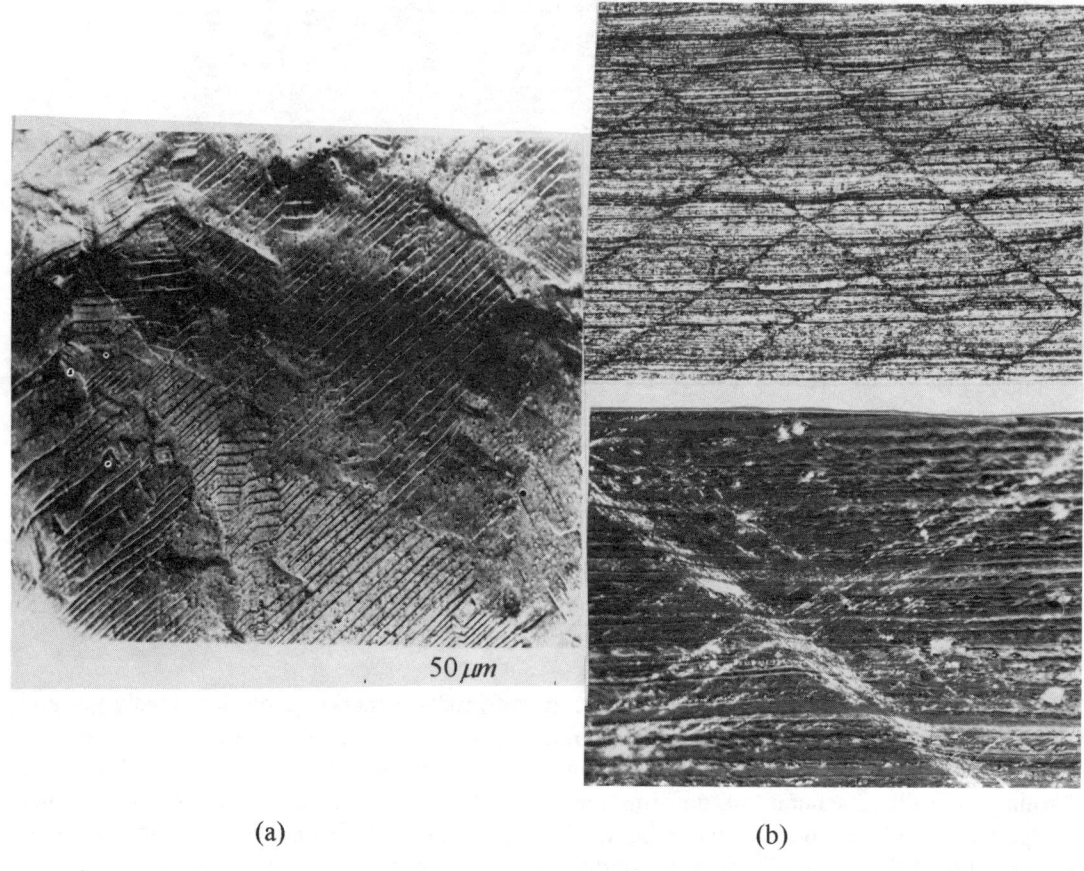

(a)   (b)

Fig. 2.16 Shear bands in copper (a) [23] and in aluminium alloy [22] after complex deformation.

cannot be ignored if one wishes to have a full pattern of plastic flow instability. Without it the understanding of the mechanism of localisation and of the governing physical rules becomes problematic. It is necessary, however, to stress at this moment that in order to explore the physical nature of localisation, we have to step down to the level of the crystal lattice, exploiting very sophisticated experimental techniques (e.g. optical metallography, electron microscopy, X-ray measurements). Only then it is possible to identify the mechanism of slip and to trace the evolution of the deformation from very local events of slip toward their organisation into a coarse slip band in a single crystal up to a macroscopic shear in polycrystalline metals. At the same time, however, we cannot lose the main track in which the geometry and the mechanics of slip are equally important. Hence, in the next chapter the attention is focused upon the properties of slip in crystals.

3. SLIP IN CRYSTALS

3.1 Slip criterion and slip geometry

Slip is defined as a relative displacement (shear) of two parts of the body across the common surface. In crystals such a surface is determined by the crystal lattice atomic plane. The number of equivalent (but not parallel) planes which may serve as slip plane depends on the crystal symmetry. The vector of the relative displacement (slip vector) is the second anisotropic property of slip in crystals. Hence, the criterion for slip in a crystal differs from that of an isotropic body in the sense that the shear stress must reach the critical value in the crystallographic slip system. The criterion for slip must, therefore, take into account the orientation of the slip plane and the slip direction with respect to the scheme of loading. Such an orientation is uniquely defined by the two unit vectors of the slip system: the vector normal to the slip plane **n** and the slip vector **b**. During uniaxial loading (tension, compression) of a single crystal, when the vectors of the slip system make with the applied stress σ_a the angles ϕ and λ, respectively, the shear stress τ (resolved shear stress) in the system is:

$$\tau = \sigma_a \cos\phi \cos\lambda \qquad (3.1)$$

and the slip in the system begins if

$$\tau > \tau_{crit} \qquad (3.2)$$

This criterion is known as the Schmid-Boas yield criterion [24]. The product: $\cos\phi \cos\lambda$, which is called the orientation factor, never exceeds the value of 0.5 but it may be smaller. Therefore, if in a crystal there are several crystallographically equivalent slip systems, the slip is initiated in this one for which the orientation factor is the highest. The criterion very selectively chooses the slip system at the onset of slip yielding, because only for „special" orientations it is met simultaneously for a few systems. If the criterion for slip is maintained for the same system throughout the deformation, then the micrograph of the crystal surface reveals one set of parallel lines (slip lines) as is shown in Fig.3.1. Such a case is called a single slip deformation. For the purpose of further analysis it is useful to distinguish the case where the slip system does not operate continuously, although it remains the most stressed by the applied load. Such a system will be called the „deformation bearing system" in order to be distinguished from the accommodating systems.

Under a much complex scheme of loading, which is represented by the applied stress tensor σ_a, the slip criterion receives a general form:

$$\tau^s = \sigma_a : \mathbf{b}^s \otimes \mathbf{n}^s > \tau_{crit} \qquad (3.3)$$

If in the course of slip some reaction or internal stresses develop, then the applied stress tensor in the formula (3.3) should be replaced by an appropriate sum of the applied σ_a and reaction σ_r stresses in order to predict in which slip system the slip has the best chance to occur:

$$\tau^s = (\sigma_a + \sigma_r) : \mathbf{b}^s \otimes \mathbf{n}^s \qquad (3.4)$$

Fig.3.1 The micrograph of slip lines on the surface of a single crystal resulting from a single system slip.

Let us suppose now that the slip criterion is met for a system leading to a homogeneous slip as it is schematically shown in Fig.3.2. This figure shows all the geometrical features of slip in crystals. To expose them, let us first notice that there are two interrelated orthogonal reference systems in the figure. One, the external or the loading scheme system of axes (x_1, x_2, x_3) coincides with the initial sample configuration (Fig.3.2a). The second is defined by the slip system unit vectors $\mathbf{b}$, $\mathbf{n}$ and ($\mathbf{b} \times \mathbf{n}$). It is easy to notice that deformation by slip is a simple shear deformation and in the slip system reference system the deformation (distortion) tensor $\mathbf{e}_{ij}$ contains only one non-zero value $e_{12} = \gamma$. Thus, the deformation of the crystal by slip is fully characterised by $\mathbf{n}$ and $\mathbf{b}$, and the simple shear γ. If the vectors $\mathbf{b}$ and $\mathbf{n}$ are defined in the external reference system, then the corresponding components of the crystalline sample distortion in the external reference system are given by:

$$e_{kl} = b^s_k n^s_l \gamma^s \qquad (3.5)$$

Structural and Mechanical Aspects

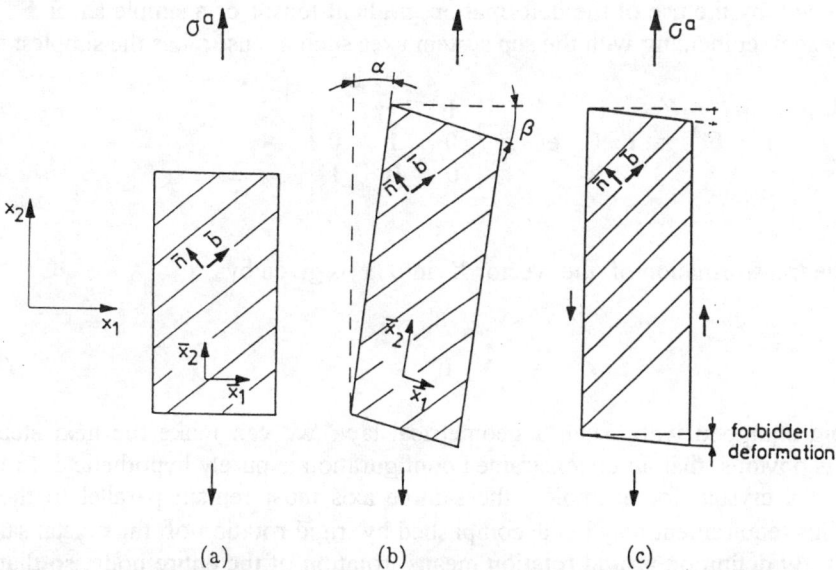

Fig.3.2 Schematic picture of the change in the crystal shape and rotation due to evenly distributed slips in a single system.

Tensor (2.5) gives the exact description of the effect of slip regardless of the value of γ, provided, however, that there are no constraints upon the new configuration of the sample (crystal shape and orientation with respect to its initial configuration) as it is shown in Fig.3.2b. The slip vectors **b** and **n** maintain the same position before (Fig.3.2a) and after slip (Fig.3.2b) and the atoms arrangement in the slip planes does not change throughout the deformation by slip. This means, in turn, that slip does not cause the crystal lattice rotation. This is a very important property of slip. In a new (unconstrained) configuration the external faces of the crystal receive, however, new orientations with respect to the crystal lattice. One may say, therefore, that there is a relative rotation of the crystal „envelope" with respect to the crystal lattice. A complete analytical account for the slip deformation must include, therefore, the rotation of the crystal faces relative to the crystal lattice. Let us show how it can be done starting with the rotation of „a material fibre" lying initially along a lattice direction **X**. Such a material fibre may be e.g. parallel to the tensile axis of the sample. A homogeneous slip γ in the system **b**, **n** brings the fibre into new (relative to the crystal lattice) position **x** such that:

$$\mathbf{x} = \mathbf{X} + \gamma (\mathbf{X} \bullet \mathbf{n}) \mathbf{b} \tag{3.6}$$

By choosing two vectors (fibres) on the crystal face and transforming them according to Eq. (3.6) a new orientation of the face (vector normal to the face) may be found from the cross product of the transformed vectors. Fig.3.2b shows the effect of slip on the position of two initially orthogonal edges of the sample with respect to the crystal lattice. Equivalent results can be obtained by the use of the deformation gradient tensor of a simple shear $\mathbf{F}^{pl}$. In the reference system coinciding with the slip system axes such a tensor has the simplest form:

$$\mathbf{F}^{pl} = \mathbf{I} + \mathbf{e} = \begin{vmatrix} 1 & 1+\gamma & 0 \\ 0 & 1 & 0 \\ 0 & 0 & 1 \end{vmatrix} \qquad (3.7)$$

and then the transformation of the vector $\mathbf{X}$ into $\mathbf{x}$ is given by:

$$\mathbf{x} = \mathbf{X}\ \mathbf{F}^{pl} \qquad (3.8)$$

Being equipped with such a geometrical law we can make the next step in the analysis. It is obvious, that an unconstrained configuration is purely hypothetical. In the case of tension of a crystal, for example, the sample axis must remain parallel to the tensile direction. This requirement may be accomplished by rigid rotation of the crystal sample (α in Fig.3.2b). By definition a rigid rotation means rotation of the entire body, so that during such a rotation also the crystal lattice rotates with respect to the external reference system (Fig.3.2c). In real deformation conditions slip in crystals is always accompanied by the lattice rotation and this last effect is due to interaction (constraints) with tools through which the load is applied. If, for example, the crystal axis slightly deviates from the tensile axis, then the pair of forces from the opposite grips provide the unbalance moment which instantly, via rigid rotation α, restores a stable configuration. Fig.3.2c shows also, that the alignment of the sample axis with the tensile axis is not the only requirement of a full matching of the sample with the tool (grip) through which the load is applied to the crystal. This is because the slip causes that the initial orthogonality of the crystal faces is lost (angles α and β in fig.3.2b in general have different values). For a full matching to be achieved the rigid rotation must be accompanied by some additional deformation which restores the right angle between the crystal edges in the case shown in Fig.3.2. More precisely, the faces of a crystal perpendicular to the tensile axis are forced to remain orthogonal. In a simplified, two dimensional picture the additional, accommodating (matching) deformation resolves itself into a homogeneous shear along the tensile direction and is driven by the reaction stress. If such an accommodation is elastic, then the crystal lattice is distorted in tact with the macroscopic distortion of the sample. Such an elastic distortion has different components than those of the (elastic) distortion caused by the applied stress. Hence, it has to be considered as resulting from the change of the stress tensor (change of the scheme of loading). The reaction induced stresses work against the slip in the system by producing the back stress. It diminishes the net value of the resolved shear. Once it happens, the yield criterion (Eq.3.4) is not fulfilled. In order to continue plastic straining the applied stress must increase either to overcome the back stress or to initiate slip in a secondary, the most stressed (Eq.3.4) slip system. This may be the case even if the crystal orientation with respect to the applied load still favours slip in the primary system. A major part of driving force (shear stress) for slip in a secondary system comes then from the reaction stresses. Thus, following the previous arguments, slip in a secondary (instantly the most stressed) system diminishes the reaction stresses, playing the role of the reaction stress relaxing system. For this reason the activity of a secondary system quickly vanishes and slip in the primary system is restored. One has to remember, however, that there iexists the elastic distortion of the lattice prior to slip in a secondary system. The elastic distortion is very small, hence, it may be, with reasonably good approximation, decomposed into the strain and the rotation tensors. It is necessary to stress here, however, that such an arithmetic decomposition of the distortion tensor into the symmetrical strain and anti-symmetrical rotation tensors is justified only if the distortions are very small. Slip in a secondary system replaces the elastic strains into plastic with no effect on the lattice rotation. In a more exact approach the rotation of the crystal lattice during tensile deformation can be approximated by a couple of elastic shears displacing the atomic planes into a configuration equivalent to the crystal lattice in a new (rotated) configuration. Stress relaxation due to a secondary slip removes the elastic strains leaving undistorted lattice in its most recent orientation. Then the atomic row along the crystal axis is exactly aligned with the tensile axis. This is not the same as the rigid rotation of the lattice by the angle calculated on the basis of antisymmetrical parts of the distortion tensor. The difference is negligible at very small strains (elastic) but appears significant for finite (plastic) distortions growing then to an important problem. It is necessary, therefore, to focus on this problem because, following Tylor proposal [25], the prediction of a crystal rotation (formation of the deformation texture in polycrystalline metals) is often based upon decomposition of the plastic distortion into strain and rotation tensors and their derivatives (plastic strain rate and plastic spin tensors) with respect to time.

3.2 Analytical account for a single and multi-system slip.

The equation (2.5) shows how to transform a simple shear deformation from the slip reference system into an external (laboratory or the loading) one. Such an external system of axes allows to summarise the effects of slip in different systems into one deformation tensor. It is also common for deformation and stress (loading scheme) tensors. Depending upon the orientation of the slip system (b,n) the deformation tensor e_{ij} may contain several non-zero value components being always non-symmetric. Regardless, however, of the slips system orientation, the sum of the diagonal components (the first invariant of the deformation tensor) is zero, which means that there is no volume change associated with slip. Now, because of the argument of the balance of moments, the stress tensor (Cauchy) has to be symmetric or $\sigma_{ij} = \sigma_{ji}$. The theory of elasticity assumes, therefore, a relationship (Hooke's law) between the symmetrical stress tensor and the symmetrical strain tensor. In order to find a constitutive law for plastic deformation, the symmetrical part has to be extracted from the deformation tensor. The simplest way of doing this is to split the deformation (distortion) tensor e_{ij} into the sum of two tensors: symmetric ε_{ij} (strain) and anti-symmetric ω_{ij} (rotation):

$$e_{ij} = \varepsilon_{ij} + \omega_{ij}, \tag{3.9}$$

such that:

$$\varepsilon_{ij} = 1/2\,(e_{ij} + e_{ji}) \tag{3.10}$$

and

$$\omega_{ij} = 1/2\,(e_{ij} - e_{ji}) \tag{3.11}$$

The term ω_{ij} determines the value of the angle of the rigid rotation of the crystal (body) around the axis $\mathbf{k} = \mathbf{i} \times \mathbf{j}$. Such rotations bring the body into a configuration in which the deformation is given by the symmetrical strain tensor. As it will be shown later, such a rotation has a doubtful physical meaning. Nevertheless, the idea to consider the deformation is such a way appeared very attractive for a simple reason. Then, due to symmetry (and the volume constancy) the deformation tensor contains only five independent components. Such a statement is heavy with consequences. It simply means that an arbitrary change of a crystal shape can be accomplished by slip in five (independent) systems. Therefore the achievement of a combination of components ε_{ij} of the strain tensor resolves itself into a solution of five equations for ε with five variables γ^s (shear in a system „s"). This is possible if the effect of slip in several systems is really additive. Then from eq.(3.5) and (3.10) one may get:

$$\varepsilon_{11} = (b_1 n_1)^1 \gamma^1 + (b_1 n_1)^2 \gamma^2 + \ldots\ldots\ldots\ldots + (b_1 n_1)^5 \gamma^5$$
$$\varepsilon_{22} = (b_2 n_2)^1 \gamma^1 + (b_2 n_2)^2 \gamma^2 + \ldots\ldots\ldots\ldots + (b_2 n_2)^5 \gamma^5$$
$$\varepsilon_{33} = -(\varepsilon_{11} + \varepsilon_{22})$$
and
$$\varepsilon_{12} = \varepsilon_{12} = 1/2\,(b_1 n_2 + b_2 n_1)^1 \gamma^1 + 1/2\,(b_1 n_2 + b_2 n_1)\gamma^2 + \ldots + 1/2\,(b_1 n_2 + b_2 n_1)^5 \gamma^5$$
$$\varepsilon_{13} = \varepsilon_{31} = 1/2\,(b_1 n_3 + b_3 n_1)^1 \gamma^1 + 1/2\,(b_1 n_3 + b_3 n_1)\gamma^2 + \ldots + 1/2\,(b_1 n_3 + b_3 n_1)^5 \gamma^5$$
$$\varepsilon_{23} = \varepsilon_{32} = 1/2\,(b_2 n_3 + b_3 n_2)^1 \gamma^1 + 1/2\,(b_3 n_2 + b_2 n_3)\gamma^2 + \ldots + 1/2\,(b_3 n_2 + b_2 n_3)^5 \gamma^5$$

$$(3.12)$$

There is a non-trivial solution of these equations provided that the determinant of the matrix of the slip systems orientation factors $M_{ij} = 1/2\,(b_k n_l + b_l n_k)$

$$\text{Det } M_{ij} \neq 0$$

Slip systems which fulfil this condition are called independent. By appropriate replacement of double indices by one index (varying from 1 to 5) one can get a simplified expression for strain components for the given γ values in the systems:

$$\varepsilon_i = M_{ij}\,\gamma_j \qquad \text{(for i and j form 1 to 5)} \tag{3.13}$$

or for γ values in the slip systems:

$$\gamma_i = M_{ij}^{-1}\,\varepsilon_j \tag{3.14}$$

or in the form similar to Eq.3.4:

$$\gamma^s = \varepsilon : \mathbf{b}^s \otimes \mathbf{n}^s \tag{3.15}$$

If a crystal possesses more than five independent slip systems (like e.g. regular crystals), then in order to predict the shears γ, needed to achieve the required deformation (strain), one has to choose these five systems first. The criterion for such a choice is a separate physical problem. We will come back to it later. If, however, such a choice has been made, then the values of shear in the system calculated upon the set of equations (3.14) determine also the components of the rotation tensor. Similarly as it was done before, we may express these components in a shorter form. Then

$$\omega_i = N_{ij}\gamma_j \qquad (3.16)$$

with $N_{ij} = 1/2 \, (b_k n_l - b_l n_k)_j$.

With the assumption that ω_i corresponds to the rigid rotations of the crystal lattice, the deformation texture can be predicted [25]. It is worth, therefore, to check whether such an assumption is true. In tension (Fig.3.2), because of gripping, the sample rotates around the axis x_3 of the external reference system by an angle α which varies with the orientation of the slip system in the sample. The value of the rotation angle is given by the dot product of the unit lattice vectors parallel to the crystal axis before and after the slip. The equation (3.11) gives the rotation around the axis x_3 (common for the external and the slip reference systems) $\omega_{12} = \gamma/2$ which does not depend upon the slip system orientation. The problem extends beyond a very particular case of the crystal behaviour in tension in which the rotation of the crystal is forced by grips (external constraints). As it was shown in Fig.2.3 the neighbouring grains play similar role. And just because of such internal constraints the crystallites rotate relative to the external form (envelope) of the sample (elastic accommodation). In result, the configuration of a grain differs from that predicted from the plastic distortion tensor, and this difference concerns the value of the grain rotation. It is easy to show that the final configuration of the grain within a polycrystalline body might be achieved solely by slip in the eight independent systems. This is because of the material volume constancy, which causes that the plastic distortion tensor has only eight independent components. Therefore, the solution of a set of eight equations, like Eq.3.6, with respect of shears in the eight slip systems ensures fulfilment of the demand of the final configuration of the crystallite within the material. But then, there is no need for a rigid rotation of the grain with respect to the external reference system. Consequently, a simultaneous operation of the eight independent systems in a crystal (in each grain of a polycrystalline aggregate) makes the crystal free from the constraints during plastic straining and moreover, free from the internal stresses and elastic accommodation of incompatible deformations. We are coming, finally, to the conclusion that in such case the orientation of the crystal lattice in each grain would remain stable in the course of straining because, as it was already shown, slip does not rotate the lattice. This is in evident conflict with the experiments. The development of a preferred orientation of crystallites in the course of deformation (formation of the deformation texture) is an important physical fact. An assumption of slip in the five systems is, therefore, a compromise between the demand of a change of the external form of the body and the lattice rotation, provided that these systems operate simultaneously. This last requirement may not follow the yield criterion (Eq. 3.3) which very selectively chooses the most stressed system. Moreover, a crystal may posses more than five crystallographically equivalent slip systems. In a face centred cubic crystals (like e.g. copper or aluminium) there are twelve equivalent slip systems and hence 792 combinations of five. This number is reduced to 364 by the demand that the systems are independent. The problem of the choice of a set of the five independent systems which as the only operates throughout the deformation, has been exposed and analysed by Taylor [25]. His criterion for yielding follows the argument that these systems operate which ensure that the deformation is accomplished by the smallest possible sum of shears. As it was shown by Bishop and Hill [26], this criterion selects these systems which are the most, although not necessary identically, stressed.

The other, not less important consequence of the decomposition of the deformation tensor concerns the evaluation of the effects of a multi- system slip. According to Eq.2.11 the sequence of the shears in operative slip systems has no influence on the final effect. This is true only if these systems operate all together at constant slip rates which, in addition, must be proportional to the contribution of the shear in a slip system in total deformation. This requirement is highly unrealistic. First of all, there does not exist such a stress state in which, in regular crystals, five slip systems would be equally stressed [26]. Secondly, even in a few special crystal orientations to the applied load (symmetrical orientations) for which the resolved shear stress is the same in a couple of systems, the chance that more than one slip system is activated at a given instant in a local area of the crystal is negligible. This is so, because an ideally symmetrical orientation is hardly probable, and if it occurs, the local conditions in the crystal (lattice imperfections) still favour the choice of a single system. It will be shown later that already very small deformation by a single system slip leads to highly anisotropic structural hardening of the crystal (latent hardening). While the slip in a system does not harden itself, except through the constraints (non-structural hardening), it hardens the other, non-coplanar systems [20]. Hence, the initial equivalence of crystallographically the same slip system in a virgin crystal is instantly lost. This fact strengthens the argument that in most cases slip systems have to operate in a sequence in the course of plastic yielding. The value of shear carried by a slip system during its temporary activity is governed by the evolution of the stress state and may vary with deformation. It is very instructive, therefore, to compare the geometrical effects of a simultaneous and a sequential slip. To simplify the analysis let us choose a crystal in the form of a cube with its edges oriented along the axes of the reference system (Fig.3.3a). The crystal undergoes deformation by shear slip in two symmetrically disposed systems $\mathbf{b}^1, \mathbf{n}^1$ and $\mathbf{b}^2, \mathbf{n}^2$. Let the shear value in each system be $\gamma^1 = \gamma^2 = 1$. According to the chosen, common reference system (x_1, x_2) the slip vector and the vector normal to the slip plane are:

$\mathbf{b}^1 \equiv [1/\sqrt{2}, 1/\sqrt{2}, 0]$, $\mathbf{n}^1 \equiv (-1/\sqrt{2}, 1/\sqrt{2}, 0)$ for the first system, and
$\mathbf{b}^2 \equiv [-1/\sqrt{2}, 1/\sqrt{2}, 0]$, $\mathbf{n}^2 \equiv (1/\sqrt{2}, 1/\sqrt{2}, 0)$ for the second.

Then, the distortions resulting from slip in the first and the second system, respectively, are:

$$e^1_{ij} \equiv \begin{vmatrix} -0.5 & -0.5 & 0 \\ 0.5 & 0.5 & 0 \\ 0 & 0 & 0 \end{vmatrix}, \quad e^2_{ij} \equiv \begin{vmatrix} -0.5 & 0.5 & 0 \\ -0.5 & 0.5 & 0 \\ 0 & 0 & 0 \end{vmatrix}$$

and their net effect:

$$e^T = e^1 + e^2 \equiv \begin{vmatrix} -1 & 0 & 0 \\ 0 & 1 & 0 \\ 0 & 0 & 0 \end{vmatrix} = \varepsilon^T$$

gives only the strain tensor and no rotation ($\omega = 0$). Therefore, the edges of the crystal (body) remain parallel to the axes of the reference system. This means that the crystal receives the form as shown in Fig.3.3b, regardless of the sequence of slip in the systems. Let us now, using the formula (3.6) or (3.8), check what happens to the crystal edges (faces) in result of the sequential slip in the first, and then in the second system.

Slip in the first, e.g. b^1, n^1, system causes that the crystal edges along the axes x_1 and x_2, whose initial length and orientation are given by the vectors [100] and [010] respectively, change the length and orientation. in tact with geometric law (Eq. 3.8): $x_i = X_i \ F^1$. The new configuration of the crystal edges (Fig.3.3c) is now given by vectors having the following components with respect to the external (or the crystal lattice) reference system:

$$x_1 = [\ 0.5, -0.5, 0] \quad \text{and} \quad x_2 = [0.5, 1,5, 0].$$

System 1: $b_1 \parallel [-1,1,0]$, $n_1 \parallel [1,1,0]$
System 2: $b_2 \parallel [1,1,0]$, $n_2 \parallel [-1,1,0]$

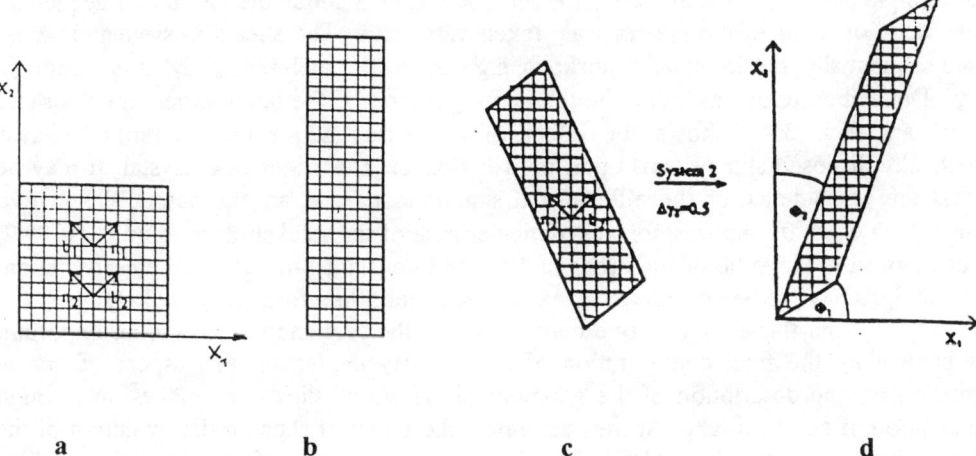

Fig.3.3 The effect of a simultaneous (a-b) and a sequential slip (a-c-d) in two symmetrically oriented systems on the final configuration of the crystal.

F^1 is the gradient deformation tensor for the first slip. Such new edges undergo a change in result of slip in the second system (F^2). Hence, a vector (an edge) x_i becomes transformed into x'_i such that $x'_i = x_i\ F^2$. Thus, the net effect of sequential slips on the length and orientation of the crystal edges, represented by the lattice vectors X_i, is given by:

$$x'_i = X_i\ F^1 F^2. \qquad (3.17)$$

In result, the edges of the crystal become transformed into:

$X_1 \Rightarrow x_1' \equiv [0.5, -0.5, 0]$ and $X_2 \Rightarrow x_2' \equiv [-0.5, 2.5, 0]$.

These new vectors reflect accurately the change of the length and orientation of the crystal edges. Therefore, they precisely define the geometrical effects of slip. A comparison of the final configurations of the crystal in Fig.3.3 b and d does not leave any doubt that they differ. Thus the slip systems do not commute and the final configuration of the body depends upon the sequence of slips in the operating systems. In a general case, in result of „n" slips in alternating systems a material fibre (the crystal edge but also an element of the crystal structure) experiences change of its length and orientation which can be exactly determined by a way of multiplication of each instant slip representing a tensor (the tensor of the deformation gradient) according to the sequence of their operation and the partial strains they carry:

$$x_i = X_i \cdot \prod_{i=1}^{n} F_i\ . \qquad (3.18)$$

In the example used to illustrate the difference between a simultaneous and a sequential slip, the value of shear in the system was taken very large. The shears in systems may be imposed sequentially, but in smaller portions, e.g. $d\gamma^1 = d\gamma^2$ such that $\Sigma_n d\gamma^1 = \gamma^1$ and $\Sigma_n d\gamma^2 = \gamma^2$. Depending upon $d\gamma$ value the final configuration of the body varies for the same total shears. Fig. 3.4 shows the effect of $d\gamma$ during sequential operation of two symmetrically disposed slip systems upon the rotation of the edges of a crystal. It may be seen that the coincidence of the effects of a simultaneous and an alternating slip in two systems ($\Phi_1 = \Phi_2 = 0$) requires, for this particular case of the total shear $\gamma^T = \gamma^1 + \gamma^2 = 2$, the increment of shear to be of the order of 10^{-4} and constant throughout the deformation. Such conditions have to be considered as very special and, therefore, very unlikely.

From this example one may learn that the sequence of slips is an important factor controlling the final configuration of a body. By neglecting this aspect of crystal deformation in the description of the geometrical effects of slips one makes very rough approximation, if not a mistake. At the same time the effect of slips on the evolution of the metal substructure is missing. At a rigorous approach to the deformation of crystalline bodies we are facing, however, a real physical problem how to predict a sequence and value of partial shears in the slip systems involved into deformation. An attempt to answer this question is made later.

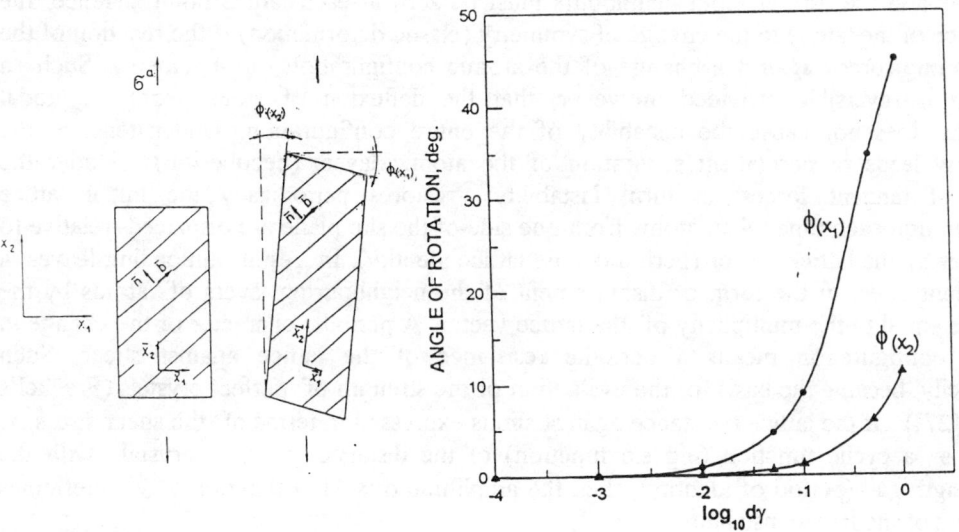

Effect of sequential slip in two systems for $\Sigma(d\gamma^1 + d\gamma^2) = 2$

Fig.3.4 The effect of the increment of shear $d\gamma$ during sequential slip in two symmetrically disposed systems upon the rotation of the edges of a hypothetical crystal.

4. MECHANISM OF SLIP

4.1 The rigid shear model

So far our attention has been focused upon the slip criterion and the geometrical effects of a single or multi-system slip. In order to have a full pattern of the plastic deformation these have to be connected with a „mechanism" of slip. This term means a manner by which one part of a crystal is displaced relative to the other across a slip plane. At first sight the mechanism of such displacement (shear) might be considered as a rigid relative motion of these two parts of the body during which all points of one part undergo the same displacement at the same time. This automatically means that the shear instantly extends across the entire body. Before possible acceptance of such a mechanism, we have to answer a couple of questions. The first one concerns the resistance of the crystal lattice to a rigid shear. It has to be followed by the question about the factors which control the distribution of slips in the crystal volume and therefore the micro-homogeneity of deformation. We wish, also, to know whether a rigid shear can change the crystal properties. The answer should be sought in the nature of inter-atomic bonds, due to which a crystalline structure exists. The problem extends far beyond the framework of the workshop used in the continuum mechanics for description of the relation between the applied load and the crystal response. This is so, because in an atomic scale the „concept of stress" has no physical meaning. The spatial arrangement of atoms in crystals results from the balance of inter-atomic forces at every atom site within the lattice. The sum of interactions between the close and the further atom neighbours must be zero at each lattice point. Hence, the resistance of the lattice to the change of symmetry (elastic deformation) is the reaction of the inter-atomic forces against a change of the atomic configuration (arrangement). Such a reaction is reversible provided, however, that the deflexion of atoms from the nodal positions does not cause the instability of the entire configuration. Under tension, the instability leads to permanent separation of the atomic layers (decohesion). Under the action of tangent forces, in turn, instability restores periodically the initial lattice configuration each time when atoms from one side of the slip plane are displaced relative to the other by the lattice vector (period of the lattice identity). In result, unloading leaves a permanent effect in the form of displacement of the neighbouring layers of atoms by the distance equal to the multiplicity of the lattice vector. A periodic character of the change in atomic configuration means a periodic resistance of the lattice against shear. Such periodicity became the basis for the evaluation of the strength of perfect crystals (Freckel's model [27]). If the lattice resistance against slip is expressed in terms of the shear stress τ, which is a cyclic function (e.g sin function) of the distance „x" in a crystal with the wavelength „a" (period of identity), then the amplitude of such a function τ_{max} determines the stress of the lattice instability:

$$\tau = \tau_{max} \sin(2\Pi/a) \tag{4.1}$$

The theoretical strength of the crystal τ_{max} may be evaluated if one takes into account that for a very small deflexion from the initial configuration x, (small shear γ) the Hook's law gives

$$\tau = \mu \cdot \gamma \tag{4.2}$$

where μ is the shear modulus (elastic constant) and $\gamma = x/a$.

From a comparison of Eq. (4.1) and (4.2) it can be found that τ_{max} is of the order of $\mu/2\Pi$. Since the elastic deformation is homogeneous in the atomic scale, hence τ_{max} reflects the strength of the crystal against a rigid, irreversible shear. Regardless how accurate function (4.1) is, it reflects the periodic structure of crystals. There is no reason, therefore, to question the order of the magnitude of the crystal strength given by such evaluation, especially that the experimentally proved strength of the very special crystal (perfect crystal in the form of a whisker) is of that order. Nevertheless, the result of such evaluation disagrees with the critical resolved shear stress of the usual single crystals being by 4-5 orders of magnitude higher. The only possible explanation to this discrepancy was to accept that slip in a crystal results from a local loss of the lattice stability and that, in a real crystal lattice, there are locations in which atomic arrangements differ from perfect. Only then, under the action of tangent forces, such local configurations may reach instability at very low forces (stress). In other words, if the lattice contains defects (dislocations), an elastic lattice distortion may destabilise configuration of atoms at the defect. Then, the defect periodically reproduces itself around the neighbouring lattice nodes. This gives rise to the movement of dislocations in the crystal. Such an idea, originated by Orovan [28], Polanyi [29] and Taylor [30,31] lay at the background of the theory of crystals plasticity

Structural and Mechanical Aspects

and attracted the attention to the role of defects in the mechanical performance of crystalline materials.

4.2 Dislocation slip and dislocation properties.

The movement of dislocations appears, therefore, as the mechanism of slip in a crystal. There are, however, some requirements concerning the properties of such a defect. A defected configuration must extend along a line (a linear defect- dislocation) and a change of it position must cause relative displacement of the parts of a crystal separated by the surface encircled by the moving dislocation. In the case of slip, the marked off surface coincides with the slip plane **n** and the displacement with the slip vector **b**. In terms of such a mechanism of slip the passage of the dislocation across the crystal gives the effect equivalent to a rigid displacement of two crystal parts across the slip plane. The value of the relative displacement is then determined by the geometrical property of the defect (Burgers vector of dislocation). It is worth noticing that dislocation movement is associated with energy dissipation. Since the destabilisation of the defect configuration occurs at the expense of the external work, the reconfiguration of the defect, which appears periodically with Burgers vector wavelength, causes that the energy dissipation takes place in every cycle of reconfiguration. This effect in the dislocation motion is often considered as „the lattice friction" effect.

Replacement of the concept of „rigid displacement" by the movement of dislocation forces us to consider slip as a set of time and space dependent events in which several dislocations are engaged. A macroscopic slip results from many such local slip events. The description of slip has to contain, therefore, the information about a number of instantly moving dislocations N and the distance in crystal Λ they pass. These (intrinsic) features quantify each event of slip in crystal. Let us look at the relationship between a global shear deformation by slip γ and the mean value of intrinsic features of slip events in a common slip system. Because dislocation is a linear defect, hence a useful parameter characterising the number of dislocations is the density of dislocations ρ. It is defined as the total length of all dislocations within a unit volume of a crystal. If a fraction ρ_m of the total dislocation density moves over an average distance Λ on the common slip plane, and each dislocation carries the displacement **b,** then a homogeneous at macroscopic approximation, shear γ in the system is:

$$\gamma = b \, \rho_m \, \Lambda \qquad (4.3)$$

In equation (4.3), known as Orovan relationship, averaging of the number of dislocations over the crystal volume makes that very local character of slip events and hence their distribution in a crystal may be lost (in some models the dislocation motion is considered in terms of the dislocation flux and analysed as a process similar to heat or mass transport). This relationship does not pretend to provide even a suggestion about the mutual relationships between the extent of slip Λ and γ, as well as between Λ and the number of dislocations engaged in the same event of slip. On the other hand, such links between the intrinsic properties of slip events appear fundamental from the point of view of slip evolution. The answer to these problems must be sought, however, in the properties of dislocations in crystals. Therefore, the most important properties of the dislocations in crystals have to be recalled. They were the subject of numerous theoretical considerations originating, as can be found in the common citations, in 1907 by the work of Volterra and applied to crystals by Burgers and Frank [32]. In the present form, they constitute a separate field of the mechanics of solids, known as the theory of dislocations (see for example [33-35]).

In the light of this last comment, it is clear that it is impossible in a short course to give sufficiently well motivated information about the nature and properties of defects in crystals. It is necessary to limit the considerations to those which are the most important from the point of view of the mechanism of plastic flow and the associated structural and mechanical effects.

Any defect of the crystal lattice causes its local elastic distortion. In case of a dislocation such a distortion remains constant along the dislocation line. The concept of the dislocation line was used to emphasise the topological features of the defects giving a highly localised distortion field which extends along a line in the mathematical sense. Such a mathematical object is, in turn, defined by the versor of the dislocation line **t**. In fact, dislocation has a finite extension in the plane perpendicular to the line which separates two parts of crystal, displaced by the vector **b**. Therefore, vectors **b** and **t** determine the geometrical properties of a dislocation within the crystal lattice. The cross product $\mathbf{b} \otimes \mathbf{t} = \mathbf{n}$ gives the vector **n** normal to the plane which contains the vector of shear (Burgers vector of dislocation) and the dislocation line. If in an actual configuration of a dislocation vector **n** coincides with the vector normal to the crystallographic slip plane (the plane of the least slip resistance), then the dislocation remains in the slip plane. Otherwise, its mobility is reduced at the most to a diffusion controlled climb. From the geometrical constraints on the slip plane it appears, that if **b** is parallel to **t**, there is more than one slip plane for such a dislocation. If, simultaneously, the dislocation line and the Burgers vector belong to a few crystalographically allowed slip planes, then the dislocation receives the property of changing the slip plane. Such a manoeuvre during a dislocation glide is called cross slip. This is an explicit property of the so-called screw dislocation.

In terms of the theory of elasticity the defect is a source of an internal stress field in a crystal. Depending on the mutual orientation of the vectors **b** and **t** such a stress field has different components. In effect, the orientation of the dislocation line relative to the Burgers vector is a controlling factor of the physical properties of the dislocation. Among the spectrum of such orientations those at $0°$ and $90°$ are distinguished as special orientations of the dislocation, often called screw and edge dislocation, respectively. Examples of the corresponding configurations of atoms and stress tensors are shown in Fig. 4.1. It must be stressed, however, that while **b** is constant for a given dislocation, its line may assume any form, except that it must be continuos within the crystal (cannot break). As it follows from the properties of dislocations (stress field), creation of such a defect needs that some work has to be done on the crystal. The work per unit length of the dislocation line is defined as the proper energy of the dislocation. It may be evaluated by summing the products of the corresponding components of the stress and strain tensors over the crystal volume. Thus, the presence of a dislocation in a crystal increases the crystal free energy. In the isotropic elasticity approximation the energy of dislocation is proportional to the shear modulus μ and the square of the Burgers vector of a dislocation:

Structural and Mechanical Aspects

$$E_d \approx \mu b^2 \qquad (4.4)$$

According to the second principle of thermodynamics the most stable state is a state of the least free energy. From these two arguments one concludes, that the most expected dislocations in a crystal are those which have the smallest Burgers vector (which cause the smallest distortion of the lattice). In crystals, the smallest **b** corresponds to the smallest interatomic distance in a perfect lattice. Dislocation with such Burgers vector is called „perfect dislocation". For example, in fcc crystals the closest atoms are along <110> lattice direction and the Burgers vector of a perfect dislocation is given by a/2<110>, where „a" is the lattice parameter. This fact explains the anisotropy of the slip direction in a crystal. The argument of the free energy causes also that the dislocation line tends to be as short as possible. This is because the free energy increase is proportional to the total length of dislocation. Hence „dislocation resists" to be curved and such a resistance, called the line tension of the dislocation (per analogy to the surface tension), is defined as **T**, such that:

$$T \approx \mu b^2 \ln(R/r_o) \qquad (4.5)$$

where R is the radius of the dislocation curvature (cut off radius) and r_o is the radius of the dislocation core ($\approx 5b$). The line tension **T** has a sense of the force applied to the ends of a dislocation segment and tangent to the dislocation line.

The other consequence of the minimum free energy argument is the interaction between dislocations in crystal. It may be deduced on the basis of the superposition of strain and the corresponding stress fields of dislocations. Such an interaction causes repulsion of dislocations, if the superposition of stress fields leads to an increase of the elastic energy in the crystal, or attraction in the opposite case. One can distinguish, therefore, high and low energy configurations of dislocations. Attraction may lead to the replacement of two superposed defects by a new one of a lower proper energy. We say, then, that there is a reaction between dislocations.

The reacting dislocations must fulfil the topological demand of continuity, known as the Frank's rule, according to which the Burgers vector of the product of the reaction $\mathbf{b}_p$ is the sum of the Burgers vectors of superposed dislocations, say $\mathbf{b}_1$ and $\mathbf{b}_2$:

$$\mathbf{b}_p = \mathbf{b}_1 + \mathbf{b}_2 \qquad (4.6)$$

or, in a more general form:

$$\Sigma_i \mathbf{b}_i = 0. \qquad (4.7)$$

The sum of the Burgers vectors of dislocations which meet at a node is zero.
The criterion for the reaction (4.6) is given by:

$$b^2_p \geq b^2_1 + b^2_2 \qquad (4.8)$$

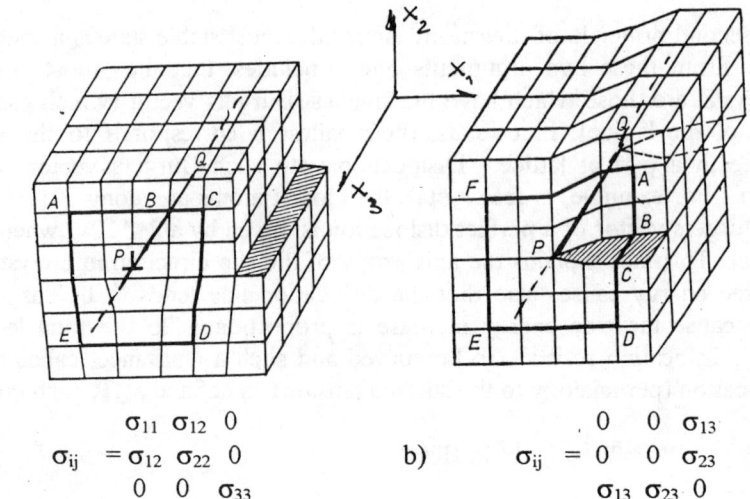

a) $\sigma_{ij} = \begin{matrix} \sigma_{11} & \sigma_{12} & 0 \\ \sigma_{12} & \sigma_{22} & 0 \\ 0 & 0 & \sigma_{33} \end{matrix}$ b) $\sigma_{ij} = \begin{matrix} 0 & 0 & \sigma_{13} \\ 0 & 0 & \sigma_{23} \\ \sigma_{13} & \sigma_{23} & 0 \end{matrix}$

with:
$\sigma_{11} = [-\mu b / 2\Pi (1-\nu)] \bullet [x_2(3x_1^2+x_2^2) / (x_1^2+x_2^2)^2]$;
$\sigma_{22} = [\mu b / 2\Pi (1-\nu)] \bullet [x_2(x_1^2-x_2^2) / (x_1^2+x_2^2)^2]$;
$\sigma_{33} = -(\sigma_{11} + \sigma_{22})$;
$\sigma_{12} = [\mu b / 2\Pi (1-\nu)] \bullet [x_1(x_1^2-x_2^2) / (x_1^2+x_2^2)^2]$;
$\sigma_{13} = (-\mu b / 2\Pi) \bullet x_2 / (x_1^2+x_2^2)$;
$\sigma_{23} = (\mu b / 2\Pi) \bullet x_1 / (x_1^2+x_2^2)$.

Fig.4.1 The model configuration of atoms around the edge (a) and screw (b) dislocations and the corresponding tensors of internal stress fields.

There are two effects of such reactions: a decrease of the crystal free energy (crystal recovery) and formation of the dislocation network (Frank's networks) with nodes which are the pinning points of the dislocation line.
   It is often useful to use the concept of „the force" on dislocation. The formula, derived for the first time by Peach and Koehler [32] bases upon the fact that in order to move a dislocation over some distance some work has to be provided to the crystal. Such a work may be done by external as well as internal stress field σ_{ij}. Because the mechanical work may be expressed as the dot product of the force **F**, and the distance passed by a dislocation, hence the formula for „the force on dislocation" may be written:

$$\mathbf{F} = \mathbf{b} \otimes \sigma_{ij} \times \mathbf{t} \qquad (4.9)$$

It defines **F** as a force acting on a unit length of the dislocation and perpendicular to the dislocation line at every point of it. If the stress tensor contains only one non-zero component, such that $\sigma_{tb} \neq 0$, then the force, $\mathbf{F} = \mathbf{b} \cdot \sigma_{tb}$, works in the plane $\mathbf{n} = \mathbf{b} \times \mathbf{t}$

(slip plane of dislocation) Hence, if dislocations form Frank's network, they bow out between the pinning point (nodes). A dislocation segment undergoes simultaneous action of the line tension **T** and the force **F**. The balance is met if:

$$b \cdot \sigma_{tb} = b \cdot \tau^s = T/R \qquad (4.10)$$

Inspection of equation (3.10) shows, in particular, that the curvature of the dislocation segment is inversely proportional to the shear stress in the slip system. Increase of the stress leads to a decrease of the curvature radius from $R = \infty$ of an initially straight segment to that given by Eq.3.10. The smallest value of R determines the highest shear stress in the slip plane. It is obvious, however, that if the distance between the pinning points is „l", the smallest bow out radius is l/2. Further expansion (glide) of dislocation causes increase of R and consequently a decrease of the stress. Thus, according to this model, the dislocation glide becomes unstable when the dislocation curvature radius reaches the value $R = l/2$. Because the force on dislocation is perpendicular to the dislocation line, the expansion is connected with lengthening of the segment and its rotation around the pinning points. The rotation brings opposite branches of the segment into direct contact. Then, due to the already mentioned topological properties, the dislocation receives the form of an expanding loop and its initial segment is reproduced between the pinning points. Such a mechanism of slip, known as the Frank-Read source, explains the most fundamental feature of plastic deformation, namely the multiplication of the dislocations in the course of slip. It shows also, that any segment of dislocation may work as a dislocation source, provided that it is locked at opposite ends. Dislocations released from a source expand in the form of loops in the common slip plane. These are identical dislocations. When they reach the free surface of a crystal they make a step which may be visible as the slip line. Figure 3.1 is an example of the slip lines at the very beginning of plastic straining of a single crystal oriented for a single system slip. Unstable emission of dislocations from a source is not, except of very special conditions, seen by the load recording system (no instability of the load), although the local nature of the slip is very well marked. The height of the step on the surface, which is several times larger than the Burgers vector of dislocation, may be used as evidence that dislocations were created at the same source and glide on the same plane. Being concentric loops (or parallel lines) they interact with one another along their length. Hence, the interaction is a long range interaction. This term emphasises that such an interaction cannot be overcome with the help of thermal fluctuation in the crystal lattice. Formation of a slip line appears, therefore, as an athermal mechanism and the movement of the dislocations emitted from a source as a collective process [36]. Those dislocations created by a source which do not leave the crystal (remain stored in the crystal) form a high energy configuration unless they rearrange or interact with the other dislocations.

Activity of a particular source (number of dislocations emitted) is fully controlled by the applied stress and interaction of the source with the neighbouring dislocations. In particular, a „back stress" on the source from the already emitted dislocations, which stack over the distance Λ (free path of the dislocations), arrests the source activity. The number of emitted dislocations N is then proportional to the shear stress in the system and inversely proportional to the free path of dislocations:

$$N \approx (\mu b/2\Pi K) \tau^s / \Lambda \tag{4.11}$$

K is the dislocation orientation dependent factor : K=1 for screw and K= (1-v) for edge oriented dislocations. v is Poisson elastic constant.

A planar arrangement of dislocations emitted from a source and arrested on an obstacle, is called dislocation pile-up because of the distribution of dislocations in the configuration. The term „pile-up" emphasises that there is a very high stress concentration at the head of the configuration. Its measure may be the force on the leading dislocation which is N times larger than that on an isolated dislocation under the applied stress only. Such a local stress concentration is sufficient to activate the other dislocation sources in its vicinity.
It is, therefore, possible to relax such a configuration by formation of an opposite Burgers vector arrangement on a parallel slip plane, as it is schematically shown in Fig.4.2. Seen in Fig.3.1 clustering of slip lines in narrow bands (slip bands), composed of several slip lines, indicates that many dislocation sources are activated in a close vicinity. A new configuration, which then forms, consists of pairs of opposite Burgers vector dislocations (dipoles). The stress field of the dipole diminishes the distance x from its centre at a rate at least as high as $1/x^2$ causing very efficient internal stress release. There is, therefore a natural tendency toward concentration of slip in a crystal which results from the demand of the minimisation of elastic energy (stored energy) in crystals.

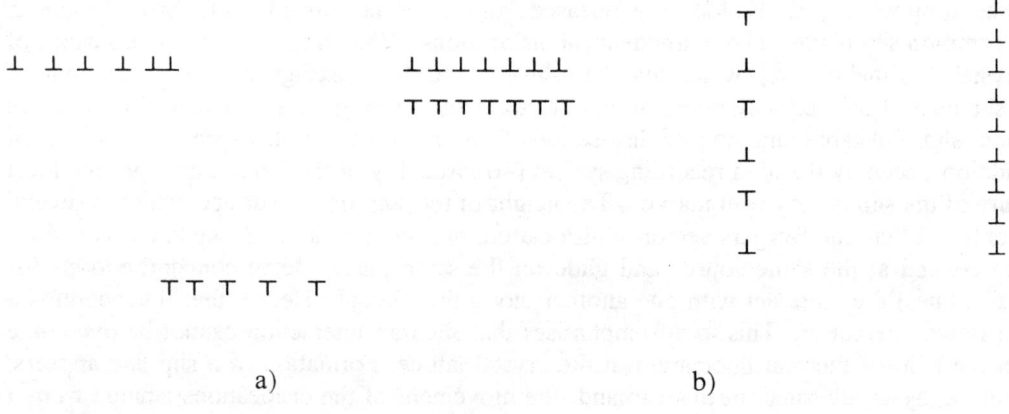

Fig.4.2 High (a) and low (b) energy configurations of dislocation in the slip plane.

Multiplication and storage of dislocations in the course of plastic straining causes that the crystal structure becomes heavily defected. While the density of dislocations in a virgin crystal is usually of the order of 10^5 - 10^6 [cm/cm^3], already a few percent elongation rises it by the factor of 10^3. Because each dislocation is a source of internal stress in crystals, hence, through relation similar to (3.4) it contributes to the shear stress in a slip system. This is the simplest way to explain the hardening of the crystal during plastic deformation. It can be shown that at the assumption of uniform distribution of dislocations the mean value of such internal stresses is proportional to the square root of the dislocation density. This is so, because the components of the dislocation stress tensor are inversely proportional to the distance from the dislocation and the mean inter-spacing of dislocation is $1/\sqrt{\rho}$. Therefore, the current flow stress of the crystal may be linked to the dislocation density by the relation:

$$\tau = \tau_o + \alpha \mu b \sqrt{\rho} \qquad (4.12)$$

where τ_o represents the resistance of the perfect lattice to the dislocation glide (friction) and α is the proportionality factor.

The assumption of an uniform distribution of dislocations is unrealistic and in a real case the dislocations form some spatial arrangements - the dislocation substructure. In result of slip in one system such substructure consists mostly of walls of dipoles in the parallel planes. Slip in a secondary system leads to formation of similar walls on the other planes and to interaction between dislocations of the primary and the secondary system. Then a more complex spatial arrangement of the substructure elements (dislocation walls or tangles) is formed. If the net Burgers vector of the dislocations forming the wall is not zero, then the wall itself gives rise to a split of a crystal into blocks with slightly misoriented (mosaic) lattice. Regardless of the extent of the rearrangement or recombination of dislocations, internal stresses within such walls are very high and make the walls hardly penetrable by the other dislocations. Dislocation substructure constitutes, therefore, the obstacles network for the dislocation glide. The proportionality of the flow stress and the square root of the dislocations is, in general, retained although the factor α varies depending on the dislocation distribution.

5. MICROSTRUCTURAL ASPECTS OF DEFORMATION

5.1 Slip pattern and mechanical characteristics of FCC single crystals.

In the light of the properties of dislocations very briefly summarised above it can be concluded, that the glide and storage of the dislocations in crystals appear as a microstructural mechanism of plastic flow which equally well satisfies the demands of geometrical changes of a body and the evolution of its mechanical properties. From the properties of the dislocations it follows also that a macroscopic effect of slip results from physical events which comprise the generation and movement of dislocations in the common plane. The extent of this elementary event - the area swept or mean free path of the dislocation glide, and the number of dislocations generated- are intrinsic features of an individual act of slip which evolve with the development of the obstacle network (substructure). Such an evolution is responsible for the rate of the dislocation storage in a crystal and, in this way, on the rate of the strain hardening. The prediction of the strain hardening rate should, therefore, be based upon the analysis of the factors influencing the intrinsic features of slip. It terms of Eq. 4.3 it can also be noticed, that a decrease of Λ works toward more slip events in the unit volume of crystal at the same macroscopic effect (γ). One may expect, therefore, that a decrease of a mesh-length of the obstacles network favours homogeneous deformation. On the contrary, an increase of Λ leads, even in a more evident way, to concentration of slip in the crystal and to decrease of the rate of the dislocation storage. According to this argument, a breakdown of the obstacle network is an instant cause of a change in the flow behaviour. A comparison of the mechanical performance of a crystal with the evolution of the slip during deformation may help to find out the factors controlling the slip intrinsic features and the rule they obey.

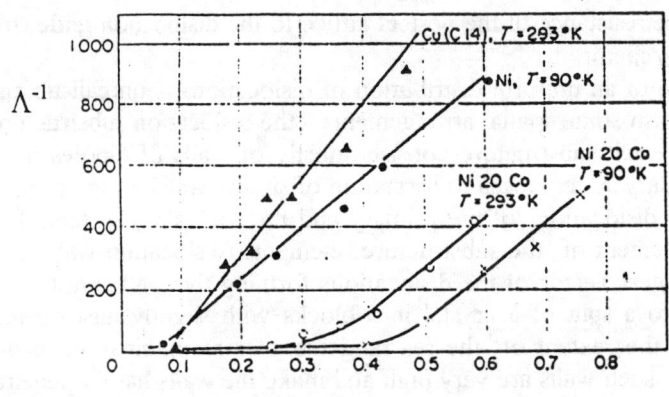

Fig.5.1 The experimental plot of the reciprocal of the slip line length Λ vs. the resolved shear deformation γ in the deformation bearing system [39].

While the number of the generated dislocations during a slip event is practically a not measurable quantity, the evolution of the extent of slip may be experimentally traced. This is so, because slip leaves a topographic replica on the crystal surface (slip line length). The early studies of slip patterns in a single crystal have shown that the length of the slip lines decreases with strain [37-39]. This change is not monotone throughout the deformation and some stages in the slip line evolution are observed as it is shown in Fig. 5.1 after Mader and Seeger [39].

It is interesting, therefore, to compare this evolution with the mechanical properties of crystal during straining. The tensile characteristic of a FCC single crystal is practically the only case for which such a comparison can be made. In Fig.5.2(a,b), quoted after Basinski [20], the resolved τ vs. γ tensile curves of a copper single crystal are shown along with the $d\tau/d\gamma$ vs. γ plot. The stress-strain curves are typical for single crystals, oriented for a single system.
They exhibit three distinct stages of different strain hardening rates. The first, of the lowest hardening, is called the easy glide stage. The extent of this stage is short in a pure metal single crystal but increases to several percents of elongation in alloys. It is followed by the

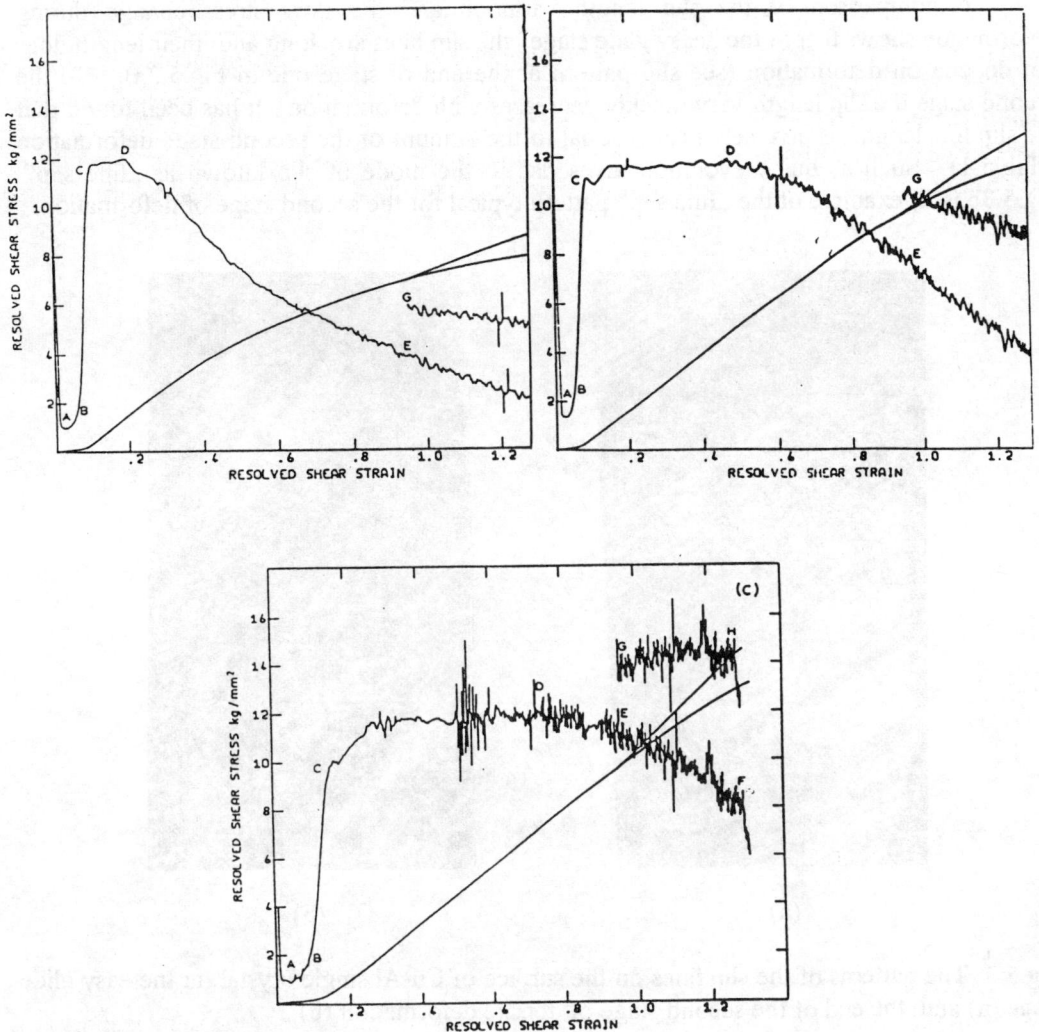

Fig.5.2 The tensile characteristics at different temperatures (a) 273K, b) 78K, c) 4K), of copper single crystals oriented for a single slip The resolved shear stress τ vs. resolved shear strain γ plots are supplemented by the strain hardening rate of the slip system $(d\tau/d\gamma)$ vs γ [20].

stage of the highest strain hardening (of the order of $10^{-2}\mu$). The extent of this stage, called a second stage or the stage of linear hardening, depends upon the material composition and temperature as may be found from Fig.5.2. A continued decrease of the strain hardening is a feature of the third stage. For this reason this stage is often called the stage of „dynamic recovery".

A comparison of the slip length variation and the flow stress change during deformation shows that in the „easy glide stage" the slip lines are long and their length does not depend on deformation (see slip pattern at the end of stage one in Fig.5.3a). In the second stage the slip length very quickly decreases with deformation . It has been found that the slip line length is inversely proportional to the amount of the second stage deformation (Fig.5.1). Such a quick evolution gives rise to the mode of slip known as „fine slip". Fig.5.3b is an example of the „fine slip" pattern typical for the second stage of deformation.

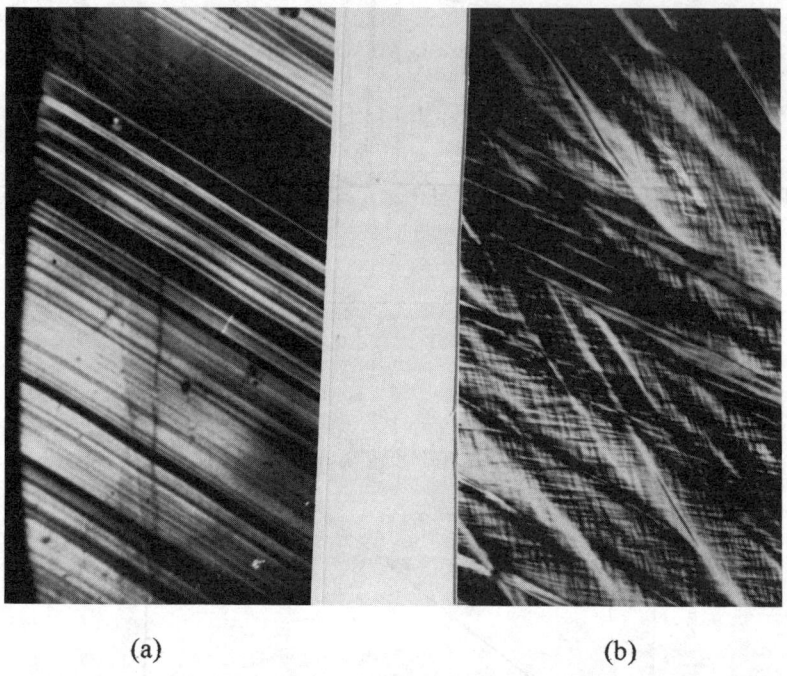

(a) (b)

Fig.5.3 The patterns of the slip lines on the surface of Cu-Al single crystal at the easy glide stage (a) and the end of the second stage of tensile deformation (b)

A comparison of the tensile characteristic with the corresponding slip pattern provides very important information. It shows that the strain hardening correlates with the development of the slip line pattern in a secondary system. The slip pattern network in Fig.5.3b proves that in the second stage of deformation, along with slip in the primary (the most stressed) system, slip takes place also in the secondary system and causes accumulation of secondary dislocations in their slip planes. In result, the obstacles for the movement of dislocation in the primary system are formed. Among different types of the obstacles which may form in result of „reaction" between primary and secondary dislocations (e.g. Lomer-Cottrell locks [33-35]), dislocations which threat the primary slip plane seem to contribute most to the strain hardening. Their configuration resembles a forest, thus the effect is called the "forest hardening" [20]. Like the primary dislocations they form sheets of accumulated dislocations (cell walls). Such a sheet is the strongest forest type obstacle. Because of small inter-spacing between forest dislocations in it, the sheet is hardly penetrable by the primary system dislocations. As long as the activity of slip in secondary systems is absent or negligibly small, the strain hardening effect is also very weak (easy glide), although the density of the dislocations produced by the primary system is high. This is one from among the arguments in favour of the „forest" model of the strain hardening. With greater activity of slip in the secondary systems the sheets of forest dislocations are more densely spaced. This leads to systematic decrease of the slip extent (slip line length). They become shorter and more densely spaced. One may conclude, therefore, that forest hardening favours the homogenous distribution of the slip in the crystal (favours homogeneous deformation).

In early attempts to explain the nature of the deformation in stage three, a decrease of the strain hardening was correlated with the cross slip manoeuvre which facilitates the reconfiguration or the annihilation of dislocations and which at this stage is easily observable. In order to explain the dependence of the onset of stage three on temperature, the assumption was made, that thermal fluctuations may assist in such manoeuvres. There are, however, several arguments against such a view of events which are responsible for the decrease of the strain hardening rate. According to the Basinski observations [20], the cross slip accompanies the plastic straining of crystals, oriented for a single system slip, from the very beginning of plastic deformation regardless of the temperature. The cross slip was even made responsible for the Luders front propagation in crystals of alloys [36], where due to a split of the perfect dislocations into the partials, the cross slip has especially little chance to occur. Moreover, cross slip in FCC crystals is controlled by the dislocation line tension and, therefore, it does not fall into the category of thermally assisted processes. And finally, a cross slip, similarly to slip in the conjugate system, produces the forest dislocations which oppose the slip in the primary system. These arguments do not discriminate the role which the slip in the cross slip plane plays in the evolution of slip features, except that in order to reveal such a role, one has to identify the difference existing in the slip patterns of stage two and three.

At the third stage of the tensile deformation of a single crystal, in addition to the fine slip, the bands of „coarse slip" are formed. They are similar to those observed at the easy glide (Fig.5.3a), except that now they intersect the early formed slip lines as can be seen in Fig.5.3c. An experimental account of the evolution of slip lines is then very difficult, nevertheless a fundamental feature of the „coarse slip" may be revealed by transmission electron microscope (TEM) observations. Fig.5.4 shows two TEM patterns of the dislocation arrangement typical for a fine (a) and a coarse slip (b) deformation. A comparison shows that a coarse slip generates a new, channel like, element of substructure which continues along the distance of the order of magnitude larger than that of the mesh length of the dislocation network (size of the dislocation cells) formed during preceding deformation. In terms of the previous arguments, the reason for such an evolution of slip must be the instability of the obstacles network. It is obvious, that a breakdown of a substructure wall must lead to a glide of the dislocations over a much longer distance. If we now take into account the already shown relationships between the strain γ and the free path of dislocations Λ (Eq.4.3), and the number of dislocations and Λ (Eq.4.11), it becomes clear that a breakdown of the obstacles network at high stress must

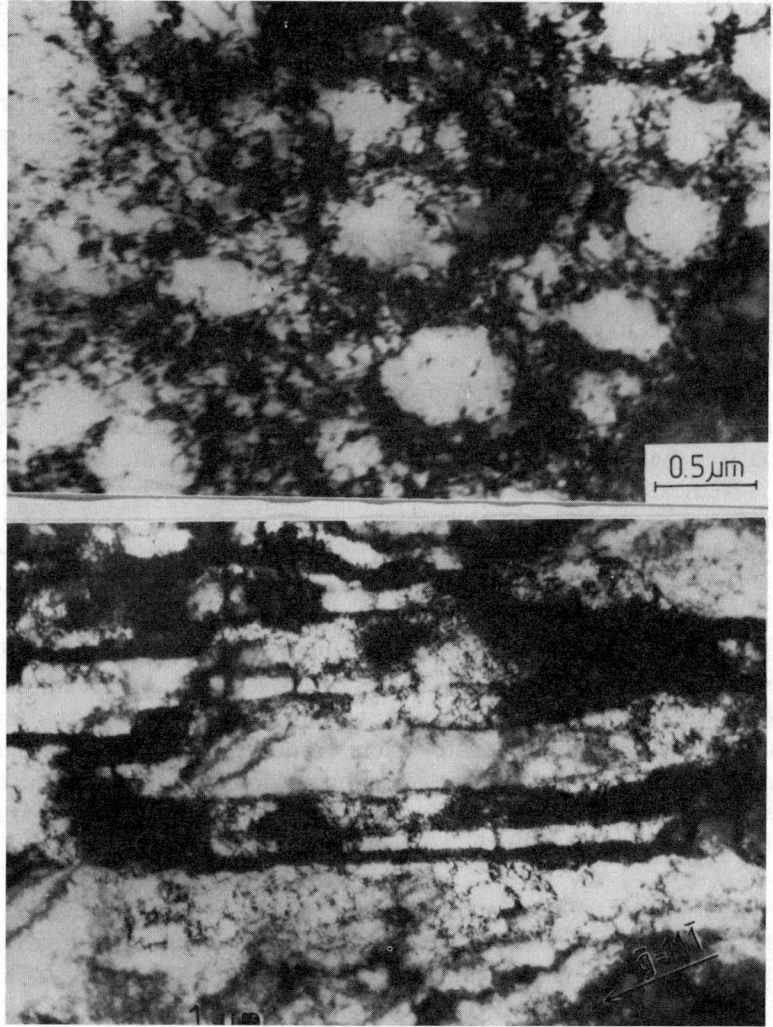

Fig. 5.4 Transmission electron microscope patterns of dislocation arrangement in copper single crystals, typical for the second (a) and the third (b) stage of deformation.

result in a local glide of very many dislocations over a large distance, or simply, in strain localisation in the form of a coarse slip. It is also obvious that with an increase of stress (breakdown of the obstacle network at higher stress) the coarse slip deformation increases, becoming a more catastrophic form of slip localisation in single crystals. A general conclusion which follows from the above argumentation is that instability of the crystal substructure is responsible for the loss of the plastic flow stability and for the localisation of deformation in crystals. Before considering when and why the substructure destabilises, it seems interesting to show that the development of the coarse slip in a crystal must lead to the decrease of the global hardening rate as it occurs at the third stage of deformation. Concentration of the deformation in a band causes that the remaining matrix is less deformed. and, therefore, less hardened than during equivalent fine slip deformation. Since slip takes place in the softest areas of crystal, hence the flow stress in softer matrix controls the global hardening of the crystal. Therefore the mode of slip should correlate with the global mechanical properties. A proof of such a correlation is shown in Table I [23]. In terms of correlation shown there it becomes obvious that the evolution of the global mechanical properties and the macroscopic flow behaviour depend upon the mode of slip. The change of the slip mode marks a threshold in such an evolution. From a purely analytical point of view such a threshold may be considered as effect of bifurcation (if e.g. a coarse slip superimposes the fine slip mode of deformation) or of a change in the governing constitutive law (if e.g. the coarse slip becomes the only operating mode of slip). From a physical point of view a smooth or sharp change of the mode of slip may result from destabilisation of the dislocation arrangements either by stress (and may be thermally assisted) or by strain (and is athermal). The experiments shed some light on this problem. When comparing the tensile curves of single crystals of pure copper (Fig.5,2) and Cu-Al alloy (Fig. 2.7a) one may see that in crystals of alloys the extent of the second stage is larger than in pure copper and continues to a much higher stress. But then it terminates abruptly by unstable localised deformation. The difference between the behaviour of copper crystals and crystals of alloys suggests that the obstacles network in alloys is more stable than that in a pure metal. And indeed, the presence of the solute atoms in the lattice of a parent metal was proved to stabilise the arrangement of dislocations. This effect is often referred to as a slow down of the recovery processes in crystals and takes roots in foreign atom-dislocation interaction. Such an interaction may lead to pinning of dislocations by atoms segregated to dislocations (Cottrell's atmosphere) or by splitting a perfect dislocation into partials (dissociation of a dislocation connected with formation of the planar lattice defect - stacking fault). In both cases the mobility of a dislocation and in particular the cross slip manoeuvre and climb are either excluded or highly reduced. With increase of the stress a chance to activate these peculiar mechanisms of the substructure rearrangement or collapse increases. The experimental fact that the onset of the third stage on the stress-strain curve shifts toward lower stress at higher temperature seems to prove the role of thermally assisted processes in the substructure destabilisation (e.g. due to climb of dipole segments). It is necessary, however, to notice a remarkable difference in crystal orientation at the onset of stage three and at the moment of a sudden loss of flow stability (Fig.2.7 a). While the change of the slope of the stress strain curve, which marks the onset of stage three, may begin at different strains depending upon the temperature and material composition and it is never sharp, the instability like that in Fig.2.7a is explicitly determined by the initial crystal orientation and closely related to the rotation of the crystal lattice relative to the applied load. In other words, the source of such instability is in the change of the scheme of stressing relative to the crystal lattice. To make this statement clear one has to trace the rotation of the crystal lattice in tension. The first systematic experimental data come from the pioneer works of Sachs and co-workers [40-42] and concern FCC single crystals of pure metals (Cu, Ag, Au) and alloys (Cu-Zn, Ag-Au). They have shown how the crystal orientation varies in the course of slip in the primary system and identified the position of the tensile axis at which in crystals of different composition and different initial orientation the secondary slip system becomes the deformation bearing one. These results pointed out the effect which was found much later as the "latent hardening" [43-48].

TABLE I

TYPE	SLIP DISTRIBUTION and deformation mode	FLOW BEHAVIOR Deformation, load Strain hardening rate	EXAMPLES
FINE SLIP (A)	Evenly distributed slips L$\downarrow$ when $\epsilon\uparrow$ multislip	homogeneous stable $\simeq 10^{-3}\mu$	Second stage of deformation in single crystals
SLIP BANDS (B)	Clusters of slip lines (parallel) single slip system	meso-heterogen. stable low > 0	easy glide
COARSE SLIP (C)	Transsubstructural long distance glide single slip system	meso-heterogen. instable zero	Lüders fronts in single crystals: (onset of yielding in alloys, overshoot and LH instability)
A + C	Fine multislip + coarse single slip complex	complex, stable decreasing with strain or low, constant	Parabolization of hardening curve (dynamic recovery) IV stage of deform.
MICRO-SHEAR BANDING (D)	Transgranular shear Pseudo single slip	macro-heterogen. instable zero	Lüders front in polycrystals Diffusing necks
SHEAR BANDING (E)	Clusters of micro-shear bands Pseudo single slip	macro-heterogen. < 0	Necking

To show these results it is helpful to use the stereographic projection of the crystal (e.g.[49]). In such a projection the crystal planes and directions are uniquely represented by points. The distribution of the points which represent crystallographically identical planes or direction follows the crystal symmetry. The plane of the projection may be arbitrarily chosen. It is useful, however, if such a projection (standard projection) is made on the plane of a high symmetry. For cubic crystals the most convenient in use is the standard

During back rotation it does not follow the same way. Now it rotates toward [110] direction, again, regardless of the initial orientation, which is the slip direction of the conjugate system. These experimental results are in exact agreement with the slip geometry of the single system deformation bearing slip.

The information which definitely violates a purely geometric argument is „overshoot". As may be seen from Fig.5.6 the tensile axis overshoots the symmetry line by several degrees, until the secondary system is finally brought into operation. The corresponding stress-strain curves are shown in Fig.5.7.

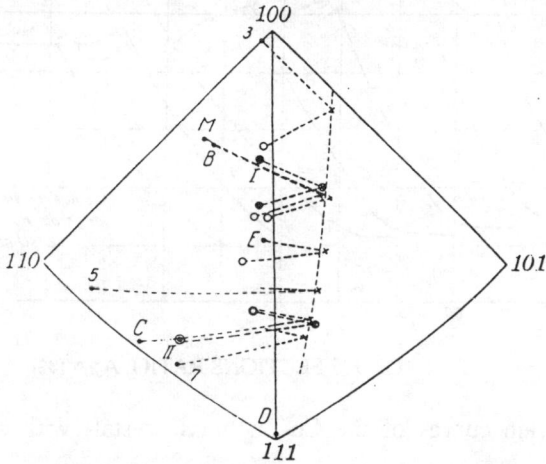

Fig.5.6 Rotation of the crystal axis during tension of Cu-Zn single crystals with different initial orientation [40].

The correlation of the overshoot strain with the corresponding tensile curve shows a very sharp change in the hardening rate toward its decrease, associated with an instant or permanent global instability of the plastic flow. To emphasise the origin of the phenomenon such instability is called the „overshoot" instability.

The very recent, very precise and systematic experiments on FCC single crystals, described in the work of Basinski, Szczerba and Embury [50], have shown, that regardless of the crystal composition the activation of the secondary system under the overshoot conditions results in strain localisation in the coarse slip band in the conjugate system. The most remarkable result of their studies is the conclusion that necking in a single crystal results from a coarse slip induced by the change of the deformation bearing system (overshoot), and that necking begins before the Considere criterion (Eq.2.2) is met. The associated decrease of the global hardening rate is already the after-effect of strain localisation.

Such a profound effect of the change in the deformation bearing slip system on strain localisation has been confirmed by the experiments on the „latent hardening" effect [44,45].

The intention of the study was to measure the strain hardening anisotropy. To do this the crystals oriented for a single system glide in tension were predeformed to different strains (different strain hardening level). Then the new sample with a different orientation of the tensile axis was machined from the parent crystal. On reloading, other than the primary system was under the highest stress, and the value of the shear stress in this system at the onset of plastic glide was measured.

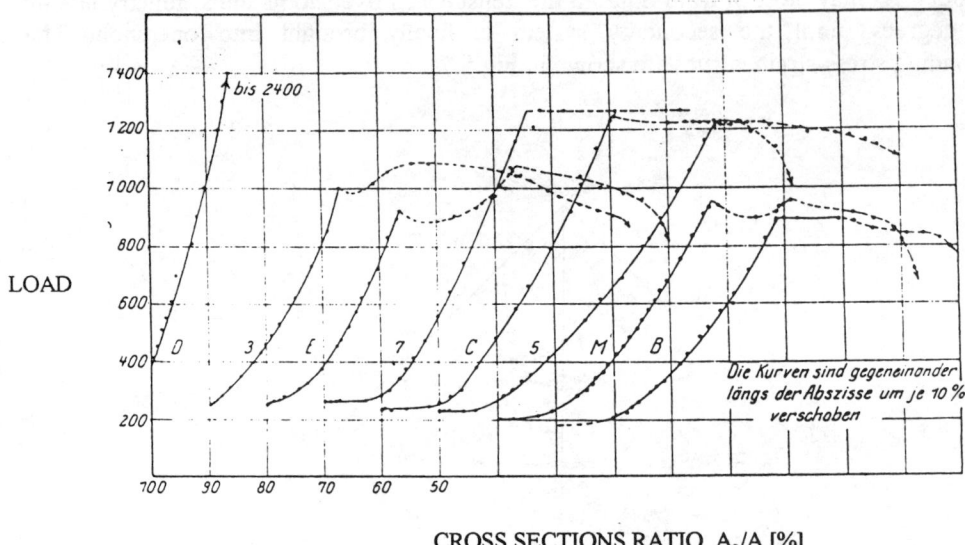

Fig.5.7 Tensile stress-strain curves of the Cu-Zn brass crystals with orientations like in figure 5.5 [40].

It has been found that the flow stress of the secondary sample is higher than that of the primary one at the end of straining. The only exception is when the secondary system was coplanar with the primary. The effect is called a latent hardening, because the flow stress in a secondary system is not seen until it becomes the deformation bearing one. The ratio of the secondary crystal flow stress to that in the primary is called the „latent hardening ratio" -LHR, respectively. Fig.5.8 shows how LHD depends upon the hardening of the primary system. In terms of these results it is obvious that already a very small slip in the primary system hardens very much (by a factor of the order of 3) the secondary systems (like e.g. the conjugate system). Therefore, the initial equivalence of flow stress in crystallographically identical systems is definitely lost. A parallel observation was that the flow stress of secondary crystals with LHR>1 was unstable and localised in Luders front. This last observation points to the role of the geometrical effects of the slip in the secondary systems as a factor controlling the stability of the dislocation .substructure.

The experiment born conclusion about the strain induced anisotropy of the shear stress practically eliminates a chance of a simultaneous glide in a few slip systems. This means, in turn, that the slip systems operate in a sequence. Thus, according to the analysis performed in Chapter 3, it is impossible to predict „a priori" which systems are brought into the operation until the evolution of the stress state, which through Eq.3.4 governs the sequential operation of the slip systems, is defined. One can ask the question, however, if the knowledge of the sequence of the slip in the systems engaged in the deformation, as well as the shear they carry, is important for the mechanical performance of the crystal. Such a question is, in a sense, similar to that in which one asks whether the behaviour of a single crystal reflects the behaviour of a grain in a polycrystalline material. The following example seems to provide the positive answer to this question.

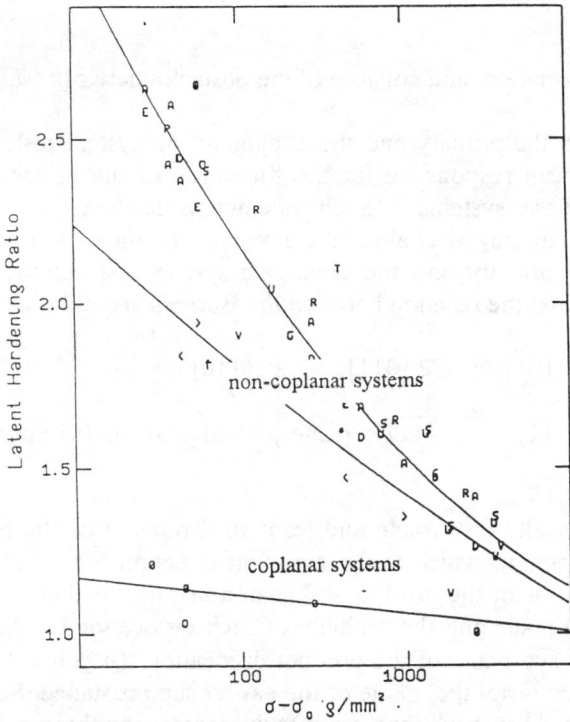

Fig.5.78 Latent hardening ratio vs. „effective stress [20].

As it was shown, the increase of the flow stress of a crystal oriented for the single system slip is caused by slip in the conjugate system. However, from the point of view of the applied stress, from the beginning of straining until the symmetrical orientation of the tensile axis is reached, the criterion for slip is fulfilled only for the primary system. Therefore the slip in the secondary system(s) must be driven by the reaction stress, and it plays the role of the stress relaxing or accommodating slip. As such, the moment of its activation and the amount of slip in it varies in tact with the rise and relaxation of the reaction stress. Therefore, the very first amount of slip in the primary system causes that the single crystal is under a complex stress state, similarly to a grain in a polycrystalline material.

The slip in the conjugate (as well as in the cross plane) system is a very clear example of accommodating and, in effect, alternating slip. Such slip may contribute to the evolution of the substructure in different ways, depending upon the value of strain during its momentary operation (operation between the slips in other system(s)). As it was shown, it favours the formation of the substructure causing the hardening of the crystal. It may, as well, lead to the substructure collapse promoting a catastrophic coarse slip. Such an apparently conflicting function of the secondary slip points to the role of the moment of its activation and the amount of shear it carries. It seems to be of value, therefore, to consider some mechanisms which may ensure such, sensitive to the amount of slip and deformation conditions (stress level., temperature), dual function of slip in the secondary system.

5.2. Mechanisms of formation and collapse of the obstacles network (substructure).

The interaction between the primary and the conjugate slip system dislocations is a good example of the mechanism responsible for the formation of the obstacles network which opposes the slip in these systems. Such interaction leads to the reaction between dislocations, provided that they meet along the direction common for both slip planes. If the Burgers vectors of the primary and the conjugate system dislocations are a/2[101] and a/2[011], respectively, then the reaction between the Burgers vectors is as follows:

$$a/2[10\bar{1}] + a/2[011] = a/2[110] \quad (5.1)$$

on: (111) ($\bar{1}\bar{1}$1) planes, all along [110] direction.

This reaction is energetically favourable and leads to formation of the product dislocation along [110] lattice direction, which is the common direction for (111) and ($\bar{1}\bar{1}$1) slip planes. The Burgers vector of the product dislocation belongs neither to the primary (111) nor the conjugate ($\bar{1}\bar{1}$1) plane and the mobility of such dislocation by slip is highly limited. This is so, because the slip plane of the product dislocation ($\mathbf{n} = \mathbf{b} \times \mathbf{t}$) is parallel to the (001) lattice plane, which is not the plane of the easiest slip resistance. Such a dislocation is called sessile dislocation. That is why the product dislocation is a barrier for the dislocations gliding in the primary and the conjugate planes (Lomer lock - Fig.5.9).

The formation of such locks explains the decrease of the free path of the glide of dislocations (slip extent). A similar reaction in crystals of alloys, where due to the low stacking fault energy dislocations split into partials, results in a more tough barrier (Lomer-Cottrell lock), composed of three partial dislocations with one sessile (stair rode dislocation). The dislocations which are arrested at such barrier form a pile-up configuration (Fig.4.2a). A high elastic energy associated with such configuration makes it unstable. A more stable (relaxed) configuration is a wall of the dislocation dipoles (Fig.4.2b), which can be formed by activation of a nearly located dislocation source on the parallel plane, as an almost instant response to the rise of the local internal stress (pile-up stress). Slip between the walls causes two effects. It provides more dislocations to the walls making them hardly penetrable, and, it gives rise to formation of new walls (finer dislocation cell structure) as it is schematically shown in Fig.5.10.

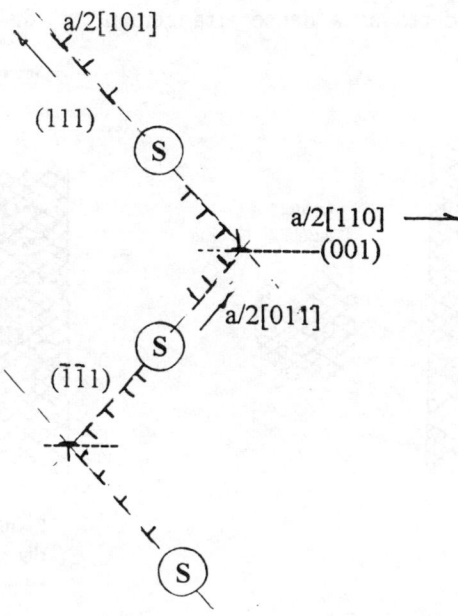

Fig. 5.9. Formation of the Lomer locks in result of the interaction of the dislocations from the primary and the secondary slip systems.

One can consider another reaction between the dislocations of the primary and the conjugate systems, in which one dislocation has the opposite Burgers vector. It might look as follow:

$$a/2[\bar{1}01] + a/2[01\bar{1}] = a/2[\bar{1}12] \tag{5.2}$$

The elastic energy of the product dislocation (proportional to b^2) is very high so that such reaction is considered as impossible. In other words, there is a strong repulsion between the reacting dislocations which plays a role similar to a barrier.

As it was already mentioned, the reaction considered so far, which leads to the formation of a sessile dislocation (lock), is possible, provided that the dislocations overlap. Thus the length of the barrier is rather short. Nevertheless, it must end with nodes (Fig.5.11a,b) which are the pinning points of the dislocations. Therefore they also contribute to the formation of the dislocation walls, because they help to arrest the dislocations in their slip planes. One can consider, therefore, that at most, dislocations are stored within the substructure walls, within which the interaction between them is the strongest. A lower density of dislocations inside the substructure cells makes these regions softer, so that slip takes place there. It would be naive, however, to assume that there is a one to one correspondence between the mesh-length of the dislocation substructure and the extent of slip. This is due to the a discontinuous structure of the wall which at the very first approximation can be considered as a dense arrangement of dislocation loops at the

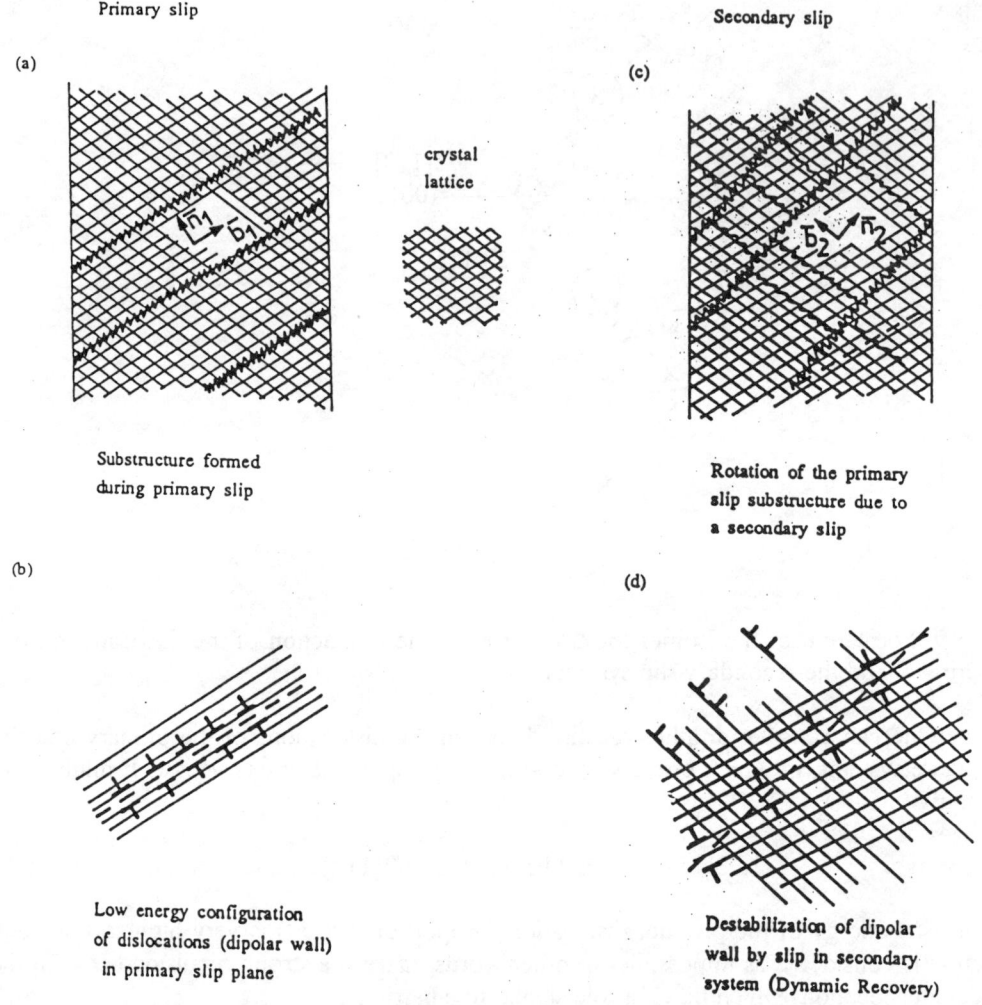

Fig. 5.10. Schematic picture of the formation and evolution of the dislocation substructure due to the interaction between the primary and the secondary slip systems.

outermost (near barriers) areas and much loosely packed dislocations in the vicinity of the source. Nevertheless, the proportionality between the free path of dislocations and the dislocation cell size is expected. Therefore, the correlation between the slip extent and the hardening behaviour of crystals (Table I) poses the question about a physically justified relationship between the flow stress and the slip line length during the homogenous deformation. In an attempt to find the answer, let us notice that the increase of the dislocation density by $d\rho$ results in the increase of the crystal free energy (elastic energy) dF such that :

$$dF = dE = E_d \, d\rho. \tag{5.3}$$

where E_d ($\approx \mu b^2$) is the proper energy of the dislocation (Eq.4.4).
As it follows from Eq.4.3 a small increment of the deformation by slip $d\gamma$, the extent of which is Λ, causes the increase of the dislocation density

$$d\rho = d\gamma / (b \Lambda). \tag{5.4}$$

Hence the rate of the increase of the free energy of crystal is:

$$dF / d\gamma = dE / d\gamma = E_d / (b \Lambda) \tag{5.5}$$

Since $dE = \tau \, d\gamma$, one can put :

$$\tau = E_d / (b \Lambda) \tag{5.6}$$

or

$$\tau \approx \mu b / \Lambda \tag{5.7}$$

At the second stage of deformation the slip line length is inversely proportional to γ. Therefore, the relationship predicts the linear strain hardening of the crystal at this stage. Taking into account the slope of the relation shown in Fig .5.1 (more accurately, the value of $d\gamma / d\Lambda$ ratio) it can be found that the hardening rate at this stage, $d\tau / d\gamma$, is of the order of $10^{-2}\mu$, in agreement with experimental measurements.

The above considerations do not pretend to compete with the other model analysis, in which the microstructure is represented by the density of the dislocations in crystals, and which yield a formula similar to that given by Eq.4.12. The main intention of such considerations was to show an alternative structural factor, which equally well determines the flow stress value and the slip behaviour. They show that an increase of the slip distance results in the decrease of the strain hardening rate even if the deformation is homogeneous in macro-scale (as at the third stage of deformation). Hence, the identification of the mechanisms which may weaken or destabilise the obstacles network, giving rise to the coarse slip, should be considered as the first step in the analysis of the processes responsible for the evolution of the mechanical performance of crystalline materials.

In terms of the interactions between the dislocations of the primary and the secondary slip systems it is impossible to explain the collapse or the breakdown of the substructure. Thus, several mechanisms were postulated as necessary for the change of the mode of the deformation from a „fine" into a coarse slip [51-53]. From the point of view of experimental observations such mechanisms must be divided into thermally assisted and athermal processes. This classification is justified by two, already cited, fundamental experimental works on crystal plasticity by Cottrell and Stokes [18] and by Basinski and Jackson [44,45].

The work of Cottrell and Stokes provides the very important information about the effect of the temperature upon the slip behaviour of single crystals of aluminium. It shows, that an increase in the temperature in the course of straining leads to unstable flow and strain localisation in bands of coarse slip which become organised into a Luders-like front, provided however, that the temperature is not increased until the stress strain curve becomes „concave to the strain axis" (until the third stage of the deformation is reached). Similar experiments on single crystals of copper [54] and copper and Cu-5%Al alloy [19] have confirmed that the change in the temperature influences neither the stability of the plastic flow (Fig. 5.12) nor the mode of the deformation until the stage three is reached. They have shown also [19] that the coarse slip develops in the primary system as is shown in Fig.5.12.. These results, therefore, provide the best proof that stage three results from a systematic replacement of fine slip by coarse slip bands. The second important information is that the mechanism responsible for the coarse slip is thermally activated. Following the suggestion expressed in the work of Cotrell and Stokes that unlocking of the barrier frees the arrested dislocations and relaxes the associated pile-up stress, it was logical to look for such a mechanism among processes which can unlock the barriers and which depend (are controlled) on the stress and temperature. An example of such a hypothetical mechanism is shown in Fig. 5.10, in order to illustrate the role of microstructural processes in the recombination of dislocations within the dislocation walls which, eventually, gives rise to a coarse slip. While such a recombination may indeed reduce the internal stress in crystal, its promoting role in the development of the coarse slip is considerably less certain. This is so, because the pile-up configuration may be quickly replaced by a dipolar wall which may be formed due to slip in the same slip system. Then, however, removal of the barrier does not lead to remobilisation of the arrested dislocations. And it does not matter whether such a removal results from the reaction between dislocations or from the slip of sessile dislocations on (001) slip plane.

The study [23] of the microstructural features of the bands of coarse slip, induced in single crystals of copper by the increase of the temperature from 78K to 300K, has shown that they are very similar to those which are observed at stage three. They penetrate the entire cross section of the crystal (Fig 5.13a in a view on the cross slip plane) being organised into a Luders-like front. They carry large deformation as it may be concluded from their discontinuous distribution and from the presence of the lines of an accommodating slip in the conjugate system. The accommodating slip occurs in the areas between sequentially formed bands of coarse slip in the primary system. The wavy character of the lines of coarse slip, seen on the other face of the crystal (Fig. 5.13b), indicates that the slip on the cross slip plane was also involved in the accommodation of the forbidden effects of the coarse slip. At the same time TEM observations showed the channelling of the substructures (Fig.5.14).

The analysis of the contrast effects associated with channelling of the structure [23] lays at the background of the model which is schematically shown in Fig.5.14. While absence of such a drastic change in the contrast across a cell wall in areas outside the band indicate that the wall consists of dislocation dipoles. Fig 5.13a shows a dipolar wall made of the primary dislocations which, due to cross slip, left the initial slip plane to form a low

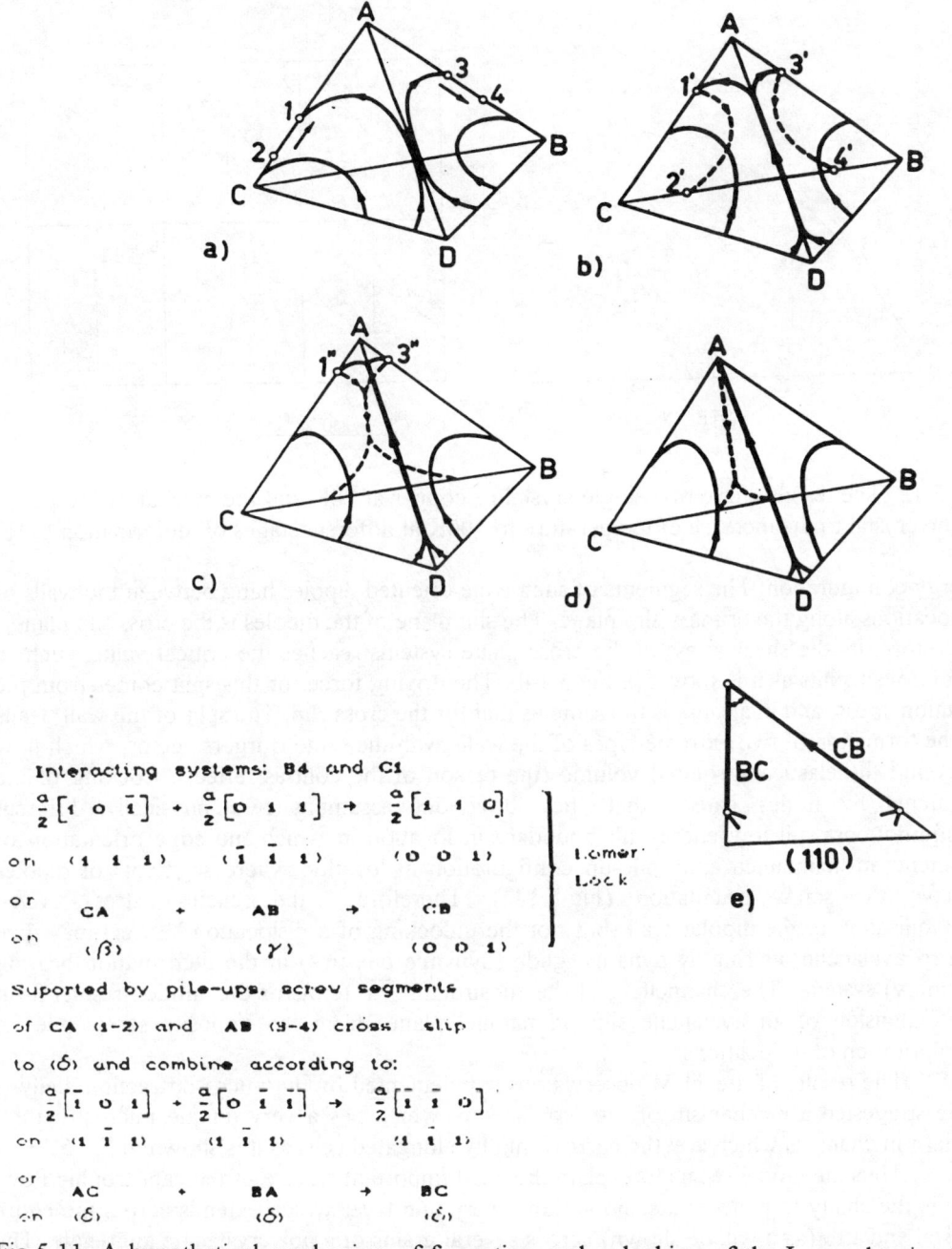

Fig.5.11. A hypothetical mechanism of formation and unlocking of the Lomer barrier **CB** along **AD** (a,b) in result of cross slip of the primary **CA** and the conjugate **AB** dislocations to the cross-system plane (δ) (c,d), formation of the obtuse dipole **CB-BC** (e) which annihilates by climb.

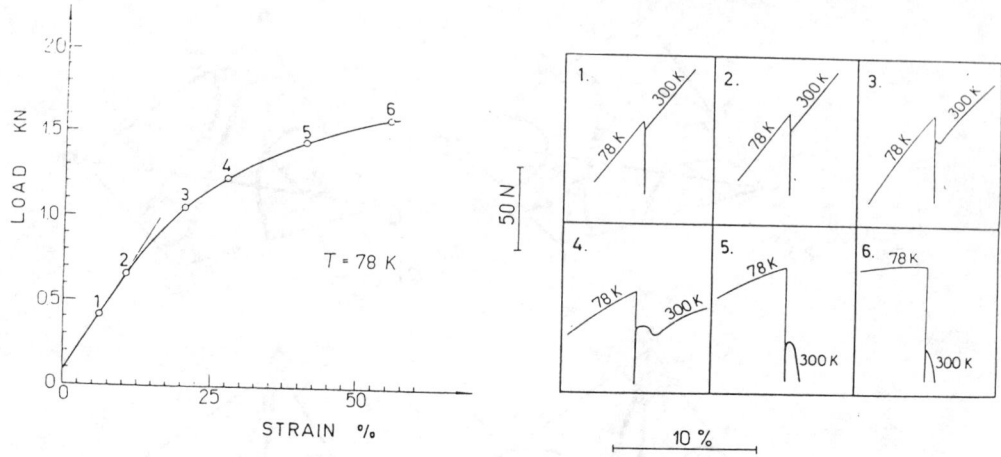

Fig.5.12 The tensile curve of a single crystal of copper at 78K and the mechanical response of the crystal to the increase of temperature to 300K at different stages of deformation [54].

energy configuration. The segments of such edge oriented dipoles hang between the walls of dislocations along the primary slip plane. The slip plane of the dipoles is the cross slip plane. Therefore, if the shear stress in the cross plane systems reaches the critical value, such a dipolar wall splits as it is shown in Fig.5.14b. The driving force for this split comes from the reaction stress, and in a sense is the same as that for the cross slip. The split of the wall leads to the formation of two extreme types of the walls with the same Burgers vector, which now surround the elastically rotated volume (the reason of the contrast effect). Locking of the segments by nodes causes that they bow out assuming two physically different configurations: still low energy tilt boundary in location in which the edge orientation of segments in maintained and pile-up configuration in location where segments of dipoles receive the screw orientation (Fig.5.14c). Therefore, the reaction stress driven destabilisation of the dipolar wall (but not the unlocking of a dislocation barrier) may give rise to avalanche, i.e. highly dynamic glide (dynamic pile ups) in the deformation bearing (primary) system. The channelling of the substructure, in terms of this model, results from the expansion of an avalanche slip on parallel planes from the opposite screw pile up configuration of dislocations.

The results of the TEM observations supplemented by the micro-diffraction analysis have suggested a mechanism of the coarse slip, which has a very unique microstructural pattern in channels which are the narrow, highly elongated cells as it is shown in Fig.5.13.

Thus the model seems to explain the most important feature of the catastrophic flow, that is the ability to concentrate slip within a very thin layer which extends across the entire crystal and also (as it will be shown) across several grains of a polycrystalline aggregate. The role of the temperature is well demonstrated by the need to have a well recovered structure of the dislocation wall (straight segments of the edge dislocations). Increase of the

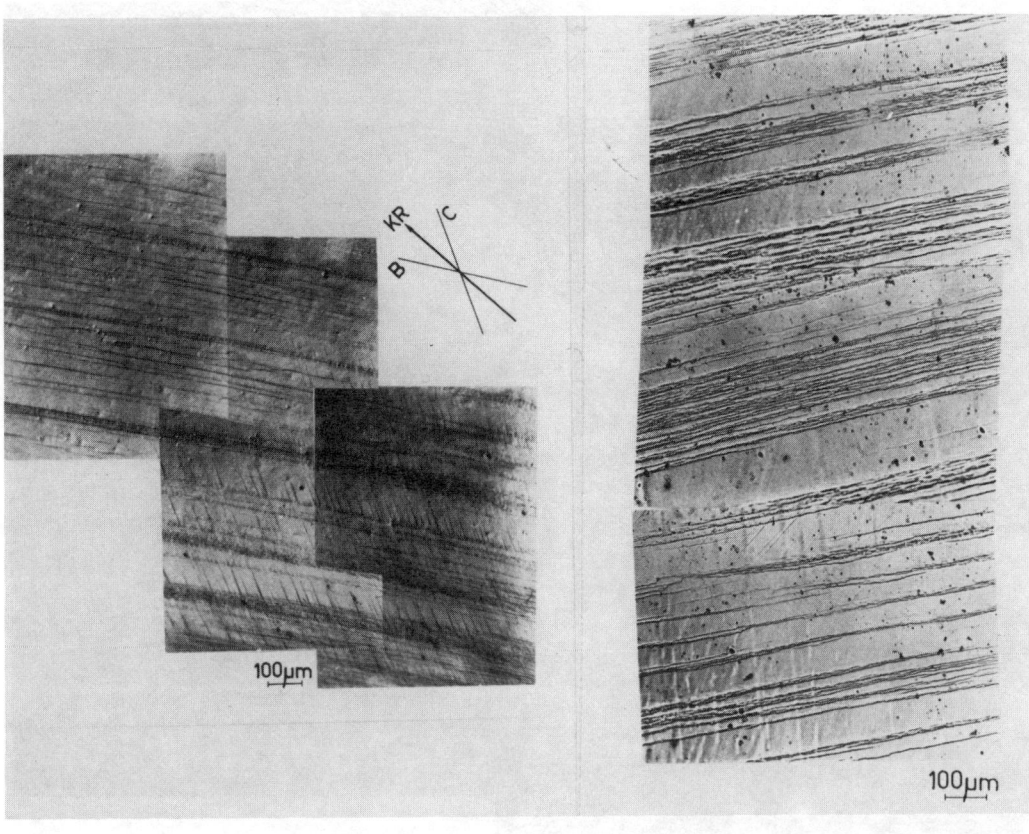

a) b)

Fig.5.13 Distribution of the line of a coarse slip at the Luders front generated in the copper single crystal due to the change of the temperature of tensile deformation (stage three, $\varepsilon=0.5$) from 78 to 300K: a) on the face parallel to the (111) lattice plane and b) on the other, parallel to the tensile direction plane.

temperature and of the load (stress) helps to form such a configuration. One may expect, therefore, that while the cross slip, which is a reaction to the geometrical constraints, can push some primary system dislocations to the cross slip plane and may as well destabilise the configuration they form, their rearrangement into „a perfect" dipolar wall depends on stress and temperature. Thus, the formation of bands of coarse slip during monotonic straining is a self induced process, which becomes intensified when the stress increases, and in contrary to the change of temperature it causes a monotone decrease of the strain hardening rate (stage three). Such a „smooth transient" from the fine slip into the coarse slip makes, however, that the change of the deformation bearing system and the associated „overshoot" instability terminate the development of the coarse slip in the primary system before it may achieve the form of macroscopic localisation (neck). It is important, therefore, to explain

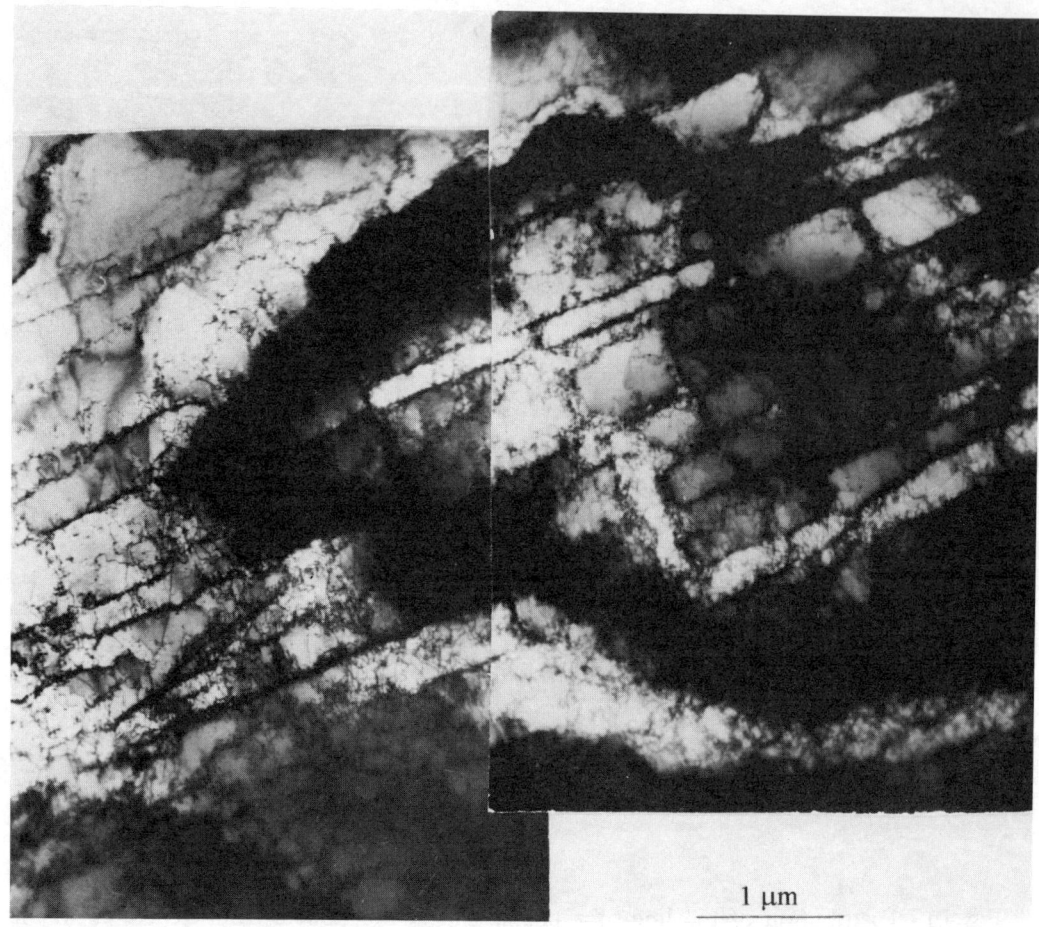

Fig.5.14 TEM picture of the dislocation arrangements in the crystal as in Fig.5.12 in the area of the cluster of bands of coarse slip.

why the change in the deformation bearing system leads to the coarse slip. Because the interaction, e.g. between the primary and the conjugate system dislocations leads to hardening, regardless of the fact which system is the deformation bearing one, the mechanism of the substructure destabilisation should be sought in the role of slip geometry. Such a role of the secondary system in the rearrangement of dislocations accumulated in the primary one is shown in Fig.5.9c. The rotation of the substructure elements (e.g. dislocation wall) with respect to the crystal lattice increases with the increase of the strain in the secondary system.. Hence, if the secondary system becomes the deformation carrying one, the rotation, which follows Eq. 3.6, may be sufficiently large to destabilise the configuration. If dislocations which are stored in the primary system form a dipolar wall, as

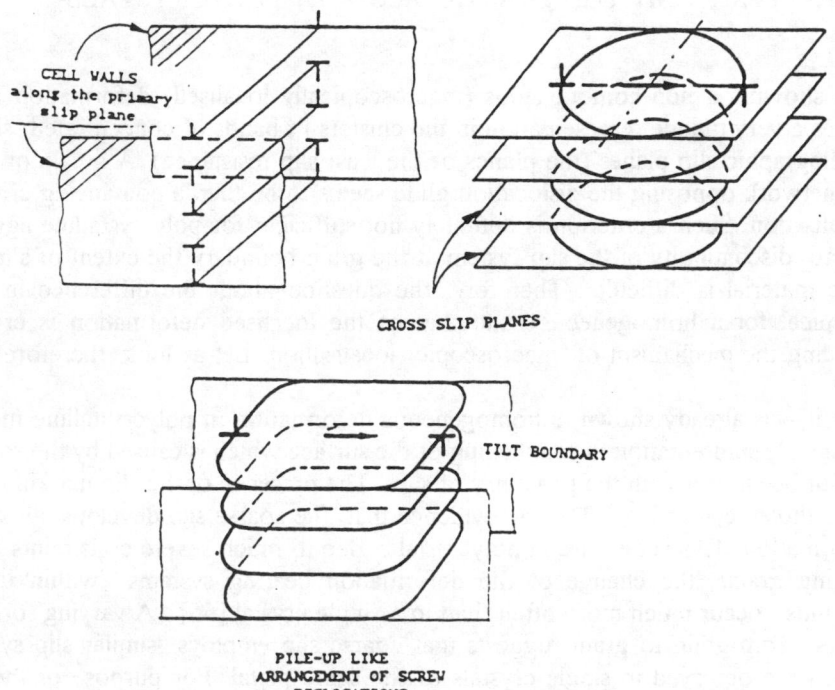

Fig.5.15. The scheme of the destabilisation of the wall of the primary dislocation dipoles on the cross slip plane and formation of an instant pile-up configuration on both sides of the channel.

shown in Fig.5.9b, then slip, e.g. in the conjugate system, may bring the opposite segments of the neighbouring dipoles to the common slip plane (Fig.5.9d). Due to strong attraction, such segments are instantly superposed and annihilated. One has to expect, therefore, that the wall collapses, giving rise to a long distance glide of the dislocations in the secondary system with all the associated consequences (dynamical pile).

In the final remarks it seems to be of value to stress that, according to the preceding considerations, the slip behaviour and the evolution of the metal substructure is very sensitive to the sequence of slips in the operative slip system and the value of deformation they carry. A monotone evolution of the substructure is expected only if the same system dominates during straining and the others are the accommodating systems. Each time when the operating system changes, the substructure undergoes a very drastic change because of substantial rearrangement and annihilation of dislocations (athermal dynamic recovery). The role of the temperature seems to be connected with its effect on the rearrangement of dislocations within the substructure (replacement of a high into a low energy configuration) but not with the kinetics of slip itself. And, at last, from the considerations it follows also that the effect of the change of the „strain path" appears as a very efficient tool in the structure control [57,58]. The practical meaning of this last remark will become more obvious in the next chapter.

6. SLIP PATTERN AND THE PLASTIC FLOW IN POLYCRYSTALS.

As it was shown, a non-homogeneous (macroscopically localised) deformation in single crystals has a very unique representation in the clusters of bands of concentrated slip along the crystallographic slip planes (the planes of the least slip resistance). A breakdown of the obstacles network opposing the dislocation glide seems to be then a convincing criterion of strain localisation. Such a criterion is definitely not sufficient for polycrystalline aggregates, since due to discontinuity of the slip system at the grain boundary the extent of slip (shear) across the material is difficult. Therefore, the question about the difference in the slip pattern typical for a homogeneous and that of the localised deformation is crucial for understanding the mechanism of macroscopic localisation. Let as look, therefore, at such difference.

As it was already shown, a homogeneous deformation in polycrystalline metals has the topological representation in roughening of the surface which is caused by the rotation of the grain surface in tact with the geometry of slip. The presence of the slip markings within grains, like those seen in Fig.2.2, is the evidence that the coarse slip develops already after small deformation. This is because in polycrystals, due to much severe constraints from the neighbouring grains, the change of the deformation bearing systems within individual crystallite must occur much more often than in a single crystal [55]. A varying orientation of slip lines from grain to grain suggests that coarse slip employs similar slip systems as those which are observed in single crystals of the same metal. For purpose of the further analysis let us call such slip systems - the easy slip systems. It is clear that the development of such coarse slip within grains may affect the rate the global hardening of polycrystalline metals sooner than in a single crystal. However, it does not lead yet to the macroscopic strain localisation in the material. The observations of the slip pattern in a zone of macroscopically localised deformation, as for example in the neck area - Fig.2.10, show that slip markings do not terminate at grain boundaries, but they expand across several grains keeping the same orientation in the material. Such a mode of deformation is called shear banding to emphasise that shear concentrates within a very thin layer (micro-shear band) which continues across the material with no respect to the orientations of grains it passes through, as it is shown in Fig.6.1. Where micro-shear bands cluster, a macroscopic shear band forms. Fig.6.2 shows a cluster of micro-shear bands within a macroscopic shear band. The orientation of the micro-shear bands in the sample is the same as the position of the macroscopic shear band. According to the TEM observations the thickness of a micro-shear band is of the order of 0.1 to 0.2 μm. If the position of the band coincides with the crystallographic slip plane then a typical for the coarse slip, channel-like penetration of the substructure is observed. (Fig.6.3). These microstructural features of micro-shear bands show, first of all, that shear must employ another then the „easy „ slip systems in those grains in which the plane of the macroscopic shear does not coincide with the plane of an easy system. Otherwise, the slip in several systems within the band is needed to accomplish the demand of the strain compatibility across the plane of macroscopic shear. There is, however no experimental evidence of a multi -system slip within such a narrow band. Instead, the discontinuity of the structural features along the micro-shear band (Figs 6.3 and 6.4) proves that deformation within a band is a simple shear.

Structural and Mechanical Aspects

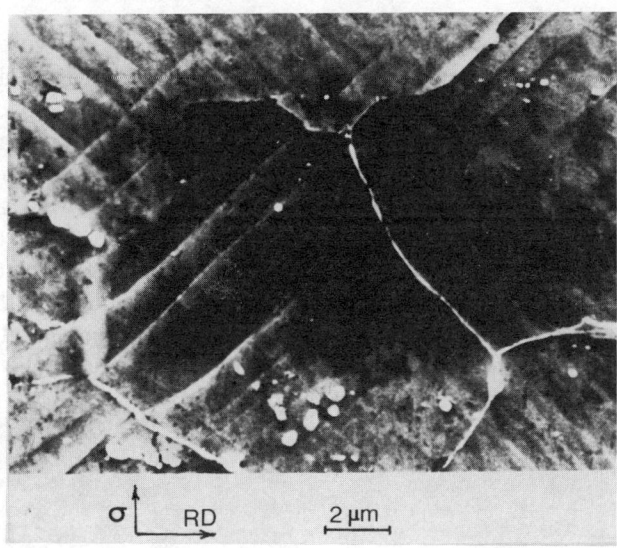

Fig.6.1 Straight markings (micro-shear bands) extending through several grains at the very front of the macroscopic band shown in Fig. 2.15 [21].

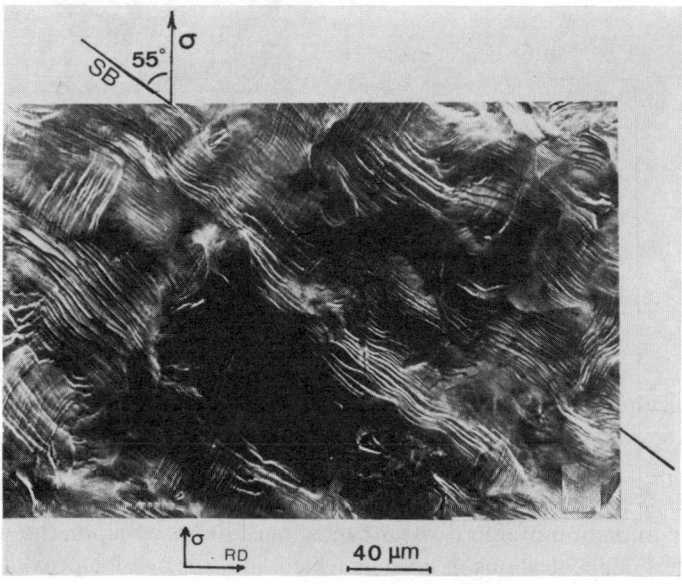

Fig.6.2 Scanning electron microscope pattern of a cluster of micro-shear bands within the zone of the macroscopic shear band shown in Fig.2.15 [21].

The mechanism of such a slip must engage dislocations with Burgers vectors oriented along the direction of the macroscopic shear. In other words, it allows for the glide of dislocations with large Burgers vectors. Although large Burgers vector dislocations are very unstable, there is a very clear thermodynamic argument in favour of a such mechanism. Continuity of the slip across the grains eliminates the need for accommodation, otherwise, incompatible the „easy systems" slip deformations of the neighbouring grains. Moreover, there are experimental evidences of slip in non-easy systems in hexagonal crystals. Therefore, such a slip in metallic materials is generally possible, provided however, the critical shear stress for slip in the non-easy system, which is much higher then that for slip in the easy systems, is reached. This statement can be consider as the verbal expression of the criterion for micro-shear banding which is a form of the trans-granular slip. Thus, the

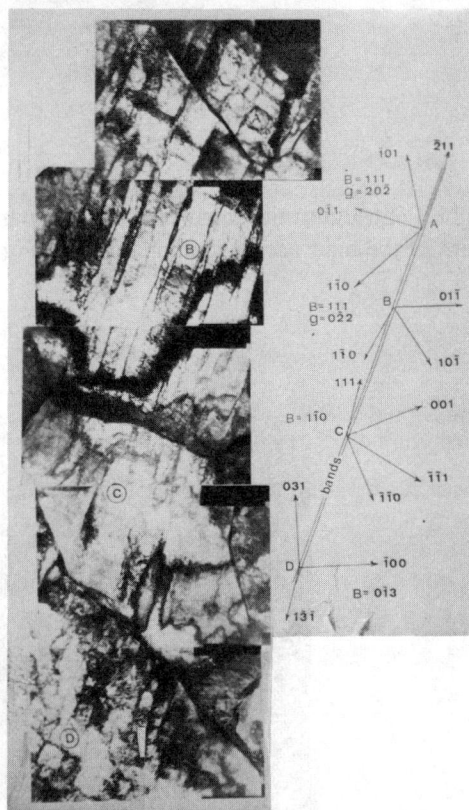

Fig. 6.3. TEM picture of micro-shear bands extending across several grains in Iron [21].

mechanism of the formation of micro-shear bands must be based upon the very high as well the very local stress concentrations in crystals which, in turn, may be provided by the coarse slip in the easy slip system. One can consider therefore, that micro-shear bands develop from the coarse slip bands. In other words, coarse slip, in agreement with observations, is necessary for micro-shear banding. Hence the factors which favour the coarse slip, favour also the formation of shear bands.

If the micro-shear banding becomes the dominating mode of the deformation of a polycrystalline aggregate, then the geometrical effects of the deformation should be similar to that of a single slip in a single crystal, and the evolution of the micro-structure should be

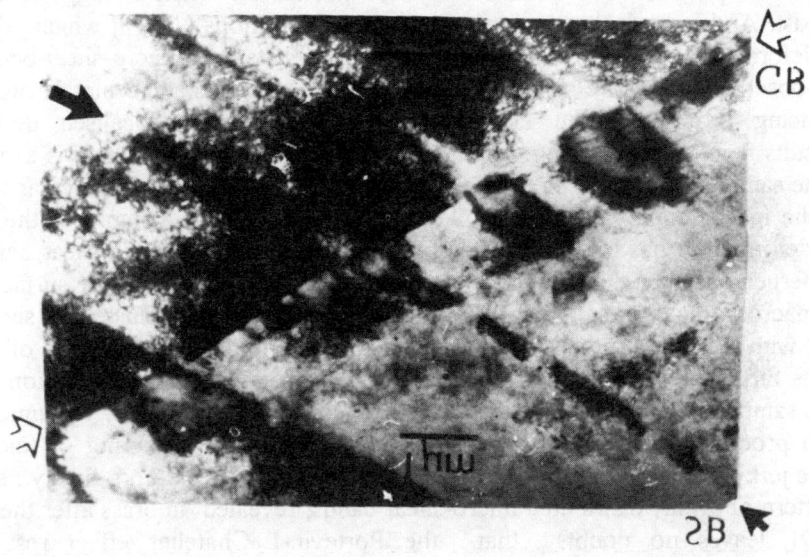

Fig.6.4 TEM micrograph of the micro-shear bands crossing the grain boundary in aluminium alloy [56].

the same as that which is caused by the coarse slip. Furthermore, if one takes into account that micro-shear bands originate at the bands of coarse slip in an easy slip system, it is reasonable to expect that the change of the deformation path will favour shear banding. This last conclusion seems to be particularly important. Let us, therefore, recall the relevant experimental data.

In a number of experimental works (summarised in [23]) it was shown that the change of the scheme of straining (change of the strain path) leads to almost instant development of micro-shear bands in polycrystalline metals. Some examples are already shown in Fig 2.16. Under such conditions micro-shear banding appears as the dominating if not the only mode of plastic flow of the material (Figs 2.15,.2.16b). During monotonic straining the onset of shear bands formation is moved toward large strains Then, however, they are considered as an unwanted mode of the deformation. This is especially true under the conditions in which micro-shear bands cluster (e.g. in tension) giving rise to macroscopic strain localisation (neck), which as an ultimate effect results in the material failure.

If we now come back to the first chapter and look at the mechanical characteristics of polycrystalline metals, one may ask the question whether different patterns of the global plastic flow instability, which are observed during monotonic deformation in different materials and under different conditions, have anything in common with micro-shear banding. Such question is fully justified by the fact that models which were postulated relate to very particular materials or deformation conditions and do not apply to the others. Therefore, the evidence that the same mode of deformation, namely micro-shear banding, is responsible for unstable, localised plastic flow, regardless of the material and the deformation conditions, would put them into common physical mechanism. Table I suggests that such evidences exist. And indeed, there are experimental observations [10,11] which show that the Luders deformation at the onset of plastic straining results from micro-shear bands.

The observations of the slip pattern in different polycrystalline metals provide the very convincing proof that the Portevin-LeChaterier unstable, localised deformation (Fig.2.8) results from shear banding. As it may be seen in Fig.6.5a, which is a magnified picture of the sample as in Fig.2.8c viewed, however, in a section perpendicular to that in Fig.2.8.c, the macroscopic fringes (Fig.2.8c) do not mark the volume of the material where the deformation has localised. They result from clusters of micro-shear bands, well seen in the perpendicular section (Fg.6.5a), which affect the flatness of the surface, giving rise to the macroscopic contrast effects. Formation of each of such clusters of shear bands is associated with the load drop (jerky flow) during tensile deformation. A drop of the load, in turn, stops further development of the band. Such a discontinuous formation of shear bands in the sample indicates that, in this case, it is controlled by a very dynamic internal stress driven process [16]. Fig.6.5b is example of slip pattern (clustering of micro-shear bands) of the jerky flow in a sample of polycrystalline titanium [58]. And, finally, shown in Fig.5.6c pattern of evenly distributed micro-shear bands, revealed in brass after the serrated yielding [59], leaves no doubts that the Portevin-LeChatelier effect results from development of micro-shear bands.

In the light of the experimental works it appears that also instability of plastic flow during deformation at high temperatures in single crystals reveals similar slip patterns. They show that in a single crystal it results from the coarse slip and in polycrystalline metals from shear banding.

In terms of such experimental evidence it seems reasonable to assume that unstable flow in polycrystals is always associated with development of a new form of glide which does not follow the anisotropy of the crystal lattice. Therefore, the criterion for shear banding must be derived from the analysis of those microstructural processes which can induce the mechanism of the transmission of slip through grain boundaries in a common „macroscopic" shear system.

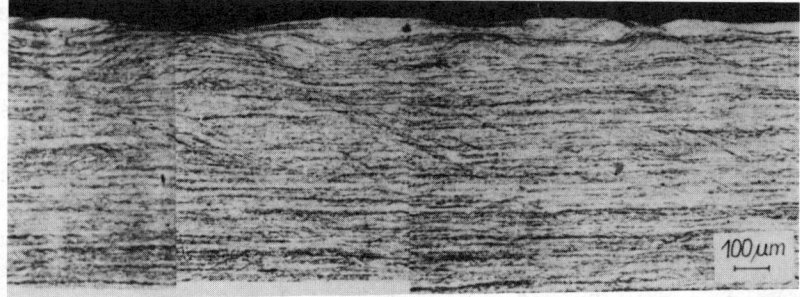

a)

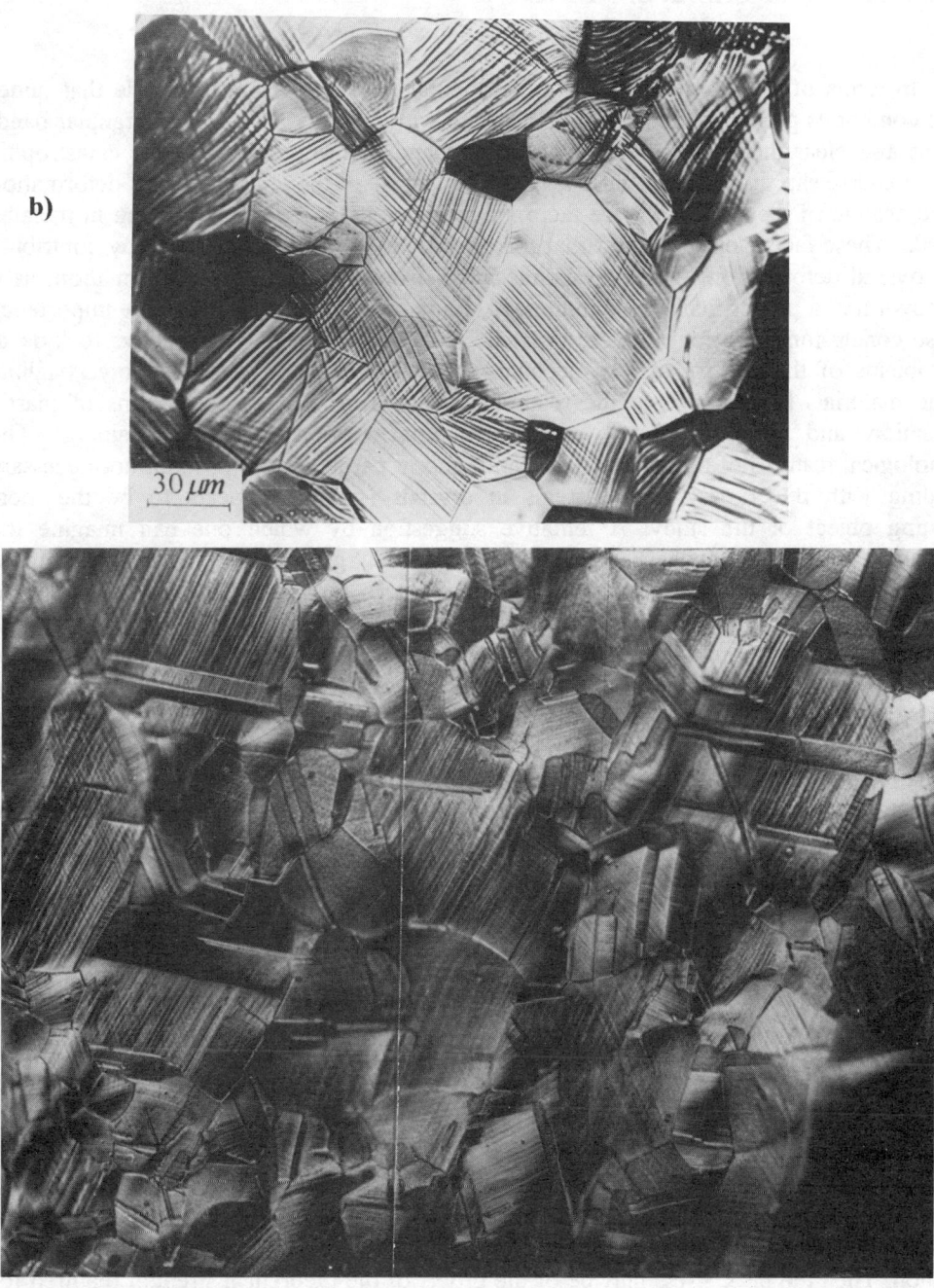

Fig.6.5 The slip patterns of the Portevin-LeChatelier deformation in different metals:
a) Al-Mg alloy [16], b) titanium [59]; c) Cu-Zn brass [60].

7. MECHANISM OF SHEAR BANDING

In terms of summarised briefly experimental results, we may conclude that under certain conditions plastic yielding may concentrate within a very narrow transgranular band. It seems also clear that tansgranular micro-shear banding has its origin in the catastrophic nature of coarse slip. It is known, finally, that the increase of the temperature of deformation and the change of the strain path are factors which favour micro shear banding in metallic materials. These facts clearly suggest that this specific mode of deformation may contribute to the overall deformation of polycrystalline metal from early stages of deformation, as it was shown for a polycrystalline aluminium [61] and austenitic steel [62]. The importance of these conclusions can not be overestimated. They provide a new perspective to look at the problems of the control of the mechanisms of plastic deformation in polycrystalline metallic materials [57,58] and at the relationship between micro mechanisms of plastic deformation and different forms of the material response during straining. The morphological features of the micro shear band, and in particular the position not necessary coinciding with the „easy" slip systems in crystals (grains), appears to be the most interesting object of the study. A tentative suggestion by which one can imagine the conversion of the avalanche-like movement of a group of dislocations into a „noncrystallographic" micro shear band has already been postulated [63]. It is based upon the argument that the mechanical of metal substructure gives rise to highly co-operative movements of dislocations [64]. Such a co-operative, highly concentrated movement of dislocations can be considered in terms of the stress profile around moving group of dislocations (dynamic pile-up) and the velocity with which the internal stress large local lattice distortion) travels within the crystal. Of particular interest is the profile attained by the dynamic pile-up at a grain boundary (a location characteristic of the discontinuity of the easy slip system) where the conversion of the slip in the easy system into micro shear band (slip in a non easy system) takes place.

The formation of the micro shear bands resolves itself into the criterion for non-dissipative (or weakly dissipative) transmission of a high amplitude pulse of internal stress across a grain boundary. The starting point of the analysis is the profile of internal stresses (or lattice distortion) associated with the pile-up formation. Two extreme momentary situations, shown in Fig.7.1 a and b, may be considered [63].

In the case (a) the pile-up stress may relax due to the activation of „easy" slip systems in a neighbouring grain (i.e. the pile-up stress assists in activation of slips in the second grain) as is schematically marked by arrows. Then, the stress pulse amplitude never reach the static pile-up stress (number of dislocations in the pile-up time the applied stress), which may be orders of the magnitude larger than the applied stress. This case can be classified as a dissipative transmission of the pulse through boundary. In the case (b) the second grain does not „see" the stress concentration, except the grain at the very last moment of the pile-up formation. Then the peak stress may be of the order of the theoretical strength required for homogeneous nucleation of dislocations (for slip in a non-easy system). A numerical distinction between these two cases can be performed on the basis of the rate of stress increase in the location on the grain boundary. The stress rate $d\sigma/dt$ may be written:

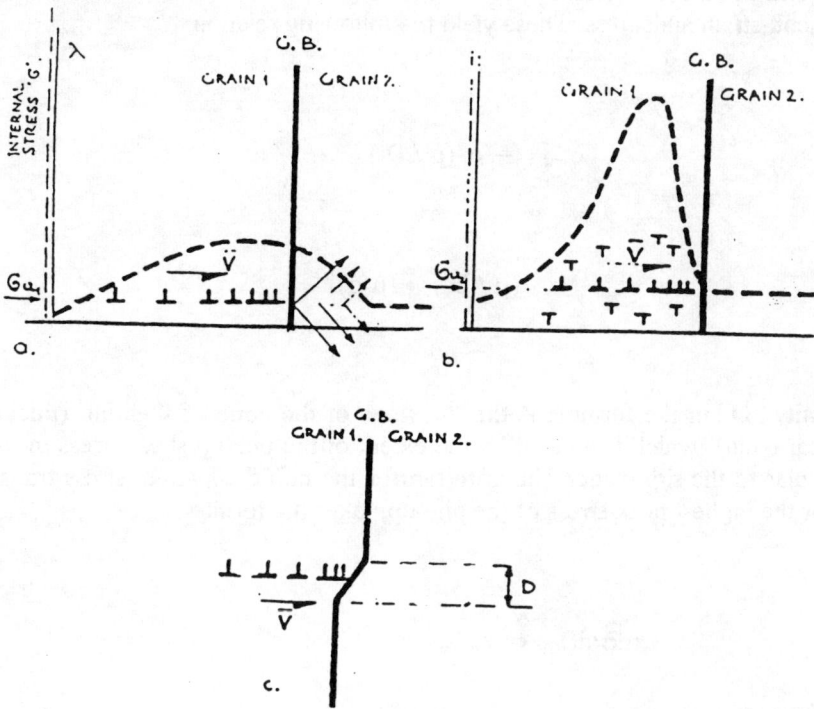

Fig.7.1. A hypothetical dissipative (a) and non-dissipative (b) transmission of the stress pulse of a group of dislocations through the grain boundary and formation of a layer of shear (micro shear band).

$$\frac{d\sigma}{dt} = \frac{d\sigma}{dx} \cdot \frac{dx}{dt} \qquad (7.1)$$

The right hand terms of the expression are the stress gradient at the grain boundary (caused by approaching group of dislocations), and the velocity of the group, respectively.

The position dependent gradient of the stress in the group of dislocations and the velocity of the group are the controlling factors for the grain boundary stressing rate, but only if there is no relaxation of stresses due to slip in the neighbouring grain. This means that non-dissipative transmission of the stress pulse through the grain boundary is possible when the rate of stressing is higher that the highest available (critical) rate of stress transmission. This critical rate of the stress transmission is, in turn, determined by the velocity of sound in a metal, which for shear stress wave is given by $c = (\mu/\rho)^{1/2}$, where μ is the shear modulus and ρ is the density, Here, „c" is the velocity of the displacement of a material point du/dt

thus, the critical rate of stressing can be found via the well known relationship between displacement, strain and stress. These yield the following relation:

$$\sigma_{crit} = c\,(\mu/D) \qquad (7.2)$$

or

$$\sigma_{crit} = (\mu/D) \bullet (\mu/\rho)^{1/2} \qquad (7.3)$$

The quantity „D" in the formula is the thickness of the zone of shearing (thickness of the micro shear band) which is defined by the extent of the pile-up shear stress in the direction perpendicular to the slip plane. The criterion for the non-dissipative stress transmission, as well as for the highest peak stress of the pile-up, takes the form:

$$(d\sigma/dt)_{GB} \bullet \mathbf{V}_{group} > \mu^{3/2}/b\,\rho^{1/2} \qquad (7.4)$$

When fulfilled, the pile-up stress may reach the theoretical value, which is transmitted with sound velocity in a neighbouring grain in the form of stress pulse. Such a stress pulse may lead to the generation of such dislocations within the volume of thickness D which are able to accommodate (to carry) the shear in the second grain to that in the first one.

Despite of numerous simplifications, the model seems to provide a coherent explanation to the observable features of micro shear banding. Its the most essential assumption is that slip may occur in slip systems in which the slip vector does not coincide with the most expected Burgers vector of dislocations and on the planes which do not resist the least of the dislocation motion. This assumption is not in conflict with the nature of the metallic bonds in crystals and receives the experimental support from the behaviour of the hexagonal crystals.

ACKNOWLEDGEMENTS:

The work was partially supported by the University of Mining and Metallurgy under the grant No: 11.180.134

REFERENCES

1. A. Korbel, L.Błaż, H.Dybiec, J.Gryziecki, J.Zasadziński, „Structure and behaviour of copper and α-brass during plastic deformation", Metals Technology, 391-397 (1979)
2. H.Tresca, „Sur l'ecoulement des corps solide soumis a de fortes pressions", C.R. Acad.Sci. Paris, 59 ,754 (1864 II)
3. K.T.Huber, „Właściwa praca odkształcenia jako miara wytężenia materiału", Czasopismo Techniczne, Lwów, 22 , 38-81 (1904)
4. R. Von Mises, „Mechanik der festen Korper im plastisch-deformablen Zustand"' Gottinger Nachtrichten, Mathematik und Physik, 582 (1913)
5. K.Pieła, A.Korbel, „Necking during the high-temperature deformation of zinc single crystals", Stregth of Materials, 7-th Japan Institute of Metals Symposium (JIMIS 7) on „ Aspects of high temperature deformations and fracture in crystalline Material", Nagoya-1993, 91 (1993)
6. A.Korbel, M.Szczerba, „ Selfinduced change of the deformation path in Cu-Al single crystals", Rev. Phys.Appl., 23, 706 (1988).
7. B.Mikułowski, Metallurgy and Foundry Practice, „Strain hardening of zinc monocrystals with additions of silver and galium", Bulleting of Academy of Mining and Metallurgy - Dissertations, Kraków, 96, (1982)
8. W.Bochniak, A.Korbel, S.Wierzbiński, „The Nature of Dynamic Recrystallization in Single and Polycrystalline FCC Metals"' Recrystallisation'90", T.Chandra (Ed.), TMS, Warrendale, 780 (1990)
9. A.Considere, „Memoire sur l'emploi du fer et de l'acier dans les constructions", Ann. des Ponts et Chaussees, 9, 574 (1885)
10. V.S.Anathan, E.O.Hall, „Microscopic shear bands at Luders fronts in mild steel", Scripta Metall., 21, 519 (1987)
11. W.Bochniak, „The microstructure of Luders band in Cu-Sn2 alloy", Scrita Metall., 23, 519 (1989)
12. B.J.Brindley, P.J.Worthington, „Yield-point phenomena in substitutional alloys", Metallurgical Review, 145, 101 (1970)
13 J.J.Jonas, C,M.Selars, J.Mc G.Tegart, „Strength and structure under hot-working conditions", Rev.Met., 14, 1 (1969)
14. S.Wierzbiński, A.Korbel, J.J.Jonas, „Structural and mechanical aspects of high temperature deformation of polycrystalline nickel", Materials Science and Technology, 8, 153-158 (1992).
15. A.Korbel, L.Błaż, „The strain localization during the hot deformation of copper", Scripta Metall., 14, 829 (1980)
16. A.Korbel, H.Dybiec, „The problem of the negative strain-rate sensitivity under the Portevin-LeChatelier deformation conditions", Acta Metall., 29, 89 (1981)
17. A.Korbel, V.S.Raghunathan, D.Teirlinck, W.Spitzig, O.Richmond, J.D.Embury, „A structural study of the influence of pressure on shear band formation",Acta Metall., 32, 511-512 (1984).
18. A.H.Cottrell, R.J.Stokes, „ Effect of temperature on the plastic properties of aluminium crystals", Proc.Roy.Soc., A233, 17 (1955).

19. A.Wusatowska-Sarnek, A.Korbel, „Low temperature work softening in Cu and C--Al single crystals oriented for single slip",Strength of Materials, Oikawa (Eds),The Japan Institute of Metal, 275 (1994).
20. Z.S.Basinski, S.J.Basinski, „Plastic deformation and work hardening", Dislocations in Solids, v.4, F.R.N. Nabarro (Ed.) North Holland Publ.Comp. (1979)
21. A.Korbel,P.Martin, „Microstructural events of macroscopic strain localization in prestrained tensile specimen", Acta Metall., 36, 2575-2586 (1988).
22. K.Piela, A.Korbel, „The effect of shear banding on spatial arrangement of the second phase particles in aluminum alloy", Materials Science Forum, 217-222, 1037-1042 (1996)
23. A.Korbel, „Mechanical instability of metal substructure - Catastrophic plastic flow in single and polycrystals, Advanced in Crystal Plasticity, Eds. D.S.Wilkinson, J.D.Embury, 43-83 (1992).
24. E.Schmid, W.Boas, Kristallpastizitat mit besonderer Berucksichtigung mit Metalle, Springer-Verlag (1935)
25. G.J.Taylor, „ Plastic Strain in Metals", Inst.Metals, 62, 218 (1938)
26. J.F.W.Bishop, R.Hill, „A theoretical derivation of the plastic properties of a polycrystalline face-centered metals", Phil.Mag., 43, 414 (1951)
27. J.I. Frenkel, „Zur Theorie der Elastizitatsgrenze und der kristallinischer Korper", Z.Phys., 37, 572 (1926)
28. E.Z.Orovan, „ Zur Kristallplastizitat. Uber den Mechanismus des Gleitnorganges", Z.Phys.,89, 605 (1934)
29. M.Z.Polanyi, „ Uber eine Art Gilterstorung die einen Kristall plastisch machen konnte", Z.Phys.,89, 660 (1934)
30. G.J.Taylor, „The mechanism of plastic deformation of crystals, Part I- Theoretical", Proc. Roy. Soc. A145, 362 (1934)
31. G.J.Taylor, „The mechanism of plastic deformation of crystals, Part II- -Comparision with observations", Proc. Roy. Soc. A145, 388 (1934)
32. F.C.Frank, Disc. Far.Soc., „The influence of dislocations on crystal growth", 5, 48 (1949)
33. J.Friedel, Dislocations, Pergamon Press, Oxford (1964)
34. F.R.N. Nabarro, Theory of Crystal Dislocations, Oxford University Press (1967)
35. J.P.Hirth, J.Lothe, Theory of Dislocations , McGrow-Hill Book Comp. (1968)
36. H.Neuhauser, „Slip-line formation and collective dislocation motion", Dislocations in Solids, F.R.N.Nabarro (Eds), North Holland Publ. Comp., Amsterdam 6, 319 (1983)
37. T.H.Blewit, R.R.Coltman, J.K.Redman, cited in [20].
38. H.Rebstock, „ Kombinierte Zug - und Torsionsverformung von Kupfer- Einkristallrohren" Zeit.f.Metallkunde, 48, 206 (1957)
39. S.Mader, H.Seeger, „Untersuchung des Gleitlinienbildes kubbisch-flachenzentriertr Einkristalle", Acta Metall., 8, 513 (1960)
40. M.Masima, G.Sachs, „ Mechanische Eigenschaften von Messingkristallen'" Zeit. f. Physik, 50, 161 (1928)
41. V.Goler, G.Sachs, „Zugversuche an Kristallen aus Kupfer und α-Messing", Zeit. f.

Physik, 55, 581 (1929)
42. G.Sachs, J.Weerts, „Zugversuche an Gold - Silberkristallen", Zeit. f. Physik, 60, 473 (1930)
43. H.W.Paxton, A.H.Cottrell, „Work -hardening in streched and twisted aluminium crystals", Acta Metall., 2, 3 (1954)
44. Z.S.Basinski, P.J.Jackson, „ Instability of Work Hardened State I- Slip in Extraenously Deformed Crystals",Phys. Stat. Sol., 9, 805 (1967)
45. Z.S.Basinski, P.J.Jackson, Z.S.Basinski, P.J.Jackson, „ Instability of Work Hardened State II - Slip in Alien Dislocation Distribution" ,Phys. Stat. Sol.,10, 45 (1965)
46 Y.Nakada, A.S.Keh, „Latent hardening in iron crystals", Acta Metall., 14 , 961 (1966)
47 E.J.H.Wessels, P.J.Jackson, „Latent Hardening in copper-aluminium alloys", Acta Metall., 17, 241 (1969)
48 P.Franciosi, M.Berveiller, A.Zaoui, „ Latent hardening in copper and aluminium crystals", Acta Metall., 28, 273 (1980)
49. A.Kelly, G.W.Groves, Crystallography and Crystal Defects, London Group Ltd.., London, 1970
50. Z.S.Basinski, M.Szczerba, D.J. Embury, „ Tensile instability in face-centered cubic materials", Phil.Mag., in press
51. J.W.Sharp, M.J.Makin, „Slip behavior in copper crystals previously deformed on another slip system",Can.Jour. Phys., 22, 519 (1967)
52. J.H.Wessels, F.R.N.Nabarro, „The hardening of latent glide systems in single crystals of copper-aluminium alloys", Acta Metall., 19, 903 (1987)
53. M.Szczerba, A.Korbel, „ Strain Softening and instability of plastic flow in Cu-Al single crystals", Acta Metall., 35, 1129 (1986)
54. M. Szczerba, not published data
55. P.J.Jackson, Z.S.Basinski, „The effect of extraneous deformation on strain hardening in Cu single crystals", Appl. Phys. Letters, 6, 148 (19645)
56. A.Korbel, P.Martin, „ Microscopic versus macroscopic aspect of shear bands deformation", Acta Metall., 34, 1905 (1986)
57. A.Korbel, „Perspectives of the control of mechanical performance of metals during forming operations" Jour. Materials Processing Technology 34, 41 (1992)
58. A.Korbel, W.Bochniak, Jour. Materials Processing Technology , 53,229 (1995)
59. A.Dziadoń, „The Role of Strain Localization in the Dynamic Strain Ageing Phenomenon of Polycrystalline Alpha Titanium", Metallurgy and Foundry Practice, Scientific Bulletings of the Academy of Mining and Metallurgy, Bulletin 146, Kraków (1993)
60. W.Bochniak, „Organization of Slip and the Portevin-LeChatelier Effect in Alpha-Brass"' Proc. 4th European Conference on Advanced Materials and Processes, Padua-Venice , 265 (1995)
61. A.Korbel, F.Dobrzański, M.Richert, „Strain hardening of aluminium at high strains", Acta Metall., 31, 293 (1983)
62. W.Oliferuk, A.Korbel, M.Grabski, „Mode of deformation and the rate of energy storage during uniaxial tensile deformation of austenitic steel", Materials Science and Engineering A220, 123 (1996)

63. A.Korbel, „The model of microshear banding in metals", Scirpta Metall. and Materialia, 24, 1229 (1990)
64. A.Pawełek, A.Korbel, „Soliton-like behaviour of moving dislocation group", Phil. Mag., B, 61, 829 (1990)

CONSTITUTIVE MODELLING OF DISSIPATIVE SOLIDS FOR LOCALIZATION AND FRACTURE

P. Perzyna
Polish Academy of Sciences, Warsaw, Poland

To the Memory of Maria

ABSTRACT

The main objective of the lectures is to survey some recent developments in the constitutive modelling of inelastic single and polycrystalline solids which can be used for the description of plastic deformation localization and fracture phenomena. Physical foundations and experimental motivations are given. Particular attention is focused on dynamic fracture (adiabatic shear band localized, spall, ductile and brittle fracture phenomena). The physical and experimental foundations for the microdamage processes are presented. The microdamage process has been treated as a sequence of nucleation, growth and coalescence of microcracks. The microdamage kinetics interacts with thermal and load changes to make failure of solids a highly rate, temperature and history dependent, nonlinear process.

The description of the kinematics of finite elasto–viscoplastic deformations is based on notions of the Riemannian space on manifolds and tangent space. A multiplicative decomposition of the deformation gradient is adopted and the Lie derivative is used to define all objective rates for introduced vectors and tensors.

A general constitutive model is developed within the thermodynamic framework of the rate type covariance structure with finite set of the internal state variables. A notion of covariance is understood in the sense of invariance under arbitrary spatial diffeomorphism. The thermodynamic theory of elasto–viscoplasticity of inelastic single crystals and damaged polycrystalline solids is developed. The relaxation time is used as a regularization parameter. By assuming that the relaxation time is equal to zero the thermo–elastic–plastic (rate independent) response for both single crystals and polycrystalline solids is accomplished.

An adiabatic inelastic flow process is analysed and the well–posedness of the Cauchy problem is investigated. The analytical methods for the investigation of plastic deformation localization phenomena are developed. The formation of the adiabatic shear band region is investigated. Criteria for adiabatic shear band localization of plastic deformation are obtained by assuming that some of eigenvalue of the instantaneous adiabatic acoustic tensor for rate independent response is equal to zero.

Numerical solution of the initial–boundary value problem (evolution problem) is discussed. Particular attention is focused on the well–posedness of the evolution problem. Convergence, consistency and stability of the discretised evolution problem are examined. The Lax equivalence theorem is formulated and the conditions of its validity are investigated. Utilizing the finite element method for regularized elasto–viscoplastic model the numerical investigation of localization and fracture phenomena is presented. The results obtained are compared with available experimental observations.

1. INTRODUCTION

In this collection of lectures emphasis is laid on experimental and physical foundations as well as on mathematical constitutive modelling for the description of localization of plastic deformation and various modes of fracture phenomena in inelastic single crystals and damaged polycrystalline solids.

The understanding of the physical origin and nature of the plastic behaviour of polycrystalline aggregates constitutes one of the major problems in modern materials science.

In recent years several models have been proposed to predict deformation textures, large plastic deformation, strain hardening and strain softening behaviour of polycrystalline solids based on the known behaviour of single crystals. The possibility of making such a prediction rests on the tacit assumption that the main mechanisms of plastic deformation in polycrystalline aggregates are substantially identical with those observed in single crystals.

Recent experimental observations and theoretical investigations have shown that the synergetic effects have great influence on the behaviour of inelastic single crystals. Particularly the adiabatic shear band localization in single crystals is affected very much by cooperative phenomena. The same conclusion can be drawn for the behaviour of damaged polycrystalline solids and particularly for fracture phenomena.

Chapter 2 presents experimental and physical foundations for both single crystals and polycrystalline solids. First, experimental observations of single crystal behaviour are discussed. Particular attention is focused on the experimental motivations of the influence of the evolution of the substructure on the behaviour of single crystals in the critical situation when the macroscopic localized shear band is formed.

Experimental observations of the macroscopic adiabatic shear band localization in single crystals performed by Chang and Asaro (1980, 1981), Spitzig ((1981) and Lisiecki et al. (1982) showed that the strain–hardening rate h_{crit} at the inception of shear band localization is positive and the direction of the localized shear band is misaligned by some small angle δ from the active slip system.

On the other hand the investigations presented by Mecking and Kocks (1975), Follansbee (1986) and Follansbee and Kocks (1988) showed the great influence of the strain rate sensitivity on the behaviour of inelastic metallic single crystals in dynamic loading processes. To describe the strain rate sensitivity effects Follansbee (1986) suggested to take into consideration the evolution of the dislocation substructure.

Experimental study of highly heterogeneous deformations in copper single crystals performed by Rashid et al. (1992) showed that the strain rate history dependence of the substructure evolution plays an important role particularly in adiabatic shear band formation phenomena.

Physical motivations of the new thermodynamic viscoplasticity theory of metallic single crystals have been presented. The discussion of various physical mechanisms of dislocation motion and particularly the interaction of the thermally activated and phonon damping mechanisms has been given. The relaxation time treated as a microstructural parameter has been introduced. It has been shown that the proposed viscoplastic model accomplishes the description of behaviour of single crystals valid for the entire range of strain rate changes and encompasses the interaction of the thermally activated and phonon damping mechanisms.

The methods of synergetics, namely the self organization conceptions, are used to explain complex macroscopic adiabatic shear band pattern.

Second, experimental justifications for the behaviour of polycrystalline solids are given. Strain rate sensitivity effect is discussed. The relaxation time for the viscoplastic model of polycrystalline solids is investigated. The localized fracture phenomenon of polycrystalline solids is experimentally motivated. The shear band formation and the micro–damage process are discussed. The thermo–mechanical coupling and anisotropy effects are analysed. Experimental observations have suggested that the shear band localization failure in dynamic loading processes is affected by complex cooperative phenomena. The intrinsic microstructure of the shear band region has been investigated and some conclusions important for the constitutive modelling have been drawn.

Chapter 3 is devoted to the description of kinematics of finite deformations and the stress tensors. The fundamental measures of total deformation are introduced. The decomposition of the strain tensor into the elastic and viscoplastic part is presented. The rates of the deformation tensor and the stress tensor are defined based on the Lie derivative.

In Chapter 4 the development of a rate dependent constitutive model within the thermodynamic framework of the rate type covariance structure with finite set of the internal state variables is presented. This constitutive model is based on the axioms as follows: (i) existence of the free energy function; (ii) invariance with respect to any diffeomorphism (any superposed motion); (iii) assumption of the entropy production inequality; (iv) assumption of the evolution equations for the internal state variables in the particular rate dependent form. For inelastic single crystals it has been assumed that, a set of the internal state variables consists of the shearings $\gamma^{(\nu)}$, the densities of mobile dislocations $\alpha^{(\nu)}$, the densities of obstacle dislocations $\beta^{(\nu)}$ and the concentrations of point defects $\xi^{(\nu)}$ ($\nu = 1, 2, \ldots, n$). The fundamental rate type constitutive equations for the Kirchhoff stress tensor τ and for temperature ϑ are formulated. These rate constitutive equations take account of the effects as follows: (i) thermomechanical coupling; (ii) influence of covariance terms, lattice deformations and rotations and plastic spin; (iii) evolution of the dislocation substructure; (iv) deviation from the Schmid rule of a critical resolved shear stress for slip; (v) rate sensitivity (viscosity). The rate independent response of single crystal is obtained under the assumption that the relaxation time $T^{(\nu)} = 0$.

For thermoviscoplasticity of damaged polycrystalline solids a set of the internal state variables is assumed to consists of the new internal vector ζ which describes the dissipation effects generated by viscoplastic flow phenomena, volume fraction porosity ξ takes account for micro-damage effects and the residual stress (the back stress) aims at the description of the kinematic hardening.

The theory developed describes the effects as follows: (i) plastic non-normality; (ii) plastic strain induced anisotropy (kinematic hardening); (iii) plastic spin; (iv) micro-damage process (softening effects generated by microkrack nucleation and growth processes); (v) influence of covariant terms; (vi) thermomechanical coupling (thermal plastic softening and thermal expansion); (vii) rate sensitivity (viscosity).

By assuming that the mechanical relaxation time is equal to zero the thermo-elastic-plastic (rate independent) response of the damaged material is accomplished.

An adiabatic inelastic flow process is formulated and investigated in Chapter 5. The conditions for the well-posedness of the Cauchy problem are examined.

Chapter 6 is devoted to the development of the analytical methods for the investigation of localization phenomena. Particular attention is focused on an analysis of acceleration waves. It has been proved that in an adiabatic process for both elastic-viscoplastic and elastic-plastic rate independent model of single crystals the acceleration discontinuity [[a]] is the solution of the appropriate eigenvalue problem. In these eigenvalue problems the instantaneous adiabatic acoustic tensors $\mathbf{A}$ and $\hat{\mathbf{A}}$ plays a fundamental role. The macroscopic shear band formation during an adiabatic process for symmetric double slip and single slip in elastic-plastic rate independent single crystal are studied. The necessary condition for a localized plastic deformation region to be formed is obtained when the determinant of the instantaneous acoustic tensor $\hat{\mathbf{A}}$ is equalled to zero. The criteria for adiabatic shear band localization in the single slip process are obtained in exact analytical form. For the symmetric double slip process these criteria have been estimated numerically. The identification procedure for material constants has been presented. Numerical estimations for the critical hardening modulus rate h_{crit} and the direction of the macroscopic shear band are given. An analysis of the influence of various effects on shear band localization criteria is presented. Particular attention is focused on the investigation of the influence of the evolution of the substructure and thermomechanical couplings. Comparison of the numerical results with available experimental data is presented. The possibility of deviations from the Schmid rule of the critical resolved shear stress is also investigated. The predicted by the theory for single slip process critical value of the hardening modulus rate h_{crit} is in accord with experimental observations while the misalignment of the shear bands from the active slip systems in the crystal's matrix is too small. For symmetric double slip process both values obtained are in accord with experimental observations. It has been found that the influence of the dislocation substructure is combined with the thermomechanical couplings, and that is why it gives distinct synergetic effect.

By using the same method criteria for adiabatic shear band localization of plastic deformation for damaged polycrystalline solids have been obtained in exact form. Particular effects have been examined. Cooperative phenomena are considered and synergetic effects are investigated. Numerical estimations and comparisons of particular effects are presented.

Numerical solutions of the initial–boundary value problem (evolution problem) are discussed in chapter 7. Mathematical formulation of the evolution problem is presented. Discretisation in space and time is proposed and convergence, consistency and stability are examined. The Lax equivalence theorem is formulated and conditions under which this theorem is valid are investigated. General regularization method is presented.

Chapter 8 is devoted to the numerical investigation of shear band localization phenomena. One particular example has been considered, namely a dynamic adiabatic process for a thin–walled steel tube. This initial boundary value problem (evolution problem) has been solved numerically by means of finite element method and ABAQUS system. Particular attention has been focused on a thin shear band region of finite width which undergoes significant deformations and temperature rise. Its evolution until occurrence of fracture has been simulated. The discussion of the results is presented and the comparison with available experimental observations is given.

A developed viscoplastic–damage type of constitutive theory for high strain rate flow processes and ductile fracture is used in Chapter 9 to model the deformation and fracture of dynamically loaded smooth cylindrical tensile bars, cf. Nemes and Eftis (1993). The analysis assumes polycrystalline materials which usually contain microvoids dispersed homogeneously throughout. It is shown that for dynamically imposed loading that produce nominal strain rates ranging between $5 \times 10^2 - 5 \times 10^3$ sec^{-1}, the inhomogeneous fields od stress and deformation caused by wave propagation and wave reflection induce necking at different locations along the gauge section, depending upon the strain rate imposed. This occurs without imposition of any geometrical or material irregularity to preposition the location of the necking.

Examples of numerical simulations of spall fracture of polycrystalline solids induced by plate impact (cf. Eftis (1996)) are also presented.

2. EXPERIMENTAL AND PHYSICAL FOUNDATIONS

2.1 Experimental observations of single crystal behaviour

The high – rate deformation of face–centered cubic (f.c.c.) metals, such as copper, aluminum, lead and nickel has been recently extensively studied (cf. review paper by Follansbee (1986)). It has been shown that the apparent strain rate sensitivity of f.c.c. metals has two origins: that associated with the finite velocity of dislocations, and that connected with the evolution of the dislocation substructure. The first of these two components – the instantaneous rate sensitivity – is related to the wait – times associated with thermally activated dislocation motion. The second component has more to do with the relative importance of dislocation generation and annihilation at different strain rates, and shall be referred to as the strain rate history effect.

Rashid et al. (1992) performed an experimental study of highly heterogeneous deformations in copper single crystals to investigate the importance of the dislocation substructure. Their experimental results are mainly qualitative in nature, and include optical photographs and micrographs of the deformed specimens, and scanning and transmission electro micrographs of the substructure. The results obtained seem to support the earlier findings related to the strain rate history dependence of the substructure evolution. The higher total dislocation density observed in the notch region of the dynamically deformed specimen, as compared to the same region in the quasi-statically deformed specimen, reflects the higher shear stress attained in the dynamic case, cf. Figs. 1 and 2.

When ductile single crystals of metals are finitely deformed, they display highly heterogeneous deformation, e.g. when crystals are stretched in tension, they can neck and then develop macroscopic bands of localized shearing.

Experimental observations of Chang and Asaro (1980, 1981) Lisiecki et al. (1982) and Spitzig (1981) for copper, aluminum – copper and nitrogenated Fe–Ti–Mn single crystals investigated in uniaxial tension tests have shown that in the first stage of the process, a crystal specimen undergoes uniform extension in single slip. At the point when the load–engineering strain trajectory reaches its maximum, a crystal specimen exhibits slight amounts of very diffuse necking, cf. Figs. 3 and 4. The neck is usually symmetric in shape indicating that double slip is operative within it, cf. Lisiecki et al.

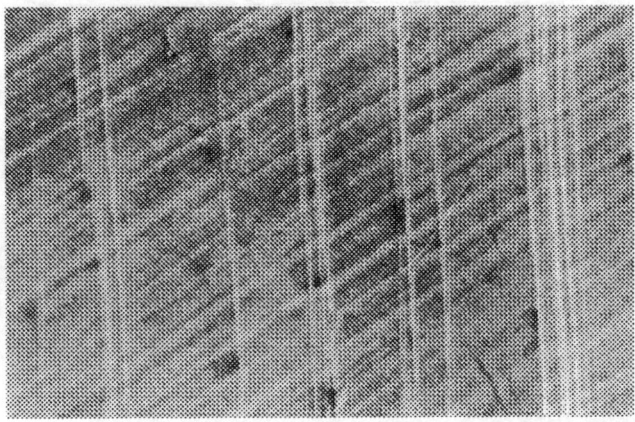

Figure 1: Scanning electron micrograph of an intensely deformed region of a dynamically deformed specimen, taken to the left of the specimen centreline. The lattice rotation is clearly evident (after Rashid et al. (1992))

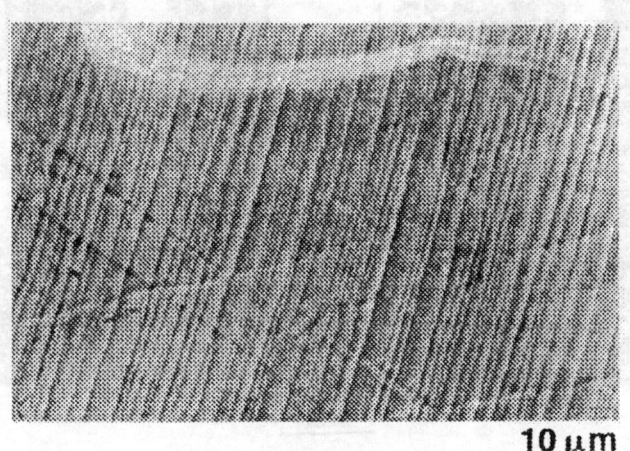

Figure 2: Scanning electron micrograph taken near the right corner of the indenter notch on the quasi-statically deformed specimen (after Rashid et al. (1992))

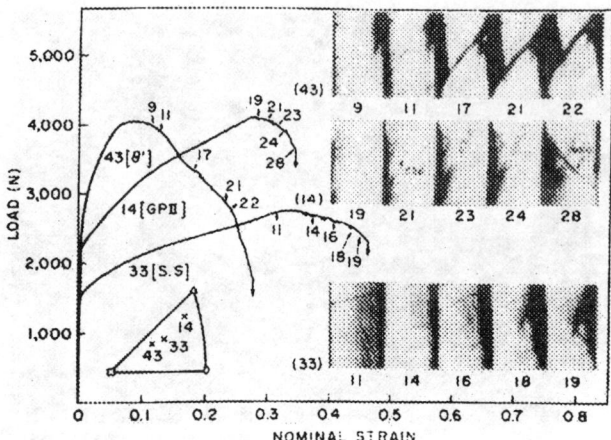

Figure 3: Load versus engineering strain curves for various ageing treatments. Numbered photos correspond to the indicated points on the load–strain curves (after Chang and Asaro (1981))

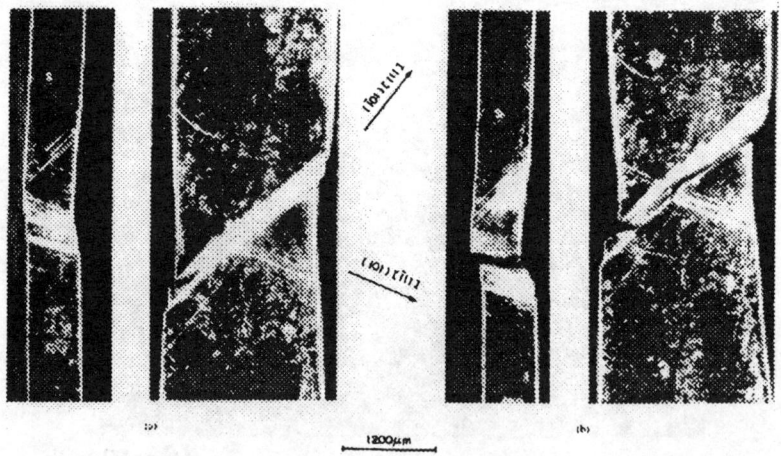

Figure 4: Propagation of localized shear band in nitrogenated Fe–Ti–Mn crystal deformed at 295 K: (a) 20 percent decrease in load maximum load, (b) 40 percent decrease in load maximum load (after Spitzig (1981))

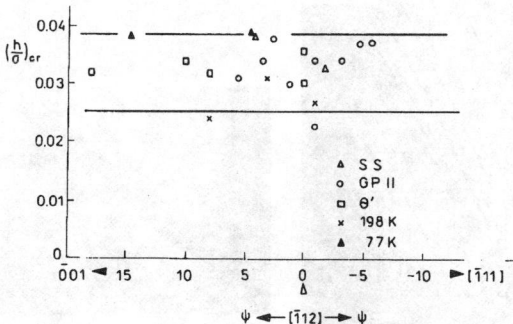

Figure 5: Critical ratio $(h/\sigma)_{crit}$ versus ψ, the angle between the tensile axis and $[\bar{1}12]$ for various ageing treatments (after Chang and Asaro (1981))

(1982). At this stage of the tensile process, the gross plastic deformations are localized to the diffusely necked region and the thermomechanical coupling effects begin to play a crucial role. That is why in this region of the specimen the tensile process has to be considered as adiabatic. With continued extension the macroscopic, adiabatic shear bands have soon developed within the diffusely necked region. This point on the shear stress – shear strain trajectory is found experimentally to lie on the increasing part of this curve very near to the maximum point, so that a critical value of the strain–hardening modulus rate $h_{crit} = \left(\frac{d\tau}{d\gamma}\right)$ is small but positive, cf. Fig. 5.

It has been experimentally observed that at the inception of the macroscopic, adiabatic shear band, the direction of the band is slightly different from the detected coarse slip bands or slip traces. In other words, the macroscopic shear bands are not aligned with the active slip systems in the crystal's matrix but are misaligned by angle δ, cf. Figs. 6 and 7.

2.2 Plastic flow mechanisms of single crystals

The rate and temperature dependence of the flow stress of metal crystals can be explained by different physical mechanisms of dislocation motion. The microscopic processes combine in various ways to give several groups of deformation mechanisms, each of which can be limited to the particular range of temperature and strain rate changes.

It will be profitable for further considerations to discuss some of these mechanisms, particularly those which lead to viscoplastic response of the crystal.

Some common thermal obstacles or mechanisms in pure metals are as follows: (i) intersection of forest dislocations; (ii) overcoming Peierls–Nabarro stress; (iii) non–conservative motion of jogs; (iv) cross–slip of screw dislocations; (v) climb of edge dislocations. Forest dislocations, the Peierls–Nabarro stress and jogs represent re-

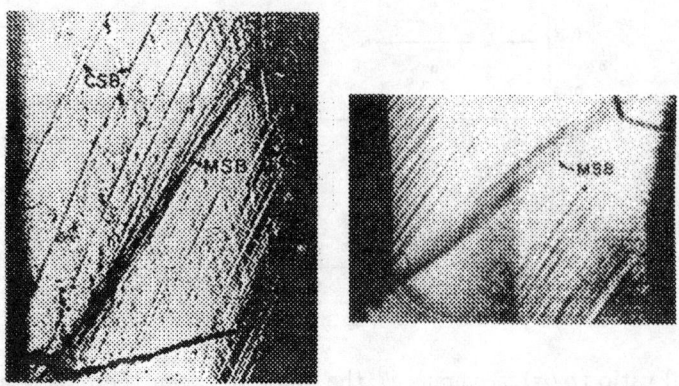

Figure 6: Coarse slip band (CSB) and macroscopic shear bands (MSB) in (a) GPII tested at 77 K and (b) a θ' strengthened crystal tested at 298 K. Note the orientation difference between CSBs and MSBs in (a), CSBs are closely aligned with the active slip systems (after Chang and Asaro (1981))

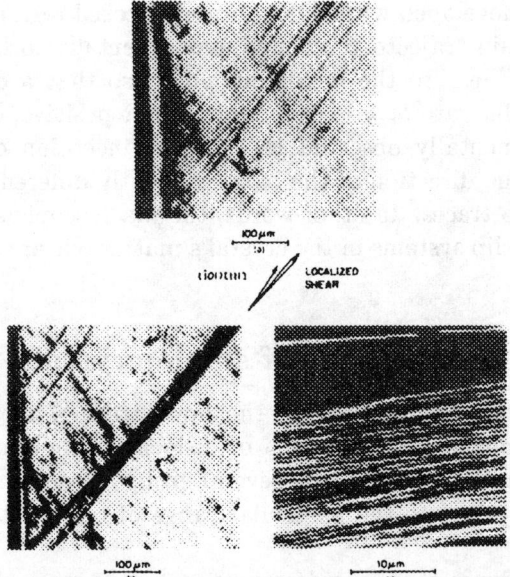

Figure 7: Slip traces and localized shear on the surface of nitrogenated Fe–Ti–Mn crystal of orientation D deformed at 295 K: (a) initial deformation until necking began, (b) subsequent deformation after removal of neck and localized shear bands from initial deformation, (c) slip traces within localized shear band in (b) (after Spitzig (1981))

sistance to the motion of dislocations in the slip plane, while cross–slip and climb represent resistance to the motion out of the slip plane. Schematic representations of the ways in which these obstacles are overcome are given in Fig. 8. In each case, thermal fluctuations assist the applied stress in getting a dislocation segment L past the barrier (cf. Conrad (1964)).

To describe theoretically all the mechanisms we have to introduce three important parameters, namely the density of mobile dislocations α, the density of obstacle dislocations β and the concentration of point defects ξ. Average density of mobile dislocations in deformed metal single crystals is of the order of 10^{15}m^{-2}, average density of obstacle dislocations is 10^{13}m^{-2}, and the average value of the concentration of point defects can be of the order of 10^{15}m^{-3}.

Since plastic flow occurs by the motion of dislocation lines, the rate at which it takes place depends on how fast the dislocations move, how many dislocations are moving in a given volume of material, and how much displacement is carried by each dislocation. The theory of crystal dislocations shows that for the single slip, the inelastic shear strain rate is as follows

$$\dot{\epsilon}^p = \alpha b v, \qquad (2.1)$$

where α is the mean density of mobile dislocations, b is the displacement per dislocation line (the Burgers vector), and v denotes the mean dislocation velocity.

2.2.1 Thermally activated mechanism

It is now generally recognized that the plastic deformation of a crystal is of dynamic nature and has been established as a thermally activated process dependent upon time, temperature and strain rate. The evolution of the activation parameters is a widely used technique for the identification of the mechanisms controlling the rate of deformation, and has been applied to b.c.c., f.c.c., h.c.p. metals, intermetallic compounds and ionic and ceramic crystals (cf. the review paper by Evans and Rawlings (1969) and books by Nabarro (1967), and Kocks et al. (1975)).

When a dislocation moves through a crystal lattice, a force is exerted upon it by obstacles present in the lattice. This force can be separated into two components, a long–range force and short–rang force.

The stress nesessary to overcome the short–range obstacles is temperature–dependent, whereas that needed to surmount fixed long–range obstacles generally depends upon temperature only through the temperature-dependence of the shear modulus. For this reason the obstacles are often referred to as thermal and athermal, respectively. When both types of obstacles are present in a lattice, the applied stress is usually composed of both thermal and athermal components

$$\tau = \tau^{\#} + \tau_\mu, \qquad (2.2)$$

where $\tau^{\#}$ is the thermal (or effective) resolved shear stress and τ_μ is the athermal stress.

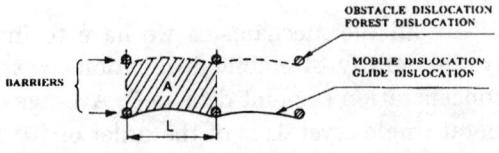
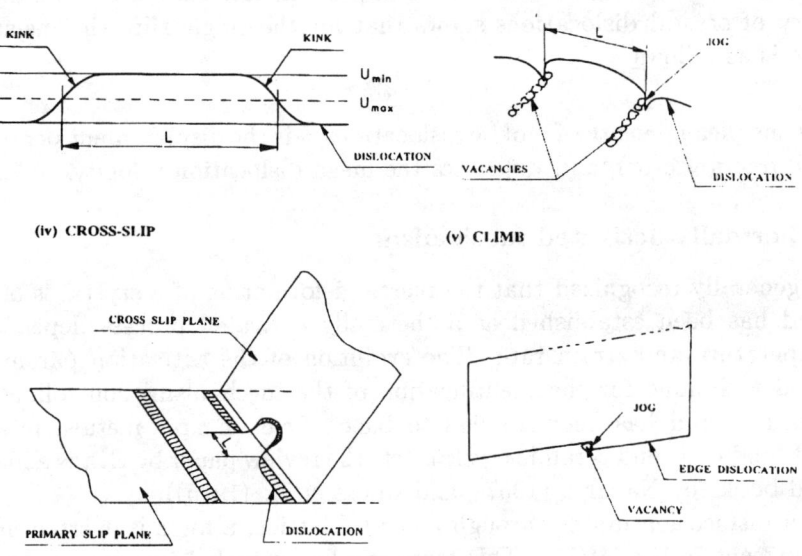

Figure 8: Schematic representation of thermal obstacles or mechanisms in pure metals (cf. Conrad (1964))

Plastic deformation occurs by the movement of a large number of dislocations through an array of obstacles. At any finite temperature, coherent atomic fluctuations can assist the applied stress in moving a dislocation past the obstacles.

The average velocity v of a dislocation that surmounts the obstacles with the assistance of thermal fluctuations is assumed to be an Arrhenius–type relationship

$$\text{v} = AL^{-1}\nu \exp\left(-\frac{U}{k\vartheta}\right), \qquad (2.3)$$

where ν is the frequency of vibration of the dislocation, AL^{-1} is the distance covered after a successful fluctuation, U is the activation energy (Gibbs free energy), k is the Boltzman constant and ϑ is actual absolute temperature.

Equations (2.1) and (2.3) give

$$\dot{\epsilon}^p = \alpha b AL^{-1}\nu \exp\left(-\frac{U}{k\vartheta}\right). \qquad (2.4)$$

Let us assume that

$$U = U[(\tau - \tau_\mu)Lb], \qquad (2.5)$$

where L is the mean cord distance between the neighboring points at which the dislocation is arrested. Expansion of the function U gives

$$U = U\big|_{\tau=\tau_\mu} + U'\big|_{\tau=\tau_\mu}(\tau - \tau_\mu)Lb + U''\big|_{\tau=\tau_\mu}\frac{(\tau - \tau_\mu)^2 L^2 b^2}{2!} + \cdots \qquad (2.6)$$

Let us denote by

$$v^* = -U'\big|_{\tau=\tau_\mu} Lb, \quad U_\circ = U\big|_{\tau=\tau_\mu}, \qquad (2.7)$$

the activation volume and the activation energy for intersection at zero effective stress, respectively.

The linear approximation to Eq. (2.4) gives the Seeger relation (cf. Seeger (1955, 1958))

$$\dot{\epsilon}^p = \alpha b AL^{-1}\nu \exp\left\{-\frac{U_\circ}{k\vartheta} + \left[(\tau - \tau_\mu)\frac{v^*}{k\vartheta}\right]\right\}, \qquad (2.8)$$

or

$$\tau = \left(\tau_\mu + \frac{U_\circ}{v^*}\right) + \frac{k\vartheta}{v^*} \ln \frac{\dot{\epsilon}^p}{\alpha b AL^{-1}\nu}. \qquad (2.9)$$

When the activation energy U is a nonlinear function of the effective stress (cf. Eq. (2.5)), the relation (2.4) yields

$$\dot{\epsilon}^p = \alpha b AL^{-1}\nu \exp\left\{-U\left[(\tau - \tau_\mu)Lb\right]/k\vartheta\right\} \qquad (2.10)$$

or

$$\tau = \tau_\mu + \frac{1}{Lb}U^{-1}\left[k\vartheta \ln\left(\alpha b AL^{-1}\nu/\dot{\epsilon}^p\right)\right]. \qquad (2.11)$$

Let us denote by

$$T_{mT} = \frac{1}{\gamma_T} = (\alpha b A L^{-1}\nu)^{-1}, \quad \tau_B = (\tau_\mu + U_o/v^*), \qquad (2.12)$$

the relaxation time for the thermally activated mechanism of dislocation motion (γ_T defines the viscosity coefficient) and the flow stress τ_B, respectively. Then the relations (2.8) and (2.9) take the form

$$\dot{\epsilon}^p = \frac{1}{T_{mT}} \exp\left[\frac{v^*}{k\vartheta}(\tau - \tau_B)\right], \quad \tau = \tau_B + (k\vartheta/v^*)\ln(T_{mT}\dot{\epsilon}^p). \qquad (2.13)$$

In this linear theory we have three intrinsic material parameters, namely the relaxation time T_{mT}, the activation volume v^* and the flow stress τ_B.

In the most general case, each of these three parameters may be considered as a function of the three independent variables ϵ^p, τ and ϑ.

In the nonlinear theory

$$\dot{\epsilon}^p = \frac{1}{T_{mT}} \exp\{-U[(\tau - \tau_\mu)Lb]/k\vartheta\} \qquad (2.14)$$

or

$$\tau = \tau_\mu + \frac{1}{Lb}U^{-1}\left[k\vartheta \ln\left(\frac{1}{T_{mT}\dot{\epsilon}^p}\right)\right] \qquad (2.15)$$

there are two intrinsic material parameters T_{mT} and τ_μ and, in addition, one response function U.

2.2.2 Damping mechanism (phonon viscosity)

With increasing dislocation velocities at high enough stress or in perfect crystal, the velocity is only governed by the phonon damping mechanism. The phonon viscosity theory has been developed by Mason (1960) (cf. Nabarro (1967)). At very high strain rates the applied stress is high enough to overcome instantaneously the dislocation barriers without any aid from thermal fluctuations. This is true for the resolved shear stress $\tau > \tau_B$, where τ_B is attributed to the stress needed to overcome the forest dislocation barriers to the dislocation motion and is called the back stress.

In this region of response, the evolution equation for the inelastic shearing has the form

$$\dot{\epsilon}^p = \frac{\alpha b^2 \tau_\mu}{B}\left[\frac{\tau}{\tau_B} - 1\right], \qquad (2.16)$$

where B is called the dislocation drag coefficient. If we introduce the denotation

$$T_{mD} = \frac{B}{\alpha b^2 \tau_B} = \frac{1}{\gamma_D} \qquad (2.17)$$

Constitutive Modelling of Dissipative Solids

for the relaxation time for the phonon damping mechanism (γ_D defines the viscosity coefficient for this region), then the evolution equation (2.16) takes the form

$$\dot{\epsilon}^p = \frac{1}{T_{mD}}\left(\frac{\tau}{\tau_B} - 1\right) \qquad (2.18)$$

or

$$\tau = \tau_B\left(1 + T_{mD}\dot{\epsilon}^p\right). \qquad (2.19)$$

For the phonon damping mechanism we have two intrinsic parameters, namely the relaxation time T_{mD} and the back stress. It is noteworthy that the dislocation drag coefficient B can be interpreted as a generalized damping parameter for phonon viscosity and electron viscosity mechanisms (cf. Gorman et al. (1969)) i.e.

$$B = B_{pv} + B_{ev}. \qquad (2.20)$$

2.2.3 Interaction of the thermally activated and phonon damping mechanisms

If a dislocation is moving through the rows of barriers, then its velocity can be determined by the expression

$$v = AL^{-1}/(t_S + t_B), \qquad (2.21)$$

where AL^{-1} is the average distance of dislocation movement after each thermal activation, t_S is the time a dislocation spent at the obstacle, and t_B is the time of travelling between the barriers.

The shearing rate in single slip is given by the relationship (cf. Kumar and Kumble (1969), Teodosiu and Sidoroff (1976) and Perzyna (1977, 1988))

$$\dot{\epsilon}^p = \frac{1}{T_{mT}}\left(\exp\left\{U[(\tau - \tau_\mu)Lb]/k\vartheta\right\} + BAL^{-1}\nu/(\tau - \tau_B)b\right)^{-1} \qquad (2.22)$$

where

$$\frac{1}{T_{mT}}\frac{b\tau_B}{BAL^{-1}\nu} = \frac{\alpha b^2 \tau_B}{B} = \frac{1}{T_{mD}}, \qquad (2.23)$$

and two effective resolved shear stresses

$$\tau_T^* = \tau - \tau_\mu \quad \text{and} \quad \tau_D^* = \tau - \tau_B \qquad (2.24)$$

are separately defined for the thermally activated and phonon damping mechanisms, respectively.

If the time t_B taken by the dislocation to travel between the barriers in a viscous phonon medium is negligible when compared with the time t_S spent at the obstacle, then

$$v = \frac{AL^{-1}}{t_S} \qquad (2.25)$$

and we can focus our attention on the analysis of the thermally activated process.

When the ratio t_B/t_S increases then the dislocation velocity (2.21) can be approximated by the expression

$$v = \frac{AL^{-1}}{t_B} \qquad (2.26)$$

for the phonon damping mechanism.

2.3 Viscoplastic model of single crystals

The main idea of the viscoplastic flow mechanism is to accomplish in one model the description of behaviour of single crystals valid for the entire range of strain rate changes. In other words, the main conception is to encompass the interaction of the thermally activated and phonon damping mechanisms.

To achieve this aim the empirical overstress function Φ has been introduced and the strain rate is postulated in the form as follows (cf. Perzyna (1988))

$$\dot{\epsilon}^p = \frac{1}{T} \langle \Phi \left[\frac{\tau}{\tau_Y(\epsilon^p, \vartheta, \beta, \zeta)} - 1 \right] \rangle \mathrm{sgn}\tau, \qquad (2.27)$$

where T is the relaxation time, $\langle \cdot \rangle$ denotes the Macauley bracket and τ_Y is the static yield stress function. In this model the static yield stress function depends on the inelastic strain ϵ^p, temperature ϑ, the density of obstacle dislocations β and the concentration of point defects ζ.

It is noteworthy that the empirical overstress function Φ can be determined basing on available experimental results performed under dynamic loading.

To describe the main experimentally observed facts connected with the macroscopic shear band localization of single crystals, namely that the strain-hardening modulus rate h_{crit} at the inception of shear band localization is positive and the direction of the localized shear band is misaligned by some angle δ from the active slip system, we intend to consider the synergetic effects resulting from taking into account spatial covariance effects and thermomechanical couplings (cf. Duszek–Perzyna and Perzyna (1993)).

To take into consideration the evolution of the substructure of crystals we introduce the density of mobile dislocations $\alpha^{(\nu)}$, the density of obstacle dislocations $\beta^{(\nu)}$ and the concentration of point defects $\zeta^{(\nu)}$ for particular slip system ν as the internal state variables.

2.4 Heuristic considerations for single crystals

From the analysis of the experimental investigations of localized shearing in single crystals performed by Chang and Asaro (1980, 1981), Spitzig (1981), Lisiecki et al. (1982) and Rashid et al. (1992) we can follow the events in the order in which things

Constitutive Modelling of Dissipative Solids

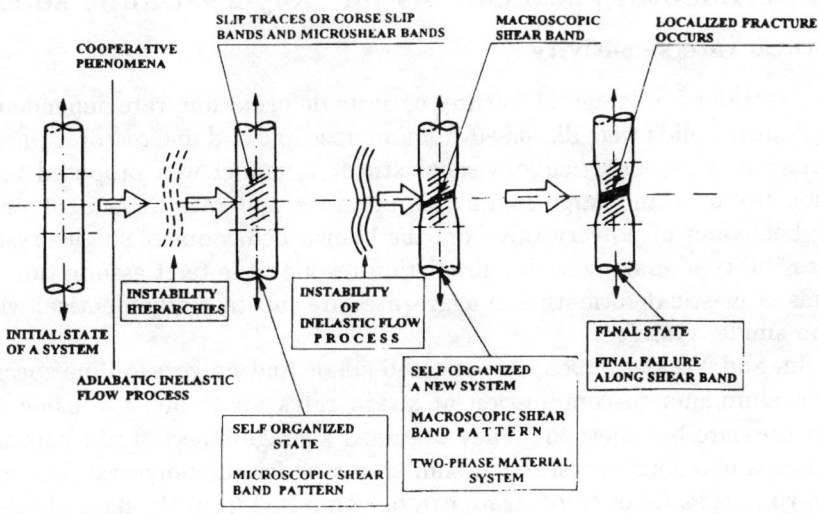

Figure 9: Subsequent states of adiabatic inelastic flow process of single crystal

naturally happen within a gauge length of the specimen during the uniaxial test, cf. Fig. 9.

In the first stage of the adiabatic inelastic flow process a crystal specimen (a system) undergoes uniform extension and slip takes place. When control parameters are changed over a wide range, our system may run through a hierarchy of instabilities and accompanying cooperative phenomena. When we look at a microscopic level we observe that a crystal is well ordered, and is self organized in microscopic shear band pattern.

At the point when the load–engineering strain trajectory reaches its maximum, i.e. when the criterion of the onset of the localization by necking mode is satisfied, a crystal specimen exhibits slight amount of very diffuse symmetric necking.

With continued extension the instability of inelastic flow process takes place and we observe on macroscopic level the formation of adiabatic shear band pattern within the diffusely necked region. A system is self organized to a new – two–phase material system (cf. Fig. 9). This is mainly due to different substructure and its evolution within the regions of adiabatic shear bands when compared with the substructure in the attached zones, cf. Fig. 7c. Final separation occurs by a ductile failure mechanism along the shear band, cf. Fig. 4.

2.5 Experimental justifications for polycrystalline solids

2.5.1 Strain rate sensitivity

In previous sections fundamental features of finite deformation, rate dependent plastic flow of crystalline solids were discussed from microscopic and macroscopic phenomenological points of view. Particular viscoplastic flow model was proposed to predict deformation textures and large strain, temperature and rate dependent and strain hardening behaviour of polycrystals from the known behaviour of single crystals.

The possibility of making such a prediction rests on the tacit assumption that the mechanisms of plastic deformation in aggregates are substantially identical with those observed in single crystals.

Lindholm and Yeakley (1965) investigated single and polycrystalline specimens of high purity aluminum in compression at strain rates up to 500 s^{-1} using the split Hopkinson pressure bar method. They obtained average stress–strain curves for the six orientations of a single crystal and similar curves for the polycrystalline material. Activation volume as function of strain can be computed from the data obtained. Results for the single and polycrystalline specimens of high purity aluminum are plotted in Fig. 10. The most interesting feature of these curves is that the activation volume for the polycrystalline material falls within the bounds and near the average of the single crystal data. This implies that the same thermally activated mechanisms control the deformation in single and polycrystals and that the distribution of the activation barriers are essentially the same in both cases. This is in agreement with the previous results obtained by Mitra and Dorn (1962) for aluminum at low temperature and those of Conrad (1964) for iron and steel.

Experimental justifications of the thermally activated and phonon damping mechanisms as well as the discussion of their range rate and temperature changes for particular materials have been given in many papers. Particular importance for our purposes have results obtained by Campbell and Ferguson (1970). In their paper an account is given of experiments in which the shear flow stress of mild steel was measured at temperature from 195 to 713 K and strain rate from 10^{-3} to $4 \cdot 10^4$ s^{-1}. The flow stress at lower yield is plotted in Fig. 11 as shear stress against the logarithm of shear strain rate, for the various temperatures used throughout the investigation.

For the purpose of the discussion which follows, it is convenient to divide the curves into three regions, each corresponding to a certain range of strain rate which is a function of the temperature. Following Rosenfield and Hahn (1966) these will be referred to as region I, II and IV. These regions are indicated in Fig. 11.

In region I the flow stress shows a small temperature and strain rate sensitivity, the latter decreasing with increasing temperature. Prestraining increases the flow stress but has little effect on the rate sensitivity of the flow stress, $(\partial \tau / \partial \ln \dot{\epsilon}^p)_\vartheta$, at room temperature (cf. Fig. 12). The dominant factor in region I seems to be the long–range internal stress fields due to dislocations, precipitate particles, grain boundaries etc.

Constitutive Modelling of Dissipative Solids

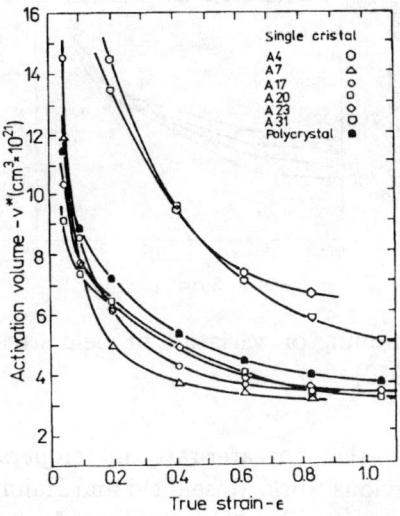

Figure 10: Activation volume versus true strain for single crystal and polycrystalline aluminium (99.995%). After Lindholm and Yeakley (1965)

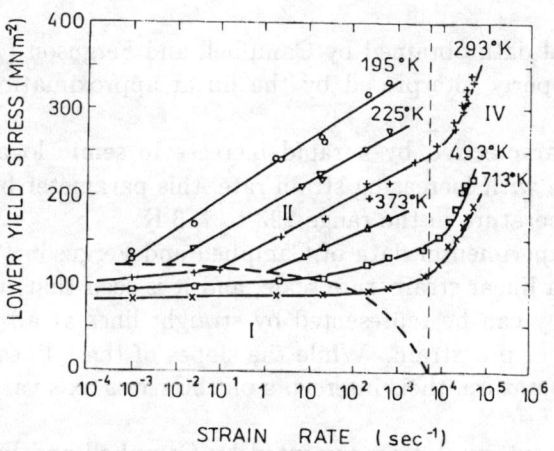

Figure 11: Variation of lower yield stress with strain rate, at constant temperature. After Campbell and Ferguson (1970)

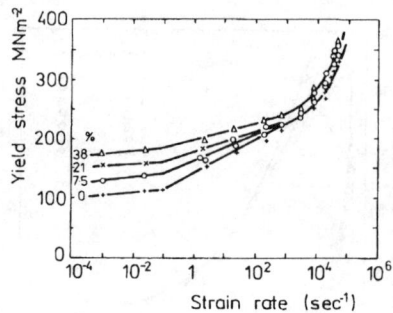

Figure 12: Effect of pre-straining on variation of yield stress with strain rate. After Campbell and Ferguson (1970)

In region II the flow stress shows greater rate and temperature sensitivities. From a survey of their own and previous work, Rosenfield and Hahn (1966) concluded that in this region the rate sensitivity $(\partial \tau / \partial \ln \dot{\in}^p)_\vartheta$ is independent of temperature and strain rate. However, the data of Campbell and Ferguson (1970) show a consistent increase in $(\partial \tau / \partial \ln \dot{\in}^p)_\vartheta$ as temperature is reduced.

It has been suggested by Campbell and Ferguson (1970) that the flow behaviour throughout region II can be explained by the thermal activation of dislocation motion.

Since the relaxation time T_{mT} is related to the dislocation structure it may be governed by the deformation history, rather than a function of the state variables $\in^p$, τ and ϑ.

The experimental data obtained by Campbell and Ferguson (1970) for mild steel in region II are properly interpreted by the linear approximation of the thermally activated theory.

Region IV is characterized by a rapid increase in semi- logarithmic rate sensitivity $(\partial \ln / \partial \ln \dot{\in}^p)_\vartheta$ with increasing strain rate, this parameter being approximately independent of temperature in the range 293 to 713 K.

In Fig. 13 the experimental data of Campbell and Ferguson (1970) for region IV are replotted using a linear strain–rate scale, and it is seen that, within the accuracy of measurement, they can be represented by straight lines at all three temperatures and all three values of pre–strain. While the slopes of these lines show only a small dependence on temperature, their intercepts on the stress axis vary greatly with temperature.

According to the interpretation presented by Campbell and Ferguson (1970), the intercepts in Fig. 13 are determined by the temperature–dependent barrier stress $\tau_\mu + U_o/v^*$ (U_o denotes the activation energy for intersection at zero effective stress and v^* the activation volume), at which the strain rate reaches $1/T_{mT}$. When the applied stress exceeds this barrier stress, the time required to activate a dislocation past

Constitutive Modelling of Dissipative Solids

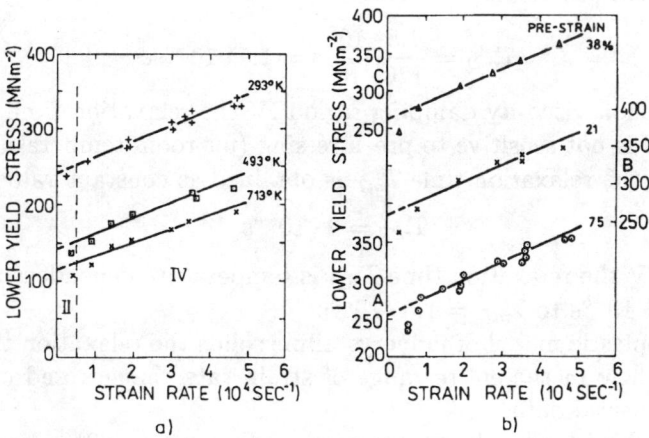

Figure 13: Variation of lower yield stress with strain rate (region IV). (a) Zero pre-strain; temperature 293, 493, 713 K. (b) Pre-strain 7.5, 21, 38%; temperature 293 K. After Campbell and Ferguson (1970)

the short-range barriers of obstacles is negligible, hence its velocity is controlled by dissipation of energy as it moves through the lattice. Assuming that this dissipation is of a linear viscous nature, the excess stress τ_D will be proportional to the strain rate $\dot{\epsilon}^p$, i.e.

$$\tau_D = \eta \dot{\epsilon}^p, \tag{2.28}$$

where η denotes the macroscopic viscosity. Equating τ_D to the difference between the applied stress and the barrier stress $\tau_B = \tau_\mu + U_o/v^*$, we obtain for region IV

$$\tau = \tau_B + \eta \dot{\epsilon}^p. \tag{2.29}$$

Comparison (2.29) with (2.17) under the condition (2.19) gives

$$\eta = T_{mD}\tau_B = \frac{B}{\alpha b^2}. \tag{2.30}$$

The values of η can be obtained from the slops of the lines of Fig. 13.

For room temperature and zero pre-strain we have the relaxation time as follows

$$T_{mD} = \frac{\eta}{\tau_B} = \frac{2.1 \cdot 10^3}{250 \cdot 10^6}\text{s} = 0.84 \cdot 10^{-5}\text{s}.$$

For 493 K and zero pre-strain

$$T_{mD} = \frac{2.0 \cdot 10^3}{150 \cdot 10^6}\text{s} = 1.3 \cdot 10^{-5}\text{s}.$$

For 713 K and zero pre–strain

$$T_{mD} = \frac{1.8 \cdot 10^3}{120 \cdot 10^6}\text{s} = 1.5 \cdot 10^{-5}\text{s}.$$

Thus, in the phonon viscosity damping region IV the relaxation time is a function of temperature and is not sensitive to pre–stressing (for room temperature).

For region II the relaxation time T_{mT} is obtained as constant value

$$T_{mT} = 2 \cdot 10^{-4}\text{s},$$

while in region IV the relaxation time T_{mD} is temperature dependent and can change from $T_{mD} = 0.8 \cdot 10^{-5}$s to $T_{mD} = 1.5 \cdot 10^{-5}$s.

For the viscoplastic model of polycrystalline solids the relaxation time T_m governs the viscoplastic flow in the entire range of strain rate changes and can be obtained based on experimental data.

Another possible idea has been presented by Perzyna (1980), namely that

$$T_m = \frac{\phi}{\gamma}, \qquad (2.31)$$

where ϕ is the control function and γ is the temperature dependent viscosity coefficient. The dimensionless control function ϕ is assumed to depend on strain rate. Thus we have

$$T_m = \frac{1}{\gamma(\vartheta)}\phi\left(\frac{\dot{\epsilon}^p}{\dot{\epsilon}^p_s} - 1\right). \qquad (2.32)$$

In Fig. 14 the theoretical results obtained by Perzyna (1980) are compared with experimental data of Campbell and Ferguson (1970) for room temperature (293 K). Taking the best fitting curve we have

$$T_m = \frac{1}{6.55 \cdot 10^3}\left(\frac{\dot{\epsilon}^p}{10^{-3}} - 1\right)^{\frac{1}{7}} (\text{s}). \qquad (2.33)$$

Figure 15 shows the plots of the relaxation time (2.33) and (2.32) for $\phi \equiv \mathcal{H}(\cdot)$, i.e. for ϕ assumed as the Heaviside function.

Dowling, Harding and Campbell (1970) investigated the strain–rate sensitivity of the yield and flow stress of aluminum, copper, brass and mild steel over a range of strain rates from 10^{-3} to $4 \cdot 10^4$ s^{-1}. These experimental data can also be used for determination of the overstress function Φ and the relaxation time T_m for these metals.

2.6 Localized fracture of polycrystalline solids

2.6.1 Shear band formation and micro–damage process

In dynamic loading processes failure may arise as a result of an adiabatic shear band localization generally attribute to a plastic instability generated by thermal softening during dynamic deformation.

Constitutive Modelling of Dissipative Solids

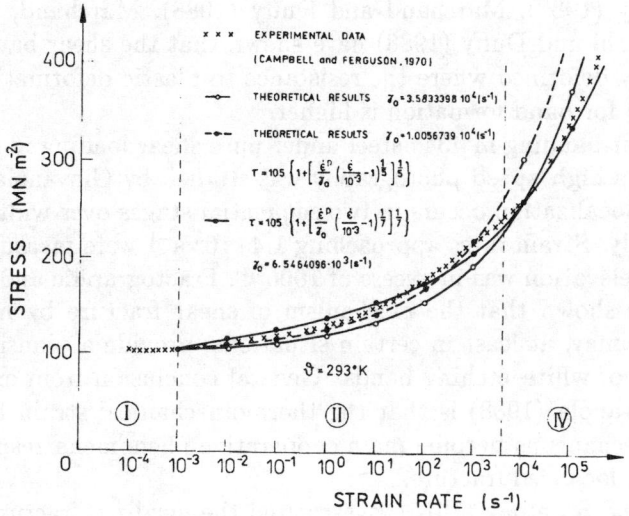

Figure 14: Comparison of theoretical description with experimental results

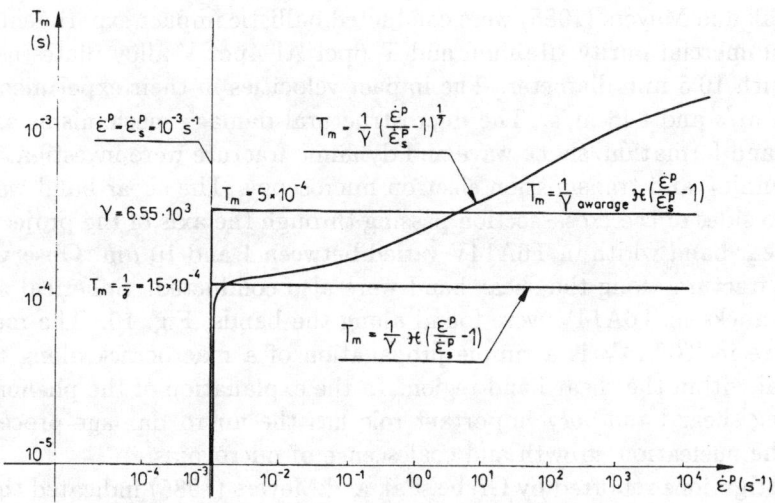

Figure 15: Variation of the relaxation time with strain rate described by Eq. (2.32) with $\phi = \mathcal{H}(\cdot)$ and by Eq. (2.33)

Recent experimental observations (cf. Grebe, Pak and Meyers (1985), Hartley, Duffy and Hawley (1987), Marchand and Duffy (1988), Marchand, Cho and Duffy (1988) and Cho, Chi and Duffy (1988) have shown that the shear band procreates in a region of a body deformed where the resistance to plastic deformation is lower and the predisposition for band formation is higher.

Adiabatic shear banding in 4340 steel under pure shear loading in split Hopkinson torsion bar using a high-speed photography was studied by Giovanola (1988). It was found that shear localization occurs in two sequential stages over width of 60 μm and 20 μm, respectively. Strain rates approaching $1.4 \cdot 10^6$ s^{-1} were measured in the band and temperature elevation was in excess of 1000^0C. Fractographic and metallographic observations have shown that the mechanism of shear fracture by microvoid nucleation and growth may, at least in certain situations, provide a plausible explanation for the formation of white-etching bands. General conclusion from experimental observations of Giovanola (1988) is that the thermomechanical strain localization and micro-damage mechanisms become main cooperative phenomena responsible for adiabatic shear band localized fracture.

Chakrabarti and Spretnak (1975) investigated the localized fracture mode for tensile steel sheet specimens simulating both plane stress and plane strain processes. The material used in their study was AISI 4340 steel. The principal variable in this flat specimen test was the width to thickness ratio. Variation in specimen geometry produces significant changes in stress state, directions of shear bands and ductility. They found that fracture propagated consistently along the shear band localized region.

Grebe, Pak and Meyers (1985) were conducted ballistic impact experiments on 12.5 mm thick commercial purity titanium and T-6pct Al-4pct V alloy plates using steel projectiles with 10.5 mm diameter. The impact velocities in their experiments varied between 578 m/s and 846 m/s. The microstructural damage mechanisms associated with shear band formation, shock wave and dynamic fracture were investigated by optical and scanning and transmission electron microscopy. The shear band were found along the two sides of the cross-section passing through the axis of the projectile. The measured shear band width in T6A14V varied between 1 and 10 μm. Observations of the onset of fracture along the shear band were also conducted. Spherical and ellipsoidal microcracks in T6A14V were found along the bands, Fig. 16. The mechanism of final failure in T6A14V is a simple propagation of a macrocrack along the damaged material within the shear band region. In the explanation of the phenomenon of fracture along shear band very important role has the micro-damage process which consists of the nucleation, growth and coalescence of microvoids.

The investigations reported by Grebe, Pak and Meyers (1985) indicated that in dynamic processes the shear band regions behave differently than adjacent zones. Within the shear band region the deformation process is characterized by very large strains (shear strains over 100%) and very high strain rates ($10^3 - 10^5$ s^{-1}). The strain rate sensitivity of a material becomes very important feature of the shear band region and

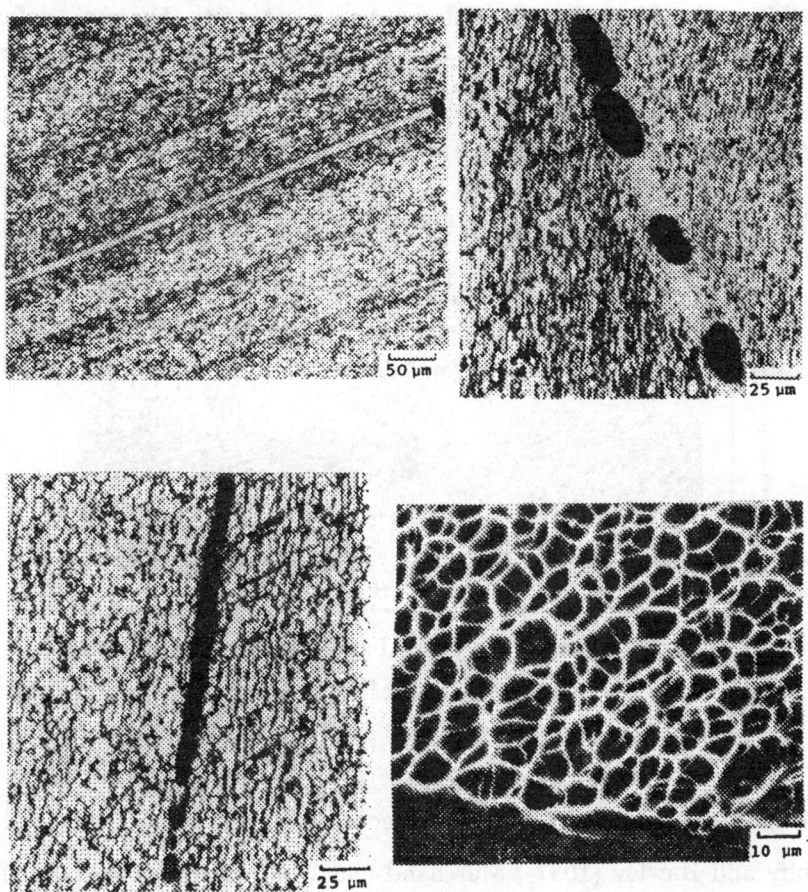

Figure 16: Shear band in Ti6Al4V target impacted at 846 m/s (After Grebe, Pak and Meyers (1985)). a) Single shear band; b) Microcracks in the shear band region; c) Elongated macrocracks along the shear band; d) Chatacteristic dimples observed in spall region

the micro–damage process is intensified.

Cho, Chi and Duffy (1988) performed microscopic observations of adiabatic shear bands in three different steels: an AISI cold rolled steel, HY-100 structural steel and AISI 4340 VAR steel subjected to two different heat treatments. Dynamic deformation in shear was imposed to produce shear bands in all the steels tested. It was found that whenever the shear band led to fracture of the specimen, the fracture occurred by a process of void nucleation, growth and coalescence. No cleavage was observed on any fracture surface, included the most brittle of the steel tested. The authors suggested that this is presumably due to softening of the shear band material that results from the local temperature rise occurring during dynamic deformation, Fig. 17.

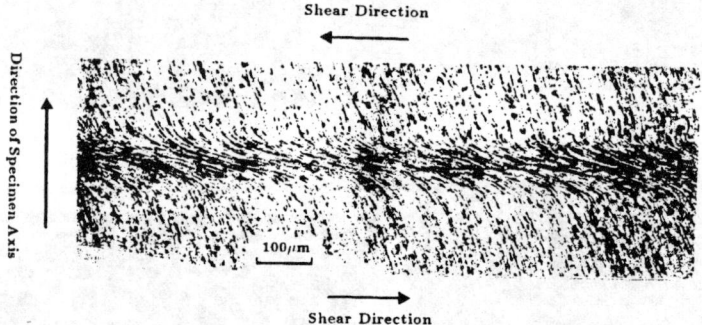

Figure 17: An optical micrograph of a shear band formed in 1018 CRS. The surface has been polished and etched. An arrested crack is shown within shear band (After Cho, Chi and Duffy (1988))

2.6.2 Thermo–mechanical coupling effects

Hartley, Duffy and Hawley (1987), Marchand and Duffy (1988) and Marchand, Cho and Duffy (1988) presented the results of experiments in which the local strain and local temperature were measured during the formation of an adiabatic shear band in an AISI 1018 cold rolled steel (CRS), and a low alloy structural steel (HY-100). In their experiments a torsional Kolsky bar was used to impose a rapid deformation rate in a short thin–walled tubular specimen. By testing a number of specimens they found that the plastic deformation process in the two steel tested can be devided into three separate stages, Fig. 18. In the first stage, the shear strain is homogeneous both in the axial and in circumferential directions. This stage ends at a nominal strain of about 15% for CRS and 25% for HY-100 steel, which corresponds approximately to the maximum stress attained during the test for each kind of steel. With continued deformation, the strain distribution is no longer homogeneous in the axial direction. During the second stage, which spans a range of nominal strains from 15% to 45% for

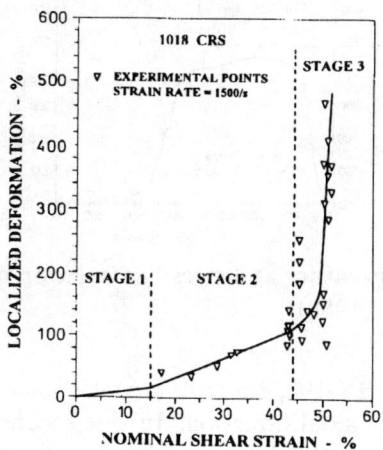

Figure 18: The maximum localized strain as a function of the nominal shear strain (After Marchand, Cho and Duffy (1988))

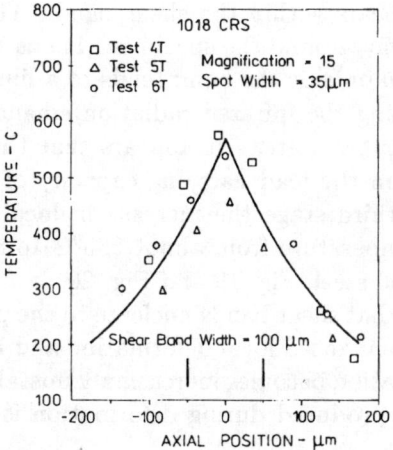

Figure 19: Measured values of the temperature as a function of axial position with respect to the centre of the shear band (After Marchand, Cho and Duffy (1988))

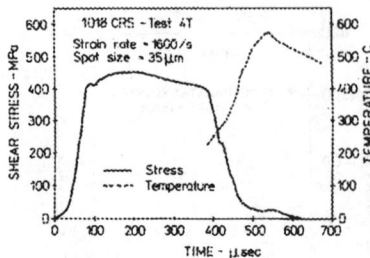

Figure 20: Shear band temperature and stress as function of time in 1018 CRS (After Hartkey, Duffy and Hawley (1988))

CRS, and 25% to 50% for HY-100, there is a continuous increase in the magnitude of the localized strain in the axial direction. In this second stage the localized strain does not vary in the circumferential direction. As the nominal strain within this second stage increases, the localized strain increases to 150% for CRS, to 170% for HY-100 steel and the width of the band decreases from about 1100 μm to 350 μm for CRS, and 600 μm to 150 μm for HY-100 steel. In this stage of deformation, the flow stress level does not vary greatly. The third stage in the deformation process in each of two steel tested involves a sharp drop in stress, i.e. a loss in the load–carrying capacity of the material. Localized strains of up to 600% for CRS, and up to 1500% for HY-steel, and a corresponding width of 100 μm and of 20 μm have been measured. The third stage continues until a crack appears within the shear band. This crack then propagates either part way or all the way around the specimen. It has been observed that, in the third stage, the deformation outside the band tends to a limit. The local temperature was determined by measuring the infrared radiation emanating from the specimen's surface, including the shear band area. It appears that the temperature rise occurs during the sharp decrease in the load–carrying capacity of the specimen for both of the two steels tested. In third stage the increase in local strain is associated with an increase of the local temperature from about 235^0C to 575^0C for CRS and about 460^0C to 900^0Cr for HY-100 steel, Fig. 19 and Fig. 20.

It is generally accepted that shear bands nucleate to the presence of a local inhomogeneity or defects, causing enhanced local deformation and heating. Once nonuniform flow procreates, the deformation becomes increasingly unstable as the dynamic process goes on if the heat that is produced during deformation is given insufficient time to be conducted away.

Experimental results have shown that localization occurs more readily in materials with a low strain hardening rate, a low strain rate sensitivity, a low thermal conductivity and a high thermal softening rate. Shear bands also form readily in high strength materials where the heat generated by plastic deformation is greater for a given plastic strain increment (cf. Hartley, Duffy and Hawley (1987)).

Along the shear band the deformation process is characterized by very intense strain and very large strain rates (cf. Grebe, Pak and Meyers (1985) and Hartley, Duffy and Hawley (1987)). Strain rate sensitivity of a material becomes very important feature of the shear band region. It causes an increase in the flow stress with a corresponding decrease in ductility.

2.6.3 Anisotropic effects

Analysis of experimental results concerning investigations of adiabatic shear band localization failure under dynamic loading suggests that there are three main reason for anisotropic effects:

(i) The strain induced anisotropy is caused by the residual type stresses which result from the heterogeneous nature of the plastic deformation in polycrystalline materials (cf. Ikegami (1982) and Phillips and Lu (1984)). Experimental evidence indicates that yield surfaces exhibit anisotropic hardening. Subsequent yield surfaces are both translated and deformed in stress space. In phenomenological description this kind of anisotropy is modelled by the shift of the yield surface in stress space. This shift of the yield surface might be described by the residual stress tensor α.

(ii) The anisotropy caused by the formation of shear bands. This effect can be described by the determination of the direction of the shear band formed.

(iii) The anisotropy induced by the micro-damage process along the shear band region. Experimental observations (cf. Yokobori JR., Yokobori, Sato and Syoji (1985), Grebe, Pak and Meyers (1985) and Hartley, Duffy and Hawley (1987)) have shown that in the micro-damage process the generated anisotropy is a consequence of rather random phenomena connected with some directional property of the formation of microcracks. This anisotropic effect is very much affected by the crystallographic structure of a material as well as by small fluctuations of main directions of the applied stress at particular point of a body during dynamic process.

To describe this kind of anisotropy one has to introduce an additional set of the internal state variables cf. Perzyna (1990).

2.6.4 Analysis of cooperative phenomena

An analysis of experimental results has clearly shown that the shear band localization failure in dynamic loading processes is affected by complex cooperative phenomena. From this analysis it is also evident that such cooperative phenomena as the thermomechanical flow process, the instability of the flow process along localized adiabatic shear bands, the micro–damage process which consists of the nucleation, growth and coalescence of microcracks and the final mechanism of failure are the most important for proper description of the fracture phenomenon under dynamic loading.

All these cooperative phenomena might be influenced by different additional effects such as the strain rate sensitivity, the induced anisotropy, the thermo-mechanical couplings and others.

It would be unrealistic to include in the description all effects observed experimentally. Constitutive modelling is understood as a reasonable choice of effects which are most important for explanation of the phenomenon described.

2.6.5 Self–organization and physical interpretation of instability hierarchies

We are interested in fracture phenomenon which is preceded by shear band localization. In this case the instability of the plastic flow process plays a fundamental role as a precursor of fracture.

Let us consider a thermodynamic plastic flow process of a system. Synergetics suggests that a system is self–organized if it acquires a spatial, temporal or functional structure without specific interference from the outside. As a result of instability of plastic flow process we observe the macroscopic shear band pattern. A system has been self–organized in a new system — the shear band pattern system. The situation is very similar to that considered for single crystals, cf. Section 2.4 and Fig. 9.

The instability phenomenon of plastic flow process can be considered at different levels. At the mezoscopic level we consider single crystals and their deformation. We describe the crystal lattice, consider movement of dislocations through the rows of barriers and take into account interactions of dislocations. At the macroscopic level by consideration of polycrystalline solids we are interested in description of the instability phenomenon of plastic flow processes. In particular we study the localization of plastic deformation along shear bands. So, we can expect the evolving macroscopic shear band pattern.

It seems that the study of instability hierarchies plays a very important role in the explanation of the interrelation between macroscopic deformation modes and dislocation structures evolved in single crystals (cf. Nakayama and Morii (1987)).

2.6.6 Intrinsic microstructure of the shear band region

Adiabatic shear band is a term used to describe the localization of plastic flow that occurs in many metals when they are deformed at high strain rates to large plastic deformations. It usually manifests itself as zones of intense shear deformation and microstructural modification of the original material up to hundreds of micrometres wide, interspersed between regions of relatively homogeneous deformation, cf. Timothy (1987).

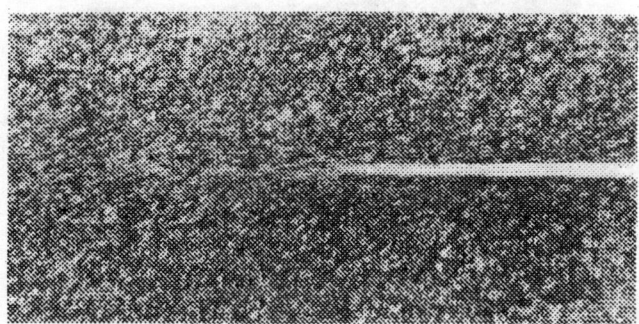

Figure 21: Transformed shear band preceded by a deformed band in AISI 1040 steel (After Rogers and Shastry (1981))

Backman and Finnegan (1963, 1973) originally proposed that shear bands in different metals could be broadly classified as either "transformed" or "deformed" on the basis of their appearance in metallographic section. A permanent change in structure is associated with the former, whereas the latter are manifested merely as zones of intense shear deformation of the original microstructure. The relative temperature rise within developing "transformed" shear zones is therefore assumed to be larger by definition.

As it has been suggested by Timothy (1987) shear bands in steels can be classified specifically on this basis, since the distinctive structure of "transformed" shear bands has been shown to be generally martensitic in nature, and they follow on from "deformed" shear bands when the adiabatic shear deformation becomes sufficiently localized.

Basing on experimental investigations performed for steels Rogers and Shastry (1981) have pointed out that under some conditions during dynamic processes a deformed shear band of some form first develops, followed by the formation of a short transformed band. Figure 21 shows a typical transformed shear band preceded by a precursor deformed band in AISI 1040 steel, generated by the impact process.

Examination of the microstructure has given evidence of a transverse structural gradient within the white–etching band. This is shown for a comparable band in AISI 4340 steel in Fig. 22. Although the band is still white over the entire transformed zone, the central zone is essentially featureless while the two outer regions appear granular in nature and have the lower hardness.

Very recently Wittman, Meyers and Pak (1990) have taken experimental investigations to identify the microstructure of a white–etching shear band in a hollow AISI 4340 steel cylinder subjected to dynamic expansion by using high–voltage transmission electron microscopy. They have determined the microstructure of an adiabatic shear

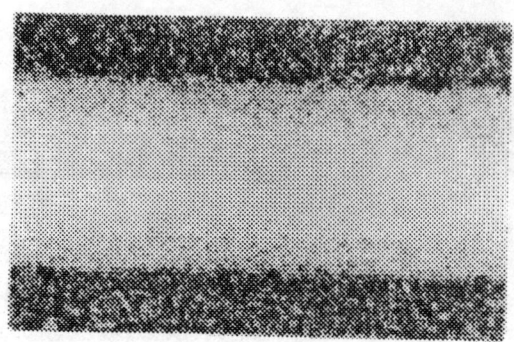

Figure 22: Transformed shear band in AISI steel, quenched and tempered at 400°C (After Rogers and Shastry (1981))

band formed at a minimum strain rate of 0.8×10^6 s^{-1} and with an accumulated shear strain of 3.92.

It has been found that the microstructure inside of the shear band is martensitic and contains carbides which exist only after a significant amount of tempering. This structure is similar to that of the surrounding matrix and has been highly deformed. There was no evidence that the material had transformed to austenite at any time during the deformation process.

Microhardeness traverses were made perpendicular to the length and along the length of the band, cf. Rogers and Shastry (1981). The average hardness value of KHN 1195 in the shear band is similar to that expected in quenched AISI 4340 microstructure.

The sample was also tested after being immersed in liquid nitrogen for 1 hour. This would transform any possible austenite to martensite. The hardness measurement of the band remained unchanged, as did the observed microstructure in the optical microscope. Thus, no evidence of austenite in the band was produced by this test.

To aid in the explanation of the microstructure of shear band observed Wittman, Meyers and Pak (1990) have modelled the thermal history of the band region by using the finite difference method.

On the basis of the thermal history analysis and the TEM observations Wittman, Meyers and Pak (1990) have concluded that the observed white etching of the band region is an artifact of the etching. The white etching is not a particular indication of a phase transformation.

Wittman, Meyers and Pak (1990) have also observed that the band often contained voids smaller microcracks. These spherical voids are thought to have been produced from tensile stresses acting within the band. The band, being at very high temperature and therefore ductile, deforms readily in tension by void nucleation and growth.

Material within the band has, by virtue of the higher temperature, a lower flow stress than the matrix. It has been pointed out that near the tip of the shear bands, these voids and microcracks were less prominent.

The reason that the results obtained by Wittman, Meyers and Pak (1990) have a great importance for the constitutive modelling of the shear band region is twofold.

(i) It has been clearly shown that the response of the material within the shear band region is different from that in the surrounding zones.

(ii) It has been proved that the phase transformation in AISI 4340 steel does not accompany the adiabatic shear band formation.

2.7 Ductile fracture

2.7.1 Experimental results and physical foundations

The most popular dynamical experiment[1] in the investigation the fracture phenomenon in metals is a plate–impact configuration system. This experimental system consists of two plates, a projectile plane plate impacts against a target plane. This is a good example of a dynamic deformation process. If impact velocity is sufficiently high the propagation of a plastic wave through the target is generated. The reflection and interaction of waves result a net tensile pulse in the target plate. If this stress pulse has sufficient amplitude and sufficient time duration, it will cause separation of the material and spalling process.

The reason for choosing this particular kind of dynamical experiment is that post-shot photomicrographic observations of the residual porosity are available, and the stress amplitude and pulse duration can be performed sufficiently great to produce substantial porosity and the spall of the target plate.

The experimental data presented by Seaman, Curran and Shockey (1976) illustrate damage phenomena and provide a common basis for considering damage criteria. They have used a plate–impact configuration system. Following the compression waves resulting from the impact, rarefaction waves have intersected near the middle of the target plate to cause damage in the form of nearly spherical voids. The heaviest damage is localized in a narrow zone, which is called the spall plane. Both the number and the size of voids decrease with distance from this zone. This type of damage is termed ductile fracture because of high ductility (ability to flow) required of the plate material, Fig. 23.

The final damage of the target plate (aluminium 1145) for a constant shot geometry but for different impact velocities has been performed by Barbee, Seaman, Crewdson

---

[1] For a thorough discussion of the experimental and theoretical works in the field of dynamic fracture and spalling of metals please consult the review paper by Meyers and Aimone (1983) and Curran, Seaman and Shockey (1987).

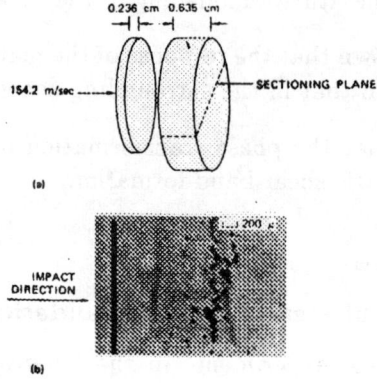

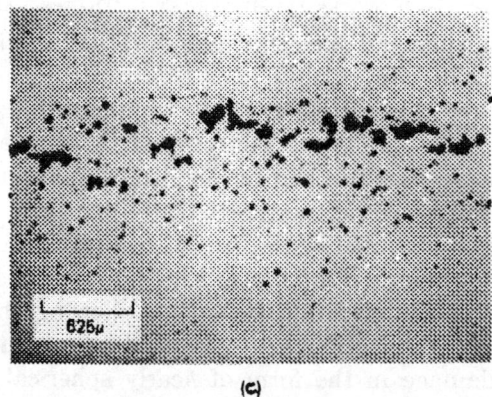

Figure 23: A cross section of an aluminum target plate that has undergone a planar impact by another aluminum plate (After Seaman, Curran and Shockey (1976))

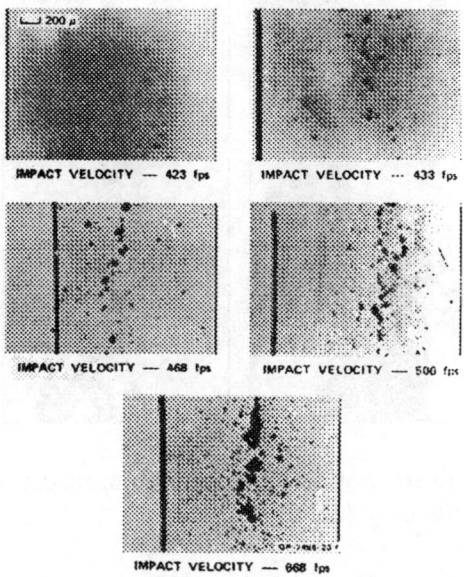

Figure 24: The final damage of the aluminum 1145 target plate for a constant shot geometry but for different impact velocities (After Barbee, Seaman, Crewdson and Curran (1972))

and Curran (1972). The results suggest dependence of spalling process on the pulse amplitude, Fig. 24. On the other hand an example of brittle fracture in armco iron (cf. Seaman, Curran and Shockey (1976)) shows dependence of damage on the tensile pulse duration. In this experimental performance an Armco iron target was impacted by a flyer plate, which was tapered on the back to provide a varying tensile wave duration across the plate. The damage, which appears as randomly oriented microcracks, varies in proportion to the tensile wave duration.

A sample of full separation is shown in Fig. 25, an aluminium target impacted by a plate has been damaged to the extent that full separation occurred near the center of the target, cf. Seaman, Curran and Shockey (1976). The authors suggested that this full separation appears as a macrocrack propagating through heavily damaged material. The macrocrack occurs as a result of coalescence of microvoids which is also visible in Fig. 25.

Similar experimental test data have been obtained for copper by Seaman, Barbee and Curran (1971)[2]. This is also a plate–impact experiment in which a 0.6 mm thick copper plate strikes as 1.6 mm copper target backed by a relatively thick plate of

---

[2]The experimental results of this unpublished report can be found in the paper by Johnson (1981).

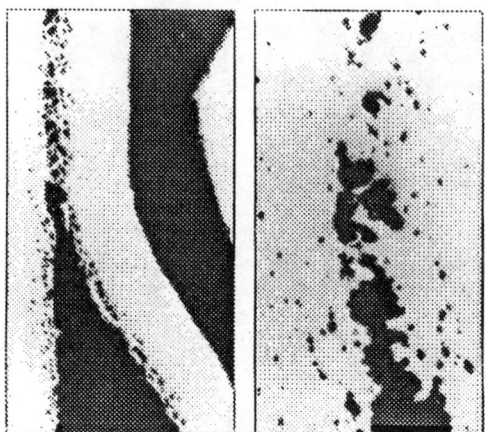

Figure 25: A sample of full separation of an aluminum target. Impact test performed by Seaman, Curran and Shockey (1976)

PMMA ((polymethylmethacrylate) in which a manganin pressure gauge is embaded. The impact velocity of 0.016 cm/μs implies a 29-kbar peak tensile stress in the copper target, and pulse duration about 0.3 μs which are sufficient to produce porosity up to 32% at the spall plane. The manganin pressure gauge record are shown in Fig. 26. The postshot photomicrographic observations of final porosity or the void volume fraction in the copper target is shown in Fig. 27.

From the experimental investigation we have the following conclusions:

(i) Damage (spalling) in ductile metals (aluminium, copper, mild steel, etc.) depends on the amplitude of tensile stress as well as on the duration of stress

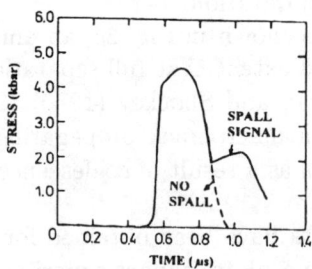

Figure 26: The pressure gauge record for copper. A plate–impact experiment performed by Seaman, Barbee and Curran (1971)

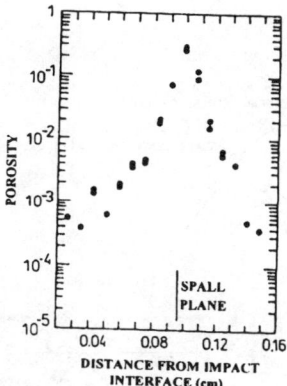

Figure 27: The postshot photomicrographic observations of final porosity in a copper target, cf. Seaman, Barbee and Curran (1971)

pulse. So, to characterize dynamic fracture one has to use the stress impulse or some other stress–time integral quantity, cf. Seaman, Curran and Shockey (1976).

(ii) As the damage occurs the stiffness of the material decreases. This softening of the material is mainly due to the nucleation, growth and coalescence of microvoids (sometimes thermal effects are also pronounced).

(iii) Full separation (fracture, fragmentation, spalling) is the result of the coalescence of microvoids and appears as a macrocrack propagating through heavily damaged material.

(iv) The propagation of the shock plastic wave induced by the impact process produces significant structural changes affect the mechanical properties. In general, one observes an increase in the flow stress with a corresponding decrease of ductility.

2.7.2 Physical mechanisms of ductile dynamic fracture

To understand better the physical mechanism of ductile dynamic fracture let us consider the variation of tensile stress with porosity or void volume fraction, cf. Fig. 28. The trajectory tensile stress–porosity represents the real dynamic process in the copper target (specimen). The process starts at the initial porosity ξ_0 and in about 0.55 μs tensile stress reaches the point at which the nucleation of microvoids can be detected. The process goes on, tensile stress peaks up at 0.72 μs and slowly breaks down to attain in 0.87 μs the point at which the coalescence of microvoids begins. At this

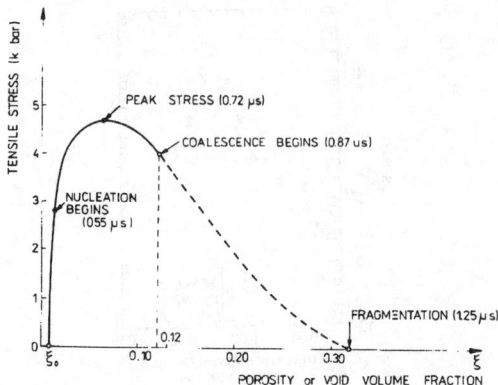

Figure 28: Tensile stress as a function of porosity in a dynamic process for copper specimen

point the fragmentation processes by the coalescence of microcracks has started. The segment of the dynamic process marked by the dashed line represents the mechanism of ductile fracture (spalling or fragmentation) which ends at zero tensile stress. The duration of the entire dynamic deformation process in the copper target (as it has been suggested by experimental observations) is approximately 1.25 μs.

Very recent experimental investigation of dynamic fracture in metals at high strain rate perfermed by Chengwei et al. (1995) and Gilath (1995) have confirmed previous results. Chengwei et al. (1995) used an electric–gun–driven plate impact (EGDPI) assembly and laser–driven shock wave (LDSW) assembly to investigate the spall strength of various metals, cf. Figs. 29 and 30. Gilath investigated spall behaviour and dynamic fracture of various metals and composite materials.

From this analysis of the dynamic deformation process one can see that the main cooperative phenomena which are most important for proper description of dynamic fracture (spalling) are as follows:

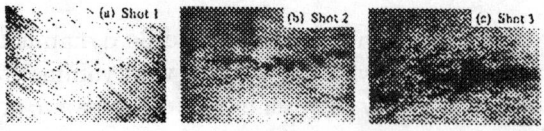

Figure 29: Microvoids in aluminum specimens loaded by EGDPI (After Chengwei, Shiming, Yanping and Cangli (1995))

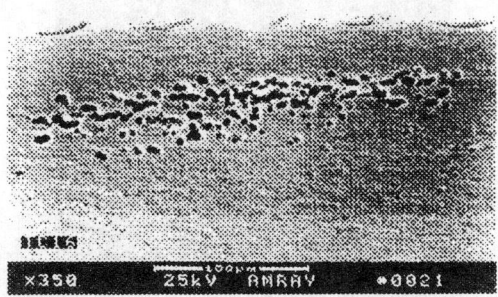

Figure 30: Microvoids in Ti–6Al–4V specimens loaded by LDSW at 3.5×10^{12} W/cm^2 (After Chengwei, Shiming, Yanping and Cangli (1995))

(i) The plastic deformation wave phenomena.

(ii) The nucleation and growth of microvoids.

(iii) The coalescence of microvoids which leads to fragmentation process.

(iv) Full separation as a result of the propagation of macrocrack through heavily damaged material.

2.8 Brittle fracture

2.8.1 Discussion of experimental results

The most popular dynamical experimental investigation of the fracture phenomena in metals as a plate–impact configuration system offers a unique opportunity for studying microvoid and microcrack kinetics under condition of extremely high tensile stress. By varying the impact velocity and target/impactor geometry it provides to change amplitude and duration of stress impulse over the range of approximately 0.1 to 10 GPa and 0.01 to 10 μs, respectively (cf. Curran, Seaman and Shockey (1981)).

An example of brittle fracture for Armco iron is presented in Fig. 31 (cf. Curran, Seaman and Shockey (1977)). It shows the polished cross section through plate impact specimen with very well visible cleavage (penny shape) microcracks. The damage, which appears as randomly oriented planar microcracks, depends on the impact velocity as well as on the duration of the tensile wave. The second property is directly observed from the results presented in Fig. 32 (cf. Curran, Seaman and Shockey (1977)). Use of a tapered flyer results in longer tensile impulses at the thicker end. As it is shown in Fig. 32 these longer pulses lead to greater damage in the Armco iron target (the inset gives to approximate durations of the tensile pulses).

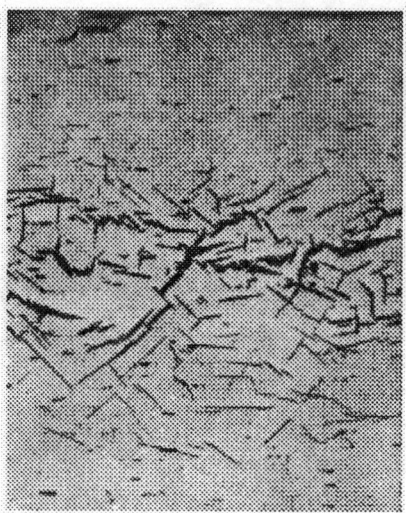

Figure 31: Internal cleavage (penny shape) microcracks caused by shock loading in the polished cross section of an Armco iron specimen (After Curran, Seaman and Shockey (1981))

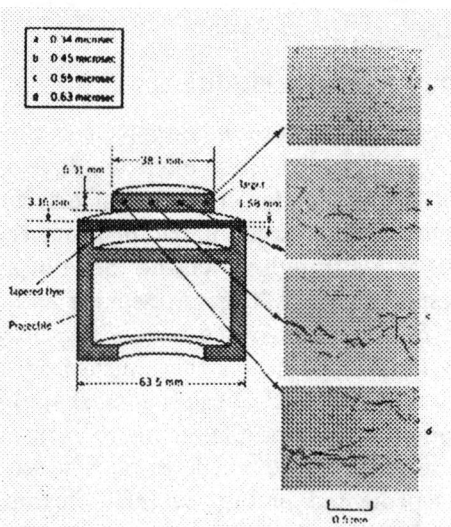

Figure 32: Tapered flyer impact experimental results for the Armco iron target (After Curran, Seaman and Shockey (1977))

Constitutive Modelling of Dissipative Solids

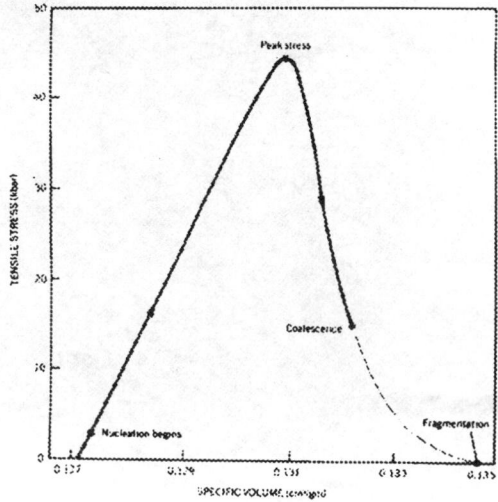

Figure 33: Stress-specific volume trajectory of Armco iron loaded to fragmentation at constant strain rate (After Curran, Seaman, Shockey (1977))

The damage observed in this experiment is termed brittle, although the microcrack growth is much slower than elastic crack velocities, indicating considerable plastic flow at micro crack tips.

2.8.2 Physical mechanism of brittle dynamic fracture

When subjected to high rate loads from impact Armco iron undergoes relatively brittle failure from nucleation, growth and coalescence of planar microcracks.

To understand better the physical mechanism of brittle dynamic fracture let us consider the variation of tensile stress with specific volume (or porosity), Fig. 33 (cf. Curran, Seaman and Shockey (1977)). The trajectory tensile stress–specific volume represents the real dynamic process in the Armco iron target (specimen) subjected to a constant strain rate of $1.3 \times 10^5 s^{-1}$. From this trajectory we can follow the events in the order in which things naturally happen during the dynamic process. The process starts at the initial specific volume of about 0.1272 and when the tensile stress reaches the threshold value for nucleation the nucleation process begins. The process goes on, the tensile stress peaks up the value of specific volume 0.1310 and dramatically breaks down to attain at 0.1323 the point at which the coalescence of microcracks begins.

If no stress relaxation were allowed, the tensile stress–specific volume trajectory would follow that determined by the constitutive laws of elastic–plastic flow theory, and the stress would increase indefinitely. However, the microcrack nucleation and

Figure 34: Polished cross section XAR 30 armor steel showing incipient coalescence of two planar macrofractures (After Shockey, Seaman and Curran (1985))

growth processes cause the stress to peak up and decay.

The segment of the dynamic process marks by the gray line represents the mechanism of brittle fracture (or fragmentation process) by microcrack coalescence which ends at zero tensile stress, at that point volume reaches the value 0.135. As it has been suggested by Seaman, Curran and Murri (1985) the physical process of coalescence occurs when the planar microcracks becomes so large that they begin to intersect other microcracks. They may intersect in the same plane, thus forming larger microcracks, and they may intersect at right angles, forming corners of fragments. Also, microcracks in the same orientation, but on different planes, may coalesce by developing crack extensions out of the plane to join nearby microcracks. Thus a family of microcracks in one orientation can coalescence and form a rough, multifaceted spall plane.

Shockey, Seaman and Curran (1985) have recently investigated the coalescence process for the XAR30 armor steel under plate impact loading conditions. Their experimental results are presented in Fig. 34, which shows two parallel but nonplanar macrofractures in the process of coalescing. A profusion of tiny microfractures has formed in a path linking the tips of macrocracks, suggesting that coalescence is a nucleation and growth process on a smaller scale.

From this analysis of the dynamic deformation process in the Armco iron target and from the analysis of the previously discussed experimental results one can see that the main cooperative phenomena which are most important for proper description of brittle dynamic fracture are as follows:

(i) The inelastic deformation wave phenomena. The propagation of the shock inelastic wave induced by the impact process produces significant structural changes affect the mechanical properties. In general, one observes an increase in the flow stress with a corresponding decrease of ductility.

(ii) The nucleation and growth processes of microcracks. Damage in brittle metals as Armco iron depends on the amplitude of tensile stress as well as on the duration of stress impulse. As the damage occurs the stiffness of the material decreases. This softening of the material is mainly due to nucleation and growth of microcracks. The nucleation and growth processes may be accompanied by thermal effects.

(iii) The coalescence of microcracks which leads to fragmentation process. As the number and sizes of microcracks increase, fragments form until the entire material disintegrates into fragments.

(iv) Full separation as a result of the propagation of a macrocrack through heavily damaged material.

3. KINEMATICS OF FINITE DEFORMATIONS AND FUNDAMENTAL DEFINITIONS

3.1 Fundamental measures of total deformation

Our notation throughout is as follows: $\mathcal{B}$ and $\mathcal{S}$ are manifolds, points in $\mathcal{B}$ are denoted by $\mathbf{X}$ and those in $\mathcal{S}$ by $\mathbf{x}$. The tangent spaces are written $T_\mathbf{X}\mathcal{B}$ and $T_\mathbf{x}\mathcal{S}$. Coordinate systems are denoted $\{X^A\}$ and $\{x^a\}$ for $\mathcal{B}$ and $\mathcal{S}$, respectively, with corresponding bases $\mathbf{E}_A$ and $\mathbf{e}_a$ and dual bases $\mathbf{E}^A$ and $\mathbf{e}^a$.

Let us take the Riemannian spaces on manifolds $\mathcal{B}$ and $\mathcal{S}$, i.e. $\{\mathcal{B}, \mathbf{G}\}$ and $\{\mathcal{S}, \mathbf{g}\}$, the metric tensors $\mathbf{G}$ and $\mathbf{g}$ are defined as follows $\mathbf{G} : T\mathcal{B} \to T^*\mathcal{B}$ and $\mathbf{g} : T\mathcal{S} \to T^*\mathcal{S}$, where $T\mathcal{B}$ and $T\mathcal{S}$ denote the tangent bundles of $\mathcal{B}$ and $\mathcal{S}$, respectively, and $T^*\mathcal{B}$ and $T^*\mathcal{S}$ their dual tangent bundles.

Let the metric tensor G_{AB} be defined by $G_{AB}(\mathbf{X}) = (\mathbf{E}_A, \mathbf{E}_B)_\mathbf{X}$, and similarly define g_{ab} by $g_{ab}(\mathbf{x}) = (\mathbf{e}_a, \mathbf{e}_b)_\mathbf{x}$, where $(\ ,\ )_\mathbf{X}$ and $(\ ,\ )_\mathbf{x}$ denote the standard inner products in $\mathcal{B}$ and $\mathcal{S}$, respectively.

Let
$$\mathbf{x} = \phi(\mathbf{X}, t) \qquad (3.1)$$
be a regular motion, then $\phi_t : \mathcal{B} \to \mathcal{S}$ is a C^1 actual configuration (at time t) of $\mathcal{B}$ in $\mathcal{S}$. The tangent of ϕ is denoted by $\mathbf{F}$ and is called the deformation gradient of ϕ; thus $\mathbf{F} = T\phi$. For $\mathbf{X} \in \mathcal{B}$, we let $\mathbf{F}(\mathbf{X})$ denote the restriction of $\mathbf{F}$ to $T_\mathbf{X}\mathcal{B}$.

Thus
$$F(X,t) : T_X\mathcal{B} \to T_{x=\phi(X,t)}\mathcal{S} \tag{3.2}$$
is a linear transformation for each $X \in \mathcal{B}$ and $t \in I \subset \mathbb{R}$. For each $X \in \mathcal{B}$ there exists an orthogonal transformation $R(X) : T_X\mathcal{B} \to T_x\mathcal{S}$ such that $F = R \cdot U = V \cdot R$. Notice that U and V operate within each fixed tangent space. We call U and V the right and left stretch tensor, respectively. For each $X \in \mathcal{B}$, $U(X) : T_X\mathcal{B} \to T_X\mathcal{B}$ and for each $x \in \mathcal{S}$, $V(x) : T_x\mathcal{S} \to T_x\mathcal{S}$.

The material (or Lagrangian) strain tensor $E : T_X\mathcal{B} \to T_X\mathcal{B}$ is defined by
$$2E = C - I, \quad (I \text{ denotes the identity on } T_X\mathcal{B}), \tag{3.3}$$
where
$$C = F^T \cdot F = U^2 = B^{-1}. \tag{3.4}$$
The spatial (or Eulerian) strain tensor $e : T_x\mathcal{S} \to T_x\mathcal{S}$ is defined by
$$2e = i - c, \quad (i \text{ denotes the identity on } T_x\mathcal{S}), \tag{3.5}$$
where
$$c = b^{-1} \quad \text{and} \quad b = F \cdot F^T = V^2. \tag{3.6}$$
The various strain tensors can be redefined in terms of pull–back and push–forward operations. For the material strain tensor E and the spatial strain tensor e we have
$$\begin{aligned} E^\flat &= \phi^*(e^\flat), & E_{AB}(X) &= e_{ab}(x)F^a_A(X)F^b_B(X), \\ e^\flat &= \phi_*(E^\flat), & e_{ab}(x) &= E_{AB}(X)(F(X)^{-1})^A_a(F(X)^{-1})^B_b, \end{aligned} \tag{3.7}$$
where the symbol $\flat$ denotes the index lowering operator.

3.2 Finite elasto–viscoplastic deformation

Motivated by the micromechanics of single crystal plasticity we postulate a local multiplicative decomposition of the form
$$F(X,t) = F^e(X,t) \cdot F^p(X,t), \tag{3.8}$$
where F^{e-1} is interpreted as the local deformation that releases the stresses from each neighborhood $\mathcal{N}(x) \subset \phi(\mathcal{B})$ in the current configuration of the body.

Let us consider a particle X, which at time $t = 0$ occupied the place X in the reference (material) configuration $\mathcal{B}$, its current place at time t in the actual (spatial) configuration $\mathcal{S}$ is $x = \phi(X,t)$, and its position in the unloaded actual configuration $\mathcal{S}'$ is denoted by y. Thus we have
$$F^e : T_y\mathcal{S}' \to T_x\mathcal{S}, \quad F^p : T_X\mathcal{B} \to T_y\mathcal{S}', \tag{3.9}$$

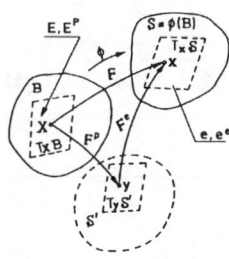

Figure 35: Schematic representation of the multiplicative decomposition of the deformation gradient

where $T_y \mathcal{S}'$ denotes the tangent space in the unloaded actual configuration $\mathcal{S}'$, cf. Fig. 35. It is noteworthy that $\mathbf{F}^e$ and $\mathbf{F}^p$ defined by (3.9) are linear transformations.

We shall treat the tangent space $T_y \mathcal{S}'$ as an auxiliary tool which helps to define the plastic strain tensors[3].

The plastic strain tensor $\mathbf{E}^p : T_\mathbf{X} \mathcal{B} \to T_\mathbf{X} \mathcal{B}$ is defined by

$$\mathbf{E}^p = \frac{1}{2}(\mathbf{C}^p - \mathbf{I}), \tag{3.10}$$

where

$$\mathbf{C}^p = \mathbf{F}^{p^T} \cdot \mathbf{F}^p = \mathbf{U}^{p^2} = \mathbf{B}^{p^{-1}} \quad \text{and} \quad \mathbf{E}^e \stackrel{\text{def}}{=} \mathbf{E} - \mathbf{E}^p. \tag{3.11}$$

Similarly the elastic strain tensor $\mathbf{e}^e : T_x \mathcal{S} \to T_x \mathcal{S}$ is defined by

$$\mathbf{e}^e = \frac{1}{2}(\mathbf{i} - \mathbf{c}^e), \tag{3.12}$$

where

$$\mathbf{c}^e = \mathbf{b}^{e^{-1}}, \quad \mathbf{b}^e = \mathbf{F}^e \cdot \mathbf{F}^{e^T} = \mathbf{V}^{e^2} \quad \text{and} \quad \mathbf{e}^p \stackrel{\text{def}}{=} \mathbf{e} - \mathbf{e}^e. \tag{3.13}$$

The plastic tensors $\mathbf{E}^p$ and $\mathbf{e}^p$ operate within each fixed tangent space; that is $\mathbf{E}^p : T_\mathbf{X} \mathcal{B} \to T_\mathbf{X} \mathcal{B}$ and $\mathbf{e}^p : T_x \mathcal{S} \to T_x \mathcal{S}$, cf. Fig. 35.

It is noteworthy to compare the relation

$$\mathbf{F} = \mathbf{R} \cdot \mathbf{U} = \mathbf{V} \cdot \mathbf{R} \tag{3.14}$$

with

$$\mathbf{F} = \mathbf{F}^e \cdot \mathbf{F}^p = \mathbf{R}^e \cdot \mathbf{U}^e \cdot \mathbf{R}^p \cdot \mathbf{U}^p = \mathbf{V}^e \cdot \mathbf{R}^e \cdot \mathbf{V}^p \cdot \mathbf{R}^p. \tag{3.15}$$

The following commutative diagrams summarize the situation.

---

[3] For precise definition of the finite elasto–plastic deformation see Perzyna (1994) and (1995). Different approach to define the finite elasto–plastic deformation has been presented by Nemat-Nasser (1992).

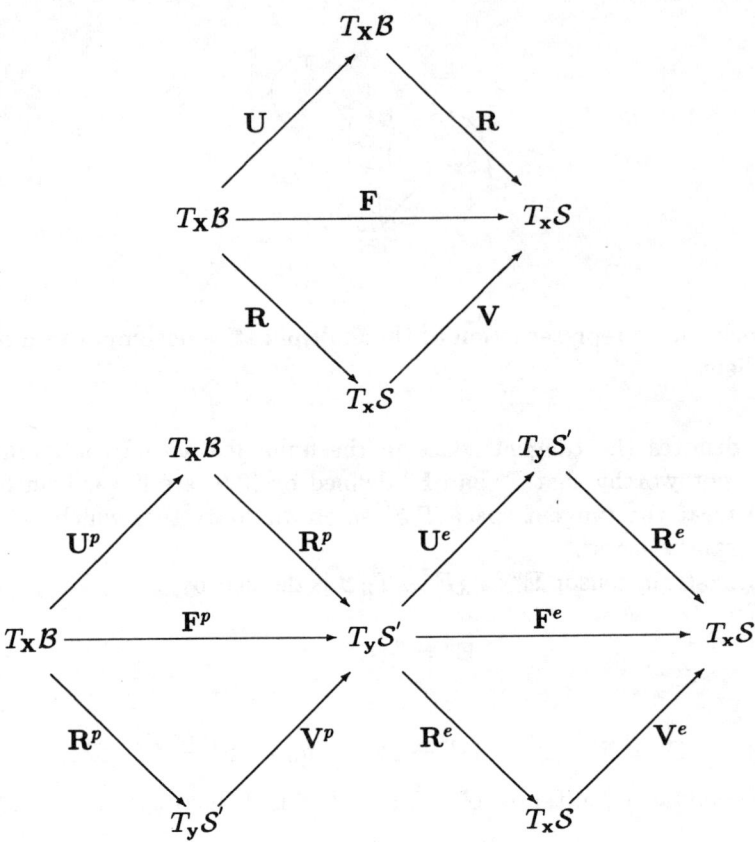

From the second diagram it is clear that the tangent space $T_y\mathcal{S}'$ is playing an auxiliary role indeed.

We can show that the following relations are valid

$$\phi_*(\mathbf{E}^{p^\flat}) = \mathbf{e}^{p^\flat}, \qquad \phi^*(\mathbf{e}^{e^\flat}) = \mathbf{E}^{e^\flat}. \tag{3.16}$$

3.3 Rates of the deformation tensor

Let $\phi(\mathbf{X},t)$ be a C^2 motion of $\mathcal{B}$. Then the spatial velocity is $v_t = \mathbf{V}_t \circ \phi_t^{-1}$, where $\mathbf{V}_t = \frac{\partial \phi}{\partial t}$ is the material velocity, i.e. $v : \mathcal{S} \times I \to T\mathcal{S}$, $I \subset \mathbb{R}$.

The collection of maps $\phi_{t,s}$ such that for each s and $\mathbf{x}$, $t \to \phi_{t,s}(\mathbf{x})$ is an integral curve of v, and $\phi_{s,s}(\mathbf{x}) = \mathbf{x}$, is called the flow or evolution operator of v, i.e.

$$\{\phi_{t,s} \mid \phi_{t,s} = \phi_t \circ \phi_s^{-1} : \phi_s(\mathcal{B}) \to \phi_t(\mathcal{B})\} \tag{3.17}$$

and

$$\phi_{t,s} \circ \phi_{s,r} = \phi_{t,r}, \qquad \phi_{t,t} = \text{identity} \tag{3.18}$$

for all $r, s, t \in I \subset \mathbb{R}$.

If **t** is a C^1 (possible time–dependent) tensor field on $\mathcal{S}$, then the Lie derivative of **t** with respect to $\boldsymbol{v}$ is defined by[4]

$$\mathbf{L}_{\boldsymbol{v}}\mathbf{t} = \left(\frac{d}{dt}\phi_{t,s}^*\mathbf{t}_t\right)|_{t=s}. \tag{3.19}$$

If we hold t fixed in $\mathbf{t}_t$, we obtain the autonomous Lie derivative

$$\mathcal{L}_{\boldsymbol{v}}\mathbf{t} = \left(\frac{d}{dt}\phi_{t,s}^*\mathbf{t}_s\right)|_{t=s}. \tag{3.20}$$

Thus

$$\mathbf{L}_{\boldsymbol{v}}\mathbf{t} = \frac{\partial \mathbf{t}}{\partial t} + \mathcal{L}_{\boldsymbol{v}}\mathbf{t}. \tag{3.21}$$

If $\mathbf{t} \in \mathbf{T}^r{}_s(\mathcal{S})$ (elements of $\mathbf{T}^r{}_s(\mathcal{S})$ are called tensors on $\mathcal{S}$, contravariant of order r and covariant of order s) then $\mathbf{L}_{\boldsymbol{v}}\mathbf{t} \in \mathbf{T}^r{}_s(\mathcal{S})$.

The spatial velocity gradient l is defined by

$$\mathbf{l} = D\boldsymbol{v} : T_{\mathbf{x}}\mathcal{S} \to T_{\mathbf{x}}\mathcal{S}, \quad \text{i.e.} \quad l_b^a = v^a|_b = \frac{\partial v^a}{\partial x^b} + \gamma_{bc}^a v^c, \tag{3.22}$$

where γ_{bc}^a denotes the Christoffel symbol for **g**.

The spatial velocity gradient l can be expressed as follows

$$\mathbf{l} = D\boldsymbol{v} = \dot{\mathbf{F}}\cdot\mathbf{F}^{-1} = \dot{\mathbf{F}}^e\cdot\mathbf{F}^{e^{-1}} + \mathbf{F}^e\cdot(\dot{\mathbf{F}}^p\cdot\mathbf{F}^{p^{-1}})\cdot\mathbf{F}^{e^{-1}} = \mathbf{l}^e + \mathbf{l}^p = \mathbf{d} + \boldsymbol{\omega} = \mathbf{d}^e + \boldsymbol{\omega}^e + \mathbf{d}^p + \boldsymbol{\omega}^p, \tag{3.23}$$

where **d** denotes the spatial rate of deformation tensor and $\boldsymbol{\omega}$ is called the spin.

Let us define the material (or Lagrangian) rate of deformation tensor **D** as follows

$$\mathbf{D}(\mathbf{X}, t) = \frac{\partial}{\partial t}\mathbf{E}(\mathbf{X}, t). \tag{3.24}$$

We have a very important relation

$$\mathbf{d}^b = \mathbf{L}_{\boldsymbol{v}}\mathbf{e}^b = \phi_*\frac{\partial}{\partial t}(\phi^*\mathbf{e}^b) = \phi_*(\frac{\partial}{\partial t}\mathbf{E}^b) = \phi_*(\mathbf{D}^b). \tag{3.25}$$

On the other hand

$$\mathbf{d}^b = \mathbf{L}_{\boldsymbol{v}}\mathbf{e}^b = \mathbf{L}_{\boldsymbol{v}}\left[\frac{1}{2}\left(\mathbf{g} - \mathbf{b}^{-1}\right)\right]^b = \frac{1}{2}\mathbf{L}_{\boldsymbol{v}}\mathbf{g} = \frac{1}{2}\left(g_{cb}v^c|_a + g_{ac}v^c|_b\right)\mathbf{e}^a \otimes \mathbf{e}^b, \tag{3.26}$$

i.e. the symmetric part of the velocity gradient l.

The components of the spin $\boldsymbol{\omega}$ are given by

$$\omega_{ab} = \frac{1}{2}\left(g_{ac}v^c|_b - g_{cb}v^c|_a\right) = \frac{1}{2}\left(\frac{\partial v_a}{\partial x^b} - \frac{\partial v_b}{\partial x^a}\right), \tag{3.27}$$

and

$$\mathbf{d}^{e^b} = \mathbf{L}_{\boldsymbol{v}}\mathbf{e}^{e^b}, \quad \mathbf{d}^{p^b} = \mathbf{L}_{\boldsymbol{v}}\mathbf{e}^{p^b}. \tag{3.28}$$

---
[4]The algebraic and dynamic interpretations of the Lie derivative have been presented by Abraham et al. (1988), cf. also Marsden and Hughes (1983).

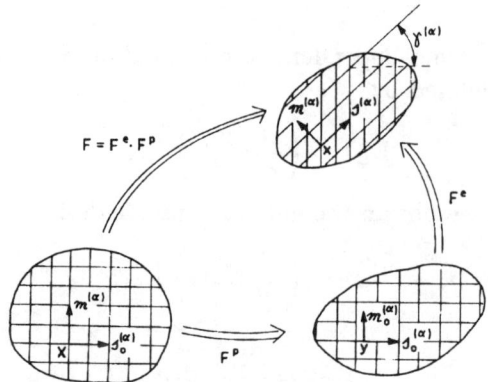

Figure 36: Decomposition of the deformation gradient for single crystal

3.4 Application to single crystal behaviour

A fundamental principle of the description of the thermomechanical, elasto-plastic behaviour of crystals had been introduced by Taylor (1938). He postulated that material flows through the crystal lattice via the dislocation motion, whereas the lattice itself, with the material embedded on it, undergoes elastic deformations and rotations. Thus there are two physically different mechanisms for deforming and reorienting material of a crystal, namely plastic slip and lattice deformation. Of course, single crystals can be generally subjected to rigid body rotations owing to boundary constraints or compatibility requirements. Then, it may be convenient (although arbitrary) to consider this as a third mechanism (cf. Asaro (1983), Peirce et al. (1982, 1983)). In this presentation we have not taken the third mechanism into account.

It is noteworthy that a local multiplicative decomposition of the form (3.8) is just motivated by the micromechanics of single crystal plasticity. It is understood that $\mathbf{F}^e$ is the lattice contribution to $\mathbf{F}$, and is associated with stretching and rotation of the crystal lattice, $\mathbf{F}^p$ describes the deformation solely due to plastic shearing on crystallographic slip systems, cf. Fig. 36.

A particular slip system α is specified by the slip vectors $\mathbf{s}_0^{(\alpha)}$, $\mathbf{m}_0^{(\alpha)}$, where $\mathbf{s}_0^{(\alpha)}$ gives the slip direction and $\mathbf{m}_0^{(\alpha)}$ is the slip plane normal. Let $\mathbf{z}_0^{(\alpha)}$ be a unit vector perpendicular to $\mathbf{s}_0^{(\alpha)}$ and $\mathbf{m}_0^{(\alpha)}$ in Fig. 36 so that $\mathbf{s}_0^{(\alpha)}$, $\mathbf{m}_0^{(\alpha)}$, $\mathbf{z}_0^{(\alpha)}$ form a right-handed triad. Thus the vectors $\mathbf{s}_0^{(\alpha)}$, $\mathbf{m}_0^{(\alpha)}$ and $\mathbf{z}_0^{(\alpha)}$ in the undeformed lattice are taken to be orthonormal.

As the crystal deforms, the vectors $\mathbf{s}^{(\alpha)}$ and $\mathbf{m}^{(\alpha)}$ are stretched and rotated according to $\mathbf{F}^e$. In the deformed lattice we have

$$\mathbf{s}^{(\alpha)} = \mathbf{F}^e \cdot \mathbf{s}_0^{(\alpha)}, \qquad \mathbf{m}^{(\alpha)} = \mathbf{m}_0^{(\alpha)} \cdot (\mathbf{F}^e)^{-1}. \tag{3.29}$$

Using Eq. (3.23) we define the elastic part of the velocity gradient as follows

$$\mathbf{l}^e = \dot{\mathbf{F}}^e \cdot \mathbf{F}^{e-1} \tag{3.30}$$

and postulate for the plastic part

$$\mathbf{l}^p = \dot{\mathbf{F}} \cdot \mathbf{F}^{-1} - \dot{\mathbf{F}}^e \cdot \mathbf{F}^{e-1} = \mathbf{F}^e \cdot \dot{\mathbf{F}}^p \cdot \mathbf{F}^{p-1} \cdot \mathbf{F}^{e-1} = \sum_{\alpha=1}^{n} \mathbf{s}^{(\alpha)} \mathbf{m}^{(\alpha)} \dot{\gamma}^{(\alpha)}, \tag{3.31}$$

where $\dot{\gamma}^{(\alpha)}$ is the rate of shearing on the slip system α.

The elastic rates of stretching and spin $\mathbf{d}^e$ and $\boldsymbol{\omega}^e$ are the symmetric and anti-symmetric parts of $\dot{\mathbf{F}}^e \cdot \mathbf{F}^{e-1}$, respectively. The plastic parts of the rate of stretching and spin are determined by the relations

$$\mathbf{d}^p = \sum_{\alpha=1}^{n} \dot{\gamma}^{(\alpha)} \mathbf{N}^{(\alpha)}, \qquad \boldsymbol{\omega}^p = \sum_{\alpha=1}^{n} \dot{\gamma}^{(\alpha)} \mathbf{W}^{(\alpha)}, \tag{3.32}$$

where

$$\mathbf{N}^\alpha = \frac{1}{2}\left[\mathbf{s}^{(\alpha)}\mathbf{m}^{(\alpha)} + \mathbf{m}^{(\alpha)}\mathbf{s}^{(\alpha)}\right], \qquad \mathbf{W}^\alpha = \frac{1}{2}\left[\mathbf{s}^{(\alpha)}\mathbf{m}^{(\alpha)} - \mathbf{m}^{(\alpha)}\mathbf{s}^{(\alpha)}\right]. \tag{3.33}$$

Of course, the schematic decomposition of the deformation gradient for single crystal shown in Fig. 36 is directly related to the appropriate decomposition represented in Fig. 35. In Fig. 36 in the points $\mathbf{X}$, $\mathbf{x}$ and $\mathbf{y}$, i.e. in the reference, actual and unloaded configurations, respectively, of the considered body, the crystalline structure of the single crystal and its deformation is explicitly presented.

3.5 Stress tensors and the resolved Schmid stress

The first Piola–Kirchhoff stress tensor P^{aA} is the two-point tensor obtained by performing a Piola transformation on the second index of the Cauchy stress tensor $\boldsymbol{\sigma}$, i.e.

$$P^{aA} = J(\mathbf{F}^{-1})^A_b \sigma^{ab}, \tag{3.34}$$

where J denotes the Jacobian of the deformation.

The second Piola–Kirchhoff stress tensor $\mathbf{S}$ is defined as follows

$$S^{AB} = (\mathbf{F}^{-1})^A_a P^{aB} = J(\mathbf{F}^{-1})^A_a (\mathbf{F}^{-1})^B_b \sigma^{ab} = (\mathbf{F}^{-1})^A_a (\mathbf{F}^{-1})^B_b \tau^{ab}, \tag{3.35}$$

i.e.

$$\mathbf{S} = \phi^*(\boldsymbol{\tau}), \tag{3.36}$$

where $\boldsymbol{\tau} = J\boldsymbol{\sigma}$ is called the Kirchhoff stress tensor.

Let us take the rate of stress working per unit reference volume

$$\boldsymbol{\tau}:\mathbf{d} = \boldsymbol{\tau}:\mathbf{d}^e + \boldsymbol{\tau}:\mathbf{d}^p = \boldsymbol{\tau}:\mathbf{d}^e + \sum_{\alpha=1}^{n} \tau^{(\alpha)}\dot{\gamma}^{(\alpha)}, \qquad (3.37)$$

where

$$\tau^{(\alpha)} = \boldsymbol{\tau}:\mathbf{N}^{(\alpha)} \qquad (3.38)$$

is the Schmid resolved stress on the slip system α.

3.6 Rates of stress tensors

The rate of the Kirchhoff stress tensor $\boldsymbol{\tau}$ is given by

$$\mathbf{L}_{\boldsymbol{v}}\boldsymbol{\tau} = \phi_*\frac{\partial}{\partial t}(\phi^*\boldsymbol{\tau}) = \phi_*(\frac{\partial}{\partial t}\mathbf{S}) = \mathbf{F}\cdot(\frac{\partial}{\partial t}\mathbf{S})\cdot\mathbf{F}^T \circ \phi_t^{-1}. \qquad (3.39)$$

Let us define

$$\boldsymbol{\tau}_1 = \tau^{ab}\mathbf{e}_a \otimes \mathbf{e}_b \in \mathbf{T}^2{}_0(\mathcal{S}), \quad \boldsymbol{\tau}_2 = \tau_a{}^b\mathbf{e}^a \otimes \mathbf{e}_b \in \mathbf{T}_1{}^1(\mathcal{S}), \quad \boldsymbol{\tau}_3 = \tau^a{}_b\mathbf{e}_a \otimes \mathbf{e}^b \in \mathbf{T}^1{}_1(\mathcal{S}).$$
$$(3.40)$$

Then

$$(\mathbf{L}_{\boldsymbol{v}}\boldsymbol{\tau}_1)^{ab} = \frac{\partial\tau^{ab}}{\partial t} + \frac{\partial\tau^{ab}}{\partial x^c}v^c - \tau^{cb}\frac{\partial v^a}{\partial x^c} - \tau^{ac}\frac{\partial v^b}{\partial x^c}. \qquad (3.41)$$

is the rate associated with the name Oldroyd (cf. Oldroyd (1950)). The Zaremba–Jaumann rate (cf. Zaremba (1903a,b) and Jaumann (1911)) is defined as follows

$$\left(\overset{\triangledown}{\boldsymbol{\tau}}\right)^{ab} = \frac{1}{2}\left[(\mathbf{L}_{\boldsymbol{v}}\boldsymbol{\tau}_3)^a{}_c g^{cb} + g^{ac}(\mathbf{L}_{\boldsymbol{v}}\boldsymbol{\tau}_2)_c{}^b\right] = \frac{\partial\tau^{ab}}{\partial t} + \frac{\partial\tau^{ab}}{\partial x^c}v^c + \tau^{ad}\omega_d{}^b - \tau^{db}\omega^a{}_d. \qquad (3.42)$$

4. CONSTITUTIVE STRUCTURES

4.1 Constitutive postulates

Let us assume that: (i) conservation of mass, (ii) balance of momentum, (iii) balance of moment of momentum, (iv) balance of energy, (v) entropy production inequality hold.

We introduce the four fundamental postulates:

(i) Existence of the free energy function. It is assumed that the free energy function is given by

$$\psi = \hat{\psi}(\mathbf{e}, \mathbf{F}, \vartheta; \boldsymbol{\mu}), \qquad (4.1)$$

where e denotes the Eulerian strain tensor, $\mathbf{F}$ is deformation gradient, ϑ temperature and μ denotes a set of the internal state variables.

The form of the free energy function $\Psi = \hat{\Psi}(\mathbf{e}, \mathbf{F}, \vartheta)$ is suggested for spatial description in thermoelasticity. To prove it let us start from material desription of thermoelastic properties of a material.

From the axiom of locality we have that the free energy function $\hat{\Psi}$ for thermoelasticity depends only on the variables $\mathbf{X}$, $\mathbf{F}$ and ϑ (cf. Marsden and Hughes (1983)), i.e.
$$\Psi = \hat{\Psi}(\mathbf{X}, \mathbf{F}, \vartheta). \tag{4.2}$$

Let us introduce the following:

Axiom of Entropy Production. For any regular motion $\mathcal{B}$, the constitutive functions for thermoelasticity are assumed to satisfy the entropy production inequality
$$\rho_{Ref}\left(\hat{N}\frac{\partial \vartheta}{\partial t} + \frac{\partial \hat{\Psi}}{\partial t}\right) - \hat{\mathbf{P}} : \frac{\partial \mathbf{F}}{\partial t} + \frac{1}{\vartheta}\hat{\mathbf{Q}} \cdot \mathrm{GRAD}\vartheta \leq 0, \tag{4.3}$$

where $\hat{N} = \hat{\eta}$ denotes the entropy constitutive function, $\hat{\mathbf{P}}$ the first Piola–Kirchhoff stress tensor constitutive function, the heat flux vector function $\hat{\mathbf{Q}} = J\mathbf{F}^{-1} \cdot \hat{\mathbf{q}}(\mathbf{x}, t)$ and the operator GRAD is computed in the material description.

Theorem. (Coleman and Noll (1963)) Suppose the axioms of locality and entropy production hold. Then we have
$$\hat{N} = -\frac{\partial \hat{\Psi}}{\partial \vartheta} \quad \text{and} \quad \hat{\mathbf{P}} = \rho_{Ref}\mathbf{g}^{\#}\frac{\partial \hat{\Psi}}{\partial \mathbf{F}} \quad \text{that is} \quad \hat{P}^A_a = \rho_{Ref}\frac{\partial \hat{\Psi}}{\partial F^a{}_A}, \tag{4.4}$$

and the entropy production inequality reduces to
$$\hat{\mathbf{Q}} \cdot \mathrm{GRAD}\vartheta \leq 0, \tag{4.5}$$

where the symbol $^\#$ denotes the index raising operator.

Taking into account (3.35) we have
$$\mathbf{S} = \mathbf{F}^{-1} \cdot \mathbf{P} = \mathbf{F}^{-1} \cdot \rho_{Ref}\mathbf{g}^{\#}\frac{\partial \hat{\Psi}}{\partial \mathbf{F}}. \tag{4.6}$$

Since
$$C_{AB} = g_{ab}F^a{}_A F^b{}_B \tag{4.7}$$

we can regard $\mathbf{C}$ as function of $\mathbf{F}$ through $\mathbf{C} = \mathbf{F}^T \cdot \mathbf{F}$. The chain rule gives

$$\mathbf{S} = \mathbf{F}^{-1} \cdot \rho_{Ref} \mathbf{g}^{\#} \frac{\partial \hat{\Psi}}{\partial \mathbf{C}} \frac{\partial \mathbf{C}}{\partial \mathbf{F}}. \tag{4.8}$$

Noting that

$$\frac{\partial C_{AB}}{\partial F^a{}_C} = g_{ab} \delta^C_A F^b_A + g_{ca} F^c_A \delta^C_B, \tag{4.9}$$

and following Marsden and Hughes (1983) we get

$$S^{DC} = \rho_{Ref} (\mathbf{F}^{-1})^D{}_d \frac{\partial \hat{\Psi}}{\partial C_{AB}} \frac{\partial C_{AB}}{\partial F^a{}_C} g^{ad} = 2\rho_{Ref} \frac{\partial \hat{\Psi}}{\partial C_{DC}}. \tag{4.10}$$

Thus we have

$$\mathbf{S} = 2\rho_{Ref} \frac{\partial \hat{\Psi}}{\partial \mathbf{C}}. \tag{4.11}$$

Since Eq. (3.3) is valid we can regard $\mathbf{C}$ as function of $\mathbf{E}$, then

$$\mathbf{S} = 2\rho_{Ref} \frac{\partial \hat{\Psi}}{\partial \mathbf{E}} \frac{\partial \mathbf{E}}{\partial \mathbf{C}} = \rho_{Ref} \frac{\partial \hat{\Psi}}{\partial \mathbf{E}}. \tag{4.12}$$

Taking into account $(3.7)_1$ we can regard $\mathbf{E}$ as a function of the point values of the deformation gradient $\mathbf{F}$ and the spatial strain tensor $\mathbf{e}$. Therefore, $\hat{\Psi}$ becomes a function of $\mathbf{X}$, $\mathbf{F}(\mathbf{X})$, $\mathbf{e}(\mathbf{x})$ and $\vartheta(\mathbf{X})$.

Set

$$\hat{\psi}(\mathbf{x}, \mathbf{F}(\mathbf{X}), \mathbf{e}(\mathbf{x}), \vartheta(\mathbf{X})) = \hat{\Psi}(\mathbf{X}, \mathbf{E}(\mathbf{F}(\mathbf{X}), \mathbf{e}(\mathbf{x})), \vartheta(\mathbf{X})). \tag{4.13}$$

By the chain rule

$$\frac{\partial \hat{\psi}}{\partial e_{ab}} = \frac{\partial \hat{\Psi}}{\partial E_{AB}} \frac{\partial E_{AB}}{\partial e_{ab}} = \frac{\partial \hat{\Psi}}{\partial E_{AB}} F^a{}_A F^b{}_B = \left(\phi_* \frac{\partial \hat{\Psi}}{\partial \mathbf{E}} \right)^{ab}. \tag{4.14}$$

Therefore (cf. Eq. (3.36))

$$\tau^{ab} = (\phi_* \mathbf{S})^{ab} = \left(\phi_* \left(\rho_{Ref} \frac{\partial \hat{\Psi}}{\partial \mathbf{E}} \right) \right)^{ab} = \rho_{Ref} \left(\phi_* \frac{\partial \hat{\Psi}}{\partial \mathbf{E}} \right)^{ab} = \rho_{Ref} \frac{\partial \hat{\psi}}{\partial e_{ab}}. \tag{4.15}$$

Thus we prove that for thermoelasticity we have

$$\psi = \hat{\psi}(\mathbf{e}, \mathbf{F}, \vartheta). \tag{4.16}$$

To extend the domain of the description of the material properties and particularly to take into consideration diferent dissipation effects we have to introduce the internal state variables represented by the vector $\boldsymbol{\mu}$.

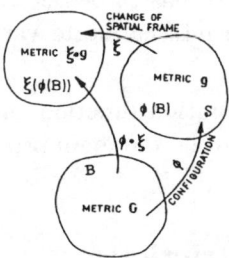

Figure 37: Schematic representation of the change of spatial frame generated by superposed spatial diffeomerphism

(ii) Axiom of objectivity (spatial covariance). The constitutive structure should be invariant with respect to any diffeomorphism (any motion) $\xi : S \to S$ (Marsden and Hughes (1983)), cf. Fig. 37. Assuming that $\xi : S \to S$ is a regular, orientation preserving map transforming $\mathbf{x}$ into $\mathbf{x}'$ and $T\xi$ is an isometry from $T_{\mathbf{x}}S$ to $T_{\mathbf{x}'}S$, we obtain the axiom of material frame indifference (cf. Truesdell and Noll (1965)).

(iii) The axiom of the entropy production. For any regular motion a body $\mathcal{B}$ the constitutive functions are assumed to satisfy the reduced dissipation inequality

$$\frac{1}{\rho_{Ref}}\boldsymbol{\tau} : \mathbf{d} - (\eta\dot{\vartheta} + \dot{\psi}) - \frac{1}{\rho\vartheta}\mathbf{q} \cdot \mathrm{grad}\vartheta \geq 0, \qquad (4.17)$$

where ρ_{Ref} and ρ denote the mass density in the reference and actual configuration, respectively, $\boldsymbol{\tau}$ is the Kirchhoff stress tensor, $\mathbf{d}$ the rate of deformation, η is the specific (per unit mass) entropy, and $\mathbf{q}$ denotes the heat flow vector field. Marsden and Hughes (1983) proved that the reduced dissipation inequality (4.17) is equivalent to the entropy production inequality first introduced by Coleman and Noll (1963) in the form of the Clausius–Duhem inequality. In fact the Clausius–Duhem inequality gives a statement of the second law of thermodynamics within the framework of mechanics of continuous media.

As it has been pointed out by Marsden and Hughes (1983) the question how one should set the basic principles of thermodynamics has given rise to much controversy (see e.g. Müller (1967) and Green and Naghdi (1977)). Alternative theories have been proposed e.g. by Green and Naghdi (1977, 1978), Müller (1967, 1975), Serrin (1979) and Day and Silhavy (1977), cf. also an exhaustive discussion on the subject given by Truesdell (1969).

(iv) The evolution equation for the internal state variable vector $\boldsymbol{\mu}$ is assumed in the form as follows

$$\mathrm{L}_{\boldsymbol{\upsilon}}\boldsymbol{\mu} = \hat{\mathbf{m}}(\mathbf{e}, \mathbf{F}, \vartheta, \boldsymbol{\mu}), \qquad (4.18)$$

where the evolution function $\hat{m}$ has to be determined based on careful physical interpretation of a set of the internal state variables and analysis of available experimental observations.

The determination of the evolution function $\hat{m}$ (in practice a finite set of the evolution functions) appears to be the main problem of the modern constitutive modelling.

4.2 Thermodynamic restrictions

Suppose the axiom of the entropy production holds. Then the constitutive assumption (4.1) and the evolution equations (4.18) lead to the results as follows

$$\tau = \rho_{Ref}\frac{\partial \hat{\psi}}{\partial \mathbf{e}}, \qquad \eta = -\frac{\partial \hat{\psi}}{\partial \vartheta},$$
$$\vartheta \hat{i} - \frac{1}{\rho\vartheta}\mathbf{q}\cdot\mathrm{grad}\vartheta \geq 0, \qquad (4.19)$$

where

$$\vartheta \hat{i} = -\frac{\partial \hat{\psi}}{\partial \boldsymbol{\mu}}\cdot\mathbf{L}_\upsilon\boldsymbol{\mu} \qquad (4.20)$$

denotes the rate of internal dissipation.

We took here advantage of the thermodynamic restrictions on materials described by internal state variables presented by Coleman and Gurtin (1967).

4.3 Constitutive description of inelastic single crystals

It has been proved by previously mentioned theoretical investigations that the main cooperative phenomena which affect the behaviour of metallic single crystals are generated by thermomechanical couplings and the evolution of the dislocation substructure.

To describe the influence of main cooperative phenomena on the behaviour of metallic single crystals, we intend to start from the development of the thermodynamic theory of single crystals with special emphasis on the investigations of thermomechanical couplings and internal dissipative effects. Then this theory is used for the investigations of the adiabatic shear band formation in single crystals under dynamic loading processes.

To take into consideration the evolution of the dislocation substructure we postulate a finite set of the internal state variables as follows[5]

$$\boldsymbol{\mu} = (\gamma^{(\nu)}, \alpha^{(\nu)}, \beta^{(\nu)}, \zeta^{(\nu)}) \qquad (4.21)$$

---

[5]The development of this idea is due to Duszek–Perzyna and Perzyna (1993, 1995, 1996), Perzyna and Korbel (1996, 1997) and Perzyna (1997). For similar propositon see Theodosiu and Sidoroff (1976).

and we interprate $\gamma^{(\nu)}$ as shearing, $\alpha^{(\nu)}$ as the density of mobile dislocations, $\beta^{(\nu)}$ as the density of obstacle dislocations and $\zeta^{(\nu)}$ as the concentration of point defects for particular slip system ν.

The evolution equations for the internal state variables are assumed in the form as follows[6]

$$\dot{\gamma}^{(\nu)} = \frac{1}{T^{(\nu)}} \langle \Phi \left[\frac{\tau^{(\nu)}}{\tau_c^{(\nu)}(\gamma,\beta,\zeta,\vartheta) + \boldsymbol{\kappa}^{(\nu)} : \boldsymbol{\tau}} - 1 \right] \rangle \mathrm{sgn}\tau^{(\nu)},$$

$$\dot{\alpha}^{(\nu)} = \sum_{\delta=1}^{n} a_1^{(\nu\delta)} \dot{\gamma}^{(\delta)} + a_2^{(\nu)} \dot{\vartheta} + \sum_{\delta=1}^{n} a_3^{(\nu\delta)} \dot{\beta}^{(\delta)} + \sum_{\delta=1}^{n} a_4^{(\nu\delta)} \dot{\zeta}^{(\delta)},$$

$$\dot{\beta}^{(\nu)} = \sum_{\delta=1}^{n} b_1^{(\nu\delta)} \dot{\gamma}^{(\delta)} + b_2^{(\nu)} \dot{\vartheta} + \sum_{\delta=1}^{n} b_3^{(\nu\delta)} \dot{\alpha}^{(\delta)} + \sum_{\delta=1}^{n} b_4^{(\nu\delta)} \dot{\zeta}^{(\delta)}, \quad (4.22)$$

$$\dot{\zeta}^{(\nu)} = \sum_{\delta=1}^{n} c_1^{(\nu\delta)} \dot{\gamma}^{(\delta)} + c_2^{(\nu)} \dot{\vartheta} + \sum_{\delta=1}^{n} c_3^{(\nu\delta)} \dot{\alpha}^{(\delta)} + \sum_{\delta=1}^{n} c_4^{(\nu\delta)} \dot{\beta}^{(\delta)},$$

where $\gamma = \sum_{\nu=1}^{n} \gamma^{(\nu)}$, $\beta = \sum_{\nu=1}^{n} \beta^{(\nu)}$, $\zeta = \sum_{\nu=1}^{n} \zeta^{(\nu)}$, $\tau^{(\nu)}$ denotes the Schmid resolved shear stress on the slip system ν, $\tau_c^{(\nu)}$ is the yield stress function on the slip system ν, $\boldsymbol{\kappa}^{(\nu)}$ is the symmetric tensor of non–Schmid effects.

For these assumptions the rate of internal dissipation is determined by

$$\vartheta \hat{i} = -\sum_{\nu=1}^{n} \frac{\partial \hat{\psi}}{\partial \gamma^{(\nu)}} \dot{\gamma}^{(\nu)} - \sum_{\nu=1}^{n} \frac{\partial \hat{\psi}}{\partial \alpha^{(\nu)}} \dot{\alpha}^{(\nu)} - \sum_{\nu=1}^{n} \frac{\partial \hat{\psi}}{\partial \beta^{(\nu)}} \dot{\beta}^{(\nu)} - \sum_{\nu=1}^{n} \frac{\partial \hat{\psi}}{\partial \zeta^{(\nu)}} \dot{\zeta}^{(\nu)}. \quad (4.23)$$

Operating with the Lie derivative on the stress relation $(4.19)_1$ and keeping the internal state variables constant, we obtain[7] (cf. Duszek–Perzyna and Perzyna (1993))

$$\mathbf{L}_\upsilon \boldsymbol{\tau} = \mathcal{L}^e : \mathbf{d} - \mathcal{L}^{th} \dot{\vartheta} - \sum_{\beta=1}^{n} \left[\mathcal{L}^e : \mathbf{N}^{(\beta)} + \mathbf{b}^{(\beta)} \right] \dot{\gamma}^{(\beta)}, \quad (4.24)$$

where

$$\mathcal{L}^e = \rho_{Ref} \frac{\partial^2 \hat{\psi}}{\partial \mathbf{e}^2}, \qquad \mathcal{L}^{th} = -\rho_{Ref} \frac{\partial^2 \hat{\psi}}{\partial \mathbf{e} \partial \vartheta},$$
$$\mathbf{b}^{(\beta)} = (\mathbf{N}^{(\beta)} + \mathbf{W}^{(\beta)}) \cdot \boldsymbol{\tau} + \boldsymbol{\tau} \cdot (\mathbf{N}^{(\beta)} - \mathbf{W}^{(\beta)}). \quad (4.25)$$

Operating on the entropy relation $(4.19)_2$ with the Lie derivative and substituting the result into the energy balance equation, we obtain

$$\rho c_p \dot{\vartheta} = -\mathrm{div}\mathbf{q} + \vartheta \frac{\rho}{\rho_{Ref}} \frac{\partial \boldsymbol{\tau}}{\partial \vartheta} : \mathbf{d} + \chi \sum_{\nu=1}^{n} \tau^{(\nu)} \dot{\gamma}^{(\nu)} + \chi^* \sum_{\nu=1}^{n} \sum_{\delta=1}^{n} (a_1^{-1})^{(\nu\delta)} \tau^{(\delta)} \dot{\alpha}^{(\nu)}$$
$$+ \chi^{**} \sum_{\nu=1}^{n} \sum_{\delta=1}^{n} (b_1^{-1})^{(\nu\delta)} \tau^{(\delta)} \dot{\beta}^{(\nu)} + \chi^{***} \sum_{\nu=1}^{n} \sum_{\delta=1}^{n} (c_1^{-1})^{(\nu\delta)} \tau^{(\delta)} \dot{\zeta}^{(\nu)}, \quad (4.26)$$

---

[6] Particular case of the evolution equations for densities of dislocations $(4.22)_{2,3}$ has been considered by Estrin and Kubin (1986) and Balke and Estrin (1994).

[7] Particular case of the rate equation (4.24) has been presented by Hill and Rice (1972).

where
$$c_p = -\vartheta \frac{\partial^2 \hat{\psi}}{\partial \vartheta^2}, \quad \chi \tau^{(\nu)} = -\rho \left(\frac{\partial \hat{\psi}}{\partial \gamma^{(\nu)}} - \vartheta \frac{\partial^2 \hat{\psi}}{\partial \vartheta \partial \gamma^{(\nu)}} \right),$$
$$\chi^* \tau^{(\nu)} = -\rho \sum_{\delta=1}^{n} a_1^{(\nu \delta)} \left(\frac{\partial \hat{\psi}}{\partial \alpha^{(\delta)}} - \vartheta \frac{\partial^2 \hat{\psi}}{\partial \vartheta \partial \alpha^{(\delta)}} \right),$$
$$\chi^{**} \tau^{(\nu)} = -\rho \sum_{\delta=1}^{n} b_1^{(\nu \delta)} \left(\frac{\partial \hat{\psi}}{\partial \beta^{(\delta)}} - \vartheta \frac{\partial^2 \hat{\psi}}{\partial \vartheta \partial \beta^{(\delta)}} \right), \quad (4.27)$$
$$\chi^{***} \tau^{(\nu)} = -\rho \sum_{\delta=1}^{n} c_1^{(\nu \delta)} \left(\frac{\partial \hat{\psi}}{\partial \zeta^{(\delta)}} - \vartheta \frac{\partial^2 \hat{\psi}}{\partial \vartheta \partial \zeta^{(\delta)}} \right).$$

A set of equations (4.22), (4.24) and (4.26) generalizes the "Duhamel–Neumann hypothesis" for inelastic single crystals, cf. Sokolnikoff (1956), p.359 and Marsden and Hughes (1983), p.204. It is noteworthy that this generalization takes account of the effects as follows: (i) thermomechanical couplings; (ii) evolution of the dislocation substructure; (iii) influence of covariance terms, lattice deformation and rotation, and plastic spin; (iv) deviation from the Schmid rule of a critical resolved shear stress for slip; and (v) rate sensitivity (viscosity).

To show synergetic effects generated by cooperative phenomena of thermomechanical couplings and the influence of the evolution of the dislocation substructure, let us take the evolution equations for the internal state variables $\alpha^{(\nu)}$, $\beta^{(\nu)}$ and $\zeta^{(\nu)}$, cf. Eqs. (4.22)$_{2-4}$. These equations can be written in the form as follows

$$\dot{\alpha}^{(\nu)} = \sum_{\delta=1}^{n} A_1^{(\nu \delta)} \dot{\gamma}^{(\delta)} + A_2^{(\nu)} \dot{\vartheta},$$
$$\dot{\beta}^{(\nu)} = \sum_{\delta=1}^{n} B_1^{(\nu \delta)} \dot{\gamma}^{(\delta)} + B_2^{(\nu)} \dot{\vartheta}, \quad (4.28)$$
$$\dot{\zeta}^{(\nu)} = \sum_{\delta=1}^{n} C_1^{(\nu \delta)} \dot{\gamma}^{(\delta)} + C_2^{(\nu)} \dot{\vartheta},$$

where
$$\mathbf{A}_1 = \overline{\mathbf{A}}^{-1} \cdot A_3^{-1} \left(A_1 + A_4 \cdot B_4^{-1} \cdot B_1 \right),$$
$$\mathbf{A}_2 = \overline{\mathbf{A}}^{-1} \cdot A_3^{-1} \left(A_2 + A_4 \cdot B_4^{-1} \cdot B_2 \right),$$
$$\mathbf{B}_1 = \overline{\mathbf{B}}^{-1} \cdot B_4^{-1} \left(B_1 + B_3 \cdot A_3^{-1} \cdot A_1 \right),$$
$$\mathbf{B}_2 = \overline{\mathbf{B}}^{-1} \cdot B_4^{-1} \left(B_2 + B_3 \cdot A_3^{-1} \cdot A_2 \right), \quad (4.29)$$
$$\mathbf{C}_1 = \mathbf{c}_1 + \mathbf{c}_3 \cdot \mathbf{A}_1 + \mathbf{c}_4 \cdot \mathbf{B}_1,$$
$$\mathbf{C}_2 = \mathbf{c}_2 + \mathbf{c}_3 \cdot \mathbf{A}_2 + \mathbf{c}_4 \cdot \mathbf{B}_2,$$

and
$$A_1 = \mathbf{a}_1 + \mathbf{a}_4 \cdot \mathbf{c}_1, \quad A_2 = \mathbf{a}_2 + \mathbf{a}_4 \cdot \mathbf{c}_2, \quad A_3 = \mathbf{1} - \mathbf{a}_4 \cdot \mathbf{c}_3, \quad A_4 = \mathbf{a}_3 + \mathbf{a}_4 \cdot \mathbf{c}_4,$$
$$B_1 = \mathbf{b}_1 + \mathbf{b}_4 \cdot \mathbf{c}_1, \quad B_2 = \mathbf{b}_2 + \mathbf{b}_4 \cdot \mathbf{c}_2, \quad B_3 = \mathbf{b}_3 + \mathbf{b}_4 \cdot \mathbf{c}_3, \quad B_4 = \mathbf{1} - \mathbf{b}_4 \cdot \mathbf{c}_4, \quad (4.30)$$
$$\overline{\mathbf{A}} = \mathbf{1} - A_3^{-1} \cdot A_4 \cdot B_4^{-1} \cdot B_3, \qquad \overline{\mathbf{B}} = \mathbf{1} - B_4^{-1} \cdot B_3 \cdot A_3^{-1} \cdot A_4.$$

Substituting (4.28) into (4.26) gives the fundamental rate equation for temperature ϑ in the form

$$(\rho c_p - \lambda)\dot{\vartheta} = -\mathrm{div}\mathbf{q} + \vartheta \frac{\rho}{\rho_{Ref}} \frac{\partial \boldsymbol{\tau}}{\partial \vartheta} : \mathbf{d} + \chi \boldsymbol{\tau} : \mathbf{d}^p + \sum_{\nu=1}^{n}\sum_{\kappa=1}^{n} \Lambda^{(\nu\kappa)} \tau^{(\delta)} \dot{\gamma}^{(\kappa)} \qquad (4.31)$$

with the denotations as follows

$$\lambda = \sum_{\nu=1}^{n} \sum_{\delta=1}^{n} \left[\chi^* \left(a_1^{-1}\right)^{(\nu\delta)} A_2^{(\nu)} + \chi^{**} \left(b_1^{-1}\right)^{(\nu\delta)} B_2^{(\nu)} + \chi^{***} \left(c_1^{-1}\right)^{(\nu\delta)} C_2^{(\nu)} \right] \tau^{(\delta)},$$

$$\Lambda^{(\delta\kappa)} = \sum_{\nu=1}^{n} \left[\chi^* \left(a_1^{-1}\right)^{(\nu\delta)} A_1^{(\nu\kappa)} + \chi^{**} \left(b_1^{-1}\right)^{(\nu\delta)} B_1^{(\nu\kappa)} + \chi^{***} \left(c_1^{-1}\right)^{(\nu\delta)} C_1^{(\nu\kappa)} \right] \qquad (4.32)$$

Let us interpret each term of Eq. (4.31). On the left hand side of Eq. (4.31) we have the term $(\rho c_p - \lambda)\dot{\vartheta}$ which represents the heat rate conversion minus the internal heating lost for the generation of new dislocations and point defects. The first term on the right hand side represents the heat conduction effects induced by the heat flux vector $\mathbf{q}$. The second term on the right hand side of Eq. (4.31) is caused by the dependence of the stress tensor $\boldsymbol{\tau}$ on temperature and has not of a dissipative nature. The third term on the right hand side represents the rate of internal dissipation due to plastic flow process, while the last term gives the contribution to the rate of internal dissipation generated by the evolution of the dislocation substructure.

This interpretation can be more understandable when we look at the character of the coefficients λ and $\Lambda^{(\delta k)}$. Both of them account for the evolution of the dislocation substructure, the first due to the transient thermal effects, the second being attributed to the influence of plastic flow phenomena.

4.4 Rate independent response of single crystals

The viscoplastic kinetic law of a single crystal (4.22)$_1$ can be written in the form

$$\tau^{(\nu)} = \left[\tau_c^{(\nu)} + \boldsymbol{\kappa}^{(\nu)} : \boldsymbol{\tau}\right] \{1 + \Phi^{-1}[T^{(\nu)} \dot{\gamma}^{(\nu)}]\}. \qquad (4.33)$$

When the relaxation time $T^{(\nu)} = 0$, then (4.33) gives

$$\tau^{(\nu)} = \tau_c^{(\nu)}(\gamma, \vartheta, \beta, \zeta) + \boldsymbol{\kappa}^{(\nu)} : \boldsymbol{\tau}. \qquad (4.34)$$

Material differentiation of (4.34) yields

$$\dot{\gamma}^{(\nu)} = \sum_{\delta=1}^{n} \hat{h}_{\nu\delta}^{-1} (\dot{\tau}^{(\delta)} - \boldsymbol{\kappa}^{(\delta)} : \dot{\boldsymbol{\tau}}) + \pi^{(\nu)} \dot{\vartheta} - \sum_{\delta=1}^{n} g_{\nu\delta} \dot{\beta}^{(\delta)} - \sum_{\delta=1}^{n} l_{\nu\delta} \dot{\zeta}^{(\delta)}, \qquad (4.35)$$

where

$$\hat{h}_{\nu\delta} = \frac{\partial \tau_c^{(\nu)}}{\partial \gamma^{(\delta)}}, \quad \pi^{(\nu)} = -\sum_{\delta=1}^{n} \frac{\partial \tau_c^{(\delta)}}{\partial \vartheta} \hat{h}_{\nu\delta}^{-1}, \quad g_{\nu\delta} = \sum_{\delta=1}^{n} \frac{\partial \tau_c^{(\nu)}}{\partial \beta^{(\delta)}} \hat{h}_{\nu\delta}^{-1}, \quad l_{\nu\delta} = \sum_{\delta=1}^{n} \frac{\partial \tau_c^{(\nu)}}{\partial \zeta^{(\delta)}} \hat{h}_{\nu\delta}^{-1}.$$

$$(4.36)$$

We interpret $\hat{h}_{\nu\delta}$ as the modulus hardening rate matrix, $\pi^{(\nu)}$ as the thermal plastic softening coefficient, $g_{\nu\delta}$ as the dislocation obstacle hardening matrix, and $l_{\nu\delta}$ as the point defect hardening matrix.

Equation (4.35) constitutes the fundamental evolution equation for shearing $\gamma^{(\nu)}$ in elastic–plastic rate independent response of single crystals.

4.5 Constitutive theory of thermoviscoplasticity of damaged polycrystalline solids

4.5.1 Fundamental assumptions

The main objective is to develop the rate type constitutive structure for an elastic–viscoplastic material in which the effects of the plastic non–normality, plastic spin, plastic strain induced anisotropy (kinematic hardening), micro–damaged mechanism and thermomechanical coupling are taken into consideration. To do this it is sufficient to assume a finite set of the internal state variables. Let us postulate

$$\boldsymbol{\mu} = (\boldsymbol{\zeta}, \xi, \boldsymbol{\alpha}), \tag{4.37}$$

where $\boldsymbol{\zeta}$ denotes the new internal state vector which describes the dissipation effects generated by viscoplastic flow phenomena, ξ is volume fraction porosity and takes account for micro–damaged effects and $\boldsymbol{\alpha}$ denotes the residual stress (the back stress) and aims at the description of the kinematic hardening effects.

Let us introduce the plastic potential function $f = f(\tilde{J}_1, \tilde{J}_2, \vartheta, \boldsymbol{\mu})$, where $\tilde{J}_1, \tilde{J}_2$ denote the first two invariants of the stress tensor $\tilde{\boldsymbol{\tau}} = \boldsymbol{\tau} - \boldsymbol{\alpha}$.

Let us postulate the evolution equations as follows

$$\mathbf{d}^p = \Lambda \mathbf{P}, \quad \boldsymbol{\omega}^p = \Lambda \boldsymbol{\Omega}, \quad L_\upsilon \boldsymbol{\zeta} = \Lambda \mathbf{Z}, \quad \dot{\xi} = \Xi, \quad L_\upsilon \boldsymbol{\alpha} = a \tilde{\boldsymbol{\tau}}, \tag{4.38}$$

where for elasto–viscoplastic model of a material we assume (cf. Perzyna (1963, 1971, 1995))

$$\Lambda = \frac{1}{T_m} \langle \Phi(f - \kappa) \rangle, \tag{4.39}$$

T_m denotes the relaxation time for mechanical disturbances, the isotropic work–hardening–softening function κ is

$$\kappa = \hat{\kappa}(\epsilon^p, \vartheta, \xi), \quad \epsilon^p = \int_0^t \left(\frac{2}{3} \mathbf{d}^p : \mathbf{d}^p\right)^{\frac{1}{2}} dt, \tag{4.40}$$

Φ is the empirical overstress function, the bracket $\langle \cdot \rangle$ defines the ramp function,

$$\mathbf{P} = \frac{1}{2\sqrt{\tilde{J}_2}} \frac{\partial f}{\partial \boldsymbol{\tau}} \bigg|_{\xi=const}, \tag{4.41}$$

Ω, Z and Ξ denote the evolution functions which have to be determined, the scalar coefficient a in the generalized Ziegler evolution law is obtained from the geometrical relation (cf. Duszek and Perzyna, 1991a)

$$(L_\upsilon \alpha - r\mathbf{d}^p) : \mathbf{Q} = 0, \tag{4.42}$$

where

$$\mathbf{Q} = \frac{1}{2\sqrt{\tilde{J}_2}} \left[\frac{\partial f}{\partial \tau} |_{\xi=const} + \left(\frac{\partial f}{\partial \xi} - \frac{\partial \hat{\kappa}}{\partial \xi} \right) \frac{\partial \xi}{\partial \tau} \right], \tag{4.43}$$

and r is a new material constant.

The relation (4.42) yields

$$a = r \frac{\mathbf{P} : \mathbf{Q}}{\tilde{\tau} : \mathbf{Q}} \Lambda. \tag{4.44}$$

4.5.2 Constitutive assumption for the plastic spin

The constitutive laws for the plastic spin[8] based on the application of the tensor function formulation have been proposed by Mandel (1971, 1973), Kratochvil (1971), Dafalias (1983, 1985, 1988) and Loret (1983, 1985). Different proposition by using generalized normality condition has been introduced by Halphen (1975), Mandel (1982), Dafalias (1984) and Van der Giessen (1989).

Let us postulate that Ω has the form (cf. Dafalias (1983) and Loret (1983))

$$\Omega = \eta_1(\alpha \cdot \mathbf{P} - \mathbf{P} \cdot \alpha), \tag{4.45}$$

where η_1 denotes the scalar valued function of the invariants of the tensors α and $\mathbf{P}$, and may depend on temperature ϑ and porosity ξ.

4.5.3 Intrinsic micro–damage process

The intrinsic micro–damage process consists of nucleation, growth and coalescence of microvoids (microcracks). Recent experimental observation results (cf. Shockey et al., 1985) have shown that coalescence mechanism can be treated as nucleation and growth process on a smaller scale. This conjecture simplifies very much the description of the intrinsic micro–damage process by taking account only of the nucleation and growth mechanisms. Then the porosity or the void volume fraction parameter ξ can be determined by

$$\dot{\xi} = \left(\dot{\xi} \right)_{nucl} + \left(\dot{\xi} \right)_{grow}. \tag{4.46}$$

---

[8] For a thorough discussion of a concept of the plastic spin and its constitutive description in phenomenological theories for macroscopic large plastic deformations please consult the critical review paper by Van der Giessen (1991).

Physical considerations (cf. Curran et al., 1987 and Perzyna, 1986) have shown that the nucleation of microvoids in dynamic loading processes which are characterized by very short time duration is governed by the thermally–activated mechanism. Based on this heuristic suggestion we postulate for rate dependent plastic flow

$$\left(\dot{\xi}\right)_{nucl} = \frac{1}{T_m} h^*(\xi, \vartheta) \left[\exp \frac{m^*(\vartheta) \mid \sigma - \sigma_N(\xi, \vartheta, \in^p) \mid}{k\vartheta} - 1\right], \qquad (4.47)$$

where k denotes the Boltzmann constant, $h^*(\xi, \vartheta)$ represents a void nucleation material function which is introduced to take account of the effect of microvoid interaction, $m^*(\vartheta)$ is a temperature dependent coefficient, $\sigma = (1/3)J_1$ is the mean stress and $\sigma_N(\xi, \vartheta, \in^p)$ is the porosity, temperature and equivalent plastic strain dependent threshold stress for microvoid nucleation.

For the growth mechanism we postulate (cf. Johnson (1981), Perzyna (1986), Perzyna and Drabik (1989, 1997), Nemes et al. (1990))

$$\left(\dot{\xi}\right)_{grow} = \frac{1}{T_m} \frac{g^*(\xi, \vartheta)}{\sqrt{\kappa}} \left[\sigma - \sigma_{eq}(\xi, \vartheta, \in^p)\right], \qquad (4.48)$$

where $T_m\sqrt{\kappa}$ denotes the dynamic viscosity of a material, $g^*(\xi, \vartheta)$ represents a void growth material function and takes account for void interaction and $\sigma_{eq}(\xi, \vartheta, \in^p)$ is the porosity, temperature and equivalent plastic strain dependent void growth threshold mean stress.

Finally the evolution equation for the porosity ξ has the form as follows

$$\dot{\xi} = \frac{1}{T_m} h^*(\xi, \vartheta) \left[\exp \frac{m^*(\vartheta) \mid \sigma - \sigma_N(\xi, \vartheta, \in^p) \mid}{k\vartheta} - 1\right] + \frac{1}{T_m} \frac{g^*(\xi, \vartheta)}{\sqrt{\kappa}} \left[\sigma - \sigma_{eq}(\xi, \vartheta, \in^p)\right]. \qquad (4.49)$$

This determines the evolution function Ξ.

4.5.4 Thermodynamic restrictions

The rate of internal dissipation is determined by

$$\vartheta \hat{i} = -\frac{\partial \hat{\psi}}{\partial \mu} \cdot \mathbf{L}_v \boldsymbol{\mu} = -\left[\frac{\partial \hat{\psi}}{\partial \zeta} \cdot \mathbf{Z} + \frac{\partial \hat{\psi}}{\partial \alpha} : r \frac{\mathbf{P}:\mathbf{Q}}{\tilde{\tau}:\mathbf{Q}} \tilde{\tau}\right] \Lambda - \frac{\partial \hat{\psi}}{\partial \xi} \Xi. \qquad (4.50)$$

It is noteworthy that the material function $\mathbf{Z}$ is intrinsically determined by the constitutive assumptions postulated. To show this it is sufficient to perform a Legendre transformation as has been presented by Duszek and Perzyna (1991b).

4.5.5 Rate type constitutive relations

Operating on the stress relation $(4.19)_1$ with the Lie derivative and keeping the internal state vector constant, we obtain (cf. Duszek–Perzyna and Perzyna (1994))

$$\mathbf{L}_\upsilon \boldsymbol{\tau} = \mathcal{L}^e : \mathbf{d} - \mathcal{L}^{th}\dot{\vartheta} - [(\mathcal{L}^e + \mathbf{g}\boldsymbol{\tau} + \boldsymbol{\tau}\mathbf{g} + \mathcal{W}) : \mathbf{P}]\frac{1}{T_m}\langle\Phi(f-\kappa)\rangle, \qquad (4.51)$$

where

$$\mathcal{L}^e = \rho_{Ref}\frac{\partial^2\hat{\psi}}{\partial \mathbf{e}^2}, \quad \mathcal{L}^{th} = -\rho_{Ref}\frac{\partial^2\hat{\psi}}{\partial \mathbf{e}\partial\vartheta}, \quad \mathcal{W} = \eta_1[(\mathbf{g}\boldsymbol{\tau} - \boldsymbol{\tau}\mathbf{g}) : (\boldsymbol{\alpha}\mathbf{g} - \mathbf{g}\boldsymbol{\alpha})]. \qquad (4.52)$$

Substituting $\dot{\psi}$ into the balance of energy equation and taking into account the results (4.20) and (4.50) gives

$$\rho\vartheta\dot{\eta} = -\text{div}\mathbf{q} + \rho\vartheta\hat{\imath}. \qquad (4.53)$$

Operating on the entropy relation $(4.19)_2$ with the Lie derivative and substituting the result into (4.53) we obtain

$$\rho c_p \dot{\vartheta} = -\text{div}\mathbf{q} + \vartheta\frac{\rho}{\rho_{Ref}}\frac{\partial\boldsymbol{\tau}}{\partial\vartheta} : \mathbf{d} + \chi\boldsymbol{\tau} : \mathbf{d}^p + \chi^{\#}\dot{\xi}, \qquad (4.54)$$

where the specific heat

$$c_p = -\vartheta\frac{\partial^2\hat{\psi}}{\partial\vartheta^2} \qquad (4.55)$$

and the irreversibility coefficients χ and $\chi^{\#}$ are determined by

$$\begin{aligned}\chi &= -\rho\left[\left(\frac{\partial\hat{\psi}}{\partial\boldsymbol{\zeta}} - \vartheta\frac{\partial^2\hat{\psi}}{\partial\vartheta\partial\boldsymbol{\zeta}}\right)\cdot\mathbf{Z} + \left(\frac{\partial\hat{\psi}}{\partial\boldsymbol{\alpha}} - \vartheta\frac{\partial^2\hat{\psi}}{\partial\vartheta\partial\boldsymbol{\alpha}}\right)r\frac{\mathbf{P}:\mathbf{Q}}{\tilde{\boldsymbol{\tau}}:\mathbf{Q}}\tilde{\boldsymbol{\tau}}\right]\frac{1}{\boldsymbol{\tau}:\mathbf{P}}, \\ \chi^{\#} &= -\rho\left(\frac{\partial\hat{\psi}}{\partial\xi} - \vartheta\frac{\partial^2\hat{\psi}}{\partial\vartheta\partial\xi}\right).\end{aligned} \qquad (4.56)$$

4.6 Rate independent plastic response for polycrystalline solids

4.6.1 Rate independent plastic response as a limit case

Let us assume that the relaxation time $T_m = 0$, then from $(4.38)_1$ and (4.39) we have the yield criterion

$$f - \kappa = 0 \qquad (4.57)$$

and the coefficient Λ in $(4.38)_1$ can now be determined from the consistency condition

$$\dot{f} - \dot{\kappa} = 0, \qquad (4.58)$$

which yields
$$\Lambda = \langle \frac{1}{H}\{\mathbf{Q} : [\dot{\boldsymbol{\tau}} - (\mathbf{d}\cdot\boldsymbol{\alpha} + \boldsymbol{\alpha}\cdot\mathbf{d})] + \pi\dot{\vartheta}\}\rangle, \tag{4.59}$$

where
$$H = H^* + H^{**}, \qquad \pi = \frac{1}{2\sqrt{J_2}}\left(\frac{\partial f}{\partial \vartheta} - \frac{\partial \hat{\kappa}}{\partial \vartheta}\right),$$
$$H^* = \frac{1}{2\sqrt{3\bar{J}_2}}\frac{\partial \hat{\kappa}}{\partial \in^p} + \frac{1}{2\sqrt{J_2}}\left(\frac{\partial f}{\partial \xi} - \frac{\partial \hat{\kappa}}{\partial \xi}\right)\frac{\partial \xi}{\partial \in^p}, \qquad H^{**} = r\mathbf{P}:\mathbf{Q}. \tag{4.60}$$

4.6.2 Rate independent intrinsic micro–damage process

For rate independent response when $T_m = 0$ the evolution equation for the porosity (4.49) yields
$$\sigma = \sigma_N(\xi, \vartheta, \in^p) = \sigma_{eq}(\xi, \vartheta, \in^p) = \sigma_Y(\xi, \vartheta, \in^p), \tag{4.61}$$
where σ_Y denotes the yield mean stress.

The last result gives the evolution equation as follows
$$\dot{\xi} = A\dot{\sigma} + B\dot{\in}^p + C\dot{\vartheta}, \tag{4.62}$$

where
$$A = \left(\frac{\partial \sigma_Y}{\partial \xi}\right)^{-1}, \quad B = -\left(\frac{\partial \sigma_Y}{\partial \xi}\right)^{-1}\frac{\partial \sigma_Y}{\partial \in^p}, \quad C = -\left(\frac{\partial \sigma_Y}{\partial \xi}\right)\frac{\partial \sigma_Y}{\partial \vartheta}. \tag{4.63}$$

The evolution equation (4.62) can be directly related to the Gurson proposition (cf. Gurson (1975)). To do this let us assume
$$A\dot{\sigma} = k_1 \dot{J}_1, \tag{4.64}$$
$$B\dot{\in}^p = k_2 \boldsymbol{\tau} : \mathbf{d}^p + k_3 \mathbf{g} : \mathbf{d}^p.$$

Introducing the results (4.64) into (4.62) gives
$$\dot{\xi} = k_1 \dot{J}_1 + k_2 \boldsymbol{\tau} : \mathbf{d}^p + k_3 \mathbf{g} : \mathbf{d}^p + C\dot{\vartheta}, \tag{4.65}$$

where k_1, k_2 and k_3 denote the new material functions.

The first two terms in (4.65) are responsible for the description of the nucleation, the first due to the cracking of the second–phase particles and the second is generated by the debonding of the second–phase particles from the matrix material as the plastic work progressively increases. The third term describes the growth process and is assumed to be controlled only by the plastic flow phenomena. The last term can be interpreted as the description of the nucleation induced by the transient change of temperature. For processes in which temperature change is small this term can be neglected (i.e. we can assume $C = 0$). This conjecture simplifies the description of the intrinsic micro–damage process by Eq. (4.65) to that proposed by Gurson (cf. Gurson (1975)). To take into consideration the kinematic hardening effects the Gurson evolution equation can be written in the form (cf. Duszek and Perzyna (1988))

$$\dot{\xi} = k_1 \dot{\tilde{J}}_1 + k_2 \tilde{\tau} : \mathbf{d}^p + k_3 \mathbf{g} : \mathbf{d}^p. \tag{4.66}$$

In what follows we shall take advantage of the evolution equation (4.66).

4.6.3 Fundamental rate type constitutive equations

The rate equation (4.51) can also be written in the form as follows

$$L_\upsilon \tau = \mathcal{L}^e : \mathbf{d} - (\mathcal{L}^e : \mathbf{P} + \mathbf{b})\Lambda - \mathcal{L}^{th}\dot{\vartheta}, \tag{4.67}$$

where

$$\mathbf{b} = (\mathbf{P} + \mathbf{\Omega}) \cdot \tau + \tau \cdot (\mathbf{P} - \mathbf{\Omega}) \tag{4.68}$$

(cf. here the results for single crystals presented by Hill and Rice (1972) and Duszek–Perzyna and Perzyna (1993)).

Substituting in (4.67) Λ and $\mathbf{\Omega}$ from Eqs. (4.59) and (4.45) we finally obtain the rate type constitutive equation for the Kirchhoff stress tensor τ in the form

$$L_\upsilon \tau = \mathcal{L} : \mathbf{d} - \mathcal{M}\dot{\vartheta}, \tag{4.69}$$

where the fundamental elasto–plastic matrix $\mathcal{L}$ and the thermal tensor $\mathcal{M}$ are defined as follows

$$\begin{aligned} \mathcal{L} &= \mathcal{L}^e - \frac{(\mathcal{L}^e + \mathbf{g}\tau + \tau\mathbf{g} + \mathcal{W}) : \mathbf{PQ} : (\mathcal{L}^e + \tilde{\tau}\mathbf{g} + \mathbf{g}\tilde{\tau})}{H + \mathbf{P} : (\mathcal{L}^e + \mathbf{g}\tau + \tau\mathbf{g}) : \mathbf{Q} + \mathbf{P} : \mathcal{W} : \mathbf{Q}}, \\ \mathcal{M} &= \mathcal{L}^{th} - \frac{(\mathcal{L}^e + \mathbf{g}\tau + \tau\mathbf{g} + \mathcal{W}) : \mathbf{P}(\mathbf{Q} : \mathcal{L}^{th} - \pi)}{H + \mathbf{P} : (\mathcal{L}^e + \mathbf{g}\tau + \tau\mathbf{g}) : \mathbf{Q} + \mathbf{P} : \mathcal{W} : \mathbf{Q}}. \end{aligned} \tag{4.70}$$

It is noteworthy to point out here that the fundamental elasto–plastic matrix $\mathcal{L}$ determined by $(4.70)_1$ is nonsymmetric. There are three reasons for the nonsymmetry of the fundamental elasto–plastic matrix $\mathcal{L}$, namely the kinematic hardening effects (i.e. the stress tensor $\tilde{\tau} = \tau - \alpha$ arises in the second bracket in the numerator of the second term instead of τ), the plastic non–normality (i.e. $\mathbf{P} \neq \mathbf{Q}$) and the plastic spin (this effect is represented by the additional term $\mathcal{W}$ in the second term). For the particular case when these three effects are neglected we have very important result, namely the fundamental elasto–plastic matrix $\mathcal{L}$ becomes symmetric.

The rate equation (4.54) for rate independent response takes the form

$$\rho c_p \dot{\vartheta} = -\text{div}\mathbf{q} + \vartheta \frac{\rho}{\rho_{Ref}} \frac{\partial \tau}{\partial \vartheta} : \mathbf{d} + \rho \chi_1 \Lambda + \rho \chi_2 [L_\upsilon \tau : \mathbf{g} + (\mathbf{g} \cdot \tilde{\tau} + \tilde{\tau} \cdot \mathbf{g}) : \mathbf{d}], \tag{4.71}$$

where the irreversibility coefficients χ_1 and χ_2 are defined as follows

$$\chi_1 = -\left\{\left(\frac{\partial \hat{\psi}}{\partial \zeta} - \vartheta\frac{\partial^2 \hat{\psi}}{\partial\vartheta\partial\zeta}\right)\cdot \mathbf{Z} + \frac{H^{**}}{\tilde{\tau}:\mathbf{Q}}\left[\left(\frac{\partial \hat{\psi}}{\partial \boldsymbol{\alpha}} \vartheta\frac{\partial^2 \hat{\psi}}{\partial\vartheta\partial\boldsymbol{\alpha}}\right):\tilde{\tau}\right.\right.$$

$$\left.\left.-\left(\frac{\partial \hat{\psi}}{\partial \boldsymbol{\xi}} - \vartheta\frac{\partial^2 \hat{\psi}}{\partial\vartheta\partial\boldsymbol{\xi}}\right)k_1\tilde{\tau}:\mathbf{g}\right] + \left(\frac{\partial \hat{\psi}}{\partial \boldsymbol{\xi}} - \vartheta\frac{\partial^2 \hat{\psi}}{\partial\vartheta\partial\boldsymbol{\xi}}\right)(k_2\tilde{\tau}:\mathbf{P} + k_3\mathbf{g}:\mathbf{P})\right\},$$

$$\chi_2 = -\left(\frac{\partial \hat{\psi}}{\partial \boldsymbol{\xi}} - \vartheta\frac{\partial^2 \hat{\psi}}{\partial\vartheta\partial\boldsymbol{\xi}}\right)k_1. \tag{4.72}$$

It is noteworthy that the rate type equations (4.69) and (4.71) take into account such effects as the plastic spin, plastic non–normality, plastic strain induced anisotropy (kinematic hardening, i.e. nonsymmetry of the fundamental matrix $\mathcal{L}$), covariant terms, micro–damage process (i.e. softening generated by microcrack nucleation and growth mechanisms), thermomechanical couplings (i.e. thermal plastic softening and thermal expansion) and of course due to cooperative phenomena the synergetic effects.

5. ADIABATIC INELASTIC FLOW PROCESS

5.1 Considerations for single crystals

5.1.1 Discussion of cooperative phenomena

For adiabatic process ($\mathbf{q} = 0$) Eq. (4.31) takes the form

$$(\rho c_p - \lambda)\dot{\vartheta} = \vartheta\frac{\rho}{\rho_{Ref}}\frac{\partial \boldsymbol{\tau}}{\partial\vartheta}:\mathbf{d} + \chi\boldsymbol{\tau}:\mathbf{d}^p + \sum_{\delta=1}^{n}\sum_{\kappa=1}^{n}\Lambda^{(\delta\kappa)}\tau^{(\delta)}\dot{\gamma}^{(\kappa)}. \tag{5.1}$$

The first term on the right–hand side of Eq. (5.1) has not a dissipative nature and is of the second order when compared with the internal dissipation terms. Its contribution to internal heating is small. This suggests that it can be neglected in some considerations like the adiabatic shear band formation. However, this nondissipative term can have important influence on the propagation of acceleration waves in an inelastic crystal.

When the nondissipative term is neglected, then Eq. (5.1) takes the form

$$(\rho c_p - \lambda)\dot{\vartheta} = \chi\boldsymbol{\tau}:\mathbf{d}^p + \sum_{\delta=1}^{n}\sum_{\kappa=1}^{n}\Lambda^{(\delta\kappa)}\tau^{(\delta)}\dot{\gamma}^{(\kappa)}. \tag{5.2}$$

From Eq. (5.2) we can compute the irreversibility coefficient χ. It gives

$$\chi = \frac{(\rho c_p - \lambda)\dot{\vartheta} - \sum_{\delta=1}^{n}\sum_{\kappa=1}^{n}\Lambda^{(\delta\kappa)}\tau^{(\delta)}\dot{\gamma}^{(\kappa)}}{\boldsymbol{\tau}:\mathbf{d}^p}. \tag{5.3}$$

For $\lambda = 0$ and $\Lambda^{(\delta k)} = 0$, i.e. when the influence of the evolution of the dislocation substructure is not taken into consideration, Eq. (5.3) takes the form

$$\chi = \frac{\rho c_p \dot{\vartheta}}{\boldsymbol{\tau} : \mathbf{d}^p}. \tag{5.4}$$

For this particular case the irreversibility coefficient χ has a simple interpretation as the heat rate conversion to plastic work rate fraction. However, Eq. (5.3) shows that the remaining work rate is attributed to the stored energy, e.g. dislocations, point defects and their interactions and is described by two additional terms, namely by $\lambda \dot{\vartheta}$ and $\sum_{\delta=1}^{n} \sum_{\nu=1}^{n} \Lambda^{(\delta \nu)} \tau^{(\delta)} \dot{\gamma}^{(\nu)}$.

Let us denote the rate of the stored energy by

$$s = \lambda \dot{\vartheta} + \sum_{\delta=1}^{n} \sum_{\nu=1}^{n} \Lambda^{(\delta \nu)} \tau^{(\delta)} \dot{\gamma}^{(\nu)}, \tag{5.5}$$

then the irreversibility coefficient χ (cf. Eq. (5.3)) takes the form

$$\chi = \frac{\rho c_p \dot{\vartheta} - s}{\boldsymbol{\tau} : \mathbf{d}^p} \tag{5.6}$$

and Eq. (5.1) can be written as follows

$$\rho c_p \dot{\vartheta} = \vartheta \frac{\rho}{\rho_{Ref}} \frac{\partial \boldsymbol{\tau}}{\partial \vartheta} : \mathbf{d} + \chi \boldsymbol{\tau} : \mathbf{d}^p + s. \tag{5.7}$$

Let us define a sum of all dissipative terms in Eq. (5.7) by

$$w = \chi \boldsymbol{\tau} : \mathbf{d}^p + \lambda \dot{\vartheta} + \sum_{\delta=1}^{n} \sum_{\nu=1}^{n} \Lambda^{(\delta \nu)} \tau^{(\delta)} \dot{\gamma}^{(\nu)}. \tag{5.8}$$

After Willems (1972) and Gurtin (1975) we can define the storage function

$$S(t) = S(0) + \int_0^t w(z) dz. \tag{5.9}$$

The storage function (5.9) plays a fundamental role in the determination of stability criteria for dynamic plastic flow processes in single crystals. It takes account of the most important cooperative phenomena coupled with thermal effects[9].

When modelling thermomechanical behaviour of materials, χ is usually assumed to be a constant in the range $0.85 - 0.95$ (a practice that dates back to the work of Taylor and Quinney (1934)).

Recent experimental investigations performed by Mason et al. (1994) by using a Kolsky (split Hopkinson) pressure bar and a high speed infrared detector array have clearly shown that this assumption may not be correct for all metals, cf. Fig. 38.

---

[9] For general methods of the description of cooperative phenomena and synergetic effects see Glansdorff and Prigogine (1977), Nicolis and Prigogine (1977) and Haken (1975, 1987, 1988).

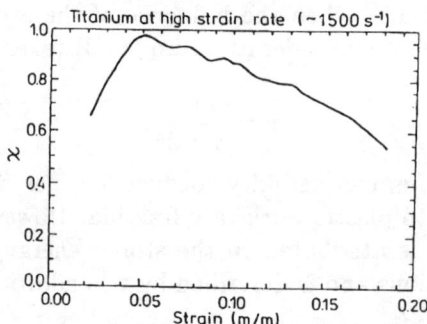

Figure 38: The irreversibility coefficient χ versus strain calculated for Ti–6Al–4V titanium using the average of the temperature of the two detectors (after Mason et al. (1994))

The reason for this considerable discrepancy is clearly visible from Eq. (5.6). The rate of the stored energy s implied by the evolution of the dislocation substructure is responsible for the reduction of χ (e.g. as it has been observed for Ti–6Al–4V deformed at high strain rates, χ decreases from 0.975 to 0.5, cf. Fig. 38).

Mason et al. (1994) observed that the irreversibility coefficient χ depends on strain and strain rate in a range of metals. Their experimental observations have significant implications in the study of the conditions preceding and governing adiabatic shear band formation and shear band growth as well as on the establishment of a criterion governing dynamic fracture mode selection in rate sensitive materials.

5.1.2 General formulation

To investigate the behaviour of rate dependent elasto–plastic single crystal during an adiabatic dynamic process and particularly, to examine the shear band formation, let us formulate the initial boundary value problem as follows. Find ϕ, $\boldsymbol{v}$, ρ, $\boldsymbol{\tau}$, $\gamma^{(\nu)}$, $\alpha^{(\nu)}$, $\beta^{(\nu)}$, $\zeta^{(\nu)}$ and ϑ as functions of t and $\mathbf{x}$ such that the following assertions are satisfied:

(i) the field equations

$$\dot{\phi} = \boldsymbol{v},$$
$$\dot{\boldsymbol{v}} = \frac{1}{\rho}\mathrm{div}(\frac{1}{J}\boldsymbol{\tau}),$$
$$\dot{\rho} = \rho\,\mathrm{div}\boldsymbol{v},$$
$$\mathbf{L}_{\boldsymbol{v}}\boldsymbol{\tau} = \mathcal{L}^e : \mathbf{d} - \mathcal{L}^{th}\dot{\vartheta} - \sum_{\nu=1}^{n}[\mathcal{L}^e : \mathbf{N}^{(\nu)} + \mathbf{b}^{(\nu)}]\dot{\gamma}^{(\nu)},$$

$$\dot\gamma^{(\nu)} = \frac{1}{T^{(\nu)}}\langle\Phi\left[\frac{\tau^{(\nu)}}{\tau_c^{(\nu)}(\gamma,\beta,\xi,\vartheta)+\kappa^{(\nu)}:\tau}-1\right]\rangle\mathrm{sgn}\tau^{(\nu)}, \qquad (5.10)$$

$$\dot\alpha^{(\nu)} = \sum_{\delta=1}^{n} A_1^{(\nu\delta)}\dot\gamma^{(\delta)} + A_2^{(\nu)}\dot\vartheta,$$

$$\dot\beta^{(\nu)} = \sum_{\delta=1}^{n} B_1^{(\nu\delta)}\dot\gamma^{(\delta)} + B_2^{(\nu)}\dot\vartheta,$$

$$\dot\zeta^{(\nu)} = \sum_{\delta=1}^{n} C_1^{(\nu\delta)}\dot\gamma^{(\delta)} + C_2^{(\nu)}\dot\vartheta,$$

$$\dot\vartheta = \frac{1}{\rho c_p - \lambda}\left[\vartheta\frac{\rho}{\rho_{Ref}}\frac{\partial\tau}{\partial\vartheta}:\mathbf{d} + \chi\tau:\mathbf{d}^p + \sum_{\nu=1}^{n}\sum_{\delta=1}^{n}\Lambda^{(\nu\delta)}\tau^{(\nu)}\dot\gamma^{(\delta)}\right];$$

(ii) the boundary conditions

  (a) displacement ϕ is prescribed on a part ∂_ϕ of $\partial\phi(\mathcal{B})$ and tractions $(\tau\cdot\mathbf{n})^a$ are prescribed on part ∂_T of $\partial\phi(\mathcal{B})$, where $\partial_\phi\cap\partial_T=0$ and $\overline{\partial_\phi\cup\partial_T}=\partial\phi(\mathcal{B})$;

  (b) heat flux $(\mathbf{q}\cdot\mathbf{n})=0$ is prescribed on $\partial\phi(\mathcal{B})$;

(iii) the initial conditions: ϕ, υ, ρ, τ, $\gamma^{(\nu)}$, $\alpha^{(\nu)}$, $\beta^{(\nu)}$, $\zeta^{(\nu)}$ and ϑ are given at $X\in\mathcal{B}$ at $t=0$.

For elasto–plastic rate independent response of crystals, to define an adiabatic flow process we have to replace Eq. $(5.10)_5$ by Eq. (4.35).

5.1.3 Rate dependent process

For an adiabatic process, the rate equation for temperature $(5.10)_9$ can be written in the form

$$\dot\vartheta = \mathcal{F}:\mathbf{d} + \sum_{\nu=1}^{n} K^{(\nu)}\dot\gamma^{(\nu)} \qquad (5.11)$$

where

$$\mathcal{F} = \frac{\rho}{\rho_{Ref}}\frac{\vartheta}{(\rho c_p - \lambda)}\frac{\partial\tau}{\partial\vartheta}, \qquad (5.12)$$

$$K^{(\nu)} = \frac{1}{\rho c_p - \lambda}\left[\chi\tau^{(\nu)} + \sum_{\delta=1}^{n}\Lambda^{(\delta\nu)}\tau^{(\delta)}\right].$$

Let us denote

$$P^{(\delta)} = \frac{1}{T^{(\delta)}}\langle\Phi\left[\frac{\tau^{(\delta)}}{\tau_c^{(\delta)}+\kappa^{(\delta)}:\tau}-1\right]\rangle\mathrm{sgn}\tau^{(\delta)}, \qquad (5.13)$$

then the evolution equations $(5.10)_{6-8}$ take the form as follows

$$\dot{\alpha}^{(\nu)} = \sum_{\delta=1}^{n} \left(A_1^{(\nu\delta)} + A_2^{(\nu)} K^{(\delta)}\right) P^{(\delta)} + A_2^{(\nu)} \mathcal{F} : \mathbf{d},$$

$$\dot{\beta}^{(\nu)} = \sum_{\delta=1}^{n} \left(B_1^{(\nu\delta)} + B_2^{(\nu)} K^{(\delta)}\right) P^{(\delta)} + B_2^{(\nu)} \mathcal{F} : \mathbf{d}, \qquad (5.14)$$

$$\dot{\zeta}^{(\nu)} = \sum_{\delta=1}^{n} \left(C_1^{(\nu\delta)} + C_2^{(\nu)} K^{(\delta)}\right) P^{(\delta)} + C_2^{(\nu)} \mathcal{F} : \mathbf{d}.$$

The fundamental rate equation for the Kirchhoff stress $(5.10)_4$ takes the form

$$L_{\boldsymbol{\upsilon}} \boldsymbol{\tau} = I\!\!E : \mathbf{d} + I\!\!F, \qquad (5.15)$$

where

$$I\!\!E = \mathcal{L}^e - \frac{\rho}{\rho_{Ref}} \frac{\vartheta}{(\rho c_p - \lambda)} \mathcal{L}^{th} \frac{\partial \boldsymbol{\tau}}{\partial \vartheta}, \qquad (5.16)$$

$$I\!\!F = -\sum_{\nu=1}^{n} \left(\mathcal{L}^e : \mathbf{N}^{(\nu)} + \mathbf{b}^{(\nu)} + \mathcal{L}^{th} K^{(\nu)}\right) P^{(\nu)}.$$

5.1.4 Rate independent process

Taking into account the equation for $\dot{\tau}^{(\nu)}$ (cf. Duszek–Perzyna and Perzyna (1996), Eq. (48))

$$\dot{\tau}^{(\nu)} = \mathbf{Q}^{(\nu)} : \mathbf{d} - \mathbf{Q}^{(\nu)} : \sum_{\delta=1}^{n} \mathbf{N}^{(\delta)} \dot{\gamma}^{(\delta)} - \mathcal{L}^{th} : \mathbf{N}^{(\nu)} \dot{\vartheta} \qquad (5.17)$$

where

$$\mathbf{Q}^{(\nu)} = \mathcal{L}^e : \mathbf{N}^{(\nu)} + \mathbf{b}^{(\nu)}, \qquad (5.18)$$

and the evolution equation (2.8) and (5.11), we obtain

$$\dot{\gamma}^{(\nu)} = \sum_{\delta=1}^{n} M_{(\nu\delta)}^{-1} \left[\mathbf{Q}^{(\delta)} - \left(\mathcal{L}^{th} : \mathbf{N}^{(\delta)} + \frac{\partial \tau_c^{(\delta)}}{\partial \vartheta} + \sum_{\kappa=1}^{n} \frac{\partial \tau_c^{(\delta)}}{\partial \beta^{(\kappa)}} B_2^{(\kappa)} \right. \right. \qquad (5.19)$$

$$\left. \left. + \sum_{\kappa=1}^{n} \frac{\partial \tau_c^{(\delta)}}{\partial \zeta^{(\kappa)}} C_2^{(\kappa)} \right) \mathcal{F} - \kappa^{(\delta)} : \mathcal{L}^e \right] : \mathbf{d}$$

where

$$M_{(\nu\delta)} = h_{\nu\delta} + \mathbf{Q}^{(\nu)} : \mathbf{N}^{(\delta)} + \left(\mathcal{L}^{th} : \mathbf{N}^{(\nu)} + \frac{\partial \tau_c^{(\nu)}}{\partial \vartheta}\right) K^{(\delta)} \qquad (5.20)$$

$$+ \sum_{\kappa=1}^{n} \left[\frac{\partial \tau_c^{(\nu)}}{\partial \beta^{(\kappa)}} \left(B_1^{(\kappa\delta)} + B_2^{(\kappa)} K^{(\delta)}\right) + \frac{\partial \tau_c^{(\nu)}}{\partial \zeta^{(\kappa)}} \left(C_1^{(\kappa\delta)} + C_2^{(\kappa)} K^{(\delta)}\right) \right]$$

$$- \kappa^{(\nu)} : \mathcal{L}^e : \mathbf{N}^{(\delta)}.$$

Substituting (5.19) and (5.11) into (5.10)$_4$ yields

$$\mathbf{L}_{\boldsymbol{v}}\boldsymbol{\tau} = \mathbb{L} : \mathbf{d} \qquad (5.21)$$

where the fundamental matrix $\mathbb{L}$ has the following form

$$\begin{aligned}\mathbb{L} &= \mathcal{L}^e - \mathcal{L}^{th}\mathcal{F} - \sum_{\nu=1}^{n}\sum_{\delta=1}^{n}\left(\mathbf{Q}^{(\nu)} + \mathcal{L}^{th}K^{(\nu)}\right) M_{(\nu\delta)}^{-1} \left\{\mathbf{Q}^{(\delta)} - \left[\mathcal{L}^{th} : \mathbf{N}^{(\delta)} + \frac{\partial \tau_c^{(\delta)}}{\partial \vartheta}\right.\right.\\ &\quad + \sum_{\kappa=1}^{n}\left(\frac{\partial \tau_c^{(\delta)}}{\partial \beta^{(\kappa)}} B_2^{(\kappa)} + \frac{\partial \tau_c^{(\delta)}}{\partial \zeta^{(\kappa)}} C_2^{(\kappa)}\right)\bigg]\mathcal{F} - \boldsymbol{\kappa}^{(\delta)} : \mathcal{L}^e \bigg\}. \end{aligned} \qquad (5.22)$$

5.2 Consideration for polycrystalline solids

5.2.1 Formulation of an adiabatic inelastic flow process

Let us define an adiabatic inelastic flow process as follows. Find ϕ, $\boldsymbol{v}$, ρ, ϑ, $\boldsymbol{\mu}$ and $\boldsymbol{\tau}$ as function of t and $\mathbf{x}$ such that

(i) the field equations

$$\begin{aligned}\dot{\phi} &= \boldsymbol{v},\\ \dot{\boldsymbol{v}} &= \frac{1}{\rho_{Ref}}\mathrm{div}\boldsymbol{\tau} + \boldsymbol{\tau}\frac{1}{\rho\rho_{Ref}}\mathrm{grad}\rho,\\ \dot{\rho} &= -\rho\,\mathrm{div}\boldsymbol{v},\\ \mathbf{L}_{\boldsymbol{v}}\boldsymbol{\tau} &= \mathcal{L}^e : \mathbf{d} - \mathcal{L}^{th}\dot{\vartheta} - [(\mathcal{L}^e + \mathbf{g}\boldsymbol{\tau} + \boldsymbol{\tau}\mathbf{g} + \mathcal{W}) : \mathbf{P}]\frac{1}{T_m}\langle\Phi(f-\kappa)\rangle,\\ \mathbf{L}_{\boldsymbol{v}}\boldsymbol{\zeta} &= \frac{1}{T_m}\langle\Phi(f-\kappa)\rangle\mathbf{Z},\\ \dot{\xi} &= \Xi;\\ \mathbf{L}_{\boldsymbol{v}}\boldsymbol{\alpha} &= r\frac{\mathbf{P}:\mathbf{Q}}{\tilde{\boldsymbol{\tau}}:\mathbf{Q}}\frac{1}{T_m}\langle\Phi(f-\kappa)\rangle\tilde{\boldsymbol{\tau}},\\ \dot{\vartheta} &= \frac{\vartheta}{c_p\rho_{Ref}}\frac{\partial\boldsymbol{\tau}}{\partial\vartheta} : \mathbf{d} + \frac{\chi}{\rho c_p}\boldsymbol{\tau} : \mathbf{d}^p + \frac{\chi^\#}{\rho c_p}\Xi,\end{aligned} \qquad (5.23)$$

(ii) the boundary conditions

(a) displacement ϕ is prescribed on a part ∂_ϕ of $\partial\phi(\mathcal{B})$ and tractions $(\boldsymbol{\tau}\cdot\mathbf{n})^a$ are prescribed on part ∂_τ of $\partial\phi(\mathcal{B})$, where $\partial_\phi \cap \partial_\tau = 0$ and $\overline{\partial_\phi \cup \partial_\tau} = \partial\phi(\mathcal{B})$;

(b) heat flux $\mathbf{q}\cdot\mathbf{n} = 0$ is prescribed on $\partial\phi(\mathcal{B})$;

(iii) the initial conditions

ϕ, $\boldsymbol{v}$, ρ, ϑ, $\boldsymbol{\mu}$ and $\boldsymbol{\tau}$ are given at each particle $X \in \mathcal{B}$ at $t = 0$;

are satisfied.

For elasto–plastic rate independent response to define an adiabatic flow process we have to replace Eqs $(5.23)_{4-8}$ by

$$\begin{aligned}
L_\upsilon \tau &= \mathcal{L}:\mathbf{d} - \mathcal{M}\dot\vartheta, \\
L_\upsilon \zeta &= \langle \frac{1}{H}\{\mathbf{Q}:[\dot{\tilde\tau} - (\mathbf{d}\cdot\boldsymbol\alpha + \boldsymbol\alpha\cdot\mathbf{d})] + \pi\dot\vartheta\}\rangle Z, \\
\dot\xi &= k_1\dot{\tilde\tau}:\mathbf{g} + (k_2\boldsymbol\tau:\mathbf{P} + k_3\mathbf{g}:\mathbf{P})\langle \frac{1}{H}\{\mathbf{Q}:[\dot{\tilde\tau} - (\mathbf{d}\cdot\boldsymbol\alpha + \boldsymbol\alpha\cdot\mathbf{d})] + \pi\dot\vartheta\}\rangle, \\
L_\upsilon \boldsymbol\alpha &= r\frac{\mathbf{P}:\mathbf{Q}}{\tilde\boldsymbol\tau:\mathbf{Q}}\langle \frac{1}{H}\{\mathbf{Q}:[\dot{\tilde\tau} - (\mathbf{d}\cdot\boldsymbol\alpha + \boldsymbol\alpha\cdot\mathbf{d})] + \pi\dot\vartheta\}\rangle\tilde\boldsymbol\tau, \\
\dot\vartheta &= \frac{\vartheta}{c_p\rho_{Ref}}\frac{\partial\boldsymbol\tau}{\partial\vartheta}:\mathbf{d} + \frac{\chi_1}{c_p}\Lambda + \frac{\chi_2}{c_p}[L_\upsilon\boldsymbol\tau:\mathbf{g} + (\mathbf{g}\cdot\tilde\boldsymbol\tau + \tilde\boldsymbol\tau\cdot\mathbf{g}):\mathbf{d}].
\end{aligned} \qquad (5.24)$$

Let us consider the Cauchy problem

$$\dot\varphi = \mathcal{A}(t,\varphi)\varphi + \mathbf{f}(t,\varphi), \quad t \in [0,t_f], \quad \varphi(0) = \varphi^0, \qquad (5.25)$$

where $\mathcal{A}$ is a spatial differential operator and $\mathbf{f}$ is a nonlinear function, both defined by the governing equations.

5.2.2 Well–posedness of the Cauchy problem

In order to examine the existence, uniqueness and well–posedness of the Cauchy problem (5.25) let us assume that the spatial differential operator $\mathcal{A}$ has domain $\mathcal{D}(\mathcal{A})$ and range $\mathcal{R}(\mathcal{A})$, both contained in a real Banach space E and the nonlinear function $\mathbf{f}$ is as follows $\mathbf{f} : E \to E$. To investigate the existence as well as the stability of solutions to (5.25) it is necessary to characterize their properties without actually constructing the solutions. This can be done by considering the properties of a nonlinear semi–group because if the operator $\mathcal{A} + \mathbf{f}(\cdot)$ generates a nonlinear semi–group $\{I\!F_t^*; t \geq 0\}$, then a solution to (5.25) starting at $t = 0$ from any element $\varphi^0 \in \mathcal{D}(\mathcal{A})$ is given by

$$\varphi(t,\mathbf{x}) = I\!F_t^* \varphi^0(\mathbf{x}) \qquad (5.26)$$

for $t \in [0, t_f]$.

We say the problem (5.25) is well posed if $I\!F_t^*$ is continuous (in the topology on $\mathcal{D}(\mathcal{A})$ and $\mathcal{R}(\mathcal{A})$ assumed) for each $t \in [0, t_f]$.

Let us postulate as follows:

(i) the strong ellipticity conditions in the form:

$$I\!E = \mathcal{L}^e - \frac{1}{c_p\rho_{Ref}}\vartheta\mathcal{L}^{th}\frac{\partial\boldsymbol\tau}{\partial\vartheta} \qquad (5.27)$$

is strongly elliptic (at a particular deformation ϕ) if there is an $\varepsilon > 0$ such that
$$\mathbb{E}^{abcd}\zeta_a\zeta_c\xi_b\xi_d \geq \varepsilon\|\zeta\|^2\,\|\xi\|^2 \tag{5.28}$$
for all vectors ζ and $\xi \in \mathbb{R}^3$;

(ii) for positive numbers λ_f^1 and λ_f^2 and for $T_m > 0$
$$\mathbf{f}(t,\varphi) \in \mathrm{E}, \quad \|\mathbf{f}(t,\varphi)\|_\mathrm{E} \leq \lambda_f^1, \quad \|\mathbf{f}(t,\varphi') - \mathbf{f}(t,\varphi)\|_\mathrm{E} \leq \lambda_f^2\|\varphi' - \varphi\|_\mathrm{E}, \tag{5.29}$$
and
$$t \to \mathbf{f}(t,\varphi) \in \mathrm{E} \quad \text{is continuous.} \tag{5.30}$$

Using the results presented by Hughes et al. (1977) it is possible to show (cf. Perzyna (1994, 1995)) that the conditions (i) and (ii) guarantee the existence of (locally defined) evolution operators $\mathbb{F}_t^* : \mathrm{E} \to \mathrm{E}$ that are continuous in all variables. In other words the solution of the Cauchy problem (5.25) in the form (5.26) exists, is unique and well–posed.

5.2.3 Cooperative phenomena (thermomechanical coupling and micro–damage mechanism)

For adiabatic process ($\mathbf{q} = 0$) Eq. (4.54) takes the form
$$\rho c_p \dot{\vartheta} = \vartheta \frac{\rho}{\rho_{Ref}} \frac{\partial \boldsymbol{\tau}}{\partial \vartheta} : \mathbf{d} + \chi \boldsymbol{\tau} : \mathbf{d}^p + \chi^\# \dot{\xi}. \tag{5.31}$$

The first term on the right–hand side of Eq. (5.31) has not a dissipative nature and is of the second order when compared with the internal dissipation terms.

The second term on the right–hand side of Eq. (5.31) represents the rate of internal dissipation due to plastic flow process while the last term gives the contribution to the rate of internal dissipation generated by the intrinsic micro–damage mechanism.

When the nondissipative term is neglected then Eq. (5.31) takes the form
$$\rho c_p \dot{\vartheta} = \chi \boldsymbol{\tau} : \mathbf{d}^p + \chi^\# \dot{\xi}. \tag{5.32}$$

From Eq. (5.32) we can compute the irreversibility coefficient χ. It gives
$$\chi = \frac{\rho c_p \dot{\vartheta} - \chi^\# \dot{\xi}}{\boldsymbol{\tau} : \mathbf{d}^p}. \tag{5.33}$$

For $\chi^\# = 0$, i.e. when the influence of the intrinsic micro–damage mechanism is not taken into consideration, Eq. (5.33) takes the form
$$\chi = \frac{\rho c_p \dot{\vartheta}}{\boldsymbol{\tau} : \mathbf{d}^p}. \tag{5.34}$$

For this particular case the irreversibility coefficient χ has a simple interpretation as the heat rate conversion to plastic work rate fraction. However Eq. (5.33) shows that the remaining work rate is attributed to the energy rate lost for micro–damage effects.

6. ANALYTICAL METHODS FOR INVESTIGATION OF LOCALIZATION PHENOMENA

6.1 Analysis of acceleration waves

6.1.1 General considerations

To investigate the intrinsic mathematical structure of the set of the field equations (5.10) and (5.23) which determine the adiabatic inelastic flow processes, let us analyse the problem of propagation of acceleration waves. We shall show that the theory of acceleration waves in the materials considered can be based on the notion of an instantaneous adiabatic acoustic tensor.

Let $\Sigma(t)$ denote a smooth surface with outward normal $\mathbf{n}$ which is moving through the solid body with velocity $\mathbf{w}(t,\mathbf{x})$. Some field quantities or their derivatives may be discontinuous across $\Sigma(t)$ which is then called a singular surface. If the surface $\Sigma(t)$ is composed of the same material points at all times, one then refers to $\Sigma(t)$ as a stationary discontinuity. Otherwise, the surface $\Sigma(t)$ is called a propagating singular surface or wave, cf. Hill (1962).

Let c denote the normal speed of propagation of $\Sigma(t)$ with respect to the material in its current configuration. It is related to the spatial velocity $\boldsymbol{v}(t,\mathbf{x})$ and to the normal wave speed $w = \mathbf{w} \cdot \mathbf{n}$, by the following equation

$$c = w - \boldsymbol{v} \cdot \mathbf{n}. \qquad (6.1)$$

Definition 1. It is said that $\Sigma(t)$ is an acceleration wave if the fields ϕ, $\boldsymbol{v}$, $\mathbf{F}$, $\boldsymbol{\mu}$ and ϑ are continuous functions of t and $\mathbf{x}$, while $\dot{\boldsymbol{v}}$, $\nabla\boldsymbol{v}$, $\dot{\mathbf{F}}$, $\nabla\mathbf{F}$, $\dot{\boldsymbol{\mu}}$, $\nabla\boldsymbol{\mu}$, $\dot{\vartheta}$, $\nabla\vartheta$ have (at most) jump discontinuities across $\Sigma(t)$ but are continuous in t and $\mathbf{x}$ jointly everywhere else ($\boldsymbol{\mu}$ denotes a set of the internal state variables).

An acceleration wave in which $\dot{\vartheta}$ and $\nabla\vartheta$ are continuous functions of t and $\mathbf{x}$ is called homothermal.

From the definition of an acceleration wave and the constitutive assumption $\psi = \hat{\psi}(\mathbf{e},\mathbf{F},\vartheta,\boldsymbol{\mu})$ we have

$$[\![\psi]\!] = [\![\boldsymbol{\sigma}]\!] = [\![\eta]\!] = 0, \qquad (6.2)$$

where $[\![\,\cdot\,]\!]$ denotes the jump of a quantity across $\Sigma(t)$ in the direction of its local normal $\mathbf{n}(t,\mathbf{x})$.

Hadamard's compatibility conditions require the jumps in velocity and stress derivatives to be related as follows (cf. Hadamard (1903)):

$$\begin{aligned}[][\![\nabla\boldsymbol{v}]\!] &= -\frac{1}{c}[\![\mathbf{a}]\!]\mathbf{n}, \\ [\![\nabla\boldsymbol{\sigma}]\!] &= -\frac{1}{c}[\![\dot{\boldsymbol{\sigma}}]\!]\mathbf{n}, \end{aligned} \qquad (6.3)$$

where ∇ denotes the spatial gradient, $\sigma = \frac{1}{J}\tau$ is the Cauchy tensor, and $\mathbf{a} = \dot{\boldsymbol{v}}$.
Balance of momentum requires that (cf. $(5.10)_2$)

$$\text{div}\,\boldsymbol{\sigma} = \rho \mathbf{a}. \tag{6.4}$$

Combining $(6.3)_2$ and (6.4) yields

$$\mathbf{n} \cdot [\![\dot{\boldsymbol{\sigma}}]\!] = -\rho c [\![\mathbf{a}]\!]. \tag{6.5}$$

From the last result it becomes clear that the existence and propagation speed of acceleration waves in solids is directly related to the assumed constitutive structure of the material.

Since ϑ is continuous across $\sum(t)$, we have

$$[\![\dot{\vartheta}]\!] = -c[\![\nabla\vartheta]\!] \cdot \mathbf{n}. \tag{6.6}$$

For an acceleration wave in an adiabatic process we have (cf. Perzyna (1994))

$$[\![\mathbf{q}]\!] = 0, \qquad [\![\dot{\mathbf{q}}]\!] = 0 \tag{6.7}$$

and

$$[\![\dot{\vartheta}]\!] \neq 0, \qquad [\![\nabla\vartheta]\!] \neq 0. \tag{6.8}$$

Hence an acceleration wave in inelastic solids for an adiabatic process is not homothermal.

This conclusion will play an important role in an analysis of acceleration waves in particular material models for adiabatic process of a crystal.

6.2 Investigation of single crystals

6.2.1 Rate dependent adiabatic process

Since $I\!F$ is continuous across $\sum(t)$ Eq. (5.15) gives

$$[\![\mathbf{L}_{\boldsymbol{v}}\boldsymbol{\tau}]\!] = I\!E : [\![\mathbf{d}]\!]. \tag{6.9}$$

Additionally we have very important jump relations

$$\begin{aligned}
[\![\dot{\gamma}^{(\nu)}]\!] &= 0, \\
[\![\dot{\alpha}^{(\nu)}]\!] &= A_2^{(\nu)}\mathcal{F} : [\![\mathbf{d}]\!], \\
[\![\dot{\beta}^{(\nu)}]\!] &= B_2^{(\nu)}\mathcal{F} : [\![\mathbf{d}]\!], \\
[\![\dot{\xi}^{(\nu)}]\!] &= C_2^{(\nu)}\mathcal{F} : [\![\mathbf{d}]\!], \\
[\![\dot{\vartheta}]\!] &= \mathcal{F} : [\![\mathbf{d}]\!].
\end{aligned} \tag{6.10}$$

It is seen (cf. Eqs. (6.10)) that all rates of the internal state variables except $\dot{\gamma}^{(\nu)}$ have jump discontinuities across $\sum(t)$. It has been proved (cf. Eqs. (6.10)$_5$) that an acceleration wave in an elastic–vicsoplastic single crystal for an adiabatic process is not homothermal.

Combining Eqs. (6.3)$_1$, (6.5) and (6.9) we can prove

Theorem 1. For an adiabatic rate dependent plastic flow process of a single crystal described by Eqs. (5.10), the acceleration discontinuity $[\![\mathbf{a}]\!]$ is the solution of the eigenvalue problem

$$\mathbf{A} \cdot [\![\mathbf{a}]\!] = \rho_{Ref} c^2 [\![\mathbf{a}]\!], \qquad (6.11)$$

where

$$\mathbf{A} = \mathbf{n} \cdot (I\!\!E \cdot \mathbf{n} + \boldsymbol{\tau} \cdot \mathbf{ng}) \qquad (6.12)$$

denotes the instantaneous adiabatic acoustic tensor.

It is noteworthy to stress that the instantaneous adiabatic acoustic tensor $\mathbf{A}$ for rate dependent response of a single crystal does depend on the evolution of the dislocation substructure. This is implied by the direct dependence of the adiabatic matrix $I\!\!E$ on the coefficient λ.

Let us assume the strong ellipticity condition in the form

$$I\!\!E^{abcd} \zeta_a \zeta_c \mu_b \mu_d \geq \epsilon \|\boldsymbol{\zeta}\|^2 \|\boldsymbol{\mu}\|^2 \qquad (6.13)$$

for all vectors $\boldsymbol{\zeta}$ and $\boldsymbol{\mu} \in I\!\!R^3$.

Then we can prove that all eigenvalues of the acoustic tensor $\mathbf{A}$ are real and positive. Thus, the Cauchy problem

$$\dot{\boldsymbol{\varphi}} = \mathcal{A}(t, \boldsymbol{\varphi})\boldsymbol{\varphi} + \mathbf{f}(t, \boldsymbol{\varphi}), \quad t \in [0, t_f], \quad \boldsymbol{\varphi}(0, \mathbf{x}) = \boldsymbol{\varphi}^0(\mathbf{x}), \qquad (6.14)$$

for the field equations (5.10) is well–posed provided some conditions for the spatial differential operator $\mathcal{A}$ and the nonlinear function $\mathbf{f}$ are satisfied[10]. This fact has very important implications for the numerical simulation of an adiabatic inelastic flow process.

It can be proved that the localization of plastic deformation in an elastic–viscoplastic crystal body may arise only as the result of the interaction and reflection of stress waves. It has a different character than that which occurs in a rate independent elasto–plastic single crystal.

Viscosity introduces implicitly a length–scale parameter into the dynamical initial-boundary value problem and hence, it implies that the localized shear band region is diffused when compared with an inviscid plastic material. Rate dependency (viscosity) allows the spatial difference operator in the governing equations to retain its ellipticity and the initial value problem is well–posed.

---

[10] For a discussion of the well–posedness of the Cauchy problem see Perzyna (1994).

Since the rate independent plastic response is obtained as the limit case when the relaxation time T tends to zero, hence the theory of viscoplasticity offers the regularization procedure for the solution of the dynamical initial–boundary value problems with localization of plastic deformation.

6.2.2 Rate independent adiabatic process

Combining Eqs. $(6.3)_1$, (6.5) and (5.21) we can prove

Theorem 2. For an adiabatic rate–independent plastic flow process of single crystal described by Eqs. $(5.10)_{1-4}$ and $(5.10)_{6-9}$ and (4.35), the acceleration discontinuity $[\![\mathbf{a}]\!]$ is the solution of the eigenvalue problem

$$\hat{\mathbf{A}} \cdot [\![\mathbf{a}]\!] = \rho_{Ref} c^2 [\![\mathbf{a}]\!], \qquad (6.15)$$

where

$$\hat{\mathbf{A}} = \mathbf{n} \cdot (\mathbb{L} \cdot \mathbf{n} + \boldsymbol{\tau} \cdot \mathbf{ng}) \qquad (6.16)$$

denotes the instantaneous adiabatic acoustic tensor.

6.2.3 Necessary conditions for macroscopic adiabatic shear band formation

Let us denote $\lambda = \rho_{Ref} c^2$, then the eigenvalue problem (6.15) takes the form

$$\hat{\mathbf{A}} \cdot [\![\mathbf{a}]\!] = \lambda [\![\mathbf{a}]\!]. \qquad (6.17)$$

The necessary and sufficient condition for (6.17) to have a non–trivial solution is

$$\det[\hat{\mathbf{A}} - \lambda \mathbb{I}] = 0, \qquad (6.18)$$

where $\mathbb{I}$ is the 3×3 unit matrix.

When zero is an eigenvalue of the instantaneous adiabatic acoustic tensor $\hat{\mathbf{A}}$, then the associated discontinuity does not propagate ($c = 0$) and we speak of a stationary discontinuity. In a quasi–static case this situation is referred to as the strain localization condition. It corresponds to a loss of hyperbolicity of the dynamical equations.

To accomplish this condition let us assume $\lambda = 0$ in Eq. (6.18). Then the necessary condition for a localized plastic deformation region to be formed is as follows

$$\det \hat{\mathbf{A}} = 0. \qquad (6.19)$$

It is noteworthy that this condition for localization is equivalent to that obtained by using the standard bifurcation method (cf. Rice (1976), Rudnicki and Rice (1975), Duszek and Perzyna (1991), Duszek et al. (1992)).

In what follows we shall neglect the influence of the point defects, i.e. we assume $\dot{\xi} = 0$ and $\xi_0 = 0$. It means that we concentrate only on the interaction of the thermally activated and phonon damping mechanisms in the case when the concentration of the point defects can be neglected, e.g. for the mechanism of the intersection of forest dislocations, cf. Fig. 8(i).

6.2.4 Necessary conditions for symmetric double slip process

Let us introduce the Cartesian coordinate system $\{x^i\}$. To obtain the direct analytical results we shall introduce the following simplifications:

(i) Let us assume
$$\mathbf{Z} = \mathbf{s} \cdot (\mathbf{s} \cdot \mathcal{L}^e). \tag{6.20}$$

(ii) Let us restrict our consideration to the linear, isotropic and homogeneous elastic properties of the crystal, i.e.
$$(\mathcal{L}^e)^{abcd} = \tau^{bd} g^{ac} + \mu(g^{ac} g^{bd} + g^{ad} g^{bc}) + \lambda g^{ab} g^{cd}, \tag{6.21}$$

where the constants μ and λ are the Lamé moduli.

The localization condition (6.19) gives the result as follows (cf. Duszek–Perzyna and Perzyna (1996) and Perzyna and Korbel (1997))
$$A\left(\frac{n_1}{n_2}\right)^4 + B\left(\frac{n_1}{n_2}\right)^3 + C\left(\frac{n_1}{n_2}\right)^2 + D\left(\frac{n_1}{n_2}\right) + E = 0, \tag{6.22}$$

where
$$\begin{aligned}
A &= (\mathbb{L}^{2222} + \tau^{22})(\mathbb{L}^{2112} + \tau^{22}), \\
B &= (\mathbb{L}^{2222} + \tau^{22})(\mathbb{L}^{1112} + \mathbb{L}^{2111}) + (\mathbb{L}^{2112} + \tau^{22})(\mathbb{L}^{1222} + \mathbb{L}^{2221}) \\
&\quad - \mathbb{L}^{2212}(\mathbb{L}^{1122} + \mathbb{L}^{2121}), \\
C &= (\mathbb{L}^{1111} + \tau^{11})(\mathbb{L}^{2222} + \tau^{22}) + (\mathbb{L}^{1222} + \mathbb{L}^{2221})(\mathbb{L}^{1112} + \mathbb{L}^{2111}) \\
&\quad + (\mathbb{L}^{1221} + \tau^{11})(\mathbb{L}^{2112} + \tau^{22}) - (\mathbb{L}^{1122} + \mathbb{L}^{2121})(\mathbb{L}^{1212} + \mathbb{L}^{2211}) - \mathbb{L}^{1121}\mathbb{L}^{2212}, \\
D &= (\mathbb{L}^{1111} + \tau^{11})(\mathbb{L}^{1222} + \mathbb{L}^{2221}) + (\mathbb{L}^{1221} + \tau^{11})(\mathbb{L}^{1112} + \mathbb{L}^{2111}) \\
&\quad - \mathbb{L}^{1121}(\mathbb{L}^{1212} + \mathbb{L}^{2211}), \\
E &= (\mathbb{L}^{1111} + \tau^{11})(\mathbb{L}^{1221} + \tau^{11}).
\end{aligned} \tag{6.23}$$

The ratio $\frac{n_2}{n_1} = \tan\beta$ determines the direction of the shear band. The localization of plastic deformations along the shear band may take place if real $\mathbf{n}$'s exist.

When the evolution of substructure and the non–Schmid effects are not taken into consideration, then the fundamental matrix $\mathbb{L}$ takes the form which has been first considered by Duszek–Perzyna and Perzyna (1996). When additionally the Lie derivative is replaced by the Zaremba–Jaumann rate and an isothermal process is assumed, then the fundamental matrix $\mathbb{L}$ is the same as that which has been considered by Peirce et al. (1982).

For symmetric double slip process it is assumed that the crystal has two active slip plane (primary and conjugate) systems, symmetrically oriented with respect to the maximum principal stress τ^{22} (the tensile axis is x^2) at the angle φ, cf. Fig. 39. Then the subscripts ν and δ take on values 1 and 2.

Constitutive Modelling of Dissipative Solids

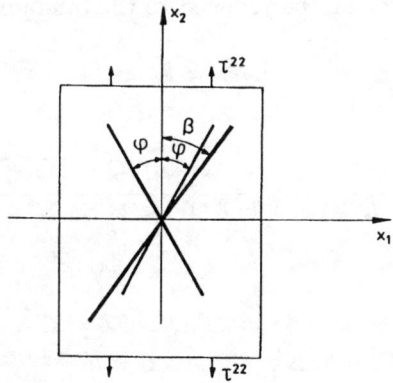

Figure 39: Schematic representation of symmetric primary–conjugate double slip systems of single crystals

Let us assume further that

$$h_{11} = h_{22} = h, \quad h_{12} = h_{21} = h_1, \quad q = \frac{h_1}{h} \quad \tau_c^{(1)} = \tau_c^{(2)} = \tau_c,$$
$$\alpha^{(1)} = \alpha^{(2)} = \alpha, \quad \beta^{(1)} = \beta^{(2)} = \beta, \qquad (6.24)$$
$$\kappa^{(1)} = \kappa^{(2)} = \begin{bmatrix} 0 & \frac{1}{2}\kappa \\ \frac{1}{2}\kappa & 0 \end{bmatrix}, \quad \kappa \sim O\left(\frac{\tau}{\mathcal{L}^e}\right) \approx 3.48 \cdot 10^{-3}.$$

6.2.5 Necessary conditions for single slip process

A rate independent constitutive structure of elasto–plastic single crystal in a single slip adiabatic process is described by the equations as follows

$$\begin{aligned}
\mathbf{L}_\upsilon \boldsymbol{\tau} &= \mathcal{L}^e : \mathbf{d} - \mathcal{L}^{th}\dot{\vartheta} - [\mathcal{L}^e : \mathbf{N} + \mathbf{b}]\dot{\gamma}, \\
\dot{\gamma} &= \frac{1}{h}(\dot{\tau} + \boldsymbol{\kappa} : \boldsymbol{\tau}) + \pi\dot{\vartheta} - g\dot{\beta}, \\
\dot{\alpha} &= a_1\dot{\gamma} + a_2\dot{\vartheta} + a_3\dot{\beta}, \\
\dot{\beta} &= b_1\dot{\gamma} + b_2\dot{\vartheta} + b_3\dot{\alpha}, \\
\dot{\vartheta} &= \frac{1}{\rho c_p}\left[\chi\tau\dot{\gamma} + \chi^*\frac{\tau}{a_1}\dot{\alpha} + \chi^{**}\frac{\tau}{b_1}\dot{\beta}\right] + \frac{\vartheta}{\rho_{Ref}c_p}\frac{\partial\boldsymbol{\tau}}{\partial\vartheta} : \mathbf{d},
\end{aligned} \qquad (6.25)$$

where

$$h = \frac{\partial \tau_c}{\partial \gamma}, \quad \pi = -\frac{1}{h}\frac{\partial \tau_c}{\partial \vartheta}, \quad g = \frac{1}{h}\frac{\partial \tau_c}{\partial \beta}. \qquad (6.26)$$

Differentiation of the resolved Schmid stress $\tau = \mathbf{s} \cdot \boldsymbol{\tau} \cdot \mathbf{m}$ gives additionally the relation (cf. Eq. (5.17))

$$\dot{\tau} = [\mathcal{L}^e : \mathbf{N} + \mathbf{b}] : (\mathbf{d} - \mathbf{d}^p) - \mathcal{L}^{th} : \mathbf{N}\dot{\vartheta}. \qquad (6.27)$$

Equations (6.25) and (6.27) can be reduced to the fundamental evolution equation as follows
$$L_\upsilon \tau = \mathbb{L} : \mathbf{d}, \qquad (6.28)$$
where
$$\mathbb{L} = \mathcal{L}^e - \Theta \frac{\iota}{\Delta} \frac{Z}{\mu} \frac{\partial \tau}{\partial \vartheta} - \frac{(\mathbf{Q} + \mathcal{L}^e : \boldsymbol{\kappa} + \Theta \frac{\tau}{\mu} \mathbf{Z})\left[\mathbf{Q} - \left(\Theta \frac{\mathbf{Z}:\mathbf{N}}{\mu} - \Pi + \Omega\right) \frac{\iota}{\Delta} \frac{\partial \tau}{\partial \vartheta}\right]}{h - \Pi \tau + \Gamma + \Omega \tau + \Theta \frac{\tau}{\mu} \mathbf{Z} : \mathbf{N} + (\mathbf{Q} + \mathcal{L}^e : \boldsymbol{\kappa}) : \mathbf{N}}, \qquad (6.29)$$
with the denotations
$$\mathbf{Q} = \mathcal{L}^e : \mathbf{N} + \mathbf{b}, \qquad \Theta = \Delta \theta \mu, \qquad \theta \mathbf{Z} = \mathcal{L}^{th}, \qquad \Pi = \Delta \pi h,$$
$$\Omega = hg\Delta \frac{b_2 + a_2 b_3}{1 - a_3 b_3}, \qquad \Gamma = hg \frac{b_1 + a_1 b_3}{1 - a_3 b_3}, \qquad (6.30)$$
$$\Delta = \frac{\chi(1-a_3 b_3) + \frac{\chi^*}{a_1}(a_1 + a_3 b_1) + \frac{\chi^{**}}{b_1}(b_1 + a_1 b_3)}{\rho c_p (1 - a_3 b_3) - \frac{\chi^*}{a_1}(a_2 + a_3 b_2) - \frac{\chi^{**}}{b_1}(b_2 + a_2 b_3)},$$
$$\iota = \frac{\vartheta \frac{\rho}{\rho_{Ref}}(1 - a_3 b_3)}{\rho c_p (1 - a_3 b_3) - \frac{\chi^*}{a_1} \tau (a_2 + a_3 b_2) - \frac{\chi^{**}}{b_1} \tau (b_2 + a_2 b_3)}.$$

Let us introduce the Cartesian coordinate system $\{x^i\}$ and restrict to the linear form of $\mathcal{L}^e$ and $\mathbf{Z}$ given by (6.20) and (6.21), respectively.

We shall study the influence of particular effects on adiabatic shear band localization. First we focus the attention on the discussion of the influence of the evolution of substructure, thermomechanical couplings, non–Schmid effects and covariance terms. This case has been considered by Perzyna and Korbel (1996, 1997). So, let us neglect in the fundamental matrix $\mathbb{L}$ the nondissipative thermal term effects. Then we have

$$\mathbb{L} = \mathcal{L}^e - \frac{(\mathbf{Q} + \mathcal{L}^e : \boldsymbol{\kappa} + \Theta \frac{\tau}{\mu} \mathbf{Z}) \mathbf{Q}}{h - \Pi \tau + \Gamma + \Omega \tau + \Theta \frac{\tau}{\mu} \mathbf{Z} : \mathbf{N} + (\mathbf{Q} + \mathcal{L}^e : \boldsymbol{\kappa}) : \mathbf{N}}. \qquad (6.31)$$

Using the necessary condition for localization of plastic deformation (6.19) and applying the perturbation procedure about $\mathbf{n} = \mathbf{m}$, we obtain

$$\mathbf{n} = \mathbf{m} + \left(\frac{\Theta \tau}{2\mu} + \frac{\tau}{4\mu\nu} + \frac{1}{2}\kappa_{ss} + \frac{2\nu - 1}{4\nu}\kappa_{zz}\right)\mathbf{s} + \kappa_{sz}\mathbf{z} + O\left(\frac{\tau}{\mathcal{L}^e}, \frac{\tau^2}{\mathcal{L}^{e2}}\right), (6.32)$$

$$h_{crit} = \Pi \tau - \Gamma - \Omega \tau + \frac{\tau^2}{\mu}\left(\nu \Theta^2 + \Theta + \frac{1}{4\nu}\right) + \tau(2\nu\Theta + 1)\kappa_{ss}$$
$$+ \tau\left[(2\nu - 1)\Theta + 1 - \frac{1}{2\nu}\right]\kappa_{zz} + \mu\nu(\kappa_{ss})^2$$
$$+ \frac{\mu(2\nu - 1)^2}{4\nu}(\kappa_{zz})^2 + \mu(2\nu - 1)\kappa_{zz}\kappa_{ss} + O\left(\frac{\tau^2}{\mathcal{L}^e}, \frac{\tau^3}{\mathcal{L}^{e2}}\right), \qquad (6.33)$$

where
$$\nu = \frac{\lambda + \mu}{\lambda + 2\mu}. \qquad (6.34)$$

Now we use the evolution equations for parameters α and β proposed by Estrin and Kubin (1986) in the form

$$\dot{\alpha} = a_1(\alpha,\beta)\dot{\gamma}, \quad a_1(\alpha,\beta) = \frac{c_1}{b^2}(\frac{\beta}{\alpha}) - c_2\alpha - \frac{c_3}{b}\sqrt{\beta},$$
$$\dot{\beta} = b_1(\alpha,\beta)\dot{\gamma}, \quad b_1(\alpha,\beta) = c_2\alpha + \frac{c_3}{b}\sqrt{\beta} - c_4\beta, \qquad (6.35)$$

where b is the Burgers vector.

Let us notice that the above equations do not describe the dependence of both densities of dislocations on temperature. Then the considered process is not a fully temperature dependent process, but it involves temperature only thorough parameter γ. The identification of the coefficients a_1 and b_1 one can find in the paper presented by Estrin and Kubin (1986), cf. Basiński and Basiński (1979).

For the particular case considered by Asaro and Rice (1977), when

$$\kappa_{ss} = \kappa_{zz} = 0, \quad \kappa_{zs} = \kappa_{sz} = \frac{1}{2}\kappa, \quad \kappa_{mz} = \kappa_{zm} = \frac{1}{2}\kappa_1, \qquad (6.36)$$

we have

$$\mathbf{n} = \mathbf{m} + \left(\frac{\Theta\tau}{2\mu} + \frac{\tau}{4\mu\nu}\right)\mathbf{s} + \frac{1}{2}\kappa\mathbf{z}, \qquad (6.37)$$

$$\left(\frac{h}{\tau}\right)_{crit} = \Pi - \frac{\Gamma}{\tau} + \left(\Theta^2\nu + \Theta + \frac{1}{4\mu}\right)\frac{\tau}{\mu} + \frac{1}{4}\kappa^2\frac{\mu}{\tau}, \qquad (6.38)$$

where

$$\Theta = \Delta\theta\mu, \quad \Pi = \Delta\pi h = -\Delta\frac{\partial\tau_c}{\partial\vartheta}, \quad \Gamma = b_1\frac{\partial\tau_c}{\partial\beta}, \quad \Delta = \frac{\chi + \chi^* + \chi^{**}}{\rho c_p}. \qquad (6.39)$$

From estimations for parameter κ done in paper of Asaro and Rice (1977) we have $\kappa = 1.1\sqrt{\frac{b}{L}}$ where L is the slip–line length. In the moment of localization we may assume that $L \approx 10^{-6}$ m.

We shall use the experimental data from the papers of Chang and Asaro (1981) for aluminum–copper single crystals tested at 298 K and Spitzig (1981) for Fe–Ti–Mn single crystals tested at 295 K.

For aluminum–copper single crystals we have the result us follows

$$\left(\frac{h}{\tau}\right)_{crit} = 0.043 + 0.0090 + 0.0010 + 0.0032 + 0.0025 + 0.0186 = 0.0773,$$

and for nitrogenated Fe–Ti–Mn crystals

$$\left(\frac{h}{\tau}\right)_{crit} = 0.0653 + 0.0100 + 0.0012 + 0.004 + 0.003 + 0.0173 = 0.1008.$$

We observe that the non–Schmid effects are about three times smaller than the thermal plastic softening effects Π, and about two times larger than the interaction between macro and microstructure $\frac{\Gamma}{\tau_{crit}}$.

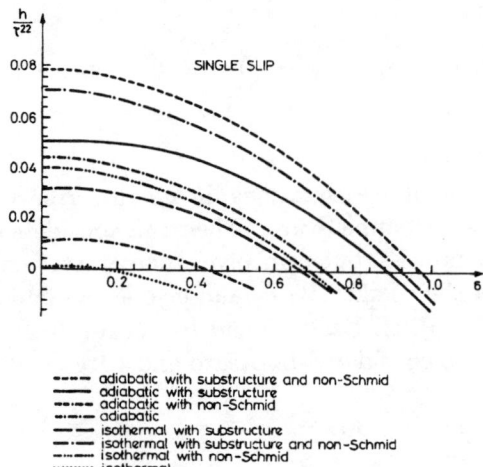

Figure 40: Numerical results for single slip process for the hardening modulus rate h/τ^{22} as function of the misalignment angle δ for Al–Cu single crystal

6.2.6 Discussion of the synergetic effects and comparison with experimental observations

We shall consider both the single slip and symmetric double slip processes. All numerical results are taken from the papers by Perzyna and Korbel (1996) and (1997).

Single slip process

For a single slip process numerical computations have been performed for the fundamental matrix $I\!L$ determined by Eq. (6.31) by using the necessary condition (6.19).

For Al–Cu single crystals some particular material parameters are taken from Chang and Asaro (1981). We consider the same example of uniaxial tension as that tested by Chang and Asaro (1981) at room temperature. The results obtained for the hardening modulus rate h/τ^{22} as function of the misalignment angle δ are plotted in Fig. 40.

For nitrogenated Fe–Ti–Mn single crystals, some particular material parameters are taken from Spitzig's (1981) experimental data. The same example as that tested by Spitzig (1981) at room temperature (295 K) has been considered. The results obtained for the hardening modulus rate h/τ^{22} as a function of the misalignment angle δ are presented in Fig. 41.

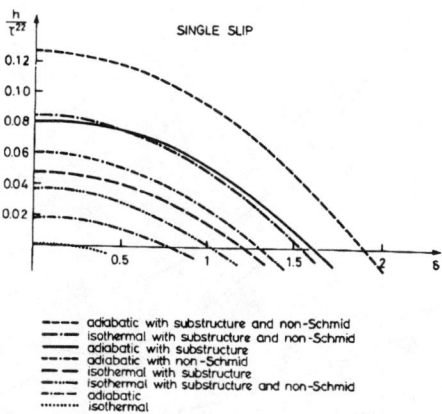

Figure 41: Numerical results for single slip process for the hardening modulus rate h/τ^{22} as function of the misalignment angle δ for Fe–Ti–Mn single crystal

Symmetric double slip process

For symmetric double slip process, numerical computations have been performed for the fundamental matrix $I\!\!L$ determined by Eq. (5.22) with simplifying assumptions (6.20), (6.21), (6.24) and by using the necessary condition (6.22).

For Al–Cu single crystals an example of uniaxial tension has been considered at room temperature with the orientation as follows: $s_1 = (\bar{1},0,1)$, $m_1 = [\bar{1},\bar{1},1]$ and $s_2 = (0,1,1)$, $m_2 = [\bar{1},\bar{1},1]$. The numerical results obtained for the hardening modulus rate h/τ^{22} as a function of the ratio $q = \frac{h_1}{h}$ are plotted in Fig. 42, and the misalignment angle δ as a function of the ratio q shown in Fig. 43.

For nitrogenated Fe–Ti–Mn single crystals the orientation is as follows: $s_1 = (0,1,1)$, $m_1 = [1,\bar{1},1]$ and $s_2 = (1,0,1)$, $m_2 = [\bar{1},1,1]$. The numerical results obtained for the hardening modulus rate h/τ^{22} as a function of the ratio q are presented in Fig. 44, and for the misalignment angle δ as a function of the ratio q are plotted in Fig. 45.

The influence of the assumed value for the irreversibility coefficient χ on the inception of the adiabatic shear band localization for Al–Cu single crystals is presented in Figs. 46 and 47.

Discussion and comparison

Comparison of the analytical theoretical results with the available experimental observations of Chang and Asaro (1981) for aluminum–copper single crystals tested at 298 K and Spitzig (1981) for nitrogenated Fe–Ti–Mn single crystals tested at 295 K clearly shows that the influence of the dislocation substructure on the critical harden-

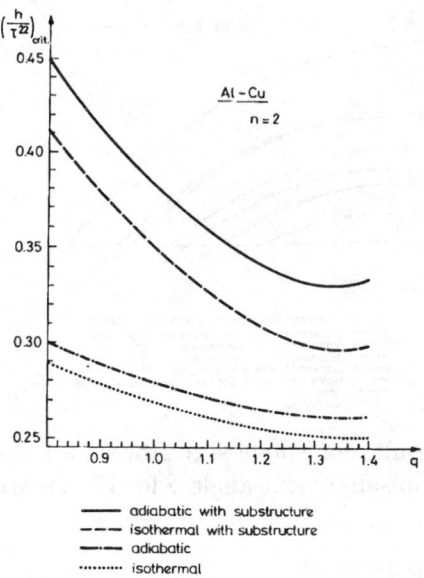

Figure 42: Numerical results for symmetric double slip process for the hardening modulus rate h/τ^{22} as function of the ratio q for Al–Cu single crystal

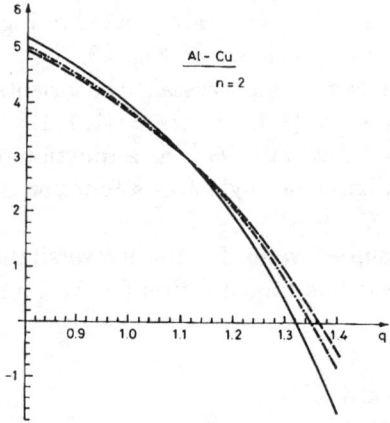

Figure 43: Numerical results for symmetric double slip process for the misalignment angle δ as function of the ratio q for Al–Cu single crystal

Constitutive Modelling of Dissipative Solids

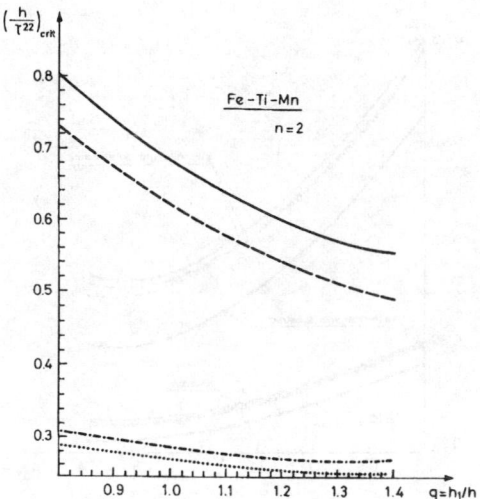

Figure 44: Numerical results for symmetric double slip process for the hardening modulus rate h/τ^{22} as function of the ratio q for Fe–Ti–Mn single crystal

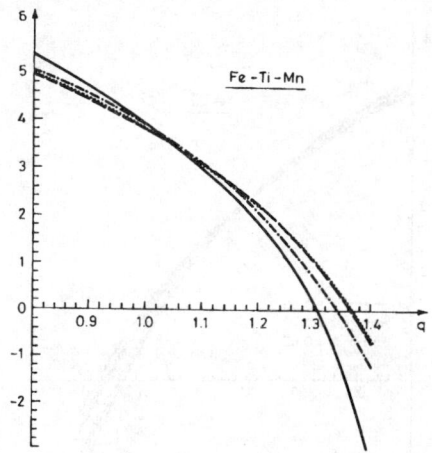

Figure 45: Numerical results for symmetric double slip process for the misalignment angle δ as function of the ratio q for Fe–Ti–Mn single crystal

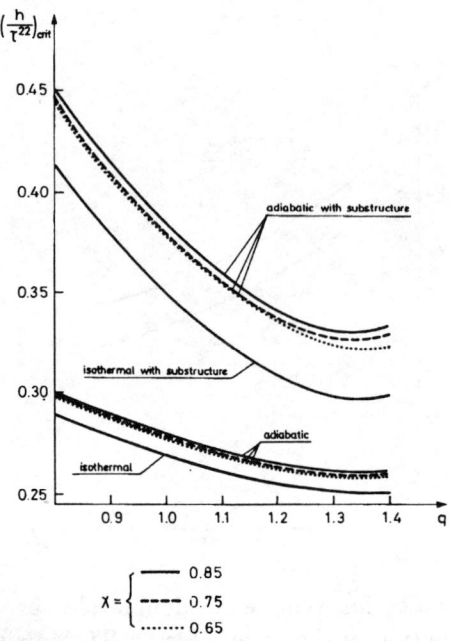

Figure 46: The influence of the assumed value for the irreversibility coefficient χ on the hardening modulus rate h/τ^{22} for Al–Cu single crystal

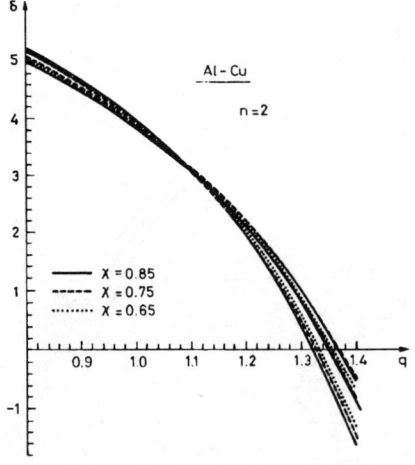

Figure 47: The influence of the assumed value for the irreversibility coefficient χ on the misalignment angle δ for Al–Cu single crystal

ing modulus rate is very pronounced. However, the misalignment of the macroscopic shear bands from the active slip systems in crystal's matrix is not very much affected by the influence of the evolution of substructure.

Comparison of the theoretical results plotted in Figs. 40–45 with those obtained experimentally by Chang and Asaro (1981) and Spitzig (1981) shows that the theoretical results for $\left(\frac{h}{\tau^{22}}\right)_{crit}$ give higher values. This seems natural since the experimental observations detected the values of the hardening modulus rate when the shear band had been already well developed, while the theoretical predictions were computed at the inception of the shear band localization.

Comparison of the theoretical results for the misalignment angle δ plotted in Figs. 43 and 45 with those obtained experimentally by Chang and Asaro (1981), Lisiecki et al. (1982) and Spitzig (1981) shows that the theoretical predictions give good agreement.

It is noteworthy that different situation takes place for a single slip process when the geometry of the deformed specimen is simplified and the misalignment angle δ computed analytically is too small.

It should be pointed out that the influence of the evolution of the dislocation substructure is combined with the thermomechanical coupling and it gives distinct synergetic effect. This synergetic effect is very well visible from the results presented in Fig. 3.15 and 43.

The changes of the assumed value for the irreversibility coefficient χ in the range of $0.65 - 0.85$ (as it has been suggested by Mason et al. (1994)) does not influence the inception of the adiabatic shear band localization too much, cf. Figs. 46 and 47.

The main features of the theory of thermodynamic viscoplasticity of single crystals developed in this paper are as follows: (i) it is invariant with respect to any diffeomorphism: (ii) it takes into considerations such important effects as thermomechanical coupling, the evolution of the dislocations substructure, the non–Schmid law, the spatial covariance and plastic spin; (iii) it describes cooperative phenomena and as the result, it takes account of synergetic effects.

The necessary criterion for adiabatic shear band localization introduced allows to discuss particular effects which can affect the localization phenomena in single slip as well as in symmetric double slip processes. It has been proved that the cooperative phenomena play very important role in the development of macroscopic shear band localization of plastic deformation in single crystals.

6.3 Investigation of polycrystalline inelastic solids

6.3.1 Elasto–viscoplastic material

Let us assume that the relaxation time $T_m > 0$ and $f - \kappa > 0$, then Eqs (5.23)$_{4-8}$ give

$$[\![L_\upsilon \tau]\!] = \mathcal{L}^e : [\![d]\!] - \mathcal{L}^{th}[\![\dot{\vartheta}]\!], \quad [\![L_\upsilon \mu]\!] = 0, \quad c_p[\![\dot{\vartheta}]\!] = \vartheta \frac{1}{\rho_{Ref}} \frac{\partial \tau}{\partial \vartheta} : [\![d]\!]. \quad (6.40)$$

Finally we have

$$[\![L_\upsilon \tau]\!] = I\!E : [\![d]\!], \quad (6.41)$$

where $I\!E$ is given by Eq. (5.27).

Combining (6.3)$_1$, (6.5) and (6.41) yields

$$\mathbf{A} \cdot [\![\mathbf{a}]\!] = \rho_{Ref} c^2 [\![\mathbf{a}]\!], \quad (6.42)$$

where

$$\mathbf{A} = \mathbf{n} \cdot (I\!E \cdot \mathbf{n} + \boldsymbol{\tau} \cdot \mathbf{ng}) \quad (6.43)$$

denotes the instantaneous adiabatic acoustic tensor for an elastic–viscoplastic model of solids. This proves

Theorem 3. In an adiabatic process for an elastic–viscoplastic model of solid body the acceleration discontinuity $[\![\mathbf{a}]\!]$ is the solution of the eigenvalue problem (6.42) with the acoustic tensor given by Eq. (6.43).

6.3.2 Rate independent plasticity

When the relaxation time $T_m = 0$ then $f = \kappa$ and Eqs (5.24) give

$$[\![L_\upsilon \tau]\!] = \mathcal{L} : [\![d]\!] - \mathcal{M}[\![\dot{\vartheta}]\!], \quad (6.44)$$

$$[\![\dot{\vartheta}]\!] = \frac{\chi_1}{c_p} \langle \frac{1}{H} \{\mathbf{Q} : [\![\![\dot{\tau}]\!]\!] - ([\![d]\!] \cdot \boldsymbol{\alpha} + \boldsymbol{\alpha} \cdot [\![d]\!])] + \pi[\![\dot{\vartheta}]\!] \} \rangle$$

$$+ \frac{\chi_2}{c_p} [\![\![L_\upsilon \tau]\!] : \mathbf{g} + (\mathbf{g} \cdot \tilde{\boldsymbol{\tau}} + \tilde{\boldsymbol{\tau}} \cdot \mathbf{g}) : [\![d]\!]\!] + \frac{\vartheta}{c_p \rho_{Ref}} \frac{\partial \tau}{\partial \vartheta} : [\![d]\!],$$

$$[\![L_\upsilon \zeta]\!] = Z \langle \frac{1}{H} \{\mathbf{Q} : [\![\![\dot{\tau}]\!]\!] - ([\![d]\!] \cdot \boldsymbol{\alpha} + \boldsymbol{\alpha} \cdot [\![d]\!])] + \pi[\![\dot{\vartheta}]\!] \} \rangle,$$

$$[\![\dot{\xi}]\!] = (k_2 \tilde{\boldsymbol{\tau}} : \mathbf{P} + k_3 \mathbf{P} : \mathbf{g}) \langle \frac{1}{H} \{\mathbf{Q} : [\![\![\dot{\tau}]\!]\!] - ([\![d]\!] \cdot \boldsymbol{\alpha} + \boldsymbol{\alpha} \cdot [\![d]\!])] + \pi[\![\dot{\vartheta}]\!] \} \rangle + k_1 [\![\dot{\tilde{\tau}}]\!] : \mathbf{g},$$

$$[\![L_\upsilon \boldsymbol{\alpha}]\!] = \frac{H^{**}}{\tilde{\boldsymbol{\tau}} : \mathbf{Q}} \tilde{\boldsymbol{\tau}} \langle \frac{1}{H} \{\mathbf{Q} : [\![\![\dot{\tau}]\!]\!] - ([\![d]\!] \cdot \boldsymbol{\alpha} + \boldsymbol{\alpha} \cdot [\![d]\!])] + \pi[\![\dot{\vartheta}]\!] \} \rangle.$$

Equation (6.44)$_2$ can be written in the form

$$[\![\dot{\vartheta}]\!] = \mathbf{M} : [\![L_\upsilon \tau]\!] + \mathbf{N} : [\![d]\!], \quad (6.45)$$

where

$$\mathbf{M} = \frac{\chi_1 \mathbf{Q} + \chi_2 H \mathbf{g}}{H c_p - \chi_1 \pi}, \tag{6.46}$$

$$\mathbf{N} = \left[\frac{\vartheta}{\rho_{Ref}} H \frac{\partial \boldsymbol{\tau}}{\partial \vartheta} + \chi_1 (\mathbf{Q} \cdot \tilde{\boldsymbol{\tau}} + \tilde{\boldsymbol{\tau}} \cdot \mathbf{Q}) + \chi_2 H (\mathbf{g} \cdot \tilde{\boldsymbol{\tau}} + \tilde{\boldsymbol{\tau}} \cdot \mathbf{g})\right](H c_p - \chi_1 \pi)^{-1}.$$

Substituting (6.45) into (6.44)$_1$ gives the fundamental relation

$$[\![\mathbf{L}_\upsilon \boldsymbol{\tau}]\!] = \mathbb{L} : [\![\mathbf{d}]\!], \tag{6.47}$$

where

$$\mathbb{L} = \left(\mathbf{I} - \frac{\mathcal{M}\mathcal{M}}{1 + \mathcal{M} : \mathbf{M}}\right) \cdot (\mathcal{L} - \mathcal{M}\mathbf{N}). \tag{6.48}$$

Combining (6.3)$_1$, (6.5) and (6.47) we obtain

$$\hat{\mathbf{A}} \cdot [\![\mathbf{a}]\!] = \rho_{Ref} c^2 [\![\mathbf{a}]\!], \tag{6.49}$$

where

$$\hat{\mathbf{A}} = \mathbf{n} \cdot (\mathbb{L} \cdot \mathbf{n} + \boldsymbol{\tau} \cdot \mathbf{n}\mathbf{g}) \tag{6.50}$$

denotes the instantaneous adiabatic acoustic tensor for an elastic–plastic model of damaged solids. This proves

Theorem 4. In an adiabatic process for an elastic–plastic damaged solid body the acceleration discontinuity $[\![\mathbf{a}]\!]$ is the solution of the eigenvalue problem (6.49) with the acoustic tensor $\hat{\mathbf{A}}$ given by Eq. (6.50).

Let us denote $\lambda = \rho_{Ref} c^2$, then the eigenvalue problem (6.42) takes the form

$$\mathbf{A} \cdot [\![\mathbf{a}]\!] = \lambda [\![\mathbf{a}]\!]. \tag{6.51}$$

The necessary and sufficient condition for (6.51) to have a non-trivial solution is

$$\det[\mathbf{A} - \lambda \mathbf{I}] = 0 \tag{6.52}$$

where $\mathbf{I}$ is the 3×3 unit matrix.

The strong ellipticity condition assumed in the form (5.28) implies that the three eigenvalues of the acoustic tensor $\mathbf{A}$ are real and strictly positive, then the set of equations (5.23) is hyperbolic.

6.3.3 Condition for localization of plastic deformation for rate independent plasticity

For rate independent plasticity it may happen that an eigenvalue of the instantaneous adiabatic acoustic tensor $\hat{\mathbf{A}}$ vanishes, then the associated discontinuity does not propagate ($c = 0$) and we speak of a stationary discontinuity. In quasi-static case this situation is referred as the strain localization condition and corresponds to a loss of hyperbolicity of the dynamical equations, cf. Section 6.2.3. Then the necessary condition for a localized plastic deformation region to be formed is as follows

$$\det[\mathbf{n} \cdot (\mathbb{L} \cdot \mathbf{n} + \boldsymbol{\tau} \cdot \mathbf{ng})] = 0. \tag{6.53}$$

Detail discussion of this condition for different constitutive assumptions of rate independent plastic materials is presented in Duszek–Perzyna and Perzyna (1994).

6.3.4 Isothermal process in undamaged solids

To obtain results in analytical form let us postulate the simplifications as follows:

(i) The flow is associated with the Huber–Mises yield criterion ($\tilde{J}_2$ – flow theory)

$$P_{ab} = Q_{ab} = \frac{\tilde{\tau}'_{ab}}{2\sqrt{\tilde{J}_2}}, \quad \tilde{J}_2 = \frac{1}{2}\tilde{\tau}'^{ab}\tilde{\tau}'^{cd} g_{ac} g_{bd}, \tag{6.54}$$

where $\tilde{\tau}'$ denotes the deviator of $\tilde{\tau}$.

(ii) Thermal effects are neglected

$$\mathcal{L}^{th} = 0, \quad \pi = 0 \implies \mathcal{M} = 0. \tag{6.55}$$

(iii) As in the infinitesimal theory of elasticity we assume linear properties of the material, i.e.

$$(\mathcal{L}^e)^{abcd} = G\left(g^{ac}g^{db} + g^{cb}g^{da}\right) + \left(K - \frac{2}{3}G\right) g^{ab}g^{cd} + \tau^{bd}g^{ac}, \tag{6.56}$$

where G and K denote the shear and bulk modulus, respectively.

Taking into account the simplifications (i), (ii) and (iii) from (6.48) and (4.70)$_1$ we obtain $\mathbb{L} = \mathcal{L}$ and

$$\mathcal{L} = \mathcal{L}^e - 2G\left[2G\tilde{\tau}'\tilde{\tau}' + 2\tilde{J}_2(\tilde{\tau}'\mathbf{g} + \mathbf{g}\tilde{\tau}') + \tilde{\tau}'\tilde{\tau}' \cdot \boldsymbol{\tau} + \tilde{\tau}' \cdot \boldsymbol{\tau}\tilde{\tau}' + \boldsymbol{\alpha} : \tilde{\tau}'(\mathbf{g}\tilde{\tau}' \right.$$
$$\left. + 3\eta_1\boldsymbol{\tau}'\tilde{\tau}')\right]\left[4\tilde{J}_2(H+G) + (\tilde{\tau}' \cdot \tilde{\tau}) : \tilde{\tau}' + 3\eta_1\boldsymbol{\alpha} : \tilde{\tau}'\tilde{\tau}' : \boldsymbol{\tau}'\right]^{-1}. \tag{6.57}$$

Let us introduce rectangular Cartesian coordinates $\{x^i\}$ in such a way that $\mathbf{n}$ is in the x^2-direction. Then taking into consideration the orientation of the plane within which the shear band localization first occurs, the necessary condition for localization (6.53) leads to the result as follows:

$$\left(\frac{H}{G}\right)_{cr} = \frac{1}{2G}\{\nu(1+\nu)\sqrt{\tilde{J}_2}\tilde{T}^3 - [G(1+\nu) - \frac{1}{3}\nu(1-2\nu)J_1 - \nu(1+\nu)\sqrt{\tilde{J}_2}\mathcal{A}$$
$$+ 3\mathcal{J}_2\eta_1(1+\nu)]\tilde{T}^2 - [2(1-\nu)\frac{1}{\sqrt{\tilde{J}_2}}(\tilde{J}_2 - \mathcal{J}_2) + 3(1+\nu)\mathcal{J}_2\eta_1\mathcal{A}]\tilde{T}$$
$$+ (1-2\nu)\sqrt{\tilde{J}_2}\mathcal{A} - \frac{3}{2}\frac{1}{\tilde{J}_2}(\tilde{J}_3 + \mathcal{J}_3) - \frac{1}{3}J_1\}, \qquad (6.58)$$

where

$$\tilde{J}_3 = \det(\tilde{\tau}), \quad \tilde{T} = \frac{\tilde{\tau}'_{II}}{\sqrt{\tilde{J}_2}}, \quad \mathcal{A} = \frac{\alpha'_{II}}{\sqrt{\tilde{J}_2}}, \quad \mathcal{J}_2 = \frac{1}{2}\alpha' : \tilde{\tau}', \quad \mathcal{J}_3 = \frac{1}{3}(\tilde{\tau}' \cdot \alpha') : \tilde{\tau}',$$
$$(6.59)$$

ν is Poisson's ratio, $\tilde{\tau}'_{II}$ and α'_{II} denote the mean principal components of $\tilde{\tau}'$ and α', respectively.

6.3.5 Thermomechanical coupling in damaged solids, spatial covariance and plastic spin are neglected

Let us assume:

(i) The yield criterion for damaged solids is postulated in the form

$$\tilde{J}_2 + (n_1 + n_2\xi)\tilde{J}_1^2 - \kappa = 0, \qquad (6.60)$$

where n_1 and n_2 are temperature dependent material functions, $\tilde{J}_2$ is defined by $(6.54)_2$ and $\tilde{J}_1 = \tilde{\tau}^{ab}g_{ab}$.

(ii) It is postulated $\mathcal{L}^{e^{-1}} : \mathcal{L}^{th} = \theta\mathbf{g}$, where θ is the thermal expansion constant.

(iii) The Lie derivative is approximated by the material derivative, that is $L_{\boldsymbol{\upsilon}}\boldsymbol{\tau} \approx \dot{\boldsymbol{\tau}}$ and $L_{\boldsymbol{\upsilon}}\boldsymbol{\mu} \approx \dot{\boldsymbol{\mu}}$.

(iv) Elastic properties are postulated linear as in the simplification (iii) in previous Section, cf. Eq. (6.56).

(v) Main contribution to internal heating is considered and additionally in the evolution equation for the jump of temperature (6.45) and (6.46) it is postulated that $\chi_2 = 0$; it means that in determination of the rate of temperature we do not take into consideration the mechanism of cracking of second phase particles.

Taking into account the simplifications (i)–(v) we obtain the fundamental matrix in the form

$$(I\!\!L)^{abcd} = (\mathcal{L}^e)^{abcd} - (H + G + 9KAB - G\Pi + 6GB\Xi)^{-1}[\frac{G}{\sqrt{\tilde{J}_2}}\tilde{\tau}'^{ab}$$

$$+(3KA + 2G\Xi)g^{ab}](\frac{G}{\sqrt{\tilde{J}_2}}\tilde{\tau}'^{cd} + 3KBg^{cd}), \qquad (6.61)$$

where

$$A = \frac{1}{\sqrt{\tilde{J}_2}}(n_1 + n_2\xi)\tilde{\tau}^{ab}g_{ab}, \quad B = A + \frac{1}{2\sqrt{\tilde{J}_2}}\left(\frac{\partial f}{\partial \xi} - \frac{\partial \hat{\kappa}}{\partial \xi}\right)\frac{\partial \xi}{\partial \mathcal{T}} : \mathbf{g},$$

$$\Pi = \frac{\chi_1 \pi}{c_p G}, \quad \Xi = \frac{3\chi_1 \theta K}{2 c_p G}. \qquad (6.62)$$

Let us proceed as in previous section, then we obtain the results as follows:

$$\left(\frac{H}{G}\right)_{cr} = -\frac{1+\nu}{2}\left(\tilde{T} + A + B + \frac{1-2\nu}{1+\nu}\Xi\right)^2 + \frac{1+\nu}{1-\nu}\left(A - B + \frac{1-2\nu}{1+\nu}\Xi\right)^2 + \Pi,$$

$$\tan^2\beta = \frac{S - \tilde{\tau}'_{III}}{\tilde{\tau}'_I - S}, \qquad (6.63)$$

where β denotes the angle between the vector $\mathbf{n}$ and the $\tilde{\tau}_{III}$-direction, and

$$S = -(1-\nu)\tilde{\tau}'_{II} + (1+\nu)(A+B)\sqrt{\tilde{J}_2} + (1-2\nu)\Xi\sqrt{\tilde{J}_2}. \qquad (6.64)$$

6.3.6 Discussion of different effects

Nonsymmetry of matrix $\mathcal{L}$. Let us examine the fundamental elasto–plastic matrix $\mathcal{L}$ (cf. Eq. (4.70)$_1$). There are three main reasons why the matrix $\mathcal{L}$ is nonsymmetric. The first is generated by the micro–damage mechanism of cracking of second–phase particles ($k_1 \neq 0 \Longrightarrow \mathbf{P} \neq \mathbf{Q}$). The second is implied by kinematic hardening and the third is caused by the plastic spin $\mathcal{W}$.

When the fundamental matrix $\mathcal{L}$ fails to be symmetric, the governing equations can admit complex solutions for c^2 that correspond to flutter–type instabilities (cf. Loret (1992)).

Plastic spin. Discussion of the influence of the plastic spin effects on the localization phenomenon can be based on the investigation of result (6.58). This result describes the influence of three cooperative phenomena, namely the plastic spin effects, the plastic strain induced anisotropy (kinematic hardening) and the geometrical covariant effects.

The result (6.58) for the critical hardening modulus rate $(\frac{H}{G})_{cr}$ as a function of the state of stress $\tilde{T}$ may be represented by the curve, as has been plotted in Fig. 48

Constitutive Modelling of Dissipative Solids

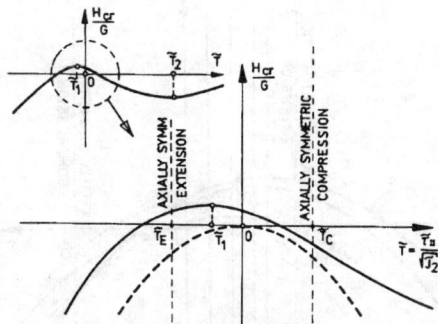

Figure 48: Schematic representation of H_{cr}/G versus $\tilde{T}$ for an isothermal process in undamaged solids (cf. Eq. (6.58)). For $\tilde{T} = 0 \Rightarrow H_{cr} > 0$ and $\frac{\partial H_{cr}}{\partial \tilde{T}} = 0 \Rightarrow \tilde{T}_E < \tilde{T}_1 < 0$ and $\tilde{T}_2 \gg \tilde{T}_C$

by a solid line. This curve when compared with the parabola given by the equation (classical $\tilde{J}_2$-flow theory)

$$\left(\frac{H}{G}\right)_{cr} = -\frac{1+\nu}{2}\tilde{T}^2 \qquad (6.65)$$

and plotted in Fig. 48 by a broken line, is translated up and shifted to the left. The translation up means that the material is more inclined to instability by localization along the shear band, and the shifting to the left shows that the inclination to instability for axially symmetric compression is different from that for the axially symmetric tension, for tension material is more sensitive to localization than for compression.

It is noteworthy that both these effects have been observed experimentally by Anand and Spitzig (1980) in the investigation of the initiation of localized shear bands in isothermal, quasi–static plane strain conditions for a maragin steel. The interaction of the plastic strain induced anisotropy (kinematic hardening) and the geometrical covariant terms causes the influence of stress triaxiality on the criterion for adiabatic shear band localization. This effect is represented in (6.58) by the terms which depend on J_1, $\tilde{J}_3$ and $\mathcal{J}_3$.

Covariant terms. Assuming that the kinematic hardening and plastic spin effects are neglected then Eq. (6.58) gives the particular result which describes only the influence of the geometrical covariant terms on shear band localization.

Kinematic hardening. The plastic strain induced anisotropy is approximated by kinematic hardening. The kinematic hardening effect represents typical cooperative phenomenon and its influence has synergetic nature.

Micro–damage mechanism (cf. Duszek–Perzyna et al (1992)). Assuming in Eq (6.63) the isothermal approximation, i.e. $\Xi = \Pi = 0$ we obtain the result for the critical hardening modulus rate as a function of the state of stress $\tilde{T}$ which may be represented by the parabola II, as it has been plotted in Fig. 49 by a broken line.

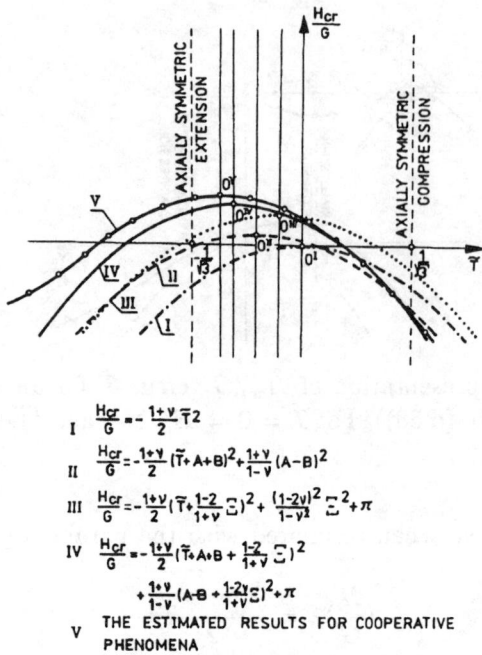

I $\quad \frac{H_{cr}}{G} = -\frac{1+\nu}{2}\tilde{T}^2$

II $\quad \frac{H_{cr}}{G} = -\frac{1+\nu}{2}(\tilde{T}+A+B)^2 + \frac{1+\nu}{1-\nu}(A-B)^2$

III $\quad \frac{H_{cr}}{G} = -\frac{1+\nu}{2}(\tilde{T}+\frac{1-2\nu}{1+\nu}\Xi)^2 + \frac{(1-2\nu)^2}{1-\nu^2}\Xi^2 + \pi$

IV $\quad \frac{H_{cr}}{G} = -\frac{1+\nu}{2}(\tilde{T}+A+B+\frac{1-2\nu}{1+\nu}\Xi)^2$
$\quad + \frac{1+\nu}{1-\nu}(A-B+\frac{1-2\nu}{1+\nu}\Xi)^2 + \pi$

V $\quad$ THE ESTIMATED RESULTS FOR COOPERATIVE PHENOMENA

Figure 49: Schematic representation of H_{cr}/G versus $\tilde{T}$ for an adiabatic process in damaged solids when spatial covariance terms and the plastic spin are neglected (cf. Eq.(6.63)) and the estimated results for cooperative phenomena

This parabola, when compared with the parabola I is translated up by $[(1+\nu)/(1-\nu)](A-B)^2$ and is shifted to the left by $A+B$. The translation up is caused by the mechanism of nucleation due to cracking of second phase particles and shifting to the left is implied by the micro–damage process (by the growth as well as nucleation mechanisms).

The micro–damage effect implies that the material is more inclined to instability by localization along the shear band and for tension is more sensitive to localization than for compression.

Thermomechanical couplings (cf. Duszek and Perzyna (1991b)). When there is no micro–damage process $(A = B = 0)$ Eq (6.63) gives the result for the critical hardening modulus rate $(H/G)_{cr}$ as a function of the state of stress $\tilde{T}$ which can be represented by the parabola III plotted in Fig. 49 by a dotted line. This parabola, when compared with the parabola I, is translated up by $[(1-2\nu)^2/(1-\nu^2)]\Xi^2 + \Pi$ and is shifted to the left by $[(1-2\nu)/(1+\nu)]\Xi$.

Considering the influence of thermal expansion and thermal plastic softening we can have shear band localization when the hardening modulus rate is positive, and we have also differences between the values of the hardening modulus rate for axially symmetric compression and tension.

Synergetic effects. The influence of different cooperative phenomena on shear band localization can be discussed basing on the results (6.58) and (6.63). The curve plotted in Fig. 48 by a solid line represents the synergetic effects of the plastic spin, kinematic hardening and the covariance terms. The parabola IV plotted in Fig. 49 by a solid line shows the influence of the micro–damage mechanism, kinematic hardening and thermomechanical couplings.

When the fundamental matrix $I\!L$ is given by (6.48) with $\mathcal{L}$ and $\mathcal{M}$ defined by (4.70) then the effects of the plastic spin, covariance terms, plastic strain induced anisotropy (kinematic hardening), micro–damage mechanism and thermomechanical coupling are taken into consideration. In this general case the influence of above–mentioned cooperative phenomena on adiabatic shear band localization can be discussed basing on the estimated qualitative results plotted in Fig. 49 as the curve V. This curve represents the synergetic effects for all cooperative phenomena considered. The effect of stress triaxiality previously mentioned has also typical synergetic nature.

7. NUMERICAL SOLUTION OF THE INITIAL–BOUNDARY VALUE PROBLEM (EVOLUTION PROBLEM)

7.1 Formulation of the evolution problem

Find φ as function of t and $\mathbf{x}$ satisfying[11]

$$\left.\begin{array}{rl} \text{(i)} & \dot{\varphi} = \mathcal{A}(t,\varphi)\varphi + \mathbf{f}(t,\varphi); \\ \text{(ii)} & \varphi(0) = \varphi^0(\mathbf{x}); \\ \text{(iii)} & \text{The boundary conditions} \\ & \text{(e.g. as have been postulated in Section 5.1.2 and 5.2.1).} \end{array}\right\} \quad (7.1)$$

A strict solution of (7.1) with $\mathbf{f}(t,\varphi) \equiv 0$ (i.e. the homogeneous evolution problem) is defined as a function $\varphi(t) \in \mathrm{E}$ (a Banach space) such that

$$\varphi(t) \in \mathcal{D}(\mathcal{A}), \quad \text{for all} \quad t \in [0, t_f], \tag{7.2}$$

$$\lim_{\Delta t \to 0} \|\frac{\varphi(t+\Delta t) - \varphi(t)}{\Delta t} - \mathcal{A}\varphi(t)\|_\mathrm{E} = 0 \quad \text{for all} \quad t \in [0, t_f].$$

The boundary conditions are taken care of by restricting the domain $\mathcal{D}(\mathcal{A})$ to elements of E that satisfy those conditions; they are assumed to be linear and homogeneous, so

---

[11] We shall follow here some fundamental results which have been discussed in Richtmyer and Morton (1967), Strang and Fix (1973), Richtmyer (1978) and Dautray and Lions (1993).

that the set **S** of all φ that satisfy them is a linear manifold; $\mathcal{D}(\mathcal{A})$ is assumed to be contained in **S**.

The choice of the Banach space E, as well as the domain of $\mathcal{A}$, is an essential part of the formulation of the evolution problem.

7.2 Well–posedness of the evolution problem

The homogeneous evolution problem (i.e. for $\mathbf{f} \equiv 0$) is called well posed (in the sense of Hadamard) if it has the following properties:

(i) The strict solutions are uniquely determined by their initial elements;

(ii) The set Y of all initial elements of strict solutions is dense in the Banach space E;

(iii) For any finite interval $[0, t_0]$, $t_0 \in [0, t_f]$ there is a constant $K = K(t_0)$ such that every strict solution satisfies the inequality

$$\|\varphi(t)\| \leq K\|\varphi^0\|, \quad \text{for} \quad 0 \leq t \leq t_0. \tag{7.3}$$

The inhomogeneous evolution problem (7.1) will be called well posed if it has a unique solution for all reasonable choices of φ^0 and $\mathbf{f}(t, \varphi)$ and if the solution depends continuously, in some sense, on those choices.

It is evident that any solution is unique, because of the uniqueness of the solutions of the homogeneous evolution problem. Namely, the difference of two solutions, for given φ^0 and given $\mathbf{f}(\cdot)$, is a solution of the homogeneous problem with zero as initial element, hence must be zero for all t.

It is possible to show (cf. Richtmyer (1978)) that strict solutions exists for sets of φ^0 and $\mathbf{f}(\cdot)$ that are dense in E and E_1 (a new Banach space), respectively.

Let $\{I\!\!F_t^*; t \geq 0\}$ be a semi-group generated by the operator $\mathcal{A} + \mathbf{f}(\cdot)$ (as it has been defined in Section 5.2) and $\{I\!\!F_t; t \geq 0\}$ be a semi-group generated by the operator $\mathcal{A}$.

Then we can write the generalized solution of the nonhomogenous evolution problem (7.1) in alternative forms

$$\begin{aligned} \varphi(t, \mathbf{x}) &= I\!\!F^*(t)\varphi^0(\mathbf{x}) \\ &= I\!\!F(t)\varphi^0(\mathbf{x}) + \int_0^t I\!\!F(t-s)\mathbf{f}(s, \varphi(s))ds. \end{aligned} \tag{7.4}$$

The generalized solution of the nonhomogenous evolution problem (7.1) in the form $(7.4)_2$ is the integral equation.

The successive approximations for $(7.4)_2$ are defined to be the functions $\varphi_0, \varphi_1, \ldots$, given by the formulas

$$\begin{aligned} \varphi_0(t) &= \varphi^0, \\ \varphi_{k+1}(t) &= I\!\!F(t)\varphi^0 + \int_0^t I\!\!F(t-s)\mathbf{f}(s, \varphi_k(s))ds, \\ k &= 0, 1, 2, \ldots; \quad t \in [0, t_f]. \end{aligned} \tag{7.5}$$

It is possible to show that these functions actually exist on $t \in [0, t_f]$ if the continuous function $\mathbf{f}$ is Lipschitz continuous with respect to the second argument uniformly with respect to $t \in [0, t_f]$. Then $(7.4)_2$ has unique solution (cf. Ionescu and Sofonea (1993)).

7.3 Discretization in space and time

We must approximate (7.1) twice. First, when E is infinite dimensional, we must replace $\mathcal{A}$ by an operator $\mathcal{A}_h$ which operates in a finite dimensional space $V_h \subset E$, where, in general, $h > 0$ represents a discretisation step in space, such that $\dim(V_h) \to \infty$ as $h \to 0$. Second, we must discretise in time, that is to say choose a sequence of moments t_n (for example $t_n = n\Delta t$, where Δt is time step) at which we shall calculate the approximate solution.

Let us introduce the following semi–discretised (discrete in space) problem.

$$\left.\begin{array}{l} \text{Find } \varphi_h \in \mathcal{C}^0([0, t_0]; V_h) \ (\mathcal{C}^0 \text{ denotes the space of functions} \\ \text{continuous on } ([0, t_0], V_h)) \text{ satisfying} \\ \frac{d\varphi_h(t)}{dt} = \mathcal{A}_h \varphi_h(t) + \mathbf{f}_h(t), \\ \varphi_h(0) = \varphi_{0,h}. \end{array}\right\} \quad (7.6)$$

The operator $\mathcal{A}_h$ for the finite element method can be obtained by a variational formulation approach. The discrete equations are obtained by the Galerkin method at particular points in the domain.

Finally, we shall define a method allowing us to calculate $\varphi_h^n \in V_h$, an approximation to $\varphi_h(t_n)$ starting from φ_h^{n-1} (we limit ourselves to a two–level scheme). Then we can write

$$\varphi_h^{n+1} = C_h(\Delta t)\varphi_h^n + \Delta t \mathbf{f}_h^n, \quad \varphi_h^0 = \varphi_{0,h} \qquad (7.7)$$

where we introduce the operator $C_h(\Delta t) \in \mathcal{L}(V_h)$ ($\mathcal{L}$ is the set of continuous linear mapping of V_h with values in V_h) and where $\mathbf{f}_h^n$ approximates $\mathbf{f}_h(t_n)$.

We shall always assume that the evolution problem (7.1) is well posed and there exists a projection R_h of E into V_h such that

$$\lim_{h \to 0} \mid R_h \varphi - \varphi \mid_E = 0 \quad \forall \varphi \in E. \qquad (7.8)$$

7.4 Convergence, consistency and stability

The first fundamental question is that of the convergence, when h and Δt tend to zero, of the sequence $\{\varphi_h^n\}$, the solution (7.7), towards the function $\varphi(t)$, the solution of (7.1). Let us restrict our consideration, for the moment, to the case where $\mathbf{f}(t) \equiv 0$.

Definition 2. The scheme defined by (7.7) will be called convergent if the condition

$$\varphi_{0,h} \to \varphi^0 \quad \text{as} \quad h \to 0 \qquad (7.9)$$

implies that

$$\varphi_h^n \to \varphi(t) \quad \text{as} \quad \Delta t \to 0, \quad n \to \infty \quad \text{with} \quad n\Delta t \to t \qquad (7.10)$$

for all $t \in]0, t_0[$, $t_0 \in [0, t_f]$, where φ_h^n is defined by (7.7) and $\varphi(t)$ is the solution of (7.1). All this holds for arbitrary φ^0.

The study of the convergence of an approximation scheme involves two fundamental properties of the scheme, consistency and stability.

Definition 3. The scheme defined by (7.7) is called **stable**, if there exists a constant $K \geq 1$ independent of h and Δt such that

$$\|(C_h(\Delta t))^n R_h\|_{\mathcal{L}(E)} \leq K \quad \forall n, \Delta t \quad \text{satisfying} \quad n\Delta t \leq t_0. \qquad (7.11)$$

In the Definition 2 and 3 there occur two parameters h and Δt. It may be that the scheme is not stable (or not convergent) unless Δt and h satisfy supplementary hypotheses of the type $\Delta t/h^\alpha \leq$ constant, $\alpha < 0$, in which case we call the scheme **conditionally stable**. If the scheme is stable for arbitrary h and Δt we say that it is **unconditionally stable**.

These schemes reflect so called explicit and implicit types of the integration procedure in a particular numerical implementation.

Definition 4. The scheme defined by (7.7) will be called **consistent** with equation (7.1) if there exists a subspace $Y \subset E$ dense in E, such that for every $\varphi(t)$ which is a solution of (7.1) with $\varphi^0 \subset Y$ (and $\mathbf{f} \equiv 0$) we have

$$\lim_{h \to 0, \Delta t \to 0} | \frac{C_h(\Delta t) R_h \varphi(t) - \varphi(t)}{\Delta t} - \mathcal{A}\varphi(t) |_E = 0. \qquad (7.12)$$

7.5 The Lax equivalence theorem

We can now state the Lax equivalence theorem (cf. Richtmyer and Morton (1967), Strang and Fix (1973) and Dautray and Lions (1993)).

Theorem 5. Suppose that the evolution problem (7.1) is well–posed for $t \in [0, t_0]$ and that it is approximated by the scheme (7.7), which we assume consistent. Then the scheme is convergent if and only if it is stable.

The proof of the Lax equivalence theorem can be found in Dautray and Lions (1993).

Remark. Let us consider the evolution problem (7.1) with

$$\mathbf{f}(t, \varphi) \neq 0 \qquad (7.13)$$

and $\varphi^0 = 0$, and also the corresponding approximation (7.7). We have

$$\varphi_h^{n+1} = \Delta t \sum_{j=1}^{n} [C_h(\Delta t)]^{n-j} \mathbf{f}_h^j. \tag{7.14}$$

If $\mathcal{A}$ is the ifinitesimal generator of a semigroup $\{I\!\!F(t)\}$ we can write

$$\varphi(t) = \int_0^t I\!\!F(t-s)\mathbf{f}(s)ds. \tag{7.15}$$

Under suitable hypotheses on the convergence of $\mathbf{f}_h^j$ to $\mathbf{f}(j\Delta t)$ we can show that expression (7.14) converges to (7.15) if the scheme is stable and consistent.

7.6 Regularization method

It may happen that for some constitutive models the nonhomogeneous evolution problem (7.1) is not well posed (e.g. for rate independent plastic model).

Then we can find a parameter $\delta \in [0, \delta^*]$ (δ^* is given) such that the new evolution problem

$$\left.\begin{array}{l}\dot{\varphi} = \mathcal{A}_\delta(t,\varphi)\varphi + \mathbf{f}_\delta(t,\varphi),\\ \varphi(0) = \varphi^0, \quad t \in [0, t_f],\\ \text{Boundary conditions are specified;}\end{array}\right\} \tag{7.16}$$

has the generalized solution

$$\varphi(t, \mathbf{x}) = I\!\!F_\delta(t)\varphi^0(\mathbf{x}) + \int_0^t I\!\!F_\delta(t-s)\mathbf{f}_\delta(s)ds \tag{7.17}$$

which satisfied the assertions as follows:

(i) The evolution operator $I\!\!F_\delta(t)$ is continuous in the topology on $\mathcal{D}(\mathcal{A}_\delta) \subset E$ assumed for each $t \in [0, t_f]$;

(ii) The function $\mathbf{f}_\delta(t)$ satisfied the conditions (cf. Eqs. (5.29) and (5.30))

$$\mathbf{f}_\delta(t,\varphi) \in E, \quad \|\mathbf{f}_\delta(t,\varphi)\|_E \le \lambda_f^1, \quad \|\mathbf{f}_\delta(t,\varphi') - \mathbf{f}_\delta(t,\varphi)\|_E \le \lambda_f^2 \|\varphi' - \varphi\|_E$$
$$\text{and} \quad t \to \mathbf{f}_\delta(t,\varphi) \in E \text{ is continuous;} \tag{7.18}$$

(iii)

$$I\!\!F_\delta(t)\mid_{\delta=0} = I\!\!F(t), \quad \mathbf{f}_\delta(t,\varphi(t))\mid_{\delta=0} = \mathbf{f}(t,\varphi(t)). \tag{7.19}$$

It means that the regularization problem (7.16) is well–posed. The parameter δ is called **the regularization parameter**.

For our case of an adiabatic rate dependent inelastic flow process (cf. Chapter 5) when $\delta = T_m$ the problem is already regularized.

7.7 Application to an adiabatic inelastic flow process

The evolution problem (7.1) describes an adiabatic inelastic flow process formulated in Section 3.1 provided

$$\varphi = \begin{bmatrix} \phi \\ v \\ \rho_M \\ \tau \\ \xi \\ \vartheta \end{bmatrix}, \quad \mathbf{f} = \begin{bmatrix} v \\ 0 \\ \frac{\rho_M}{1-\xi}\Xi \\ \left[\left(\frac{\chi^*}{\rho_M(1-\xi)c_p}\mathcal{L}^{th}\tau + \mathcal{L}^e + \mathbf{g}\tau + \tau\mathbf{g}\right):\mathbf{P}\right]\frac{1}{T_m}\langle(\frac{f}{\kappa}-1)^m\rangle - \frac{\chi^{**}\Xi}{\rho_M(1-\xi)c_p}\mathcal{L}^{th} \\ \Xi \\ \frac{\chi^*}{\rho_M(1-\xi)c_p}\tau:\mathbf{P}\frac{1}{T_m}\langle(\frac{f}{\kappa}-1)^m\rangle + \frac{\chi^{**}}{\rho_M(1-\xi)c_p}\Xi \end{bmatrix},$$

$$\mathcal{A} = \begin{bmatrix} 0 & 0 & 0 & 0 & 0 & 0 \\ 0 & 0 & \frac{1}{\rho_M^0(1-\xi_0)}\frac{\tau}{\rho_M}\text{grad} & \frac{\text{div}}{\rho_M^0(1-\xi_0)} & \frac{\tau\text{grad}}{\rho_M^0(1-\xi_0)(1-\xi)} & 0 \\ 0 & -\rho_M\text{div} & 0 & 0 & 0 & 0 \\ 0 & \mathbb{E}:\text{sym}\frac{\partial}{\partial \mathbf{x}} + 2\text{sym}(\tau:\frac{\partial}{\partial \mathbf{x}}) & 0 & 0 & 0 & 0 \\ 0 & 0 & 0 & 0 & 0 & 0 \\ 0 & \frac{\vartheta}{c_p\rho_{Ref}}\frac{\partial \tau}{\partial \vartheta}:\text{sym}\frac{\partial}{\partial \mathbf{x}} & 0 & 0 & 0 & 0 \end{bmatrix}.$$

(7.20)

It is noteworthy that the spatial operator $\mathcal{A}$ has the same form as in elastodynamics of damaged material while all dissipative effects generated by viscoplastic flow phenomena influence the process through the nonlinear function $\mathbf{f}$.

Let us consider first undamaged material (i.e. we assume $\Xi \equiv 0$). For this case the spatial operator $\mathcal{A}$ has strictly the form as in elastodynamics. Then, for the proof of the well–posedness of the homogeneous evolution problem (for $\mathbf{f} \equiv 0$) we can use the results obtained in elastodynamics. Next, we can extent the results to elasto-viscoplasticity by considering the nonhomogeneous evolution problem (when $\mathbf{f} \neq 0$) and by superposing suitable smoothness assertions for the nonlinear function $\mathbf{f}$ (cf. Ionescu and Sofonea (1993) nad Perzyna (1994, 1995)).

It is sufficient to assume for the nonlinear function $\mathbf{f}$ the assertions as have been postulated by (5.29) and (5.30). These conditions are satisfied provided we assume:

(i) The yield function $f = f(J_1, J_2, \xi, \vartheta)$ is smooth in the stress space and depends continuously on ξ and ϑ. The particular form of the yield function (6.60) satisfies this assertion if the material functions $n_1 = n_1(\vartheta)$ and $n_2 = n_2(\vartheta)$ are continuous.

(ii) The isotropic hardening material function $\kappa = \hat{\kappa}(\in^p, \xi, \vartheta)$ is a continuous function in all variables.

(iii) The overstress viscoplastic function $\Phi = \Phi(f - \kappa)$ is absolutely continuous.

(iv) The evolution function Ξ is absolutely continuous. This assertion is satisfied by the nucleation and growth terms (4.47) and (4.48) provided the material functions h^*, m^*, σ_N, g^* and σ_{eq} are sufficiently smooth.

This analysis clearly shows how the constitutive modelling of the material is directly related to the well–posedness of the evolution problem (i.e. the initial boundary value problem).

8. LOCALIZATION PHENOMENA

8.1 Particular constitutive assumptions

Let us introduce the plastic potential function for damaged material in two alternative forms

(i) cf. Shima and Oyane (1976) and Perzyna (1984, 1986)

$$f = \frac{J_2}{\kappa_0^2} + n\xi \left(\frac{J_1}{\kappa_0}\right)^2; \tag{8.1}$$

(ii) cf. Gurson (1975) and Tvergaard and Needleman (1986)

$$f = 3\frac{J_2}{\sigma_M^2} + 2q_1\xi \cosh\left(q_2 \frac{J_1}{\sigma_M}\right); \tag{8.2}$$

where κ_0 denotes the yield stress of the matrix material, $n = n(\vartheta)$ is the temperature dependent material function, σ_M denotes the effective tensile yield stress in the matrix material, q_1 and q_2 are constant parameters, $\sigma_M = \sigma_M(\epsilon_M^p, \vartheta)$, i.e. σ_M depends on the equivalent plastic strain of the matrix material ϵ_M^p and temperature ϑ and **g** denotes the metric tensor in $\mathcal{S}$.

The isotropic hardening–softening material function κ is assumed in two alternative forms

(i) cf. Perzyna (1984, 1986) and Nemes, Eftis and Randles (1990)

$$\kappa = \kappa_0^2 \{q + (1-q) \exp[-h(\vartheta)\, \epsilon^p]\}^2 \left[1 - \left(\frac{\xi}{\xi^F}\right)^{\frac{1}{2}}\right], \tag{8.3}$$

(ii) cf. Gurson (1975) and Tvergaard and Needleman (1986)

$$\kappa = 1 + q_3\xi^2; \tag{8.4}$$

where $q = \frac{\kappa_1}{\kappa_0}$, κ_0 and κ_1 denote the yield and saturation stress of the matrix material (both can be temperature dependent functions), respectively, $h = h(\vartheta)$ is the temperature dependent strain hardening function for the matrix material, $\epsilon^p = \int_0^t (\frac{2}{3}\mathbf{d}^p : \mathbf{d}^p)^{\frac{1}{2}} dt$ is the equivalent plastic deformation, ξ^F denotes the value of porosity at which the incipient fracture occurs and q_3 is a constant parameter. The overstress viscoplastic function Φ is postulated in the form (cf. Perzyna (1963, 1971))

$$\Phi(f - \kappa) = (f - \kappa)^m, \quad \text{where } m = 1, 3, 5, \ldots \quad (8.5)$$

To make possible numerical investigation of the three-dimensional dynamic adiabatic deformations of a body for different ranges of strain rate we introduce some simplifications of the constitutive model.

(i) By analogy with the infinitesimal theory of elasticity we postulate linear elastic properties of the material, cf. Eq. (6.56).

(ii) It is assumed that
$$\mathcal{L}^{e^{-1}} : \mathcal{L}^{th} = \theta \mathbf{g}, \quad (8.6)$$
where θ is the thermal expansion coefficient in the elastic range.

(iii) Kinematic hardening is neglected.

It is noteworthy that the influence of the evolution of microvoids on elastic properties of the material is not taken into account.

8.2 Fracture criterion

We base the fracture criterion on the evolution of the porosity internal state variable.

Let us assume that for $\xi = \xi^F$ catastrophe takes place (cf. Perzyna (1984)), that is

$$\kappa = \hat{\kappa}(\epsilon^p, \vartheta, \xi)|_{\xi=\xi^F} = 0. \quad (8.7)$$

It means that for $\xi = \xi^F$ the material looses its stress carrying capacity. The condition (8.7) describes the main feature observed experimentally that the load tends to zero at the fracture point. It is noteworthy that the isotropic hardening–softening material function $\hat{\kappa}$ proposed in particular form (8.3) satisfies the fracture criterion (8.7). To satisfy the fracture criterion (8.7) for the case (8.4) it suffices to assume $q_3 = -(\xi^F)^{-2}$.

8.3 Formulation of an adiabatic inelastic flow process

Let us define an adiabatic inelastic flow process as follows (cf. Perzyna (1994, 1995)). Find ϕ, $\boldsymbol{v}$, ρ_M, $\boldsymbol{\tau}$, ξ and ϑ as function of t and $\mathbf{x}$ such that

(i) the field equations

$$\dot{\phi} = \boldsymbol{v},$$

$$\dot{\boldsymbol{v}} = \frac{1}{\rho_M^0(1 - \xi_0)} \left(\frac{\boldsymbol{\tau}}{\rho_M} \mathrm{grad}\rho_M + \mathrm{div}\boldsymbol{\tau} - \frac{\boldsymbol{\tau}}{1 - \xi} \mathrm{grad}\xi \right),$$

Constitutive Modelling of Dissipative Solids

$$\dot{\rho}_M = \frac{\rho_M}{1-\xi}\Xi - \rho_M \text{div}\boldsymbol{v},$$

$$\dot{\boldsymbol{\tau}} = \left[\mathcal{L}^e - \frac{1}{c_p\rho_{Ref}}\vartheta\mathcal{L}^{th}\frac{\partial\boldsymbol{\tau}}{\partial\vartheta}\right]:\text{sym}D\boldsymbol{v} + 2\text{sym}\left(\boldsymbol{\tau}:\frac{\partial\boldsymbol{v}}{\partial\mathbf{x}}\right)$$

$$- \left[\left(\frac{\chi}{\rho_M(1-\xi)c_p}\mathcal{L}^{th}\boldsymbol{\tau} + \mathcal{L}^e + \mathbf{g}\boldsymbol{\tau} + \boldsymbol{\tau}\mathbf{g}\right):\mathbf{P}\right]\frac{1}{T_m}\langle(f-\kappa)^m\rangle$$

$$- \frac{\lambda\mathcal{L}^{th}}{\rho_M(1-\xi)c_p}\Xi, \qquad (8.8)$$

$$\dot{\xi} = \Xi,$$

$$\dot{\vartheta} = \frac{\vartheta}{c_p\rho_{Ref}}\frac{\partial\boldsymbol{\tau}}{\partial\vartheta}:\text{sym}D\boldsymbol{v} + \frac{\chi}{\rho_M(1-\xi)c_p}\boldsymbol{\tau}:\mathbf{P}\frac{1}{T_m}\langle(f-\kappa)^m\rangle$$

$$+ \frac{\lambda}{\rho_M(1-\xi)c_p}\Xi;$$

(ii) the boundary conditions

   (a) displacement ϕ is prescribed on a part ∂_ϕ of $\partial\phi(\mathcal{B})$ and tractions $(\boldsymbol{\tau}\cdot\mathbf{n})^a$ are prescribed on a part ∂_T of $\partial\phi(\mathcal{B})$, where $\overline{\partial_\phi\cap\partial_T} = 0$ and $\overline{\partial_\phi\cup\partial_T} = \partial\phi(\mathcal{B})$;

   (b) heat flux $\mathbf{q}\cdot\mathbf{n} = 0$ is prescribed on $\partial\phi(\mathcal{B})$;

(iii) the initial conditions

   $\phi, \boldsymbol{v}, \sigma_M, \vartheta, \xi$ and $\boldsymbol{\tau}$ are given at each particle $X \in \mathcal{B}$ at $t = 0$;

are satisfied.

In the field equations (8.8) ρ_M and ρ_M^0 denote the actual and reference mass density of the matrix material, respectively, ξ_0 is the initial porosity of a material and $D\boldsymbol{v}$ denotes the spatial velocity gradient.

8.4 Formulation of the initial–boundary value problems (evolution problems)

Let us consider an adiabatic dynamic process for thin steel tube. Cho, Chi and Duffy (1988) tested the specimens machined in the shape of thin-walled tubes with integral hexagonal flanges for gripping. Torsional loading at high strain rates was applied in a torsional Kolsky bar (split-Hopkinson bar).

We idealize the initial-boundary value problem (cf. Batra and Zhang (1994)) by assuming the specimen in the shape of thin-walled tube.

The initial conditions are taken in the form

$$\phi(\mathbf{x},0) = 0, \quad \boldsymbol{v}(\mathbf{x},0) = 0, \quad \rho(\mathbf{x},0) = \rho_{Ref} = \rho_M^0(1-\xi_0),$$
$$\boldsymbol{\tau}(\mathbf{x},0) = 0, \quad \xi(\mathbf{x},0) = \xi_0, \quad \vartheta(\mathbf{x},0) = \vartheta_0 = \text{constant in } \mathcal{B}. \qquad (8.9)$$

That is, the body is initial at rest, is stress free at a uniform temperature ϑ_0 and the initial porosity at every material point is ξ_0.

For the boundary conditions, we assume

$$\boldsymbol{\tau} \cdot \mathbf{n} = 0 \text{ on the inner and outer surfaces of the tube,}$$
$$\mathbf{q} \cdot \mathbf{n} = 0 \implies \text{grad}\vartheta \cdot \mathbf{n} = 0 \text{ on all bounding surfaces,}$$
$$\boldsymbol{v}(x_1, x_2, 0, t) = 0, \quad \boldsymbol{v}(x_1, x_2, L, t) = \omega^*(t)\left(x_1^2 + x_2^2\right)^{\frac{1}{2}} \mathbf{n}^*, \qquad (8.10)$$

where $\mathbf{n}$ is a unit outward normal to the respective surfaces, $\omega^*(t)$ is the angular speed of the end surface $x_3 = L$ of the tube, and $\mathbf{n}^*$ is a unit vector tangent to the surface $x_3 = L$. It is assumed that

$$\omega^*(t) = \begin{cases} \omega_0^* t/20, & 0 \leq t \leq 20\mu s, \\ \omega_0^*, & t > 20\mu s. \end{cases} \qquad (8.11)$$

The rise time of 20 μs is typical for torsional tests done in a split Hopkinson bar (cf. Batra and Zhang (1994)).

The following values for various material parameters are assumed (AISI 1018 cold rolled steel)

$$\rho_M = 7860 \text{ kg/m}^3, \quad c_p = 460 \text{ J/kg}^0\text{C}, \quad G = 80 \text{ GPa},$$
$$K = 210 \text{ GPa}, \quad \kappa_0 = 237 \text{ MPa}, \quad \kappa_1 = 1.2 \cdot \kappa_0, \quad \vartheta_0 = 20\ ^0\text{C},$$
$$h = const = 5.15, \quad n = const = 1.25, \quad \chi^* = 0.90,$$
$$m = 5 \text{ (for } \vartheta = 0\text{)}, \quad T_m = 2.5 \cdot 10^{-2} \text{ s (for } \vartheta = 0\text{)},$$
$$m = 4.7 \text{ (for } \vartheta = 80^0\text{C)}, \quad T_m = 1.0 \cdot 10^{-2} \text{ s (for } \vartheta = 80^0\text{C)}.$$

The tube has been twisted at nominal shear strain rates ranging $10^3 - 10^4$ s^{-1}.

For the particular example considered it has been assumed $\dot{\xi} = 0$ (no influence of the micro–damage mechanism) and $\omega_0^* = 253$ s^{-1}.

8.5 Numerical computation results

The aforestated initial–boundary value problem has been solved by using the wide spectrum of ABAQUS possibilities (cf. Łodygowski et al. (1994) and Łodygowski and Perzyna (1996)).

The half of the specimen (the thin walled cylinder, cf. Fig. 50) is modelled via both multilayer shell and three-dimensional brick elements with 4 layers in radial direction. In circumferential direction the model consists of 24 segments with 10 elements on the depth, cf. Fig. 51.

To avoid the reflection of waves and to model the influence of the rest of the specimen (cf. Fig. 50) it has been postulated that the additional spring and mass elements are taken into consideration. The evolution of the deformation of the mesh for the

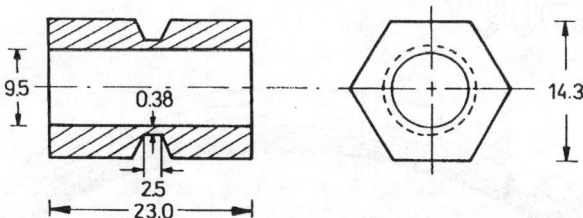

Figure 50: Details of specimen with hexagonal mounting flanges used in the torsional Kolsky bar experiment (all dimensions are in millimeters), cf. Hartley, Duffy and Hawley (1987)

segment is shown in Fig. 52. In Fig. 53 one can observe the evolution of the equivalent plastic deformation in different laminates and clear plastic strain localization effect around the mid cross section. The predicted width of this localized region is 0.2 − 0.3 mm, cf. Fig. 54. The evolution of temperature in different laminates is presented in Fig. 55 and for the segment in Fig. 56. The maximum strain is of the order of 200%, and the temperature rise is of the order of $150^0 C$.

8.6 Discussion of the results and comparison with experimental observations

To be sure that the numerical scheme is convergent the conditions (7.11) and (7.12) have been estimated numerically for each evolution problem separately.

The main result of these numerical simulations is the determination of a thin shear band region of finite width which undergoes significant deformations and temperature rise.

Comparison of the numerical results obtained with experimental observations can have only qualitative nature.

In the case of adiabatic dynamic process for thin steel tube a thin shear band region of finite width along the circumference of the tube is propagated. The predicted width of the determined shear band, the changes of temperature and deformation are in sufficiently good agreement with the experimental observations of Hartley, Duffy and Hawley (1987), Marchand and Duffy (1988), Marchand, Cho and Duffy (1988) and Cho, Chi and Duffy (1988).

It has been also found that the width of the shear band region and the temperature rise vary with the nominal strain rate as well as with the relaxation time assumed.

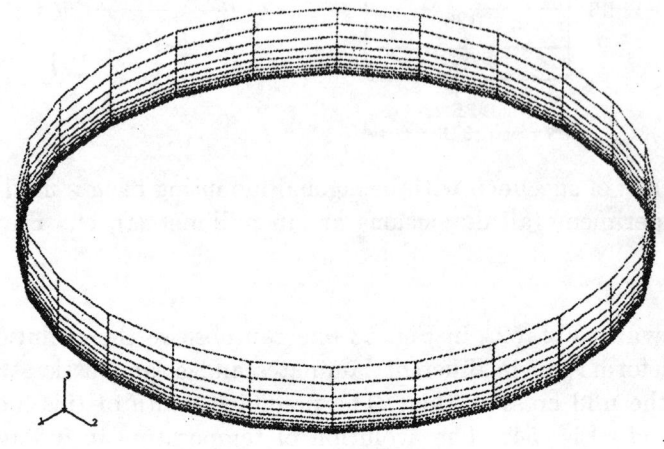

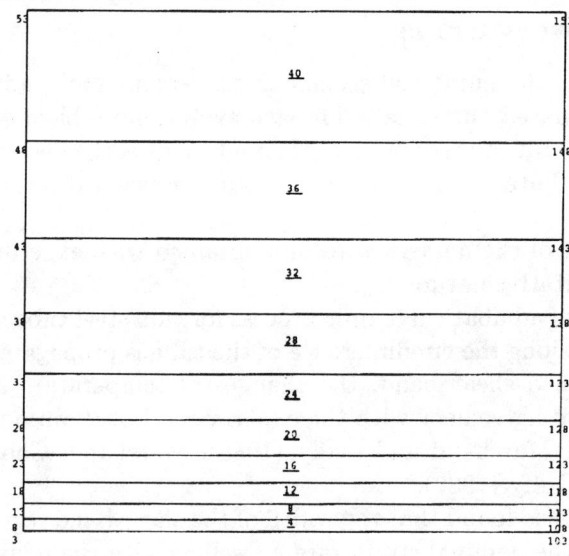

Figure 51: Layer shell divided into 24 segments with 10 elements on the depth

Constitutive Modelling of Dissipative Solids

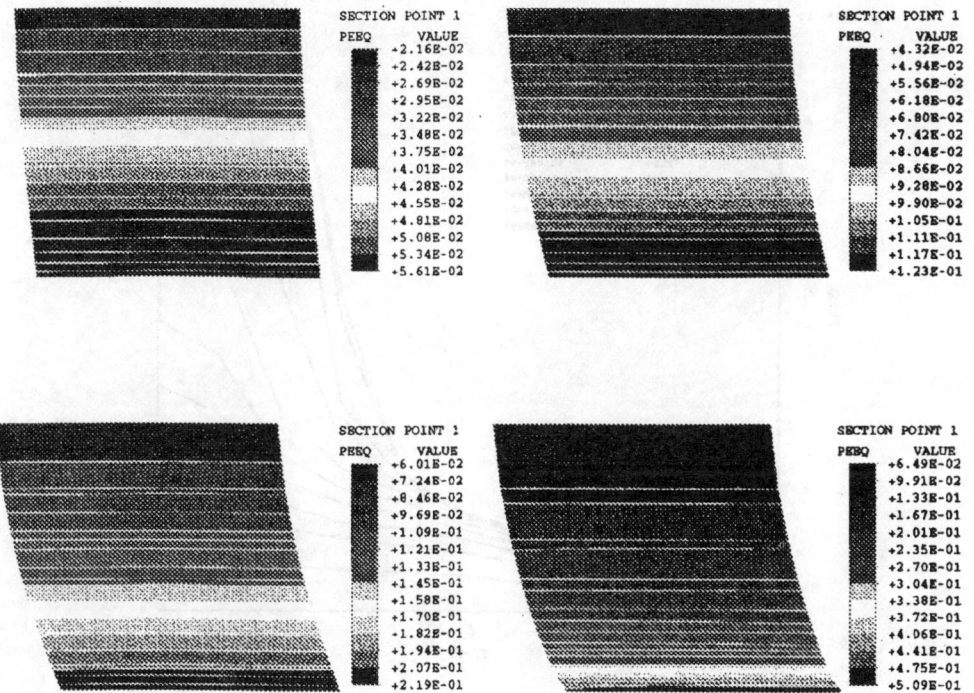

Figure 52: Evolution of the deformation of the mesh for the segment

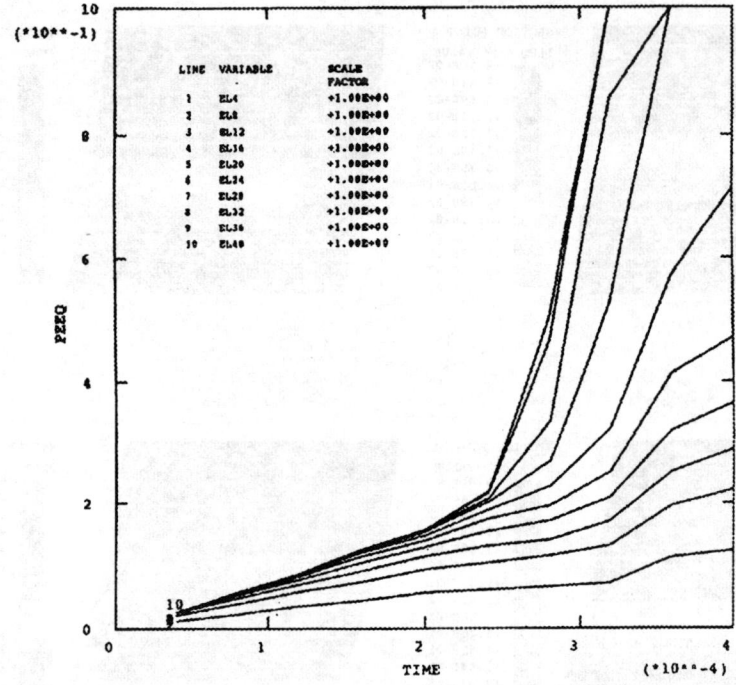

Figure 53: Evolution of the plastic equivalent strain in different laminates

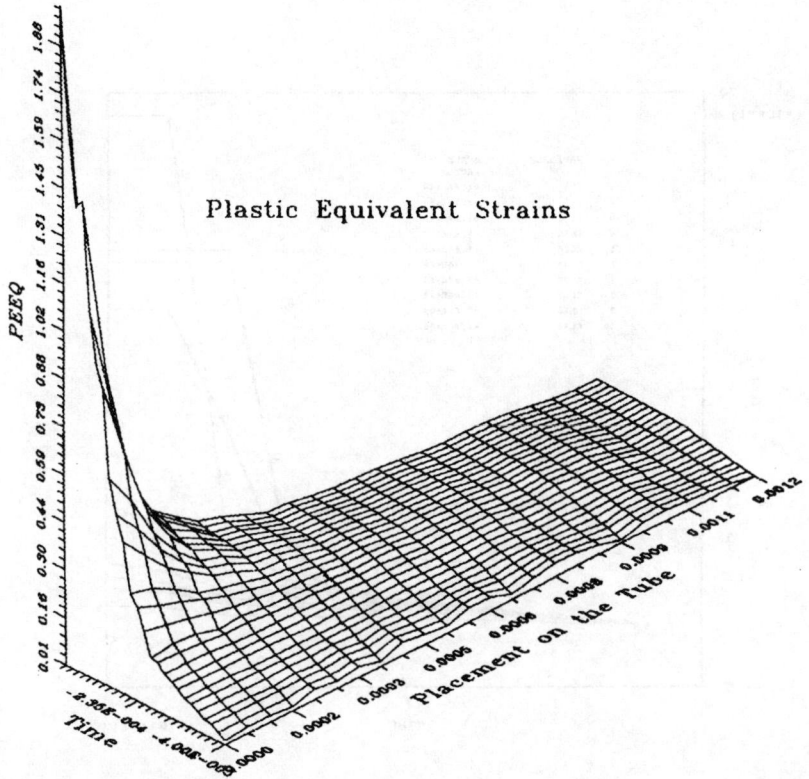

Figure 54: Plastic equivalent strain as function of time and placement on the tube

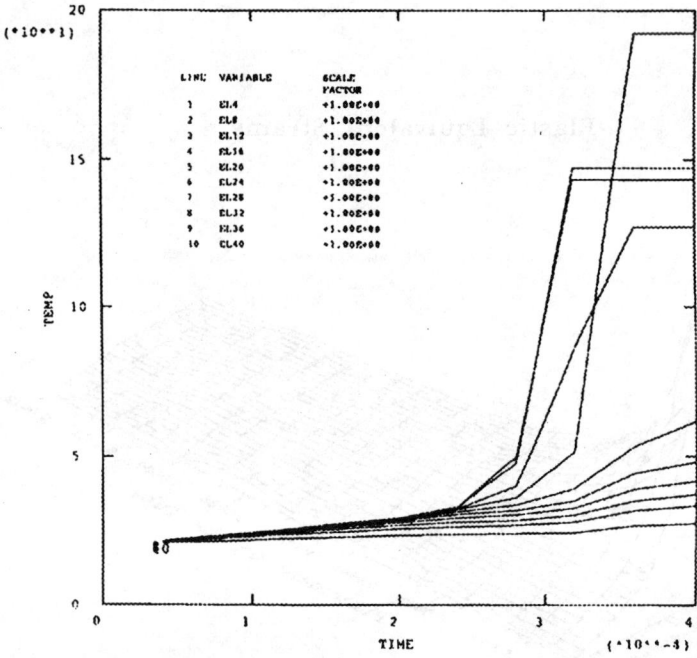

Figure 55: Evolution of the temperature rise in different laminates

Constitutive Modelling of Dissipative Solids

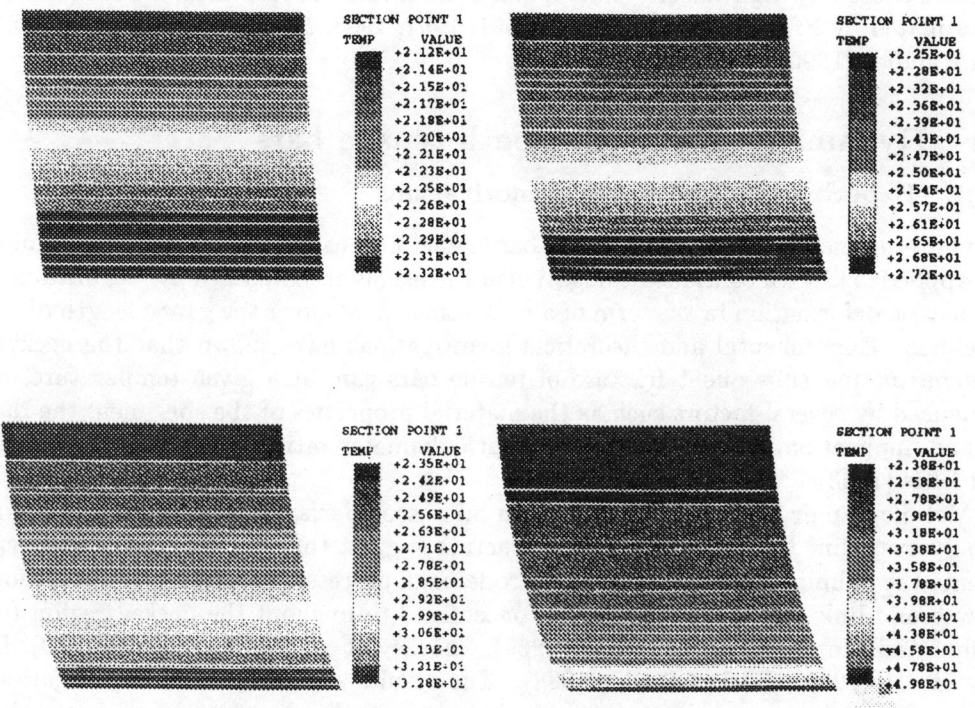

Figure 56: Evolution of temperature for the segment

9. FRACTURE PHENOMENA

To show application of the developed viscoplastic–damage type constitutive theory for high strain rate flow processes and ductile fracture we shall consider two examples, namely the problem of strain localization and fracture of dynamically loaded cylindrical tensile bars experiencing strain rates between $10^2 - 10^4$ sec^{-1} and the problem of spall fracture caused by high velocity plate impact. Both these problems have been recently investigated by Eftis and Nemes (1990, 1991, 1992, 1993, 1995), cf. also Nemes, Eftis and Randles (1990, 1991).

9.1 Dynamic fracture of smooth tensile bars

9.1.1 Experimental and physical motivations

The onset of instability of a cylindrical bar loaded in tension, that is, the reduction of the applied load with continued elongation of the bar, is accompanied by localization of the plastic deformation in the form of a neck somewhere along the gauge length of the specimen. Experimental and theoretical investigations have shown that the necking phenomena and subsequent fracture of tensile bars can, at a given temperature, be influenced by several factors such as the material properties of the specimen, the rate of load application and the geometry (length/diameter ratio) of the bar, cf. Nemes and Eftis (1993).

Metallographic, ultrasonic and electron microscopic examination of necked region of polycrystalline bars tensile loaded to fracture suggest that ductile type failure represents the culmination of a series of microdamage processes consisting of nucleation, growth and linkage of microvoids that concentrate throughout the necked region (cf. Bluhm and Morrissey (1965), Wray (1969), Shockey et al. (1980), Fisher (1980), Le Roy et al. (1981), Becker et al. (1988). The results shown by these investigations clearly establish the correlation of microdamage with localized strain concentration and the non–stable portion of the load–extension curve.

In a tensile test of a uniform bar at low strain rate (less than 10^{-2} sec^{-1}) inertial effects are negligible and prior to the onset of instability the stress may be shown to be essentially uniaxial (cf. Regazzoni, Johnson and Follansbee (1986), with homogeneous deformation throughout the length of the bar. However at the higher strain rates (10^3 sec^{-1} and above) inertial effects are important because of the propagation of stress waves that cause heterogeneity in the distribution of stress and deformation at the initial stages of the test. High strain rate tensile bar tests for several structural steels, aluminum alloy, tantalum and OHFC copper at room temperature have been performed by Regazzoni and Montheillet (1984) using a high velocity tensile test machine, and by Rajendren and Bless (1985) using a split Hopkinson bar, producing tensile strain rates of order $10^3 - 10^4$ sec^{-1}. These tests have shown that the strains at which the tensile necking instability initiates, and at which fracture takes place, can be affected by the strain rate imposed and by the nature of the crystal lattice structure of the material. At about the same time analytical/numerical studies of high strain rate behaviour of shallow notched bar tensile specimen using different viscoplastic constitutive models have been carried out for several materials, including copper, by Rajendren and Bless (1985) and by Regazzoni, Johnson and Follansbee (1986).

The conventional tensile at low to moderate strain rates the necking localization that occurs arises principally because of heterogeneity of deformation caused by geometric transitions, by the grip constrains at the ends of the bar, and possibly by a local material inhomogeneity. In the dynamic tensile test the formation of a neck does not necessarily require the presence of a geometrical transition or a material defect. The heterogeneity of the propagation stress and deformation fields immediately following application of the high strain rate loading can serve as the initiator, or 'defect', in an otherwise homogeneous test specimen.

Modelling the actual tensile bar geometry, localization is allowed to occur without recourse to artificial geometric or material imperfections to predetermine the localization of the necking. Under dynamic conditions the propagating stress and deformation establish heterogeneous distributions of stress and strain along the gage section of the specimen, causing variations of the position of localization as the imposed strain rate is changed.

In the constitutive–damage model employed for this study the possible effects of temperature change are not included. In this regard it is noted that in the dynamic tensile test study conducted by Regazzoni, Johnson and Follansbee (1986), their calculations found that the temperature changes induced by adiabatic deformation were not significant because of the low temperature dependence of the flow stress for copper. It is recognized, however, that for other materials the temperature effects could be important in leading to decreased ductility.

9.1.2 Geometry of the bar and simulation of the dynamic loading

The geometry of the cylindrical tensile bar, presumed to be made of OFHC copper, is shown in Fig. 57. Because of the circular symmetry extending the length of the bar only one half need be modelled. Te design of the finite element discretization is also shown, containing approximately 100 elements. More refined meshes containing 4000 elements were used in this study.

Dynamic loading conditions were simulated by applying specified velocities to the upper nodal points along the outer diameter of the bar using a ramp history in which the applied velocity $U = U_0(t/t_0)$ for $t < t_0$, $U = U_0$ for $t \geq t_0$, with a rise time of $t_0 = 50\mu s$ as shown in Fig. 58. The nodal points at the outer diameter of the lower end of the bar are held to zero displacements (fixed end). Simulations were conducted using values of U_0 equal to 10, 25, 50 and 100 m/s, which result in nominal strain rates over the gage length of 500, 1250, 2500 and 5000 s^{-1}, respectively. The range of imposed strain rates coincides with the experimentally observed range at which

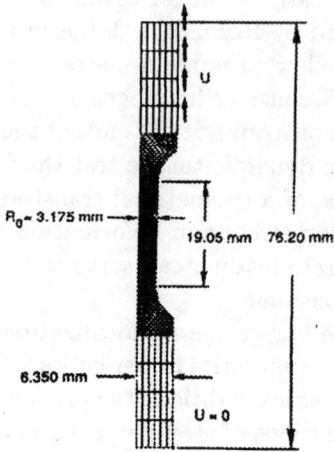

Figure 57: Geometry and finite element discretization of the tensile bar

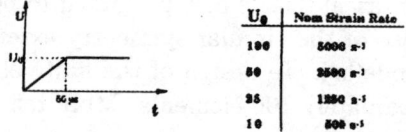

Figure 58: Imposed velocities and nominal strain rates

pure copper begins to show marked increase in strain rate sensitivity, and also the range of strain rates over which the strain required for onset of necking instability, and for fracture, begin to show appreciable change, Regazzoni and Montheillet (1984), Follansbee and Kocks (1988). It was also observed that at strain rates at 5×10^2 s^{-1} and below the calculated load–displacement curves no longer show signs of inertial effects (wave propagation).

9.1.3 Material parameter values for OFHC copper

For the strain rates considered in the dynamic tensile test simulations it was found that the tensile mean stresses that develop in the neck region of the bar are not sufficient to overcome the void nucleation threshold stress $\sigma_N = 500$ MPa that was used for the plate impact spall fracture studies of copper. This suggests that at the mean stress levels associated with the imposed nominal strain rates for the dynamic tensile bar test, the micromechanisms leading to void nucleation can perhaps be better modeled in terms of strain rather than by mean tensile stress. Thus in the results described below void nucleation does not contribute to the increase in void volume fraction as we choose, initially at least, to consider only void growth from the average initial void volume fraction ξ_0. Of course in so doing we underestimate somewhat the rate of damage evolution. However this should not compromise the main result to be shown by this study, namely, the influence of the heterogeneity of the initial dynamic stress and strain in determining where necking instability initiates along the bar.

Values for the material parameters $G, K, \xi_0, \xi_F, n, n_1, \alpha$, initial density ρ_R, and the material function $g(\xi) = e^{\alpha \xi}$ for OFHC copper are taken from previous spall fracture studies. The remaining parameters $m_1, m, \gamma_0, \kappa_0, I_2^s, q, \beta$, have been determined by optimal curve fit of compression test data for pure copper for strain rates ranging from $10^{-4} - 10^4$ s^{-1} as reported by Follansbee and Kocks (1988).

The viscosity of polycrystalline metals subject to high strain rates at room temperature usually ranges between $10^3 - 10^5$ P, falling to as low as 10^{-2} P at melting temperature Carroll, Kim and Nesterenko (1986). In the numerical simulation of plate impact spallation tests for copper at room temperature, the effects of temperature change induced by inelastic deformation was not considered and the microviscosity parameter value $\eta = 10$ P, suggested by Johnson (1981) was assumed, giving good correlation between predicted and experimental spall fracture data. (cf. Eftis et al. (1991a)). The low value for η is attributed to highly localized heating effects that develop at and around the plasticity expanding microvoid walls, creating local 'hot spots' that can rise to a substantial fraction of the melting temperature.

For the dynamic tensile bar test under consideration the maximum mean is about an order of magnitude smaller, while the nominal strain rates are approximately two orders magnitude smaller, than the mean stress and strain rates that are encountered in high velocity plate impact tests. Accordingly for the tensile bar test the microviscosity parameter is assumed to have the higher value $\eta = 120$ P. Illustrations of the effects

Table 1: Material parameters for OFHC copper

G	=	4.48×10^4	γ_0	=	2.18×10^5 s^{-2}
K	=	14.00×10^4 MPa	I_s^2	=	1.21×10^{-4} s^{-2}
q	=	202 MPa	κ_0	=	23.1 MPa
β	=	5.34	n	=	0.25
ρ_R	=	8.93 gm/cm^3	m	=	0.167
ξ_0	=	3×10^{-4}	ξ_F	=	0.32
η	=	120 Poise	n_1	=	1.77
g	=	$e^{\alpha\xi}$	α	=	20
m_1	=	2			

that varying values of η can have on the material stress–strain behaviour and on the evolution of the material damage can be found in Eftis and Nemes (1991a,b) and Nemes and Eftis (1991). A summary of the material parameter values for OFHC copper used in the numerical calculations discussed below are provided in Table 1.

9.1.4 Boundary–initial value problem

The problem to be solved seeks to determine the stress and deformation of a cylindrical bar subject to prescribed velocities applied to one end of the bar while the other end is stationary. Referring to cylindrical coordinates, any particle of the body which in the reference configuration has position $\mathbf{X} = (X^1, X^2, X^3) = (R, \Theta, Z)$, under the induced motion $\mathbf{x} = \chi(\mathbf{X}, t)$ has the spatial location $\mathbf{x} = (x^1, x^2, x^3) = (r, \theta, z)$ in the current configuration. Since the imposed conditions are axisymmetric all components of the field variables will be independent of the angle variable θ, and correspondingly the stress and rate of deformation tensor will have the non–zero components σ_{11}, σ_{22}, σ_{33}, $\sigma_{13} = \sigma_{31}$, d_{11}, d_{22}, d_{33}, $d_{13} = d_{31}$.

A set of the field equations are as follows

$$\frac{\rho_{Ref}}{\rho} = \det \mathbf{F} = \left[\frac{\partial \mathbf{x}}{\partial \mathbf{X}}\right],$$

$$\frac{\partial \sigma_{rr}}{\partial r} + \frac{\partial \sigma_{rz}}{\partial z} + \frac{1}{r}(\sigma_{rr} - \sigma_{\theta\theta}) = \rho \frac{dv_r}{dt} = \rho \left(\frac{\partial v_r}{\partial t} + v_r \frac{\partial v_r}{\partial r} + v_z \frac{\partial v_r}{\partial z}\right),$$

$$\frac{\partial \sigma_{rz}}{\partial r} + \frac{\partial \sigma_{zz}}{\partial z} + \frac{1}{r}\sigma_{rz} = \rho \frac{dv_z}{dt} = \rho \left(\frac{\partial v_z}{\partial t} + v_r \frac{\partial v_z}{\partial r} + v_z \frac{\partial v_z}{\partial z}\right),$$

$$\frac{1}{2G}\left[\stackrel{\nabla}{\sigma}_{rr} - \frac{\bar{\nu}}{1+\bar{\nu}}\left(\stackrel{\nabla}{\sigma}_{rr} + \stackrel{\nabla}{\sigma}_{\theta\theta} + \stackrel{\nabla}{\sigma}_{zz}\right)\right] = d_{rr} - \frac{\gamma_0}{\phi}\langle\Phi(\hat{F})\rangle\frac{1}{\kappa_0}\left(2n\xi J_1 + \sigma'_{rr}\right), \quad (9.1)$$

$$\frac{1}{2G}\left[\stackrel{\nabla}{\sigma}_{\theta\theta} - \frac{\bar{\nu}}{1+\bar{\nu}}\left(\stackrel{\nabla}{\sigma}_{rr} + \stackrel{\nabla}{\sigma}_{\theta\theta} + \stackrel{\nabla}{\sigma}_{zz}\right)\right] = d_{\theta\theta} - \frac{\gamma_0}{\phi}\langle\Phi(\hat{F})\rangle\frac{1}{\kappa_0}\left(2n\xi J_1 + \sigma'_{\theta\theta}\right),$$

$$\frac{1}{2G}\left[\stackrel{\nabla}{\sigma}_{zz} - \frac{\bar{\nu}}{1+\bar{\nu}}\left(\stackrel{\nabla}{\sigma}_{rr} + \stackrel{\nabla}{\sigma}_{\theta\theta} + \stackrel{\nabla}{\sigma}_{zz}\right)\right] = d_{zz} - \frac{\gamma_0}{\phi}\langle\Phi(\hat{F})\rangle\frac{1}{\kappa_0}\left(2n\xi J_1 + \sigma'_{zz}\right),$$

$$\frac{1}{2G}\overset{\nabla}{\sigma}_{rz} = d_{rz} - \frac{\gamma_0}{\phi}\langle\Phi(\hat{F})\rangle\frac{1}{\kappa_0}\sigma'_{rz},$$

$$\frac{\partial\xi}{\partial t} + v_r\frac{\partial\xi}{\partial r} + v_z\frac{\partial\xi}{\partial z} = \frac{h(\xi)}{1-\xi}\left[\exp\left(\frac{m_2(\sigma-\sigma_N)}{k\theta}\right)-1\right] + \frac{1}{\eta}g(\xi)F(\xi,\xi_0)(\sigma-\sigma_G),$$

where

$$\Phi(\hat{F}) = \left[\frac{J'_2 + n\xi J_1^2}{[q+(\kappa_0-q)e^{-\beta\epsilon^p}]^2[1-n_1\xi^{1/2}]^2} - 1\right]^{m_1},$$

$$\phi = \left(\frac{I_2}{I_2^s} - 1\right)^m, \tag{9.2}$$

and

$$[d] = \begin{bmatrix} \frac{\partial v_r}{\partial r} & 0 & \frac{1}{2}\left(\frac{\partial v_r}{\partial z} + \frac{\partial v_z}{\partial r}\right) \\ 0 & \frac{v_r}{r} & 0 \\ \frac{1}{2}\left(\frac{\partial v_r}{\partial z} + \frac{\partial v_z}{\partial r}\right) & 0 & \frac{\partial v_z}{\partial z} \end{bmatrix}. \tag{9.3}$$

The invariants J_1, J'_2 and I_2 have the explicit forms

$$J_1 = \sigma_{rr} + \sigma_{\theta\theta} + \sigma_{zz}, \tag{9.4}$$

$$J'_2 = \frac{1}{3}\left(\sigma_{rr}^2 + \sigma_{\theta\theta}^2 + \sigma_{zz}^2\right) + \sigma_{rz}^2 - \frac{4}{9}\left(\sigma_{rr}\sigma_{\theta\theta} + \sigma_{rr}\sigma_{zz} + \sigma_{\theta\theta}\sigma_{zz}\right),$$

$$I_2 = \left|\left\{-\left[\frac{1}{3}\left(d_{rr}^2 + d_{\theta\theta}^2 + d_{zz}^2\right) - \frac{4}{9}\left(d_{rr}d_{\theta\theta} + d_{rr}d_{zz} + d_{\theta\theta}d_{zz}\right) + d_{rz}^2\right]\right\}^{\frac{1}{2}}\right|.$$

Because of the presence of microvoids the elastic behaviour is modified through degradation of the elastic material response, cf. Perzyna (1986). This is expressed in the manner developed by Mac Kenzie (1950) whereby the elastic shear and bulk moduli, $\overline{G}$ and $\overline{K}$ respectively, are dependent upon porosity ξ in the following manner:

$$\overline{G} = G(1-\xi)\left(1 - \frac{6K+12G}{9K+8G}\xi\right),$$

$$\overline{K} = \frac{4GK(1-\xi)}{4G+3K\xi}, \tag{9.5}$$

and the Poisson ratio

$$\overline{\nu} = \frac{1}{2}\frac{3\overline{K} - 2\overline{G}}{3\overline{K} + \overline{G}}. \tag{9.6}$$

Referring to Fig. 57 we designate the initial length of the bar as $L_0 = 76.20$ mm. The initial gage length is $L_g = 19.05$ mm. The initial radius of the bar is R_0 and has the values $R_0 = 6.35$ mm at the end sections and $R_0 = 3.175$ along the gage section. The upper and lower section are taken to be gripped over a fixed length $L_f = 19.85$ mm along the outer radius. The transition lengths are designed by L_t. The initial values for the problem require that at $t = 0$ for $r \in [0, R_0]$ and $z \in [0, L_0]$

$$\mathbf{v}(r,z,0) = 0, \quad \sigma(r,z,0) = 0, \quad \xi(r,z,0) = \xi_0, \tag{9.7}$$

which assume an initial average porosity that is uniform throughout the bar.

The rate type boundary conditions applied to be the axisymmetric body are as follows. The circumferential surface of the gage and transition section of the bar and the bar ends are traction–free during the entire load–deformation process. Thus with $\mathbf{t} = \boldsymbol{\sigma} \cdot \mathbf{n}$, where $\mathbf{t}$ is the surface stress vector and $\mathbf{n}$ the unit normal vector, for all $t \geq 0$ and all points $(r, z, t) = \chi(R_0, z \in [0, L_0], t)$, $(r, z, t) = \chi(R \in [0, R_0], 0, t)$ and $(r, z, t) = \chi(R \in [0, R_0], L_0, t)$, we have

$$\dot{t}_r = \dot{t}_\theta = \dot{t}_z = 0. \tag{9.8}$$

The upper and lower end sections of the bar have the respective velocities

$$v_z(r, z, t) = U = U_0(t/t_0),\ 0 \leq t \leq t_0$$
$$v_z(r, z, t) = 0,\ t > t_0 \qquad (r, z, t) = \chi(R_0, z \in [L_0-(L_f+2L_t+L_g), L_0], t), \tag{9.9}$$

and

$$v_z(r, z, t) = 0, \quad t \geq 0 \quad (r, z, t) = \chi(R_0, z \in [0, L_f], t). \tag{9.10}$$

Along the longitudinal axis $r = 0$ symmetry conditions require that

$$\begin{aligned} v_z(0, z, t) &= 0 \\ t_z(0, z, t) &= 0 \end{aligned} \quad t \geq 0, \quad (r, z, t) = \chi(0, z \in [0, L_0], t). \tag{9.11}$$

The numerical calculations simulating dynamic deformation and fracture of the tensile bar requires determination of four components of stress, two components od velocity, the current mass density and the current void volume fraction. The eight unknowns are determined from the set of eight equations (9.1) in conjunction with the set of initial conditions (9.7) and boundary conditions (9.8)–(9.11).

9.1.5 Numerical results

(i) Load–rate effect on location of necking

The conditions pertaining to the stress and deformation of a tensile bar at high strain rate differs considerably from that at low strain rates. For the latter situation the inertial effects of particle acceleration and wave propagation are negligible. This leads to the assumption of quasi–static equilibrium according to which the longitudinal force in the bar is constant at each cross–section, with corresponding uniform uniaxial stress along the gage length. These conditions do not hold at high strain rates. For example for the tensile bar shown in Fig. 57 with length 76.2 mm, in a test at nominal strain rate of 5000 s^{-1} the elastic wave initiated at the mobile end reaches the fixed end approximately 21 μs later, assuming a longitudinal wave speed of 3.67 mm/μs. Meanwhile the test bar has undergone an overall elongation of about five percent, with

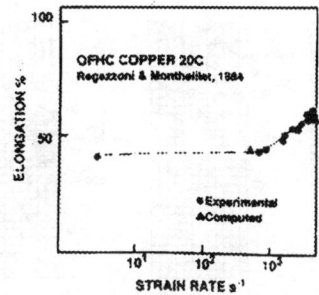

Figure 59: Computed and measured elongation at tensile instability. After Nemes and Eftis (1993)

the mobile end of the bar accommodating most of the deformation. At the initial state of the test the upper reaches of the bar can be deforming elasto–viscoplastically while the lower end is still undeformed. Wave reflections from the cylindrical surface further complicate the stress–deformation pattern along the length of the specimen. Thus during the early stages of a dynamic tensile test the stress and deformation are not uniformly distributed, and this kind of heterogeneity can lead to strain localization in the absence of geometrical or material irregularities.

The position of strain localization along the gage section is strongly influenced by the loading rate, indicating thereby the effect of the heterogeneous deformation field established by the dynamic loading. At a nominal strain rate of 5000 s^{-1} localization occurs 7.04 mm above the specimen centerline, whereas for the imposed strain rate 2500 s^{-1} the necking appears at 5.38 mm below centerline. At 1250 s^{-1} strain rate two neck form symmetrically at 3.31 mm above and below the centerline. Only at the lowest strain rate considered, at 500 s^{-1}, did localization take place at the center of the gage section indicating, perhaps, negligible dynamic effects.

(ii) Effect of strain–rate on the strain at tensile instability

The elongation ($\Delta l/l$) at initiation of instability is determined by considering the change of length (Δl) of the gage section length (l) at the point of maximum load. The computed values for the elongation corresponding to the strain rates at 5000 s^{-1} and 500 s^{-1} are shown in Fig. 59, and compare well with the experimental data obtained by Regazzoni and Montheillet (1984). For OFHC material (fcc crystal lattice structure) the strain at which necking instability begins increases as the strain rate increases, with the effect becoming pronounced at the higher rates.

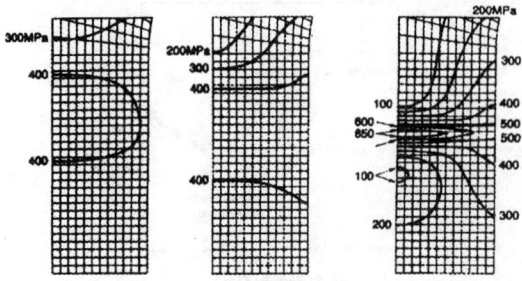

Figure 60: Contours for the axial stress component σ_{zz} at 150 μs, 200 μs and 250 μs after load initiation. After Nemes and Eftis (1993)

(iii) Localized intensification of stress and plastic strain without material damage

Figures 60 and 61 show contours for the stress and equivalent plastic strain that are drawn at 150 μs, 200 μs and 250 μs after load initiation. The imposed nominal strain rate is 5000 s^{-1} and the location of the necking instability is 7.04 mm above the gage length center, which is the region shown in the figures. Figure 60 illustrates the distribution of the axial stress component, σ_{zz} drawn over the undeformed configuration of the bar, where a number of the horizontal mesh elements of the finite element mesh have not been drawn in order to enhance clarity. During the 100 μs interval they show the stress increasing at the center of the bar at first, moving to the circumferential edge before localizing to a narrow band of intense stress. Contour lines for equivalent plastic strain are shown in Fig. 61, plotted on the deformed configuration of the specimen where the necking occurs. The pattern of the plastic strain increase is generally similar to that for stress, being most intense initially at the center before radiating across to the circumferential edge, with the severest strain located across the smallest cross section of the necked region.

These plots also give some indication of the temporal element associated with the spatial localization, illustrating the dramatic change that takes place in the dynamically loaded bar during the 50 μs time interval from 200 μs to 250 μs after start of the loading process.

(iv) Void growth and fracture

For the calculations that produced Figs. 60 and 61 the void volume fracture within the material was maintained at the initial average value in order to focus on the transient

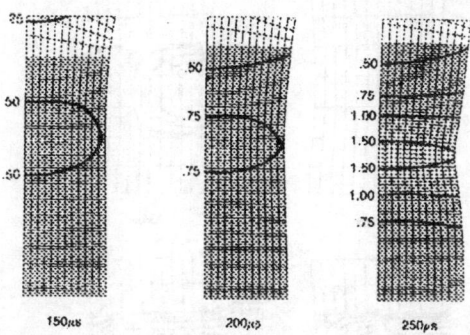

Figure 61: Contours for the equivalent stress $\in^p$ at 150 µs, 200 µs and 250 µs after load initiation. After Nemes and Eftis (1993)

aspect of the localization. However as a outgrowth of the heterogeneity caused by the dynamic nature of the tensile loading, within the region of induced necking instability triaxial tensile stress are developed that exceed the void growth threshold mean stress and initiate the void growth process. With the tensile bar deforming at 5×10^3 s^{-1} nominal strain rate where the necking instability is located 7.04 mm above the center of the bar, Fig. 62 show contours of the void volume fraction for the necked region at four instants of time. Let us note that the finite element mesh for the bar that was used in the calculations producing Fig. 62 numbered approximately 1000 elements. In these numerical computations no provision is made for separation of the material in the sense that the constitutive model issues no instructions for the servering of the mesh lines that cross the potential plane of fracture. Numerically the material continues to be modeled as a continuous body in which extensive material damage takes place across the plane of minimum cross-section, appearing as excessive elongation of the mesh elements.

The predicted fracture surface can be modelled however in some detail by employing a finer mesh (4000 elements) for the finite element grid covering the bar, and by making use of the element deletion capability in the PRONTO code. Upon reaching a prescribed void volume fraction the stresses in the finite element are set to zero over several time steps and the element is deleted from further calculations. Attention is drawn to the fact that as the critical value for the microvoid volume fraction is reached, i.e., as $\xi \to \xi_F$, it is clear that the rate of inelastic deformation $\mathbf{D}^p \to \infty$. In the numerical simulations this condition can be approached but never actually reached. For this reason the critical value for the void volume fraction that is used in the finite element deletion technique for calculation of the fracture surface is chosen at 0.25, somewhat below the experimentally determined value of $\xi_F = 0.32$ for OFHC

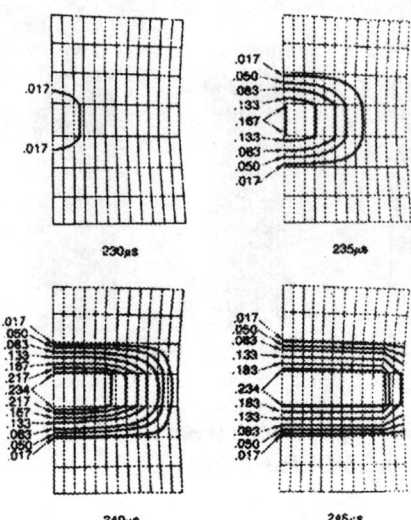

Figure 62: Contours for tvoid volume fraction at 230 μs, 235 μs, 240 μs and 245 μs after load initiation. After Nemes and Eftis (1993)

copper. The result is shown by Fig. 63. The fracture surface profile now exhibits a non–planar feature near the circumferential edge that might be interpreted as a 'shear lip', common to ductile tensile fracture of round bars. A more detailed resolution of the features of the fracture would require a much finer finite element mesh to cover the region of interest than was used here.

The elongation of the gage section at fracture is computed and shown along with the experimental values measured by Regazzoni and Montheillet (1981) in Fig. 64. The data shows that for the high strain rate range copper experiences a strong increase of ductility as the strain rate in increased. As with the elongation at tensile instability, the influence of strain rate on the elongation at fracture can be quite different for different materials. Kawata, Hashimoto and Kurokawa (1978) observe a systematic behaviour related to the lattice structure of the material. For fcc solids the elongation at fracture increases as the strain rate is increased from low to high values, that is, from 10^{-4} s^{-1} to 10^3 s^{-1}. For bcc solids over the same range of strain rate this trend is reversed. The predicted values for the elongation at fracture for the fcc copper material is seen to be considerably larger than the experimental values. They do however shown an increasing trend as the strain rate increased from 5×10^2 s^{-1} to 5×10^3 s^{-1}, in accord with this feature of the experimental data. The excessive values for the predicted elongations may possibly be due to omission of the void nucleation contribution to the void growth rate, thereby underestimating the extent of material damage and overestimating the material ductility.

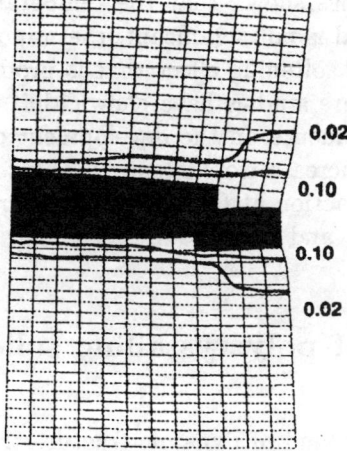

Figure 63: Deleted elements and void volume fraction conturs at fracture. After Nemes and Eftis (1993)

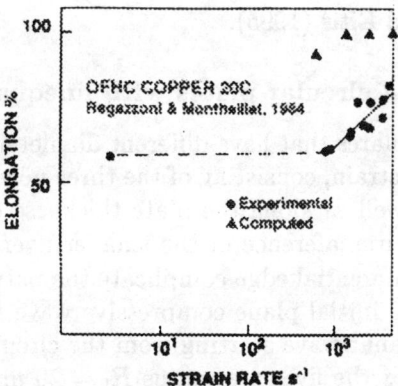

Figure 64: Elongation at fracture. After Nemes and Eftis (1993)

Another important physical aspect not considered is the adiabatic thermal softening associated with high strain rate inelastic deformation. Although a previous dynamic tensile test study has shown that the temperature changes caused by deformation for copper material is not significant, the importance of induced temperature change and the thermal softening effect that is produced cannot, in general, be overlooked. Our spall fracture studies have shown that the material microviscosity parameter η has a strong influence on the development of the void growth damage. Any material temperature increase would tend to reduce the value of the viscosity, which for the void volume fraction at fracture would correspondingly tend to reduce the level of associated strain, and thus the ductility at fracture (cf. Eftis and Nemes (1991b)).

9.2 Spall fracture of polycrystalline solids induced by plate impact

Spall fracture is a particular kind of dynamic fracture that results from the tension produced by the interaction of propagating waves of rarefaction. In the case of plate impact, the initial compressive stress wave travelling across the target reflects back at the free surface as a propagating decompression wave. The superposition of the reflected decompression wave fronts, if sufficient intensity and time duration, can caused partial or complete separation of the material along a plane perpendicular to the direction of the traveling wave fronts. The plate impact spall experimental results have been discussed in Chapter 2, Sections 2.7 and 2.8.

Numerical simulation of plate impact spall fracture has been presented by Eftis and Nemes in series of papers, cf. Eftis et al. (1991), Nemes and Eftis (1991, 1992), Eftis and Nemes (1991) and Eftis (1995).

9.2.1 Normal impact of circular plates with unequal diameter

Normal impact of circular plates that have different diameters generates multi-dimensional axisymmetric stress/strain, consisting of the three normal components that vary in the radial direction, as well as along the plate thickness direction. Because of the edge effect caused by the circumference of the smaller flyer plate, non-planar waves originating from the circumferential edge complicate the pattern of wave deformation.

Figure 65 illustrates the initial plane compressive wave fronts propagating across each plate, and the non-planar wave starting from the circular edge of the flyer.

The plate dimensions for the flyer are radius $R_1 = 20$ mm, thickness $L_1 = 3$ mm. For the target, radius $R_2 = 40$ mm and thickness $L_2 = 6$ mm, cf. Fig. 66. Because of the smaller diameter of the flyer, spall initiates only over the central region of the target. When the impact velocity of the flyer is sufficiently large the induced tensile mean stress in the target reach intensities high enough to cause spall fracture. The post-spall behaviour of the partially fragmented target was modelled by carrying the

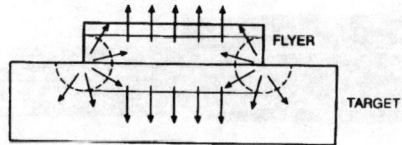

Figure 65: Impact of a circular flyer with a larger diameter target

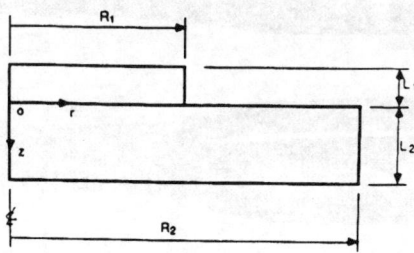

Figure 66: Flyer and target plate dimensions

numerical calculations to a post–impact time of 50 μs. The calculations required use of the transient finite–element computer code PRONTO, and were carried out on a Cray X–MP computer.

9.2.2 Numerical results

The effect of increasing the flyer impact velocity in three steps from 100 m/s to 350 m/s, are shown graphically by Fig. 67(a)–(c), which are the plate configurations 5 μs after impact (the computed deformations have been magnified by a factor of three). From these test simulations it was possible to estimate the damage threshold impact velocity at approximately 50 m/s for OFHC copper. This estimate is in approximate agreement with experimental data. The corresponding contours of the void volume fraction across the target plate thickness are shown by Fig. 68(a)–(c). Both sets of figures illustrate the pronounced effect of increasing the flyer impact velocity.

With the flyer impact at 350 m/s, several numerical simulations were performed with the parameter η assigned values of 10 P, 120 P and 500 P. The corresponding plate configurations and the void volume fraction contours at 5 μs post–impact are shown by Figs. 69 and 70, respectively, illustrating the controlling influence of this material parameter on the development of microvoid growth and material damage. Since $\eta = T_m \kappa_0$ the result obtained can be interpreted as the examination of the influence of the relaxation time T_m on the constitutive modelling. As the viscosity η is increased the extent of damage that can develop within the material after a 5 μs

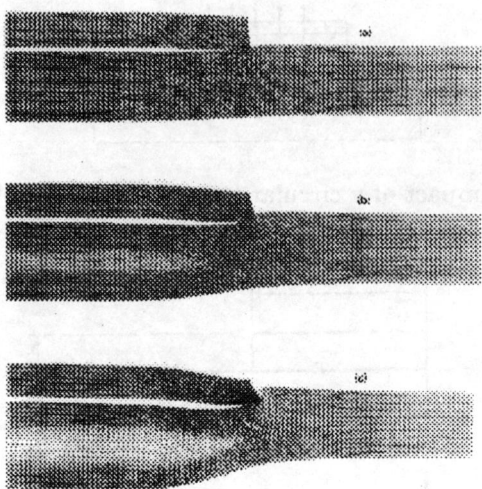

Figure 67: Deformed geometry at 5 μs post–impact for impact velocity at (a) 100 m/s, (b) 200 m/s, (c) 350 m/s. After Eftis (1996)

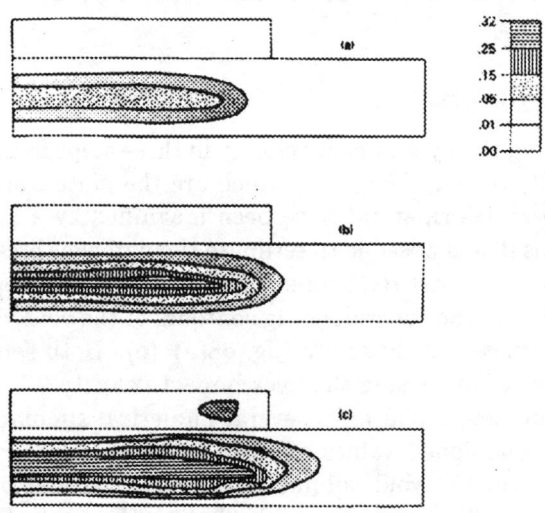

Figure 68: Contours for void volume fraction for impact velocity at (a) 100 m/s, (b) 200 m/s, (c) 350 m/s. After Eftis (1996)

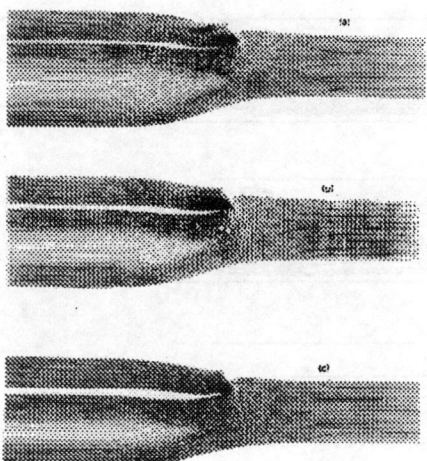

Figure 69: Deformed geometry at 5 μs post–impact for viscosity parameter η at (a) 10 P, (b) 120 P, (c) 500 P. After Eftis (1996)

interval is reduced considerably, while the material, which has correspondingly smaller microvoid volume, experiences greater strain hardening.

To study the post–spall behaviour the calculations have been performed for larger duration of the process. The calculated deformation of the plates at 5, 15, 25 and 50 μs after impact by the flyer traveling at 350 m/s are shown by Figs. 71(a)–(d). For these figures the deformation has not been magnified. Figure 72(a) displays the void volume distribution at 50 μs where, as expected, the highest values are adjacent to the spall plane. The equivalent plastic strain distribution across the plates appearing in Fig. 72(b) indicates that the regions of highest plastic strain are those where the edge effects are present due to larger shear deformation that occurs.

10. FINAL COMMENTS

The main features of the theory of thermodynamic viscoplasticity of dissipative solids developed in these lectures are as follows: (i) it is invariant with respect to any diffeomorphism; (ii) it takes into considerations many important effects and phenomena which affect the behaviour of single crystals and polycrystalline materials; (iii) it describes cooperative phenomena and as the result takes account of synergetic effects.

To accomplish this purpose the theory has been developed within the thermodynamic framework of the rate type covariance constitutive structure with finite set of

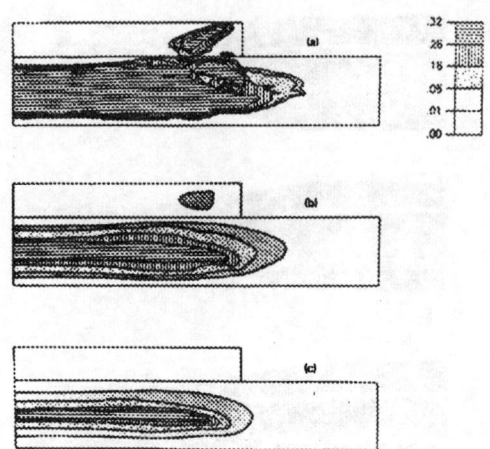

Figure 70: Contours for void volume fraction for viscosity parameter η at (a) 10 P, (b) 120 P, (c) 500 P. After Eftis (1996)

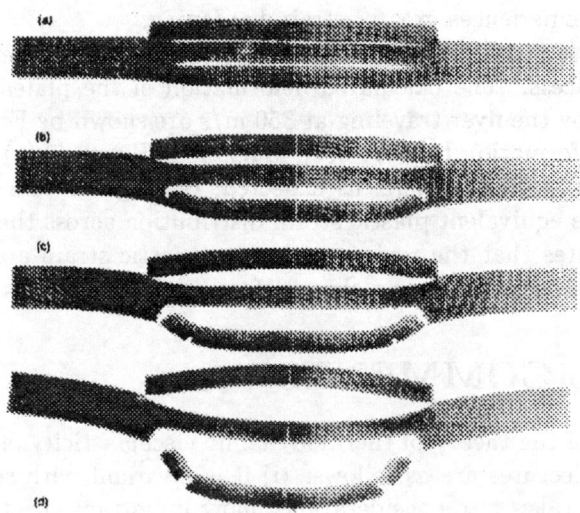

Figure 71: Deformed geometry for a 3.5 mm flyer with 350 m/s impact velocity at (a) 5 μs, (b) 15 μs, (c) 25 μs, (d) 50 μs post–impact. After Eftis (1996)

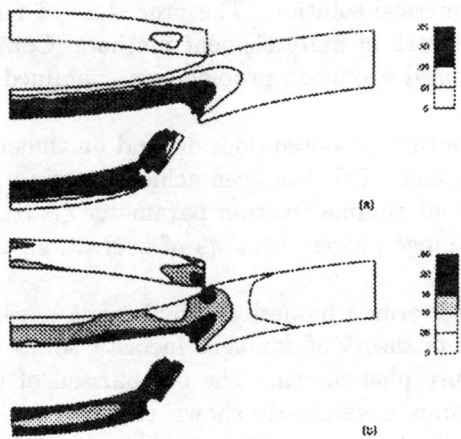

Figure 72: Contours due to 350 m/s impact velocity at 50 μs post–impact, (a) Void fraction, (b) Equivalent plastic strain. After Eftis (1996)

the internal state variables. The crucial idea in this theory is the very efficient physical interpretation of the internal state variables.

This interpretation has been given for both single crystals and polycrystalline solids. For single crystals the assumption that $\alpha^{(\nu)}$ and $\beta^{(\nu)}$ are interpreted as densities of mobile and obstacle dislocations in particular slip system ν, respectively, and $\xi^{(\nu)}$ as the concentration of point defects permitted to base all considerations on good physical foundations and to use the available experimental observations.

Similarly for polycrystalline solids the void volume porosity ξ and the residual stress (the back stress) α have very precise physical interpretation.

This theory has been inspired by recent theoretical and experimental investigations for both single crystals and polycrystalline solids. Many recent experimental investigations, as has been shown in Chapter 2, have brought deep understanding of real features of the deformation process and have given many important measurements needed for the development of the theoretical descriptions.

The instability of inelastic flow process of solids observed on macroscopic level as the formation of adiabatic shear band pattern is treated as the self–organization of a system to a new two–phase material system. To investigate the adiabatic shear band formation in inelastic solids under dynamic loading processes besides the theoretical methods based on the analysis of acceleration waves the methods of synergetics are used and the development of the numerical procedure is proposed. The problem is reduced to the examination of well–posedness of the evolution problem and to the construction of its numerical solution. The procedure of numerical solution is developed within the framework of finite element method. Convergence, consistency and stability of the discretised evolution problem are examined and the Lax equivalence theorem is formulated.

The criterion of fracture proposed does depend on the entire evolution of the constitutive structure of solids. This has been achieved mainly due to careful analysis of the evolution of the void volume fraction parameter ξ. It has been postulated that the intrinsic micro-damage process consists of nucleation, growth and coalescence of microcracks.

Examples of the performed numerical simulations have proved the usefulness of the thermoviscoplasticity theory of damaged inelastic solids to the investigation of the localization and fracture phenomena. The comparison of the results obtained with experimental observations have clearly shown that particularly the localized ductile fracture mode is described properly. The main feature of the developed theoretical model is the possibility of the investigation of the entire process considered. The ductile fracture is treated as a final stage of the entire inelastic flow process.

It has been proved that the localization of plastic deformation in an elastic-viscoplastic solid body can arise only as the result of the reflection and interaction of waves. It has different character than that which occurs in a rate independent elastoplastic solid body. Viscosity introduces implicitly a length-scale parameter into the dynamical initial-boundary value problem and hence it implies that the localization region is diffused when compared with an inviscid plastic material. In dynamical initial-boundary value problem the stress and deformation due to wave reflections and interactions are not uniformly distributed, and this kind of heterogeneity can lead to strain localization in the form of shear band regions or necking regions in the absence of geometrical or material irregularities (superposed imperfections).

These regions vary with the imposed strain rate and are affected by the assumed value of the relaxation time (or viscosity parameter).

It has been also proved that the insensitivity of the results to mesh refinement is apparent whether material softening is present or not.

Examples of spallation as a particular manifestation of dynamic fracture and fragmentation phenomena showed great technological usefulness of the developed theoretical model.

REFERENCES

1. Abraham, R. and Marsden, J.E.: Foundations of Mechanics, Second Edition, Addison–Wesley, Reading Mass., 1978.
2. Abraham, R., Marsden, J.E. and Ratiu, T.: Manifolds, Tensor Analysis and Applications, Springer, Berlin 1988.
3. Agah–Tehrani, A., Lee, E.H., Mallett, R.L. and Onat, E.H.: The theory of elastic–plastic deformation at finite strain with induced anisotropy modelled as combined isotropic–kinematic hardening, J. Mech. Phys. Solids, 35 (1987), 519–539.
4. Anand, L. and Spitzig, W.A.: Initiation of localized shear band in plane strain, J. Mech. Phys. Solids, 28 (1980), 113–128.
5. Asaro, R.J.: Crystal plasticity, J. Appl. Mech., 50 (1983), 921–934.
6. Asaro, R.J.: Micromechanics of crystals and polycrystals, Adv. Appl. Mech., 23 (1983), 1–115.
7. Asaro, R.J. and Needleman, A.: Texture development and strain hardening in rate dependent polycrystals, Acta Metall., 33 (1985), 923–953.
8. Asaro, R.J. and Rice, J.R.: Strain localization in ductile single crystals, J. Mech. Phys. Solids, 25 (1977), 309–338.
9. Asay, J.R. and Kerley, G.I.: The response of materials to dynamic loading, Int. J. Impact Engng., 5 (1987), 69–99.
10. Ashby, M.F. and Frost, H.J.: in : Constitutive Equation in Plasticity, (Ed. A.S. Argon), MIT Press, Cambridge, Mass., 1975, 117–147.
11. Ayres, R.A.: Thermal gradients, strain rate and ductility in sheet steel tensile specimen, Metall. Trans., 16A (1985), 37–43.
12. Balke, H. and Estrin, Y.: Micromechanical modelling of shear banding in single crystals, Int. J. Plast., 10 (1994), 133–147.
13. Barbee, T.W., Seaman, L., Crewdson, R. and Curran, D.: Dynamic fracture criteria for ductile and brittle metals, J. Mater., 7 (1972), 393–401.
14. Basiński, S.J. and Basiński, Z.S.: Plastic deformation and work hardening, in: *Dislocations in Solids*, Vol.4 Dislocations in Metallurgy, (Ed. F.R.N. Nabarro), Nort–Holland, Amsterdam 1979, 261–362.
15. Bassani, J.L.: Plastic flow of crystals, Adv. Appl. Mech., 30 (1994), 191–258.
16. Batra, R.C.: Effect of material parameters on the initiation and growth of adiabatic shear bands, Int. J. Solids Structures, 23 (1987), 1435–1446.
17. Batra, R.C. and Ko, K.I.: Analysis of shear bands in dynamic axisymmetric compression of a thermoviscoplastic cylinder, Int. J. Engr. Sci., 31 (1993), 529–547.
18. Batra, R.C. and Zhang, X.: On the propagation of a shear band in a steel tube, J. Engng. Materials and Technology, 116 (1994), 155–161.

19. Becker, R., Needleman, A., Richmond, O. and Tvergaard, V.: Void growth and failure in notched bars, J. Mech. Phys. Solids, 36 (1988), 317–351.
20. Bingham, E.C.: Fluidity and Plasticity, McGraw Hill, New York 1922.
21. Bluhm, J.I. and Morrissey, R.J.: Fracture in a tension specimen, in: Proc. First Int. Conf. on Fracture, Sendai, (Eds. T. Yokoburi et al.), Japan, Sendai 1965, 1739–1780.
22. Bridgeman, P.W.: Studies in Large Plastic Flow and Fracture, McGraw Hill, New York 1952.
23. Campbell, J.D.: Dynamic plasticity macroscopic and microscopic aspects, Material Sci. Engng., 12 (1973), 3–12.
24. Campbell, J.D. and Ferguson, W.G.: The temperature and strain-rate dependence of the shear strength of mild steel, Phil. Mag., 81 (1970), 63–82.
25. Carroll, M.M. and Holt, A.C.: Static and dynamic pore-collapse relations for ductile solids, J. Appl. Phys., 43 (1972), 1626–1636.
26. Carroll, M.M., Kim, K.T. and Nesterenko, V.F.: The effect of temperature on viscoplastic pore collapse, J. Appl. Phys., 59 (1986), 1962–1967.
27. Chakrabarti, A.K. and Spretnak, J.W.: Instability of plastic flow in the direction of pure shear, Metallurgical Transactions, 6A (1975), 733–747.
28. Chang, Y.W. and Asaro, R.J.: Lattice rotations and shearing in crystals, Arch. Mech., 32 (1980), 369–388.
29. Chang, Y.W. and Asaro, R.J.: An experimental study of shear localization in aluminum-copper single crystals, Acta Metall., 29 (1981), 241–257.
30. Chengwei, S., Shiming, Z., Yanping, W. and Cangli, L.: Dynamic fracture in metals at high strain rate, in: High-Pressure Shock Compression of Solids, II. Dynamic Fracture and Fragmentation, (Eds. L. Davison, D.E. Grady and M. Shahinpoor), Springer-Verlag, New York 1996, 71–89.
31. Cho, K., Chi, Y.C. and Duffy, J.: Microscopic observations of adiabatic shear bands in three different steels, Brown University Report, 1989.
32. Coleman, B.D. and Gurtin, M.E.: Thermodynamics with internal state variables, J. Chem. Phys., 47 (1967), 597–613.
33. Coleman, B.D. and Noll, W.: The thermodynamics of elastic materials with heat conduction and viscosity, Arch. Rational Mech. Anal., 13 (1963), 167–178.
34. Conrad, H.: Thermally activated deformation of metals, J. Metals, 16 (1964), 582–588.
35. Cox, T.B. and Low, J.R.: An investigation of the plastic fracture of AISI 4340 and Nickel-200 grade maraging steels, Met. Trans., 5 (1974), 1457–1470.
36. Curran, D.R., Seaman, L. and Shockey, D.A.: Dynamic failure in solids, Physics Today, January (1977), 46–55.
37. Curran, D.R., Seaman, L. and Shockey, D.A.: Linking dynamic fracture to microstructural processes, in: Shock Waves and High-Strain Rate Phenomena in

Metals: Concepts and Applications, (Eds. M.A. Meyers and L.E. Murr), Plenum Press, New York 1981, 129–167.
38. Curran, D.R., Seaman, L. and Shockey, D.A.: Dynamic failure of solids, Physics Reports, 147 (1987), 253–388.
39. Dafalias, Y.F.: Corotational rates for kinematic hardening at large plastic deformations, J. Appl. Mech., 50 (1983), 561–565.
40. Dafalias, Y.F.: The plastic spin, J. Appl. Mech., 52 (1985), 865–871.
41. Dafalias, Y.F.: Issues on the constitutive formulation at large elastoplastic deformations, Part 1: Kinematics, Acta Mechanica, 69 (1987), 119.
42. Dafalias, Y.F.: Issues on the constitutive formulation at large elastoplastic deformations, Part 2: Kinetics, Acta Mechanica, 73 (1988), 121.
43. Dautray, R. and Lions, J.-L.: Mathematical Analysis and Numerical Methods for Science and Technology, Vol. 6. Evolution Problems II, Springer, Berlin 1993.
44. Day, W.A. and Silhavy, M.: Efficiency and the existence of entropy in classical thermodynamics, Arch. Rational Mech. Anal., 64 (1977), 205–219.
45. Dowling, A.R., Harding, J. and Campbell, J.D.: The dynamic punching of metals, J. Inst. of Metals, 98 (1970), 215–224.
46. Duszek, M.K. and Perzyna, P.: Plasticity of damaged solids and shear band localization, Ing. Arch., 58 (1988a), 330–392.
47. Duszek, M.K. and Perzyna, P.: Influence of the kinematic hardening on the plastic flow localization in damaged solids, Arch. Mech., 40 (1988b), 595–609.
48. Duszek, M.K. and Perzyna, P.: On combined isotropic and kinematic hardening effects in plastic flow processes, Int. J. Plasticity, 7 (1991a), 351–363.
49. Duszek, M.K. and Perzyna, P.: The localization of plastic deformation in thermoplastic solids, Int. J. Solids Structures, 27 (1991b), 1419–1443.
50. Duszek–Perzyna, M.K., Korbel, K. and Perzyna, P.: Adiabatic shear band localization in single crystals under dynamic loading processes, Arch. Mechanics, 49 (1997), (in print).
51. Duszek–Perzyna, M.K. and Perzyna, P.: Adiabatic shear band localization in elastic–plastic single crystals, Int. J. Solids Structures, 30 (1993), 61–89.
52. Duszek–Perzyna, M.K. and Perzyna, P.: Analysis of the influence of different effects on criteria for adiabatic shear band localization in inelastic solids, in: Material Instabilities: Theory and Applications, ASME Congress, Chicago, 9–11 November 1994 (Eds. R.C. Batra and H.M. Zbib), AMD–Vol. 183/MD–Vol.50, ASME, New York 1994, 59–85.
53. Duszek–Perzyna, M.K. and Perzyna, P.: Acceleration waves in analysis of adiabatic shear band localization, IUTAM Symposium on Nonlinear Waves in Solids, August 15–20, 1993, Victoria, Canada; (Eds. J.L. Wegner and F.R. Norwood), ASME 1995, 128–135.

54. Duszek-Perzyna, M.K. and Perzyna, P.: Adiabatic shear band localization of inelastic single crystals in symmetric double slip process, Archive of Appled Mechanics, 66 (1996), 369–384.
55. Duszek-Perzyna, M.K. and Perzyna, P.: Analysis of anisotropy and plastic spin effects on localization phenomena, Arch. Appl. Mechanics, (1997) (in print).
56. Duszek-Perzyna, M.K., Perzyna, P. and Stein, E.: Adiabatic shear band localization in elastic-plastic damaged solids, Int. J. Plasticity, 8 (1992), 361–384.
57. Eftis, J.: Constitutive modelling of spall fracture, in: High-Pressure Shock Compression of Solids, II. Dynamic Fracture and Fragmentation, (Eds. L. Davison, D.E. Grady and M. Shahinpoor), Springer-Verlag, New York 1996, 399–451.
58. Eftis, J. and Nemes, J.A.: Constitutive modelling of spall fracture, Arch. Mech., 43 (1991a), 399–435.
59. Eftis, J. and Nemes, J.A.: Evolution equation for the void volume growth rate in a viscoplastic-damage constitutive model, Int. J. Plasticity, 7 (1991b), 275–293.
60. Eftis, J. and Nemes, J.A.: Modelling of impact-induced spall fracture and post spall behaviour of a circular plate, Int. J. Fracture, 53 (1992), 301–324.
61. Eftis, J., Nemes, J.A. and Randles, P.W.: Viscoplastic analysis of plate–impact spallation, Int. J. Plasticity, 7 (1991), 15–39.
62. Eisenberg, M.A. and Yew, C.F.: The anisotropic deformation of yield surfaces, J. Eng. Mat. Tech., 106 (1984), 355–360.
63. Estrin, Y. and Kubin, L.P.: Load strain hardening and nonuniformity of plastic deformation, Acta Metall., 34 (1986), 2455–2464.
64. Evans, A.G. and Rawlings, R.D.: The thermally activated deformation of crystalline materials, Phys. Stat. Sol., 34 (1969), 9–31.
65. Fisher, J.R.: Viod Nucleation in Spheroidized Steels During Tensile Deformation, Ph.D. Dissertation, Brown University 1980.
66. Flanagan, D.P. and Belytschko, T.: A uniform strain hexahedron and and quadrilateral with orthogonal hourglass control, Int. J. Numerical Methods in Engr., 17 (1981), 679–706.
67. Flanagan, D.P. and Taylor, L.M.: An accurate numerical algorithm for stress integration with finite rotations, Computer Methods in Appl. and Engng., 62 (1987), 305–320.
68. Follansbee, P.S.: Metallurgical Applications of Shock – Wave and High-Strain-Rate Phenomena, (Eds. L.E. Murr, K.P. Staudhammer, M.A. Meyeres MA), Marcel Dekker, New York 1986, 451–480.
69. Follansbee, P.S. and Kocks, U.F.: A constitutive description of the deformation of copper based on the use of the mechanical treshold stress as an internal state variable, Acta Met., 36 (1988), 81–93.
70. Fressengeas, C. and Molinari, A.: Inetria and thermal effects on the localization of plastic flow, Acta Metall., 33 (1985), 387–396.

71. Gilath, I.: Laser-induced spallation and dynamic fracture at ultra high strain rate, in: High-Pressure Shock Compression of Solids, II. Dynamic Fracture and Fragmentation, (Eds. L. Davison, D.E. Grady and M. Shahinpoor), Springer-Verlag, New York 1996, 90–120.
72. Gilbert, J.E. and Knops, R.J.: Stability of general systems, Arch. Rat. Mech. Anal., 25 (1967), 271–284.
73. Giovanola, J.H.: Adiabatic shear banding under pure shear loading, Mechanics of Materials, 7 (1988), 59–87.
74. Glansdorff, P. and Prigogine, I.: Thermodynamic Theory of Structure, Stability and Fluctuations, Wiley-Interscience, London 1977.
75. Gorman, J.A., Wood, D.S. and Vreeland, T.: Mobility of dislocation in aluminium, J. Appl. Phys., 40 (1969), 833–841.
76. Grebe, H.A., Pak, H.R. and Meyer, M.A.: Adiabatic shear band localization in titanium and Ti-6PctAl-4PctV alloy, Met. Trans., 16A (1985), 761-775.
77. Green, A.E. and Naghdi, P.M.: A general theory of an elastic–plastic continuum, Arch. Rat. Mech. Anal., 18 (1965), 251–281.
78. Green, A.E. and Naghdi, P.M.: Some remarks on elastic–plastic deformation at finite strain, Int. J. Engn. Sci., 9 (1971), 1219–1229.
79. Green, A.E. and Naghdi, P.M.: On thermodynamics and the nature of the second law, Proc. R. Soc. Lond., A357 (1977), 253–270.
80. Green, A.E. and Naghdi, P.M.: On thermodynamics and the nature of the second law for mixtures of interacting continua, Quart. J. Mech. Appl. Maths., 31 (1978), 265–293.
81. Gurson, A.L.: Plastic flow and fracture behaviour of ductile materials incorporating void nucleation, growth, and interaction, PhD Thesis, Brown University, 1975.
82. Gurson, A.L.: Continuum theory of ductile rupture by void nucleation and growth – Part I – Yield criteria and flow rules for porous ductile media, J. Eng. Mater. Technology, 99 (1977), 2–15.
83. Gurtin, M.E.: Thermodynamic and stability, Arch. Rational Mech. Anal., 59 (1975), 63–96.
84. Hadamard, J.: Lecons sur la propagation des ondes et les equations de l'hydrodynamique, Chap. 6, Paris 1903.
85. Haken, H.: Cooperative phenomena in systems far from thermal equilibrium and in nonphysical systems, Reviews of Modern Physics, 47 (1975), 67–121.
86. Haken, H.: Advanced Synergetics, Springer, Berlin 1987.
87. Haken, H.: Information and Self-Organization, Springer, Berlin 1988.
88. Hale, K.J.: Dynamic systems and stability, J. Math. Anal. Appl., 26 (1969), 39–59.

89. Hartley, K.A., Duffy, J. and Hawley, R.H.: Measurement of the temperature profile during shear band formulation in steels deforming at high strain rates, J. Mech. Phys. Solids, 35 (1987), 283–301.
90. Hauser, F.E., Simmons, J.A. and Dorn, J.E.: Strain rate effects in plastic wave propagation, in: Response of Metals to High Velocity Deformation, Wiley (Interscience), New York 1961, 93–114.
91. Hill, R.: Acceleration wave in solids, J. Mech. Phys. Solids, 10 (1962), 1–16.
92. Hill, R.: The essential structure of constitutive laws for metal composities and polycrystals, J. Mech. Phys. Solids, 15 (1967), 255–262.
93. Hill, R.: The Mathematical Theory of Plasticity, last ed., Oxford University Press, Oxford 1983.
94. Hill, R.: Aspects of invariance in solids mechanics, Adv. Appl. Mech., 18 (1987), 1–75.
95. Hill, R. and Rice, J.R.: Constitutive analysis of elastic–plastic crystals at arbitrary strain, J. Mech. Phys. Solids, 20 (1972) 401–413.
96. Hill, R. and Rice, J.R.: Elastic potentials and the structure of inelastic constitutive laws, SIAM J. Appl. Math., 25 (1973), 448–461.
97. Hohenemser, K. and Prager, W.: Uber die Ansatze der Mechanik isotroper Kontinua, ZAMM, 12 (1932), 216–226.
98. Hughes, T.J.R., Kato, T. and Marsden, J.E.: Well-posed quasilinear second order hyperbolic system with application to nonlinear elastodynamics and general relativity, Arch. Rat. Mech. Anal., 63 (1977), 273–294.
99. Hughes, T.J.R. and Winget, J.: Finite rotation effects in numerical integration of rate constitutive equations arising in large–deformation analysis, Int. J. Numerical Methods in Engng., 15 (1989), 1862–1967.
100. Hutchinson, J.W.: Bounds and self–consistent estimates for creep of polycrystalline materials, Proc. R. Soc. Lond., A348 (1976), 101–127.
101. Ikegami, K.: Experimental plasticity on the anisotropy of metals, in: Proc. Euromech Colloquium 115, Mechanical Behaviour of Anisotropic Solids, (Ed. J.P. Biehler), 1982, 201–242.
102. Ionescu, I.R. and Sofonea, M.: Functional and Numerical Methods in Viscoplasticity, Oxford 1993.
103. Iwakuma, T. and Nemat-Nasser, S.: Finite elasto–plastic deformation of polyctystalline metals, Proc. R. Soc. Lond., A23(394) (1984), 87–119.
104. Jaumann, G.: Geschlossenes System physikalischer und chemischer Differentialgesetze. Sitzgsber. Akad. Wiss. Wien (IIa), 120 (1911), 385–530.
105. Johnson, J.N.: Dynamic fracture and spallation in ductile solids, J. Appl. Phys., 52 (1981), 2812–2825.
106. Johnson, G.C. and Bamman, D.J.: A discussion of stress rates in finite deformation problems, Int. J. Solids & Struct., 20 (1984), 725–737.

107. Kawata, K., Hashimoto, S. and Kurokawa, K.: Analyses of high velocity tension of bars of finite length of bcc and fcc metals with their own constitutive equation, in: High Velocity Deformation of Solids, (Eds. H.S. Kawata and J. Shioiri), Springer Verlag, New York 1978.
108. Knops, R.J. and Wilkes, E.W.: Theory of elastic stability, in: Handbuch der Physik, VI a/3, Springer, Berlin, Heidelberg, New York 1973.
109. Kocks, U.F., Argon, A.S. and Ashby, M.F.: Thermodynamics and Kinetics of Slip, Pergamon Press 1975.
110. Kratochvil, J.: Finite-strain theory of crystalline elastic-inelastic materials, J. Appl. Phys., 42 (1971), 1104.
111. Kumar, A., Hauser, F.E. and Dorn, J.E.: Viscous drag on dislocations in aluminium at high strain rates, Acta Met., 16 (1968), 1189-1197.
112. Kumar, A. and Kumble, R.G.: Viscous drag on dislocations at high strain rates in copper, J. Appl. Physics, 40 (1969), 3475-3480.
113. Le Roy, G., Embury, J.D., Edward, G. and Ashby, M.F.: A model of ductile fracture based on the nucleation and growth of voids, Acta Met., 29 (1981), 1509-1522.
114. Lee, E.H.: Elastic-plastic deformations at finite strains, J. Appl. Mech., 35 (1969), 1-6.
115. Lee, E.H. and Liu, D.T.: Finite-strain elastic-plastic theory partucularly for plane wave analysis, J. Appl. Phys., 38 (1967).
116. Lindholm, U.S.: Some experiments with the split Hopkinson pressure bar, J. Mech. Phys. Solids, 12 (1964), 317-335.
117. Lindholm, U.S.: in: Mechanical Behaviour of Materials under Dynamic Loads, (Ed. U.S. Lindholm), Springer Verlag 1968, 77-95.
118. Lisiecki, L.L., Nelson, D.R. and Asaro, R.J.: Lattice rotations, necking and localized deformation in f.c.c. single crystals, Scripta Met., 16 (1982), 441-449.
119. Loret, B.: On the effect of plastic rotation in the finite deformation of anisotropic elastoplastic materials, Mech. Mater., 2 (1983), 287-304.
120. Loret, B.: On the effects of plastic rotation on the localization of anisotropic elastoplastic solids, in: Plastic Instability, (Eds. J. Salencon et al.), Presses Ponts et Chausees, Paris 1985, 89-100.
121. Loret, B.: Does deviation from deviatoric associativity lead to the onset of flutter instability?, J. Mech. Phys. Solids, 40 (1992), 1363-1375.
122. Loret, B. and Dafalias, Y.F.: The effects of anisotropy and plastic spin on fold formations, J. Mech. Phys. Solids, 40 (1992), 417-439.
123. Lodygowski, T.: On avoiding of spurious mesh sensitivity in numerical analysis of plastic strain localization, CAMES, 2 (1995), 231-248.
124. Lodygowski, T., Lengnick, M., Perzyna, P. and Stein E.: Viscoplastic numerical analysis of dynamic plastic strain localization for a ductile material, Archives of Mechanics, 46 (1994), 1-25.

125. Łodygowski, T. and Perzyna, P.: Localized fracture of inelastic polycrystalline solids under dynamic loading processes, I. J. Damage Mechanics, (1997), (in print).
126. Łodygowski, T. and Perzyna, P.: Numerical modelling of localized fracture of inelastic solids in dynamic loading processes, Int. J. Num. Meth. Engng., 40 (1997), (in print).
127. Mac Kenzie, J.H.: The elastic constants of a solid containing spherical holes, Proc. Phys. Soc., 63B (1950), 2–11.
128. Malvern, L.E.: The propagation of longitudinal waves of plastic deformation in a bar of material exhibiting a strain–rate effects, J. Appl. Mech., 18 (1951), 203–208.
129. Mandel, J.: Plasticité Classique et Viscoplasticite, CISM Lecture Notes No. 97, Udine, Springer–Verlag, Vien 1971.
130. Mandel, J.: Equations constitutives et directeurs dans les milieux plastiques et viscoplastiques, Int. J. Solids Structures, 9 (1973), 725.
131. Mandel, J.: Thermodynamics and plasticity, in: Foundations of Continuum Thermodynamics, (Eds. J.J. Delgado et al.), MacMillan, New York 1974.
132. Mandel, J.: Définition d'un repère privilégé pour l'etude des transformations anélastiques du polycrystal, J. Méc. Théo. Appl., 1 (1982), 7.
133. Marchand, A., Cho, K. and Duffy, J.: The formation of adiabatic shear bands in an AISI 1018 cold–rolled steel, Brown University Report 1988.
134. Marchand, A. and Duffy, J.: An experimental study of the formation process of adiabatic shear bands in a structural steel, J. Mech. Phys. Solids, 36 (1988), 251–283.
135. Marsden, J.E. and Hughes, T.J.R.: Mathematical Foundations of Elasticity, Prentice–Hall, Englewood Cliffs, New York 1983.
136. Mason, W.P.: Phonon viscosity and its effect on acoustic wave attenuation and dislocation motion, J. Acoustical Soc. Amer., 32 (1960), 458–472.
137. Mason, J.J., Rosakis, A.J. and Ravichandran, R.: On the strain and strain rate dependence of the fraction of plastic work converted to heat: an experimental study using high speed infrared detectors and the Kolsky bar, Mechanics of Materials, 17 (1994), 135–145.
138. Matic, P., Kirby II, G.C. and Jolles, M.I.: The relation of tenssile specimen size and geometry effects to unique constitutive parameters for ductile materials, Proc. R. Soc. Lond., A417 (1988), 309–333.
139. Mear, M.E. and Hutchinson, J.E.: Influence of yield surface curvature on flow localization in dilatant plasticity, Mech. Mater., 4 (1985), 395–407.
140. Mecking, H. and Kocks, U.F.: Kinetics of flow and strain–hardening, Acta Metall., 29 (1981), 1865–1875.
141. Mehrabadi, M.M. and Nemat–Nasser, S.: Some basic kinematical relations for finite deformations of continua, Mechanics of Materials, 6 (1987), 127–138.

142. Meyers, M.A. and Aimone, C.T.: Dynamic fracture (spalling) of metals, Prog. Mater. Sci., 28 (1983), 1–96.
143. Müller, I.: On the entropy inequality, Arch. Rational Mech. Anal., 26 (1967), 118–141.
144. Müller, I.: Thermodynamics of mixtures of fluids, J. Mécanique, 14 (1975), 267–303.
145. Nabarro, F.R.N.: Theory of Crystal Dislocations, Oxford 1967.
146. Nadai, A.: Theory of Flow and Fracture of Solids, Vol. 1, McGraw Hill, New York 1950.
147. Needleman, A. and Rice, J.R.: Limits to ductility set by plastic flow localization, in: Mechanics of Sheet Metal Forming, (Ed. D.P. Koistinen and N.M. Wang), Plenum, New York 1978, 237–267.
148. Nemat-Nasser, S.: Phenomenological theories of elastoplasticity and strain localization at high strain rates, Appl. Mech. Rev., 45 (1992), S19–S45.
149. Nemat-Nasser, S.: Decomposition of strain measures and their rates in finite deformation elastoplasticity, Int. J. Solids Structures, 15 (1979), 155–166.
150. Nemat-Nasser, S., Chung, T.D. and Taylor, L.M.: Phenomenological modelling of rate-dependent plasticity for high strain rate problems, Mechanics of Materials, 7 (1989), 319–344.
151. Nemat-Nasser, S. and Obata, M.: Rate-dependent, finite elasto-plastic deformation of polycrystals, Proc. R. Soc. Lond., A407 (1986), 343–375.
152. Nemes, J.A. and Eftis, J.: Several features of a viscoplastic study of plate-impact spallation with multidimensional strain, Computers and Structures, 38 (1991), 317–328.
153. Nemes, J.A. and Eftis, J.: Pressure-shear waves and spall fracture described by a viscoplastic-damage constitutive moddel, Int. J. Plasticity, 8 (1992), 185–207.
154. Nemes, J.A. and Eftis, J.: Constitutive modelling on the dynamic fracture of smooth tensile bars, Int. J. Plasticity, 9 (1993), 243–270.
155. Nemes, J.A., Eftis, J. and Randles, P.W.: Viscoplastic constitutive modelling of high strain-rate deformation, material damage and spall fracture, J. Appl. Mech., 57 (1990), 282–291.
156. Nicolis, G. and Prigogine, I.: Self-Organization in Nonequilibrium Systems, Wiley-Interscience, New York 1977.
157. Oldroyd, J.: On the formulation of rheological equations of state, Proc. R. Soc. Lond., A200 (1950), 523–541.
158. Peirce, D., Asaro, J.R. and Needleman, A.: An analysis of nonuniform and localized deformation in ductile single crystals, Acta Metall., 30 (1982), 1087–1119.
159. Peirce, D., Asaro, J.R. and Needleman, A.: Material rate dependence and localized deformation in crystalline solids, Acta Metall., 31 (1983), 1951–1976.

160. Perzyna, P.: The constitutive equations for rate sensitive plastic materials, Quart. Appl. Math., 20 (1963), 321–332.
161. Perzyna, P.: Fundamental problems in viscoplasticity, Advances in Applied Mechanics, 9 (1966), 243–377.
162. Perzyna, P.: Thermodynamic theory of viscoplasticity, Advances in Applied Mechanics, 11 (1971), 313–354.
163. Perzyna, P.: Thermodynamics of a unique material structure, Arch. Mechanics, 27 (1975), 791–806.
164. Perzyna, P.: Coupling of dissipative mechanisms of viscoplastic flow, Arch. Mechanics, 29 (1977), 607–624.
165. Perzyna, P.: Modified theory of viscoplasticity. Application to advanced flow and instability phenomena, Arch. Mechanics, 32 (1980), 403–420.
166. Perzyna, P.: Thermodynamics of dissipative materials, in: Recent Developments in Thermodynamics of Solids, (Eds. G. Lebon and P. Perzyna), Springer, Wien 1980, 95–220.
167. Perzyna, P.: Stability phenomena of dissipative solids with internal defects and imperfections, in: Proc. XV-th IUTAM Congress, Toronto, August 1980, Theoretical and Applied Mechanics, (Eds. F.P.J. Rimrott and B. Tabarrok), North-Holland, Amsterdam 1981, 369–376.
168. Perzyna, P.: Stability problems for inelastic solids with defects and imperfections, Arch. Mechanics, 33 (1981), 587–602.
169. Perzyna, P.: Application of dynamical system methods to flow processes of dissipative solids, Arch. Mechanics, 34 (1982), 523–539.
170. Perzyna, P.: Stability of flow processes for dissipative solids with internal imperfections, ZAMP, 35 (1984a), 848–867.
171. Perzyna, P.: Constitutive modelling of dissipative solids for postcritical behaviour and fracture, ASME J. Eng. Materials and Technology, 106 (1984b), 410-419.
172. Perzyna, P.: Dependence of fracture phenomena upon the evolution of constitutive structure of solids, Arch. Mechanics, 37 (1985), 485–501.
173. Perzyna, P.: Constitutive modelling for brittle dynamic fracture in dissipative solids, Arch. Mechanics, 38 (1986), 725–738.
174. Perzyna, P.: Internal state variable description of dynamic fracture of ductile solids, Int. J. Solids Structures, 22 (1986), 797-818.
175. Perzyna, P.: Temperature and rate dependent theory of plasticity of crystalline solids, Revue Phys. Appl., 23 (1988), 445–459.
176. Perzyna, P.: Influence of anisotropic effects on micro–damage process in dissipative solids, in: Proc. IUTAM/ICM Symposium on Yielding, Damage and Failure of Anisotropic Solids, Villerd–de–Lance, August 1987, Mech. Eng. Publ. Limited, London 1990, 483–507.

177. Perzyna, P.: Constitutive equations for thermoplasticity and instability phenomena in thermodynamic flow processes, in: Progress in Computational Analysis of Inelastic Structures, (Ed. E. Stein), Springer–Verlag, Wien 1993, 1–78.
178. Perzyna, P.: Adiabatic shear band localization fracture of solids in dynamic loading proceses, Proc. of Int. Conference on Mechanical and Physical Behaviour of Materials under Dynamic Loading, Oxford, September 26–30, 1994; (Ed. J. Harding), Les Editions de Physique Le Ulis 1994, 441-446.
179. Perzyna, P.: Instability phenomena and adiabatic shear band localization in thermoplastic flow processes, Acta Mechanica, 106 (1994), 173–205.
180. Perzyna, P.: Interactions of elastic–viscoplastic waves and localization phenomena in solids, IUTAM Symposium on Nonlinear Waves in Solids, August 15–20, 1993, Victoria, Canada; (Eds. J.L. Wegner and F.R. Norwood), ASME 1995, 114–121.
181. Perzyna, P.: Thermodynamic theory of inelastic single crystals, Continuum Mechanics and Thermodynamics, (1997), (Submitted for publication).
182. Perzyna, P.: Thermodynamics and synergetics of inelastic single crystals, Mathematics and Mechanics of Solids, (1997), (Submitted for publication).
183. Perzyna, P. and Drabik, A.: Description of micro–damage process by porosity parameter for nonlinear viscoplasticity, Arch. Mechanics, 41 (1989), 895–908.
184. Perzyna, P. and Drabik, A.: Micro–damage mechanism in adiabatic processes, Int. J. Plasticity, (1997) (in print).
185. Perzyna, P. and Duszek–Perzyna, M.K.: Phenomenological modelling of adiabatic shear band localization fracture of solids in dynamic loading processes, in: Localized Damage III; Computer – Aided Assessment and Control, (Eds. M.H. Aliabadi, A. Carpinteri, S. Kalisky, D.J. Cartwright), Computational Mechanics Publications., Southampton 1994, 579–588.
186. Perzyna, P. and Duszek–Perzyna, M.K.: Constitutive modelling of inelastic single crystals for localization phenomena, in: *Constitutive Laws: Experiments and Numerical Implementation*, (Eds. A.M. Rajendran, R.C. Batra), CIMME, Barcelona 1995, 70–83.
187. Perzyna, P. and Korbel, K.: Analysis of the influence of substructure of crystal on the localization phenomena of plastic deformation, Mechanics of Materials, 24 (1996), 141–158.
188. Perzyna, P. and Korbel, K.: Analysis of the influence of variuos effects on criteria for adiabatic shear band localization in single crystals, Acta Mechanica, (1997) (in print).
189. Philips, A. and Lu, W.Y.: An experimental investigation of yield surface and loading surface of pure aluminium with stress–controlled and strain–controlled paths of loading, ASME J. Eng. Mater. Technol., 106 (1984), 349–354.
190. Prager, W.: The theory of plasticity: a survey of recent achievements, (J. Clayton Lecture), Proc. Inst. Mech. Eng., 169 (1955), 41.

191. Prager, W.: Introduction to Mechanics of Continua, Gin and Co., New York 1961.
192. Qin, Q. and Bassani, J.L.: Non–Schmid yield behavior in single crystals, J. Mech. Phys. Solids, 40 (1992), 813–833.
193. Qin, Q. and Bassani, J.L.: Non–associated plastic flow in single crystals, J. Mech. Phys. Solids, 40 (1992), 835–862.
194. Rajendren, A.M. and Bless, S.J.: High strain rate material behaviour, in: AFWALTR–85–4009, Materials Laboratory, Air Force Wright Aeronautical Laboratories, Wright–Patterson Air Force Base, Ohio, 45433, 1985.
195. Rashid, M.M., Gray, G.T. and Nemat–Nasser, S.: Heterogeneous deformations in copper single crystals at high and low strain rates, Philosophical Magazine A, 65 (1992), 707–735.
196. Regazzoni, G., Johnson, J.N. and Follansbee, P.S.: Theoretical study of the dynamic tensile test, ASME J. Appl. Mech., 53 (1986), 519–528.
197. Regazzoni, G. and Montheillet, F.: Influence of strain rate on the flow stress and ductility of copper and tantalum at room temperature, in: Third Int'l. Conf. on the Mechanical Properties of Materials at High Rates, (Ed. J. Harding), The Institute of Physics, 70, pp. 63–70, 1984.
198. Rice, J.R.: On the structure of stress–strain relations for time-dependent plastic deformation in metals, J. Appl. Mech., 37 (1970), 728–737.
199. Rice, J.R.: Inelastic constitutive relations for solids: an internal–variable theory and its application to metal plasticity, J. Mech. Phys. Solids, 19 (1971), 433–455.
200. Rice, J.R.: Continuum mechanics and thermodynamics of plasticity in relation to microscale deformation mechanisms, in: Constitutive Equations in Plasticity, (Ed. A.S. Argon), The MIT Press, Cambridge 1975, 23–75.
201. Rice, J.R.: The localization of plastic deformation, in: *Theoretical and Applied Mechanics*, (Ed. W.T. Koiter), North–Holand 1976, 207–220.
202. Rice, J.R. and Rudnicki, J.W.: A note on some features of the theory of localization of deformation, Int. J. Solids Structures, 16 (1980), 597–605.
203. Rice, J.R. and Thomson, R.: Ductile versus brittle behaviour of crystals, Phil. Mag., 29 (1974), 73–97.
204. Richtmyer, R.D.: Principles of Advance Mathematical Physics, Vol. I, Springer, New York 1978.
205. Richtmyer, R.D. and Morton, K.W.: Difference Methods for Initial–Value Problems, John Wiley, New York 1967.
206. Ritchie, R.O., Knott, J.F. and Rice, J.R.: On thee relationship between critical tensile stress and fracture toughness in mild steel, J. Mech. Phys. Solids, 21 (1973), 395–410.
207. Rogers, H.C. and Shastry, C.V.: Material factors in adiabatic shearing in steels, in: Shock Waves and High–Strain–Rate Phenomena in Metals, (Eds. M.A. Meyers and L.E. Murr), Plenum, New York 1981, 285–298.

208. Rudnicki, J.W. and Rice, J.R.: Conditions for the localization of deformation in pressure–sensitive dilatant materials, J. Mech. Phys. Solids, 23 (1975), 371–394.
209. Saje, M., Pan, J. and Needleman, A.: Void nucleation effects on shear localization in porous plastic solids, Int. J. Fracture, 19 (1982), 163–182.
210. Seaman, L., Curran, D.R. and Shockey, D.A.: Stanford Res. Inst. Tech. Rep., No. AFWL-TR-71-156, Dec. 1971.
211. Seaman, L., Curran, D.R. and Murri, W.J.: A continuum model for dynamic tensile microfracture and fragmentation, J. Appl. Mech., 52 (1985), 593–600.
212. Seeger, A.: The generation of lattice defects by moving dislocations and its application to the temperature dependence of the flow–stress of f.c.c. crystals, Phil. Mag., 46 (1955), 1194–1217.
213. Seeger, A.: Kristalplastizitat, in: *Handbuch der Physik VII/2*, (Ed. S. Flugge), Springer–Verlag 1958, 1–208.
214. Serrin, J.: Conceptual analysis of the classical second laws of thermodynamics, Arch. Rational Mech. Anal., 70 (1979), 355–371.
215. Shawki, T.G. and Clifton, R.J.: Shear band formation in thermal viscoplastic materials, Mechanics of Materials, 8 (1989), 13–43.
216. Shield, R.T. and Ziegler, H.: On Prager's hardening rule, ZAMP, 9a (1958), 260–276.
217. Shima, S. and Oyane, M.: Plasticity for porous solids, Int. J. Mech. Sci, 18 (1976), 285–291.
218. Shockey, D.A., Seaman, L. and Curran, D.R.: in: Metallurgical Effects at High Strain Rates, (Eds. R.W. Rohde, B.M. Butcher, J.R. Holland, and C.H. Karbes), Plenum Press, New York 1973, 473.
219. Shockey, D.A., Seaman, L. and Curran, D.R.: The microstatistical fracture mechanics approach to dynamic fracture problem, Int. J. Fracture, 27 (1985), 145–157.
220. Shockey, D.A., Seaman, L. Dao, K.C. and Curran, D.R.: Kinetics of void development in fracturing A533B tensile bars, Trans. ASME J. Pressure Vessel. Tech., 102 (1980), 14–21.
221. Simo, J.C.: A framework for finite strain elastoplasticity based on maximum plastic dissipation and the multiplicative decomposition: Part I. Continuum formulation, Comput. Meths. Appl. Mech. Engrg., 66 (1988), 199–219.
222. Simo, J.C.: A framework for finite strain elastoplasticity based on maximum plastic dissipation and the multiplicative decomposition: Part II. Computational aspects, Comput. Meths. Appl. Mech. Engrg., 66 (1988), 1–31.
223. Simo, J.C. and Taylor, R.L.: Consistent tangent operators for rate–dependent plasticity, Comput. Meths. Appl. Mech. Engrg., 48 (1985), 101–118.
224. Sluys, L.J., Bolck, J. and de Borst, R.: Wave propagation and localization in viscoplastic media, Proc. Third Inter. Conference on Computational Plasticity, Fundamentals and Applications, Barcelona, April 6–10, 1992, (Eds. D.R.J. Owen, E. Onate and E. Hinton), Pineridge Press, Swansea 1992, 539–550.

225. Sokolnikoff, I.S.: The Mathematical Theory of Elasticity, (2d. ed.), Mc Graw-Hill, New York 1956.
226. Spitzig, W.A.: Deformation behaviour of nitrogenated Fe–Ti–Mn and Fe–Ti single crystals, Acta Metall., 29 (1981), 1359–1377.
227. Strang, G. and Fix, G.J.: An Analysis of the Finite Element Method, Prentice-Hall, Englewood Cliffs 1973.
228. Taylor, G.I.: Analysis of plastic strain in a crystal, in: *Stephen Timoshenko 60th Anniversary Volume*, (Ed. J.M. Lessels), Macmillan, New York 1938.
229. Taylor, G.I. and Quinney, H.: The latent energy remaining in a metal after cold working, Proc. R. Soc. Lond., A143 (1934), 307–326.
230. Teodosiu, C. and Sidoroff, F.: A theory of finite elastoplasticity of single crystals, Int. J. Engng. Sci., 14 (1976), 165–176.
231. Thomas, T.Y.: Theoretical effect of large hydrostatic pressures on the tensile strength of materials, Proc. Nat. Acad. Sci., 57 (1967a), 1195–1197.
232. Thomas, T.Y.: Effect of pressure on the ductility od solids, Proc. Nat. Acad. Sci., 58 (1967b), 1274–1278.
233. Thomas, T.Y.: Theory of the physical characteristics of tensile fracture under pressure, Proc. Nat. Acad. Sci., 59 (1968), 700–704.
234. Truesdell, C.: Rational Thermodynamics, Mc Graw–Hill, New York 1969.
235. Truesdell, C. and Noll, W.: The Non–Linear Field Theories of Mechanics, in: *Handbuch der Physik III/3*, (Ed. S. Flűgge), Springer-Verlag, Berlin 1965.
236. Tvergaard, V.: Effects of yield surface curvature and void nucleation on plastic flow localization, J. Mech. Phys. Solids, 35 (1987), 43–60.
237. Tvergaard, V. and Needleman, A.: Analysis of the cup–cone fracture in a round tensile bar, Acta Metall., 32 (1984), 157–169.
238. Tvergaard, V. and Needleman, A.: Effect of material rate sensitivity on failure modes in the Charpy V–notch test, J. Mech. Phys. Solids, 34 (1986), 213–241.
239. Tvergaard, V. and Needleman, A.: Ductile failure modes in dynamically loaded notched bars, in: Damaga Mechanics in Engineering Materials, (Eds. J.W. Ju, et al.), ASME, New York 1990, 117–128.
240. Tvergaard, V. and Van der Giessen, E.: Effects of plastic spin on localization predictions for a porous ductile material, J. Mech. Phys. Solids, 39 (1991), 763–781.
241. Van der Giessen, E.: Continuum models of large deformation plasticity, Part I: Large deformation plasticity and the concept of a natural reference state, Eur. J. Mech., A/Solids, 8 (1989), 15.
242. Van der Giessen, E.: Continuum models of large deformation plasticity, Part II: A kinematic hardening model and the concept of a plastically induced orientational structure, Eur. J. Mech., A/Solids, 8 (1989), 89.
243. Van der Giessen, E.: Micromechanical and thermodynamic aspects of the plastic spin, Int. J. Plasticity, 7 (1991), 365–386.

244. Willems, J.C.: Dissipative dynamical systems, Arch. Rat. Mech. Anal., 45 (1972), 321–393.
245. Wray, P.J.: Strain–rate dependence of the tensile failure of polycrystalline material at elevated temperatures, J. Appl. Phys., 46 (1969), 4018–4029.
246. Yokobori, A.T., Yokobori, T.Jr., Sato, K. and Syo, K.: Fatigue crack growth under mixed modes I and II, Fatigue Fract. Engng. Mater. Struct., 8 (1985), 315–325.
247. Zaremba, S.: Sur une forme perfectionnée de la théorie de la relaxation, Bull. Int. Acad. Sci. Cracovie, (1903), 594–614.
248. Zaremba, S.: Le principe des mouvements relatifs et les équations de la mécanique physique, Bull. Int. Acad. Sci. Cracovie, (1903), 614–621.
249. Zbib, H.M. and Jurban J.S.: Dynamic shear banding: A three–dimensional analysis, Int. J. Plasticity, 8 (1992), 619–641.
250. Ziegler, H.: A modification of Prager's hardening rule, Quart. Appl. Math., 17 (1959), 55–65.

COMPUTATIONAL MODELLING OF LOCALISATION AND FRACTURE

L.J. Sluys
Delft University of Technology, Delft, The Netherlands

ABSTRACT

The computational modelling of failure in a broad class of materials is analysed. Beyond a critical load level deformation may localise in small bands or cracks dependent on the load condition and the properties of the material. The localisation process and fracture have been studied in a finite element framework. The virtual work equation that describes the motion of an inelastic body is discretised and nonlinear solution techniques have been treated. This computational procedure can be used in combination with simple models to describe failure as plasticity, damage and crack models. Localisation is analysed by means of a simple bar analysis and the problem of mesh sensitivity has been explained. To solve this problem enhancement of the plasticity theory is proposed by means of higher-order time derivatives (viscoplastic model) or higher-order spatial derivatives (gradient model). An extensive algorithmic elaboration of all models has been given.

1. INTRODUCTION

A large number of engineering materials including metals, polymers, soils, concrete and rock are classified as *softening* materials. These materials show a reduction of the load-carrying capacity accompanied by increasing localised deformations after reaching the limit load, i.e. the load-displacement characteristic exhibits a descending branch. For a continuum description of this type of structural behaviour the standard procedure to derive stress-strain relations, which consists of an affine mapping from the measured load-displacement relation onto a stress-strain diagram, i.e. stress and strain are computed as the quotients of the force and the virgin load-carrying cross-section, and of the displacement and the length of the specimen, respectively, then leads to a negative slope of the stress-strain diagram. This is commonly called *strain softening*.

Beyond a critical load level softening causes all further deformation to localise in small bands, which are often a precursor to failure. Roughly, we can observe two types of localisation. Firstly, if the cohesive properties of the material are more critical than the frictional properties localisation of deformation takes place in fracture zones (mode-I localisation). However, in the opposite case localisation becomes manifest along shear bands (mode-II localisation).

The straightforward use of the strain-softening model in a classical continuum generally does not result in a mathematically well-posed problem. The field equations that describe the motion of the body lose hyperbolicity and become elliptic as soon as strain softening occurs. In fact, the domain is split up into an elliptic part, in which the waves have imaginary wave speeds and are not able to propagate (standing waves), and into a hyperbolic part with propagating waves. The initial value problem becomes ill-posed and can no longer be a proper description of the underlying physical problem. Because of the inability of the standing waves to propagate, localisation zones stay confined to a line with zero thickness (or a discrete plane in a three-dimensional continuum). Spurious wave reflections occur on these localisation zones with zero thickness and the energy consumed in the failure zones is zero. These results are in contradiction with experimental data, which for mode-I as well as for mode-II localisation show finite widths of the localisation zone and finite values for the energy consumption andthe wave reflection. The finite element solution tries to capture the localisation zone of zero thickness which results in a *mesh sensitivity*.

For proper failure analyses the above observations are unacceptable and we must therefore rephrase our continuum description of the softening solid such that it is able to capture zones of highly localised deformation. The main goal of the present study is to scrutinise approaches which may remedy the deficiencies. For a proper mathematical modelling of the softening solid extra or higher-order derivative terms are necessary in the continuum description. Such enriched, or higher-order continua do not necessarily lose hyperbolicity at the onset of strain softening and admit a solution with real wave speeds. In these lecture notes we suggest two so-called regularisation techniques to solve mesh sensitivity, namely (i) the addition of viscous, or higher-order time derivatives and (ii) the addi-

Box 1.1 Mathematical preliminaries.

Inner product :	$\mathbf{u}^T \mathbf{v}$
Cross (outer) product :	$\mathbf{u}\mathbf{v}^T$
Matrix transpose :	$\mathbf{B} = \mathbf{A}^T \;\rightarrow\; b_{ij} = a_{ji}$
Matrix inverse :	$\mathbf{A}^{-1} \quad \mathbf{A}\mathbf{A}^{-1} = \mathbf{I} \quad \mathbf{A}^{-1}\mathbf{A} = \mathbf{I}$
norm (Euclidian or L_2) :	$|\mathbf{v}|_2 = (\mathbf{v}^T \mathbf{v})^{\frac{1}{2}}$
Gradient of a scalar function :	$\mathbf{b} = \dfrac{\partial f}{\partial \mathbf{a}}$
Sherman-Morrison formula :	$(\mathbf{A} + \mathbf{u}\mathbf{v}^T)^{-1} = \mathbf{A}^{-1} - \dfrac{\mathbf{A}^{-1}\mathbf{u}\mathbf{v}^T\mathbf{A}^{-1}}{1 + \mathbf{v}^T \mathbf{A}^{-1}\mathbf{u}}$
Einstein convention :	$c_{ij} = \sum\limits_{k=1}^{n} a_{ik} b_{kj} = a_{ik} b_{kj}$
Kronecker delta δ_{ij} :	$\delta_{ij} = 0 \quad \text{if} \quad i \neq j$
	$\delta_{ij} = 1 \quad \text{if} \quad i = j$
Gauss divergence theorem :	$\int_V \operatorname{div} \mathbf{v}\, dV = \int_S \mathbf{n}^T \mathbf{v}\, dS$
or (tensor notation) :	$\int_V \dfrac{\partial v_i}{\partial x_i}\, dV = \int_S n_i v_i\, dV$
	$\operatorname{div} \mathbf{v} = \dfrac{\partial v_1}{\partial x_1} + \dfrac{\partial v_2}{\partial x_2} + \dfrac{\partial v_3}{\partial x_3}$

tion of higher-order deformation gradients. A proper modelling of the softening solid is obtained if the numerical results converge to a finite size of the localisation zone upon mesh refinement with unique properties with respect to energy consumption and wave reflection (in dynamics). Below a certain discretisation level at which accuracy of the results is obtained a refinement of the mesh as well as a change of orientation of the mesh lines should not affect the numerical outcome.

In Chapter 2 the basic equations for the motion of an inelastic body will be set-up. By means of the finite element representation of the virtual work equation the problem is discretised. In Chapter 3 nonlinear solution techniques that can be used in combination with the plasticity, damage and crack models treated in Chapter 4 are discussed. In Chapter 5 the computational problems that come up with the modelling of localisation processes will be explained. Finally, in Chapter 6 two methods will be described to solve the numerical problem of mesh dependence as described in Chapter 5. Firstly, viscoplastic approaches will be discussed and, secondly, a modelling that involves the inclusion of higher-order strain gradients is treated. For the notation both matrix-vector and tensor format will be used. In Box 1.1 some mathematical preliminaries of these notes are outlined.

2. FORMULATION OF THE BOUNDARY VALUE PROBLEM

In this Chapter the equations of motion of the inelastic body will be derived. In the derivation a restriction is made to small displacement gradients. We use the finite element representation of the virtual work equation to discretise the problem.

2.1 Preliminary equations

In a general three-dimensional continuum the equations of motion of an elementary volume V without damping can be written as

$$\mathbf{L}^T \boldsymbol{\sigma} + \mathbf{p} = \mathbf{R}\ddot{\mathbf{u}}, \qquad (2.1)$$

in which $\boldsymbol{\sigma}$ is a vector containing the stress components $(\sigma_{xx}, \sigma_{yy}, \sigma_{zz}, \sigma_{xy}, \sigma_{yz}, \sigma_{zx})$, while in the vector $\mathbf{u}$ the displacement components are assembled (u_x, u_y, u_z). A superimposed dot denotes differentiation with respect to time and a superimposed double dot implies that a quantity is differentiated twice, which means that $\ddot{\mathbf{u}}$ is the acceleration vector. The density matrix $\mathbf{R}$ is equal to $diag[\rho, \rho, \rho]$ with density ρ. In the vector $\mathbf{p}$ the body forces are assembled. The kinematic equations

$$\boldsymbol{\varepsilon} = \mathbf{L}\mathbf{u}, \qquad (2.2)$$

set the relations between the strain components $(\varepsilon_{xx}, \varepsilon_{yy}, \varepsilon_{zz}, 2\varepsilon_{xy}, 2\varepsilon_{yz}, 2\varepsilon_{zx})$, assembled in the vector $\boldsymbol{\varepsilon}$ and the displacement components. In eq.(2.1) and (2.2) the differential operator matrix $\mathbf{L}$ is defined as

$$\mathbf{L} = \begin{bmatrix} \frac{\partial \cdot}{\partial x} & 0 & 0 \\ 0 & \frac{\partial \cdot}{\partial y} & 0 \\ 0 & 0 & \frac{\partial \cdot}{\partial z} \\ \frac{\partial \cdot}{\partial y} & \frac{\partial \cdot}{\partial x} & 0 \\ 0 & \frac{\partial \cdot}{\partial z} & \frac{\partial \cdot}{\partial y} \\ \frac{\partial \cdot}{\partial z} & 0 & \frac{\partial \cdot}{\partial x} \end{bmatrix}, \qquad (2.3)$$

and the superscript T is the transpose symbol. The constitutive equations can be given in the general rate format

$$\dot{\boldsymbol{\sigma}} = \mathbf{D}_i \dot{\boldsymbol{\varepsilon}}, \qquad (2.4)$$

with matrix $\mathbf{D}_i$ containing the tangent stiffness moduli. In the nonlinear calculations treated in these lecture notes, we apply a decomposition of the total strain rate $\dot{\boldsymbol{\varepsilon}}$ into the elastic strain rate $\dot{\boldsymbol{\varepsilon}}_e$ and the inelastic strain rate $\dot{\boldsymbol{\varepsilon}}_i$ according to

$$\dot{\varepsilon} = \dot{\varepsilon}_e + \dot{\varepsilon}_i \,. \tag{2.5}$$

The stress rate must satisfy

$$\dot{\sigma} = \mathbf{D}_e \dot{\varepsilon}_e \,, \tag{2.6}$$

with matrix $\mathbf{D}_e$ containing the elastic stiffness moduli according to

$$\mathbf{D}_e = \frac{E}{(1+\nu)(1-2\nu)} \begin{bmatrix} 1-\nu & \nu & \nu & 0 & 0 & 0 \\ \nu & 1-\nu & \nu & 0 & 0 & 0 \\ \nu & \nu & 1-\nu & 0 & 0 & 0 \\ 0 & 0 & 0 & \tfrac{1}{2}(1-2\nu) & 0 & 0 \\ 0 & 0 & 0 & 0 & \tfrac{1}{2}(1-2\nu) & 0 \\ 0 & 0 & 0 & 0 & 0 & \tfrac{1}{2}(1-2\nu) \end{bmatrix} \tag{2.7}$$

with E the Young's modulus and ν Poisson's ratio. The shear modulus $G = E/(2(1+\nu))$. Substitution of eq.(2.5) into (2.6) leads to

$$\dot{\sigma} = \mathbf{D}_e (\dot{\varepsilon} - \dot{\varepsilon}_i) \,, \tag{2.8}$$

which can be used for the derivation of the tangent stiffness matrix. If the inelastic strain rate $\dot{\varepsilon}_i$ can be written in an explicit format substitution into eq.(2.8) yields the matrix $\mathbf{D}_i$. At the boundary S of the body it is required that either

$$\mathbf{t} - \sigma \bar{\mathbf{n}} = 0 \,, \tag{2.9}$$

with $\mathbf{t}$ the boundary traction and $\bar{\mathbf{n}}$ the outward normal to the surface of the body, or

$$\mathbf{u}_{\bar{n}} = \mathbf{u}_s \,, \tag{2.10}$$

with $\mathbf{u}_{\bar{n}}$ the displacements at the boundary and $\mathbf{u}_s$ the prescribed displacements.

2.2 Weak formulation of the boundary value problem

While eq.(2.1) describes the motion of the body in a strong sense, a weak form of these equations is obtained by setting

$$\int_V \delta \mathbf{u}^T [\mathbf{L}^T \sigma - \mathbf{R}\ddot{\mathbf{u}} + \mathbf{p}] \, dV = 0 \,, \tag{2.11}$$

in which δ denoting the variation of a quantity. With aid of Green's theorem

$$\int_V \delta \mathbf{u}^T [\mathbf{L}^T \sigma] \, dV = -\int_V \delta \varepsilon^T \sigma \, dV + \int_S \delta \mathbf{u}^T [\sigma \bar{\mathbf{n}}] \, dS \,, \tag{2.12}$$

eq.(2.11) can be transformed into

$$\int_V \delta \mathbf{u}^T [\mathbf{R}\ddot{\mathbf{u}}] \, dV + \int_V \delta \varepsilon^T \sigma \, dV = \int_V \delta \mathbf{u}^T \mathbf{p} \, dV + \int_S \delta \mathbf{u}^T \mathbf{t} \, dS \,, \tag{2.13}$$

in which eq.(2.9) has been substituted. Note that in the derivation of eq.(2.13) no assumptions have been made with regard to the material behaviour.

If eq.(2.13) is considered to be valid at time $t+\Delta t$ the evolution of the stress follows from

$$\sigma^{t+\Delta t} = \sigma^t + \int_t^{t+\Delta t} \dot{\sigma}\, d\tau \,. \tag{2.14}$$

If the constitutive relation (2.4) is substituted in eq.(2.14) we can rewrite the virtual work equation (2.13) into

$$\int_V \delta \mathbf{u}^T [\mathbf{R}\ddot{\mathbf{u}}^{t+\Delta t}]\, dV + \int_V \delta \boldsymbol{\varepsilon}^T \int_t^{t+\Delta t} \mathbf{D}_i \dot{\boldsymbol{\varepsilon}}\, d\tau\, dV = \tag{2.15}$$

$$\int_V \delta \mathbf{u}^T \mathbf{p}^{t+\Delta t}\, dV + \int_S \delta \mathbf{u}^T \mathbf{t}^{t+\Delta t}\, dS - \int_V \delta \boldsymbol{\varepsilon}^T \boldsymbol{\sigma}^t\, dV \,.$$

2.3 Discretisation of the boundary value problem

We shall consider the finite element representation of the virtual work equation (2.15) for the dynamic motion of the inelastic body. First, we will discuss the spatial discretisation of the problem and briefly the time integration is treated. The body can be divided into a finite number of elements. For each element the continuous displacement field $\mathbf{u}$ can be interpolated by

$$\mathbf{u} = \mathbf{H}\mathbf{a} \,, \tag{2.16}$$

and the continuous acceleration field by

$$\ddot{\mathbf{u}} = \mathbf{H}\ddot{\mathbf{a}} \,, \tag{2.17}$$

in which the matrix $\mathbf{H}$ contains the interpolation polynomials and $\mathbf{a}$ and $\ddot{\mathbf{a}}$ the nodal displacements and nodal accelerations, respectively. Combining eqs.(2.2) and (2.16) and introducing the strain-nodal displacement matrix

$$\mathbf{B} = \mathbf{L}\mathbf{H} \,, \tag{2.18}$$

the relation between the strains and the nodal displacements is obtained as

$$\boldsymbol{\varepsilon} = \mathbf{B}\mathbf{a} \,, \tag{2.19}$$

or in rate format as

$$\dot{\boldsymbol{\varepsilon}} = \mathbf{B}\dot{\mathbf{a}} \,. \tag{2.20}$$

We can now substitute eqs.(2.16), (2.17), (2.19) and (2.20) into the virtual work expression (2.15), which yields

$$\delta \mathbf{a}^T \int_V \mathbf{H}^T \mathbf{R} \mathbf{H} \ddot{\mathbf{a}}^{t+\Delta t}\, dV + \delta \mathbf{a}^T \int_V \int_t^{t+\Delta t} \mathbf{B}^T \mathbf{D}_i \mathbf{B} \dot{\mathbf{a}}\, d\tau\, dV =$$

$$\delta \mathbf{a}^T \int_V \mathbf{H}^T \mathbf{p}^{t+\Delta t}\, dV + \delta \mathbf{a}^T \int_S \mathbf{H}^T \mathbf{t}^{t+\Delta t}\, dS - \delta \mathbf{a}^T \int_V \mathbf{B}^T \boldsymbol{\sigma}^t\, dV \,. \tag{2.21}$$

Here, we consider eq.(2.21) on a structural level and the solution method for the nonlinear set of algebraic equations determines the time integral. In these notes we use the incremental iterative Newton-Raphson solution method (see Chapter 3). For the determination of $\int \mathbf{B}^T \mathbf{D}_i \mathbf{B} \dot{\mathbf{a}} \, d\tau$ both explicit and implicit integration schemes have been used (see Chapter 4 and 6). In the zero-th iteration we start from the stress-strain matrix at time t, i.e. $\mathbf{D}_i^t$. Furthermore, we define the incremental nodal displacement vector as

$$\Delta \mathbf{a} = \mathbf{a}^{t+\Delta t} - \mathbf{a}^t = \int_t^{t+\Delta t} \dot{\mathbf{a}} \, d\tau . \tag{2.22}$$

Since we assume that identity (2.21) must hold for any admissible $\delta \mathbf{a}$ we obtain for the zero-th iteration at time $t + \Delta t$

$$\int_V \mathbf{H}^T \mathbf{R} \mathbf{H} \ddot{\mathbf{a}}^{t+\Delta t} \, dV + \int_V \mathbf{B}^T \mathbf{D}_i^t \mathbf{B} \Delta \mathbf{a} \, dV = \int_V \mathbf{H}^T \mathbf{p}^{t+\Delta t} \, dV + \int_S \mathbf{H}^T \mathbf{t}^{t+\Delta t} \, dS - \int_V \mathbf{B}^T \boldsymbol{\sigma}^t \, dV \tag{2.23}$$

If we introduce the notations

$$\mathbf{M} = \int_V \mathbf{H}^T \mathbf{R} \mathbf{H} \, dV \tag{2.24}$$

for the mass matrix,

$$\mathbf{K}^t = \int_V \mathbf{B}^T \mathbf{D}_i^t \mathbf{B} \, dV \tag{2.25}$$

for the tangential stiffness matrix,

$$\mathbf{f}_e^{t+\Delta t} = \int_V \mathbf{H}^T \mathbf{p}^{t+\Delta t} \, dV + \int_S \mathbf{H}^T \mathbf{t}^{t+\Delta t} \, dS \tag{2.26}$$

for the external load vector and

$$\mathbf{f}_i^t = \int_V \mathbf{B}^T \boldsymbol{\sigma}^t \, dV \tag{2.27}$$

for the internal force vector, we can rewrite eq.(2.23) into

$$\mathbf{M} \ddot{\mathbf{a}}^{t+\Delta t} + \mathbf{K}^t \Delta \mathbf{a} = \mathbf{f}_e^{t+\Delta t} - \mathbf{f}_i^t . \tag{2.28}$$

Eq.(2.28) represents the semi-discrete nonlinear equation of motion governing the response of a system of finite elements.

2.4 Time discretisation

In dynamic problems equation (2.28) needs to be discretised further by means of a time integration scheme. Explicit and implicit schemes can be used (see Hughes (1987)). The explicit scheme in combination with a lumped mass matrix results in a fully decoupled system of equations that can be solved directly. The implicit time integration results in a coupled set of equations which is of the same type as eq.(2.28) with a modified mass matrix **M** and stiffness matrix **K**. The solution procedures treated in the next Chapter can be applied to this system of equations and a solution can be found.

3. SOLUTION TECHNIQUES

3.1 Incremental-iterative solution techniques

When equation (2.28) is solved for the static case the inertia term is omitted which yields

$$\mathbf{K}^t \Delta \mathbf{a} = \mathbf{f}_e^{t+\Delta t} - \mathbf{f}_i^t \ . \tag{3.1}$$

The time t represents the pseudo-time. When a structure is in a state of equilibrium the difference between the internal forces and the external loads vanishes. As a consequence the vector which assembles the displacement increments, $\Delta \mathbf{a}$, also vanishes. In principle it would be possible to impose the entire external load $\mathbf{f}_e$ in a single step. Often, this is not very sensible. In the first place the iterative procedure usually has a hard time to iterate towards a properly converged solution for very large load steps. Since the convergence radius is limited for most commonly used iterative procedures (e.g. the Newton-Raphson process which will be treated next), the possibility of divergence of the iterative procedure used to solve the set of non-linear algebraic equations definitely exists. Secondly, experiments show that most structural materials are path-dependent, which means that different values for the stress will be obtained depending on the strain path that is followed. For instance, the resulting stress will be different when we first apply tension on a panel followed by a shear strain increment or when the same strain increments are imposed in the reversed order. It is obvious that the true structural behaviour can only be predicted accurately if the strain increments are relatively small, so that the true strain path is followed closely.

Accordingly, it is recommended to apply the total external load in a number of small loading steps (or increments). Such a procedure is usually called an *incremental procedure* and is shown graphically in Figure 3.1. In this method the total external load is decomposed in a contribution $\mathbf{f}_e^t$ that is already present at time t, i.e. at the beginning of the time or load step, and an increment $\Delta \mathbf{f}_e$. We can rewrite eq. (3.1) as

$$\mathbf{K}^t \Delta \mathbf{a} = \Delta \mathbf{f}_e + \mathbf{f}_e^t - \mathbf{f}_i^t \ . \tag{3.2}$$

At the beginning of each step the difference between external loads and internal forces should be zero. This is indeed the case if a properly converged solution is obtained after each loading step. In practice, perfect convergence in each loading step is only seldom accomplished and, therefore, omission of the contribution $\mathbf{f}_e^t - \mathbf{f}_i$ from the equilibrium equations (3.2) may entail significant errors. In particular, any unbalance that has been left behind at the end of the previous loading step will be carried along in all subsequent loading steps. This implies that we get an accumulation of errors, which causes a significant drift from the true load-deflection path the structure is following (see Figure 3.1). Ultimately, this drifting tendency will result in a considerable overshoot of the collapse load of the structure. Another cause for the drifting tendency in a pure incremental method is the fact that the tangential stiffness matrix is derived through a linearisation of the (non-linear) set of equations *at the beginning* of a loading step. Strictly speaking, this stiffness matrix is tangential to the equilibrium path in the force-displacement space only during an infinitesimally small increment after the beginning of the loading step, but certainly not during the

Computational Modelling of Localisation and Fracture

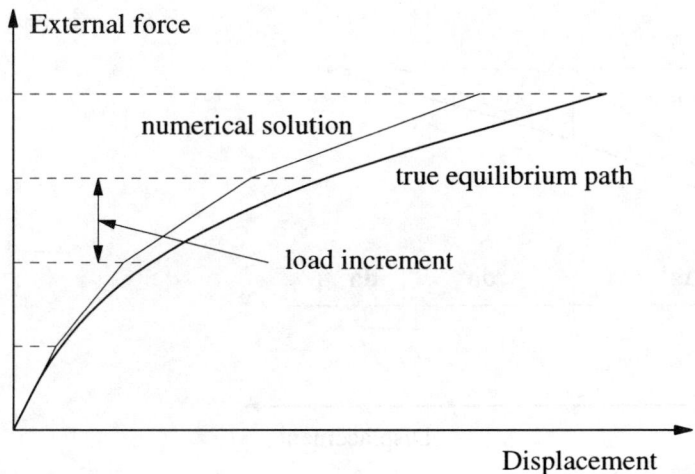

Figure 3.1 Purely incremental solution procedure: drifting tendency of the numerical solution.

entire loading step. Consequently, eq. (3.1) is only valid in an exact sense at the beginning of the loading step, while it is just an approximation thereafter.

The gradual departure of the numerical solution from the true solution can be prevented, or at least be made smaller, by adding equilibrium iterations within each loading step. Now, we obtain an *incremental-iterative procedure* instead of a pure incremental procedure. For explanation of the method we will use the superscript for the iteration number and no longer for the time step which implies that

$$\mathbf{K}^t = \mathbf{K}^0 , \tag{3.3}$$

$$\mathbf{f}_i^t = \mathbf{f}_i^0 \tag{3.4}$$

at the beginning of the step. The loading

$$\mathbf{f}_e^{t+\Delta t} = \mathbf{f}_e , \tag{3.5}$$

is fixed during iteration. The converged state after a sufficient number of iterations belongs to time $t + \Delta t$. In an incremental-iterative solution method which is known as the *Newton-Raphson procedure* a first estimate for the displacement increment $\Delta \mathbf{a}$ is made through

$$\Delta \mathbf{a}^1 = [\mathbf{K}^0]^{-1}[\mathbf{f}_e - \mathbf{f}_i^0] , \tag{3.6}$$

where the superscript 1 of the $\Delta \mathbf{a}$ signifies that we deal with the estimate in the first iteration for the incremental displacement vector. Likewise, the superscript 0 of the internal force vector relates to the fact that this vector is calculated using the stresses at the beginning of the loading step $\sigma^t = \sigma^0$,

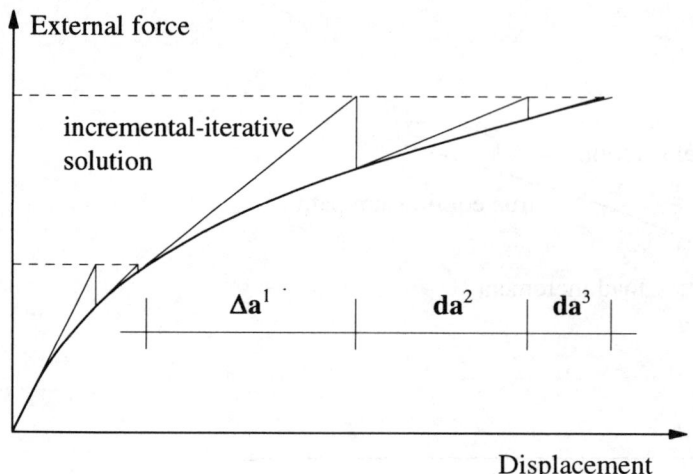

Figure 3.2 Iterative-incremental solution procedure.

$$\mathbf{f}_i^0 = \int_V \mathbf{B}^T \sigma^0 dV \,. \tag{3.7}$$

From the incremental displacement vector $\Delta \mathbf{a}^1$ a first estimate for the strain increment $\Delta \varepsilon^1$ can be calculated, whence, using the stress-strain law the stress increment $\Delta \sigma^1$ can be computed according to

$$\Delta \sigma^1 = \mathbf{D}_i^1 \Delta \varepsilon^1 \tag{3.8}$$

assuming that an implicit stress update scheme is used to determine $\mathbf{D}_i^1$ (see Chapter 4 and 6). The total stresses after the first iteration are then given by:

$$\sigma^1 = \sigma^0 + \Delta \sigma^1 \,. \tag{3.9}$$

Generally, the internal force vector $\mathbf{f}_i^1$, that is computed on basis of the stresses σ^1 is not in equilibrium with the external loads $\mathbf{f}_e$ that have been added up to and including this loading step and an unbalance vector $\mathbf{r}$ remains which is equal to

$$\mathbf{r}^1 = \mathbf{f}_e - \mathbf{f}_i^1 \,. \tag{3.10}$$

For the second iteration first a better tangential stiffness matrix can be computed. From the updated stress-strain law we calculate

$$\mathbf{K}^1 = \int_V \mathbf{B}^T \mathbf{D}_i^1 \mathbf{B} \, dV \,. \tag{3.11}$$

A correction to the displacement increment is necessary and denoting this correction by $\mathbf{da}^2$ we have

$$\mathbf{da}^2 = [\mathbf{K}^1]^{-1}[\mathbf{f}_e - \mathbf{f}_i^1] \,. \tag{3.12}$$

The displacement increment after the second iteration in the loading step follows from

$$\Delta \mathbf{a}^2 = \Delta \mathbf{a}^1 + \mathbf{da}^2 \,. \tag{3.13}$$

In a similar fashion as with the calculation of the strain and stress increment in the first iteration the quantities $\Delta \varepsilon^2$ and $\Delta \sigma^2$ are now computed, while using the latter quantity an improved approximation for the stress at the end of the loading step, σ^2 can be made. Repetition of this process, which can be formulated mathematically as

$$\mathbf{da}^{j+1} = [\mathbf{K}^j]^{-1}[\mathbf{f}_e - \mathbf{f}_i^j] \,, \tag{3.14}$$

$$\Delta \mathbf{a}^{j+1} = \Delta \mathbf{a}^j + \mathbf{da}^{j+1} \,, \tag{3.15}$$

$$\Delta \varepsilon^{j+1} = \mathbf{B} \Delta \mathbf{a}^{j+1} \,, \tag{3.16}$$

$$\Delta \sigma^{j+1} = \mathbf{D}_t^{j+1} \Delta \varepsilon^{j+1} \,, \tag{3.17}$$

$$\sigma^{j+1} = \sigma^0 + \Delta \sigma^{j+1} \,, \tag{3.18}$$

and with the internal and unbalance forces

$$\mathbf{f}_i^{j+1} = \int_V \mathbf{B}^T \sigma^{j+1} dV \,, \tag{3.19}$$

$$\mathbf{r}^{j+1} = \mathbf{f}_e - \mathbf{f}_i^{j+1} \,, \tag{3.20}$$

respectively. This iterative process ultimately results in stresses that are in equilibrium with each other and with the applied external loading to within some user-prescribed convergence tolerance. A graphical explanation is given in Figure 3.2.

The procedure summarised in eqs. (3.14)-(3.20) is sometimes referred to as the total incremental method. Every iteration the total displacement increment within the step is computed and on basis of this total displacement increment the total strain increment and the total stress increment are computed. Then, the new stresses are found as the sum of the stresses at the beginning of the step and the total stress increment. As an alternative approach, we might continue to work with corrections. Rather than first adding the correction to the total displacement increment obtained in the previous iteration, we can also proceed by calculating a correction to the strain increment $\mathbf{d}\varepsilon^{j+1}$, which can be used to compute a correction to the stress increment $\mathbf{d}\sigma^{j+1}$. Instead of eqs. (3.14)-(3.20) we then get the iteration sequence:

$$\mathbf{da}^{j+1} = [\mathbf{K}^j]^{-1}[\mathbf{f}_e - \mathbf{f}_i^j] \,, \tag{3.21}$$

$$\mathbf{d}\varepsilon^{j+1} = \mathbf{B}\mathbf{da}^{j+1} \,, \tag{3.22}$$

$$\mathbf{d}\sigma^{j+1} = \mathbf{D}_t^{j+1} \mathbf{d}\varepsilon^{j+1} \,, \tag{3.23}$$

$$\sigma^{j+1} = \sigma^j + \mathbf{d}\sigma^{j+1} \,. \tag{3.24}$$

Usually, this so-called delta-incremental update methodology is less robust than the first algorithm. This is particularly so when we have materially non-linear models in which we have different behaviour in loading than in unloading, e.g. plasticity or damage. Then, the second algorithm may result in pseudo-unloading which has severe consequences for numerical stability.

The case of load control is used here but we can also prescribe that one or more displacements grow. This so-called displacement control procedure causes a stress development within the specimen, which in turn results in nodal forces at the nodes where the displacements are prescribed. Summation of these forces gives the total reaction force (which equals the external load) on the prescribed displacements. When there is no preference for either load or displacement control from a physical point of view, the latter method is nearly always to be preferred. The reasons for the preference for displacement control are twofold:

- The tangent stiffness matrix is better conditioned for displacement control than for load control. This will result in a faster convergence behaviour of the iterative procedure.
- Under load control, the tangent stiffness matrix becomes singular at a limit point in the load-deflection diagram, not only when global failure occurs, but also when we have a local maximum along this curve. The tangent stiffness matrix of the displacement controlled problem on the other hand does not become singular.

Sometimes, typical structural behaviour (e.g. snap-back behaviour) is still not traceable with a displacement control procedure. The most elegant procedure that can be used to analyse these kind of problems properly is known as the 'arc-length control' method (see Riks (1979), Crisfield (1991)).

3.2 Convergence criteria

In order to be able to assess whether an iterative procedure has converged a so-called convergence criterion is needed. Such a criterion requires that some quantity, e.g. a force or a displacement, must be approximated within some user-specified tolerance. If the error does not become smaller than the pre-set tolerance the iterative process is said to be not converged. If the quantity that is being monitored becomes unbounded the process diverges. To prevent that the computer program continues to search for a solution when either of the latter two possibilities occur, a maximum number of iterations must be specified by the user. It is difficult to provide guidelines to which number of iterations the iterative procedure should be limited, since this is not only problem-dependent, but also depends on the type of iterative procedure, e.g. Newton-Raphson, modified Newton (no update of the tangent stiffness matrix in every iteration), elastic stiffness, that is employed and on the quantity that is being monitored. Nonetheless, a general recommendation could be to set the maximum number of iterations equal to 8 for the full Newton-Raphson process, equal to 20 for the modified Newton-Raphson process and equal to 30 for elastic stiffness approaches.

The above recommendations are primarily valid for so-called global convergence criteria, in which a global quantity is monitored. Here, one may think of a norm of the unbalanced forces, a norm of the displacement increments or the energy of the system. When the interest mainly lies in obtaining an impression of the global behaviour of the structure, e.g. the ultimate bearing capacity, such a global criterion is always sufficient. Also, in analyses in which attention is focused on local structural behaviour one can usually rely on a global convergence criterion. Yet, in critical cases it may be wise to adopt a local convergence criterion in which for instance the unbalanced forces of a number of nodes identically have to be zero to within some tolerance.

Since the majority of the finite element packages only offer options for monitoring the convergence behaviour in a global sense, we will restrict the treatment to such criteria. In particular, a force criterion, a displacement criterion and an energy criterion will be discussed, since these criteria are the most popular in existing finite element software.

The most demanding criterion is usually the force norm. With this criterion equilibrium iterations are added till the change in the norm of the unbalanced force vector is smaller than the prescribed convergence tolerance ε times the value of the norm in the first iteration of that loading step. As a rule, the L_2-norm is used to measure the unbalanced force vector and iterations are terminated when:

$$|\mathbf{f}_e - \mathbf{f}_i^j|_2 \leq \varepsilon \times |\mathbf{f}_e - \mathbf{f}_i^1|_2 . \tag{3.25}$$

A reasonable balance between accuracy and consumption of computer time is usually achieved if the tolerance ε is set equal to 10^{-3}. Another criterion that is used frequently in non-linear finite element analysis is the energy criterion. Now, the iterations are terminated when:

$$[\mathbf{f}_i^j]^T \mathbf{da}^j \leq \varepsilon \times [\mathbf{f}_i^1]^T \Delta \mathbf{a}^1 . \tag{3.26}$$

Experience shows that this convergence criterion is easier to satisfy than the preceding criterion. To achieve the same accuracy the tolerance may often not be chosen higher than 10^{-4}. The norm of the incremental displacements is relatively easy to satisfy. If convergence is achieved if

$$|\mathbf{da}^j|_2 \leq \varepsilon \times |\Delta \mathbf{a}^1|_2 . \tag{3.27}$$

then the ε-parameter may not exceed 10^{-6} in order that a reasonably accurate solution is obtained.

The choice of a convergence criterion and the associated convergence tolerance ε must be done with great care. A simple example is pure relaxation. For this problem the incremental displacements have the correct value immediately after the first iteration, whereas it may need several more iterations to let the stresses, and consequently the internal forces, relax to their proper values. It will be clear that in such a case any convergence criterion that involves the incremental displacement vector $\mathbf{da}$, including the energy criterion and the norm of incremental displacements, will erroneously identify the process as converged after the first iteration. In such cases only a force norm does not result in a premature termination of the iterative procedure.

Also the value of the convergence tolerance ε must be chosen carefully. On one hand

Box 3.1 Flow of nonlinear finite element code.

FOR EACH LOAD OR TIME STEP :

1. Iteration $j = 0$, $\Delta \mathbf{a}^0 = \mathbf{0}$
2. Compute new external force vector : $\mathbf{f}_e = \mathbf{f}_e^{t+\Delta t}$
3. Compute tangent stiffness matrix : $\mathbf{K}^j = \int_V \mathbf{B}^T \mathbf{D}_i^j \mathbf{B} \, dV$
4. Solve, e.g. by LDU-decomposition the linear system : $\mathbf{K}^j \mathbf{da}^{j+1} = \mathbf{f}_e - \mathbf{f}_i^j$
5. Add correction $\mathbf{da}^{j+1}$ to the increm. displacement vector : $\Delta \mathbf{a}^{j+1} = \Delta \mathbf{a}^j + \mathbf{da}^{j+1}$
6. Compute strain increment $\Delta \varepsilon^{j+1} = \mathbf{B} \Delta \mathbf{a}^{j+1}$ for each int. point
7. Stress update : $\Delta \sigma^{j+1} = \mathbf{D}_i^{j+1} \Delta \varepsilon^{j+1}$ for each int. point

   (implicit integration of the constitutive equations is assumed)
8. Add stress increment to total stress $\sigma^{j+1} = \sigma^0 + \Delta \sigma^{j+1}$ for each int. point
9. Compute internal force vector : $\mathbf{f}_i^{j+1} = \int_V \mathbf{B}^T \sigma^{j+1} \, dV$
10. Check convergence : $[\mathbf{f}_e - \mathbf{f}_i^{j+1}] <$ tolerance

    if YES, go to next step

    if NO, $j = j + 1$ go to 3.

---

a too loose convergence may result in inaccurate and unreliable answers. On the other hand, a too strict convergence tolerance sometimes hardly improves the results while costing a lot of expensive computer time. One percent improvement in accuracy may cause a doubling of the required computer time. Here, one should also realise that convergence is a relative matter. Neither of the convergence criteria discussed above warrants that the error be smaller than some prescribed value in an absolute sense.

Finally, while it is clear that a diverged solution is unreliable, it is more difficult to assess whether a non-converged solution must also be abandoned. For instance, it usually makes perfect sense to continue a solution from the end of a loading step in which the change of the internal energy in the last iteration is equal to let's say 10 times the convergence tolerance.

3.3 Summary

The flow of computation in a nonlinear finite element code as discussed above is summarised in Box 3.1. For this a load-controlled set-up with a total incremental approach and a force norm is used. The next Chapter focusses on the determination of the nonlinear stress-strain laws with matrix D_i.

4. PLASTICITY, DAMAGE AND FRACTURE

Failure in softening materials can be modelled using a phenomenological approach or a micromechanical approach. For the micromechanical approach exact knowledge about the microstructural changes in the softening zone is necessary. For instance, for concrete fracture data about particles, matrix and interface should be described in force-displacement relations. Especially for large scale structures there is a need for models at a macro level and for this reason phenomenological failure models will be treated here. Examples of phenomenological strain-softening models are plasticity models with yield limit degradation, continuous damage models, based on a degradation of the stiffness determined by a damage parameter and crack models, based on a total or decomposed strain concept. Here, we will first treat plasticity models, next damage models and finally smeared crack models formulated in a fixed crack concept.

4.1 Computational plasticity

The plastic response of a material becomes manifest as soon as a combination of the stress components reaches a characteristic value. For isotropic hardening or softening plasticity this characteristic value is governed by a yield function f of the form

$$f(\sigma, \kappa) = 0 , \qquad (4.1)$$

in which κ is a scalar valued hardening or softening parameter which is dependent on the strain history. In rate-independent plasticity inelastic deformations occur if the stress point is on the yield surface. Stress states outside the yield contour are not possible. For this reason, the stress point must remain on the yield contour during plastic flow. This leads to a second condition for plastic deformation

$$\dot{f}(\sigma, \kappa) = 0 , \qquad (4.2)$$

which is commonly referred to as Prager's consistency condition. In plasticity theory the strain rate vector is decomposed into an elastic $\dot{\varepsilon}_e$ and a plastic $\dot{\varepsilon}_p$ part according to

$$\dot{\varepsilon} = \dot{\varepsilon}_e + \dot{\varepsilon}_p . \qquad (4.3)$$

The elastic strain rate is related to the stress rate by the bijective relation

$$\dot{\sigma} = \mathbf{D}_e \dot{\varepsilon}_e , \qquad (4.4)$$

therefore the stress-strain relation can be written in a rate form as

$$\dot{\sigma} = \mathbf{D}_e (\dot{\varepsilon} - \dot{\varepsilon}_p) . \qquad (4.5)$$

The plastic strain rate vector is written as the product of a non-negative scalar $\dot{\lambda}$ and a vector $\mathbf{m}$, representing the magnitude and the direction of the plastic flow, respectively

$$\dot{\varepsilon}_p = \dot{\lambda} \mathbf{m} . \qquad (4.6)$$

The vector $\mathbf{m}$ is often assumed to be the gradient of the plastic potential function g_p

$$\mathbf{m} = \frac{\partial g_p}{\partial \boldsymbol{\sigma}}. \tag{4.7}$$

The plastic potential function g_p is equal to the the yield function f in case of associative flow. The consistency equation can be elaborated as

$$\mathbf{n}^T \dot{\boldsymbol{\sigma}} + \frac{\partial f}{\partial \kappa} \dot{\kappa} = 0, \tag{4.8}$$

in which the vector $\mathbf{n}$ is the gradient to the yield surface

$$\mathbf{n} = \frac{\partial f}{\partial \boldsymbol{\sigma}}. \tag{4.9}$$

If we define the softening modulus h as

$$h = -\frac{1}{\dot{\lambda}} \frac{\partial f}{\partial \kappa} \dot{\kappa} \tag{4.10}$$

an explicit expression for the magnitude of the plastic flow can be derived by premultiplying eq.(4.5) by $\mathbf{n}^T$. Combination with eqs.(4.6), (4.8) and (4.10) then yields

$$\dot{\lambda} = \frac{\mathbf{n}^T \mathbf{D}_e \dot{\boldsymbol{\varepsilon}}}{h + \mathbf{n}^T \mathbf{D}_e \mathbf{m}}. \tag{4.11}$$

We can now obtain the relation between stress rate and strain rate by substitution of eq.(4.11) in eq.(4.5) and (4.6)

$$\dot{\boldsymbol{\sigma}} = \left[\mathbf{D}_e - \frac{\mathbf{D}_e \mathbf{m} \mathbf{n}^T \mathbf{D}_e}{h + \mathbf{n}^T \mathbf{D}_e \mathbf{m}} \right] \dot{\boldsymbol{\varepsilon}}. \tag{4.12}$$

The expression between brackets is called the continuum tangent stiffness matrix.

A yield function is used to distinguish stress points that are in a plastic state ($f = 0$) from stress points that are in the elastic domain of the stress space ($f < 0$). If $f = 0$, plastic straining can occur, $\dot{\boldsymbol{\varepsilon}}^p = \dot{\lambda} \mathbf{m}$ with direction $\mathbf{m}$ an experimentally determined quantity and $\dot{\lambda}$ a non-negative plastic multiplier, which gives the magnitude of the plastic flow and which is determined from the consistency condition $\dot{f} = 0$. $\dot{\lambda}$ and f are constrained by the the discrete Kuhn-Tucker conditions

$$f \le 0, \quad \dot{\lambda} \ge 0, \quad f \dot{\lambda} = 0. \tag{4.13}$$

The latter condition becomes clear by considering that during plastic flow $\dot{\lambda} > 0$ while simultaneously $f = 0$, but that in the elastic domain $f < 0$, but that then $\dot{\lambda} = 0$.

The hardening/softening parameter κ is typically dependent on the strain history and grows monotonically. Here we assume strain-hardening/softening in which the evolution of the hardening/softening parameter is postulated to be equal to

$$\dot{\kappa} = \sqrt{2/3 (\dot{\boldsymbol{\varepsilon}}_p)^T \dot{\boldsymbol{\varepsilon}}_p}, \tag{4.14}$$

which is basically the second invariant of the plastic strain vector (Note that $\boldsymbol{\varepsilon}_p =$

$(\varepsilon_{pxx}, \varepsilon_{pyy}, \varepsilon_{pzz}, \varepsilon_{pxy}, \varepsilon_{pyz}, \varepsilon_{pzx}))$. The hardening/softening parameter can be integrated in time via

$$\kappa = \int \dot{\kappa} \, dt \ . \tag{4.15}$$

Because the yield function f is dependent on a scalar κ, the yield surface can only expand (hardening) or shrink (softening), which means that the theory is limited to isotropic type of hardening/softening.

The abovementioned derivation is general and can be elaborated for a specific choice of the yield function $f(\boldsymbol{\sigma}, \kappa)$. For instance, for the pressure-independent criterion of von Mises we define

$$f(\boldsymbol{\sigma}, \kappa) = \sqrt{3J_2} - \bar{\sigma}(\kappa) \ , \tag{4.16}$$

in which

$$J_2 = \tfrac{1}{2}(s_1^2 + s_2^2 + s_3^2) \tag{4.17}$$

is the second invariant of the deviatoric stresses $s_i = \sigma_i - p$, with $p = \tfrac{1}{3}(\sigma_1 + \sigma_2 + \sigma_3)$ the hydrostatic pressure. $\bar{\sigma}$ is the yield stress which is a function of the softening parameter. Typically, metals and normally-consolidated clays under relatively rapid loading conditions (undrained behaviour) satisfy the pressure-insensitive criterion of von Mises. Another criterion is the Drucker-Prager yield function which includes a dependence on the hydrostatic pressure p. The Drucker-Prager yield function is defined by

$$f(\boldsymbol{\sigma}, \kappa) = \sqrt{3J_2} + \alpha \, p - k \ , \tag{4.18}$$

with α and k constants given by

$$\alpha = \frac{6 \sin \phi}{3 - \sin \phi} \quad \left(\alpha_\psi = \frac{6 \sin \psi}{3 - \sin \psi} \right) \tag{4.19}$$

and

$$k = \frac{6c(\kappa) \cos \phi}{3 - \sin \phi} \ , \tag{4.20}$$

in which ϕ signifies the internal friction angle and c the cohesion of a material. The cohesion c can be made a function of the softening/hardening parameter κ. We can invoke the concept of non-associative flow by defining the plastic potential function g_p equal to f but with the dilatancy angle ψ substituted for the friction angle ϕ. The Drucker-Prager criterion is well suited to describe the inelastic behaviour of sands, drained clays, rocks and concrete under compressive loading. The yield functions of von Mises and Drucker-Prager can be used for a mode-II dominated failure pattern. On the other hand, the principal stress yield criterion of Rankine is suited to predict a mode-I failure pattern. The yield function according to Rankine is defined as

$$f(\boldsymbol{\sigma}, \kappa) = \sigma_i - \bar{\sigma}(\kappa) \ , \tag{4.21}$$

in which $\sigma_i = \max(\sigma_1, \sigma_2, \sigma_3)$. This model can be used to model fracture of brittle materials with a plasticity model instead of a smeared crack, discrete crack or damage model.

The stress-strain relation for plasticity (eq.(4.12)) involves a careful integration along the loading path. The most straightforward method to integrate eq.(4.12) is a one-point Euler forward integration rule. Such a scheme is fully explicit: the stresses and the value of the hardening parameter are known at the beginning of the strain increment so that the tangent stiffness at the beginning of the strain increment can be evaluated directly. If the initial stress point σ_0 is on the yield contour the stress increment can be computed as:

$$\Delta \sigma = \left[\mathbf{D}_e - \frac{\mathbf{D}_e \mathbf{m}^0 (\mathbf{n}^0)^T \mathbf{D}_e}{h^0 + (\mathbf{n}^0)^T \mathbf{D}_e \mathbf{m}^0} \right] \Delta \varepsilon \ . \tag{4.22}$$

where the superscript 0 at the flow direction $\mathbf{m}$, the normal to the yield surface $\mathbf{n}$ and the hardening modulus h refers to the fact that these quantities are evaluated for $\sigma = \sigma^0$. The new stress, i.e. the stress state at the end of the loading step, is then computed as

$$\sigma^n = \sigma^0 + \Delta \sigma \ . \tag{4.23}$$

If the stress point is initially inside the yield contour the total strain increment must first be subdivided into a purely elastic part, i.e. a part that is needed to make the stress point reach the yield surface ($\Delta \varepsilon_A$ in Figure 4.1), and a part that involves elasto-plastic straining. Now, the stress increment is computed as

$$\Delta \sigma = \mathbf{D}_e \Delta \varepsilon_A + \left[\mathbf{D}_e - \frac{\mathbf{D}_e \mathbf{m}^c (\mathbf{n}^c)^T \mathbf{D}_e}{h^c + (\mathbf{n}^c)^T \mathbf{D}_e \mathbf{m}^c} \right] \Delta \varepsilon_B \ , \tag{4.24}$$

where the superscript c now denotes that the respective quantities are evaluated at $\sigma = \sigma^c$. Here, a major disadvantage of the explicit Euler method becomes apparent: the contact stress must be calculated explicitly.

The whole procedure may also be viewed as follows. First, the stress increment

$$\Delta \sigma^{trial} = \mathbf{D}_e \Delta \varepsilon \tag{4.25}$$

is calculated. For this calculation it is quite irrelevant whether the initial location of the stress is inside or on the current yield surface. This stress increment may be conceived as a trial stress increment, which rests upon the assumption of pure elastic behaviour during the load increment. Possible plastic straining is not considered during this trial step. Then, the total trial stress σ^{trial} is set up as the sum of the stress at the beginning of the loading step, σ^0, and the trial stress increment:

$$\sigma^{trial} = \sigma^0 + \Delta \sigma^{trial} = \sigma^0 + \mathbf{D}_e \Delta \varepsilon \ . \tag{4.26}$$

If the trial stress σ^{trial} appears to violate the yield condition, $f(\sigma^{trial}) > 0$, a correction is applied. The direction and the magnitude of this correction are inferred from eq.(4.24):

$$\sigma^n - \sigma^{trial} = - \frac{(\mathbf{n}^c)^T \mathbf{D}_e \Delta \varepsilon_B}{h^c + (\mathbf{n}^c)^T \mathbf{D}_e \mathbf{m}^c} \mathbf{D}_e \mathbf{m}^c \tag{4.27}$$

for the most general case that the initial stress point is located within the yield surface.

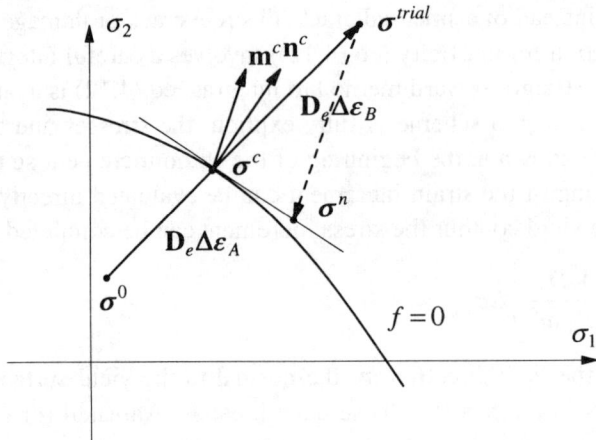

Figure 4.1 Explicit integration: the total strain increment is divided into an elastic and a plastic part. The plastic part is integrated with an Euler forward rule.

Eq.(4.27) shows that in the explicit Euler method, the flow direction **m**, the gradient to the yield surface **n** and the hardening modulus h are computed either on basis of the initial stress state σ^0, or on basis of the stresses at the contact or intersection point of the 'elastic' stress path with the yield contour σ^c if the initial stress state is within the yield contour. Combining eqs.(4.25)-(4.27) shows that the entire 'elastic predictor-plastic corrector' process has the following format:

$$\sigma^n = \sigma^{trial} - \Delta\lambda \mathbf{D}_e \mathbf{m}^c \qquad (4.28)$$

where $\Delta\lambda$ is the (finite) amount of plastic flow within this loading step:

$$\Delta\lambda = \frac{(\mathbf{n}^c)^T \mathbf{D}_e \Delta\boldsymbol{\varepsilon}_B}{h^c + (\mathbf{n}^c)^T \mathbf{D}_e \mathbf{m}^c} \,. \qquad (4.29)$$

Eq.(4.28) can also be interpreted as follows. First, a trial stress state is computed assuming fully elastic behaviour. Then, the trial stress is mapped back, i.e. projected in the direction of the yield surface. Therefore, the name return-mapping algorithms has become the vogue for this type of integration methods.

It can be read from Figure 4.1 that, while the correction for plastic straining is governed by the direction of the flow vector at σ^c, the final or new stress point σ^n is found at the intersection of the hyperplane that is tangent to the yield surface at σ^c and the return direction $\mathbf{m}^c$ (in the absence of hardening). A formal proof thereof will not be given here. Apparently, the forward Euler method does not guarantee a rigorous return to the yield surface. An error is committed, the magnitude of which depends upon the local curvature of the yield surface. A strongly curved yield surface gives rise to larger errors than an almost flat yield contour. Especially when relatively large loading steps are used, the accumulation

Box 4.1 Euler backward method - return mapping scheme.

FOR EACH INTEGRATION POINT :

1. Iteration $k = 0$, $\Delta\lambda^0 = 0$, κ^0, σ^0

2. Compute trial stress state : $\sigma^{trial} = \sigma^0 + \mathbf{D}_e \Delta\varepsilon$

3. Plasticity ? YES, if $f(\sigma^{trial}, \kappa^0) \geq 0$ ELSE, $\sigma = \sigma^{trial}$ and go to 10.

4. Calculate $\mathbf{m}^k$ and $\mathbf{n}^k$ for current stress σ^k ($=\sigma^{trial}$ in first iteration)

5. Compute $\Delta\lambda^{k+1} = \Delta\lambda^k + f(\sigma^k, \kappa^k)/[h^k + (\mathbf{n}^k)^T \mathbf{D}_e \mathbf{m}^k]$

6. Correction of stress : $\sigma^k = \sigma^{trial} - \Delta\lambda^{k+1} \mathbf{D}_e \mathbf{m}^k$

7. Calculate κ^{k+1} on basis of hardening/softening hypothesis

8. Calculate h^{k+1} and $\bar{\sigma}^{k+1}$ for updated κ^{k+1}

9. Check convergence : $f(\sigma^{k+1}, \kappa^{k+1}) <$ tolerance

    if YES, go to 10.

    if NO, $k = k + 1$ go to 4.

10. Next integration point

---

of errors may become quite significant, and may even lead to numerical instability of the algorithm.

A good and still relatively simple algorithm that warrants for numerical stability irrespective of the step size is the implicit Euler backward method. Formally, this algorithm is also given by (4.28), but for the fact that all stress-dependent quantities are now evaluated for $\sigma = \sigma^n$

$$\sigma^n = \sigma^{trial} - \Delta\lambda \mathbf{D}_e \mathbf{m}^n \ . \tag{4.30}$$

Comparing eqs.(4.28) and (4.30) we observe that in the fully implicit Euler backward algorithm there is no need to determine the contact stress σ^c. However, the equations are now implicit in the sense that neither $\Delta\lambda$, nor σ^n (and therefore also not $\mathbf{m}^n$ and $\mathbf{n}^n$) can be calculated directly. To determine $\Delta\lambda$ and σ^n we must use the requirement that the yield condition is complied with at the end of the loading step

$$f(\boldsymbol{\sigma}^n, \kappa^n) = 0 \ . \tag{4.31}$$

Eqs.(4.30) and (4.31) constitute a total of seven non-linear equations in seven unknowns, namely the six stress components of $\boldsymbol{\sigma}_n$ and the plastic multiplier $\Delta\lambda$. An iterative procedure, e.g. a Newton-Raphson procedure, must be set up at integration point level - within each global iteration on structural level as discussed in Chapter 3 ! - to solve this non-linear system of equations. During iteration k a first-order Taylor series expansion around the current stress point gives

$$f^{k+1} = f^k + \left(\frac{\partial f^k}{\partial \boldsymbol{\sigma}}\right)^T \Delta\boldsymbol{\sigma} + \frac{\partial f^k}{\partial \kappa} \Delta\kappa \ , \tag{4.32}$$

which can be rewritten in

$$f^{k+1} = f^k + (\mathbf{n}^k)^T \Delta\boldsymbol{\sigma} - h^k \Delta\lambda \ . \tag{4.33}$$

For the increment of stress between two iterations eq.(4.30) reduces to

$$\Delta\boldsymbol{\sigma} = -\Delta\lambda \mathbf{D}_e \mathbf{m}^k \ . \tag{4.34}$$

Assuming that $f = 0$ (eq.(4.31)) in the new state $k + 1$ we have derived a delta incremental value for the plastic multiplier

$$\Delta\lambda = \frac{f^k}{h^k + (\mathbf{n}^k)^T \mathbf{D}_e \mathbf{m}^k} \ . \tag{4.35}$$

The stresses can now be updated and iteratively a converged solution can be found in which the stresses are mapped backed on the yield surface according a certain tolerance. In Box 4.1 an elaboration of the method is given in an algorithmic way. Finally, it is emphasized that the implicit Euler backward algorithm is superior to the explicit variant especially if the curvature of the yield surface becomes large and if large strain increments are taken.

Within a loading step eq.(4.30) that sets the dependence of the stress increment on the prescribed strain increment is basically a total stress-strain relation. Accordingly, we have a deformation type plasticity theory within a loading step rather than a flow theory when a return-mapping algorithm is used. In this spirit the tangential relation between stress rate and strain rate that is required when implicit solution strategies are used at a global (structural) level, bears much resemblance to the tangential operators that result from a deformation theory of plasticity. In particular, eq.(4.30) can be differentiated to yield

$$\dot{\boldsymbol{\sigma}} = \mathbf{D}_e \dot{\boldsymbol{\varepsilon}} - \dot{\lambda} \mathbf{D}_e \mathbf{n} - \Delta\lambda \mathbf{D}_e \frac{\partial^2 f}{\partial \boldsymbol{\sigma}^2} \dot{\boldsymbol{\sigma}} \ . \tag{4.36}$$

where for sake of simplicity an associated flow rule and ideal plasticity have been assumed. However, this restriction is by no means necessary. When we introduce the auxiliary matrix $\mathbf{H}$,

$$\mathbf{H} = \left(\mathbf{I} + \Delta\lambda \mathbf{D}_e \frac{\partial^2 f}{\partial \sigma^2} \right)^{-1} \mathbf{D}_e ,\qquad(4.37)$$

eq.(4.36) can be rewritten to give a familiar form of the rate equation

$$\dot{\sigma} = \mathbf{H}(\dot{\varepsilon} - \dot{\lambda}\mathbf{n}) .\qquad(4.38)$$

Via the usual argument involving the consistency condition the tangent stiffness operator can be derived as

$$\dot{\sigma} = \left[\mathbf{H} - \frac{\mathbf{H}\mathbf{n}\mathbf{n}^T\mathbf{H}}{\mathbf{n}^T\mathbf{H}\mathbf{n}} \right] \dot{\varepsilon} .\qquad(4.39)$$

This tangent stiffness matrix is called the 'consistent tangent stiffness matrix'. It differs from the conventional tangent stiffness matrix (eq.(4.12)) in that the elastic rigidity matrix is replaced by the **H**-matrix which includes effects of plastic flow. Especially for large plastic strain increments, e.g. due to large loading steps, **H** can differ significantly from $\mathbf{D}_e$. When used within a full Newton-Raphson procedure, the use of a consistent tangent stiffness matrix is highly advantageous compared with the conventional tangent stiffness matrix. The amount of iterations necessary to obtain a properly converged solution is then typically reduced by a factor two. Note, however, that use of a consistent tangent stiffness matrix is meaningful only when a full Newton-Raphson procedure is employed to solve the set of non-linear equations at a structural level. This is because the magnitude of the plastic strain within the loading step enters the tangent stiffness matrix ($\Delta\lambda$ in eq.(4.37)).

4.2 Computational damage

For an elasticity-based damage model we assume that

$$\sigma = (1-\omega)\mathbf{D}_e \varepsilon ,\qquad(4.40)$$

with $\mathbf{D}_e$ given by eq.(2.7) and ω the scalar-valued internal variable, which reflects the amount of damage which the material has experienced. Sometimes ω is split up into a value ω_1 for volumetric behaviour and ω_2 for deviatoric behaviour. Here we assume that $\omega_1 = \omega_2 = \omega$. As in plasticity we must distinguish between further 'loading' and 'unloading'. In a multiaxial generalisation damage growth occurs if the damage loading function

$$f(\tilde{\varepsilon},\kappa) = \tilde{\varepsilon} - \kappa \qquad(4.41)$$

vanishes as well as its derivative

$$f = 0 \quad \text{and} \quad \dot{f} = 0 ,\qquad(4.42)$$

with κ a history-dependent parameter, which reflects the loading history, i.e. from an initial value κ_0 it grows, as it memorises the largest value ever attained of the equivalent strain $\tilde{\varepsilon}$. It takes an assumption to define $\tilde{\varepsilon}$. In a fashion similar to plasticity, any invariant measure of the total strains or of the elastic energy per unit mass (the so-called free energy ψ) could be employed. Possible choices are via the elastic energy

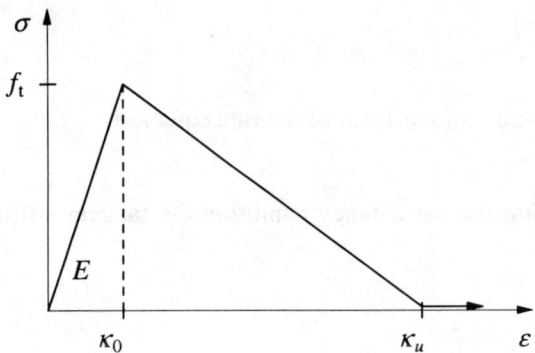

Figure 4.2 Elastic-linear damaging material behaviour.

$$\tilde{\varepsilon} = \sqrt{1/2}\varepsilon^T \mathbf{D}_e \varepsilon \tag{4.43}$$

or via the principal strains as proposed by Mazars for concrete and ceramics

$$\tilde{\varepsilon} = \sqrt{\sum_{i=1}^{3}(<\varepsilon_i>)^2}, \tag{4.44}$$

with ε_i the principal strains, and $<\varepsilon_i> = \varepsilon_i$ if $\varepsilon_i > 0$ and $<\varepsilon_i> = 0$ otherwise. Any dependence of $\tilde{\varepsilon}$ on the strain tensor ε must be postulated and later verified experimentally. Normally, the relation will differ for each material.

At all instances, the damage loading function f and the rate of the history parameter $\dot{\kappa}$ have to satisfy the discrete Kuhn-Tucker conditions

$$f \leq 0 \;,\; \dot{\kappa} \geq 0 \;,\; f\dot{\kappa} = 0. \tag{4.45}$$

In a spirit similar to plasticity models damage theory is completed by defining an evolution law for the damage variable ω as a function of the history parameter κ

$$\omega = \omega(\kappa). \tag{4.46}$$

We have $0 \leq \omega \leq 1$, the initial value $\omega = 0$ representing the state where we have fully intact material, and the final value $\omega = 1$ representing the state where we have a total loss of coherence. This relation can be inferred from one-dimensional tensile tests. For instance, suppose that we have measured a stress-strain behaviour that can be characterised by a bilinear relation, in the sense that we have a linear elastic relation with a Young's modulus E up to a peak value of the stress $\sigma = f_t$, and that after the peak we approximately have a linear descending branch of the stress-strain curve that ends at an axial strain $\varepsilon = \kappa_u$, at which there is complete loss of coherence (Figure 4.2). Damage starts upon the first departure from linear elasticity. For the present relation, that occurs when the tensile strength is reached. The uniaxial strain is then $\varepsilon = f_t/E$. We now assume that we have a choice for the history parameter κ that reduces to the uniaxial strain in case of uniaxial loading, so that then $\varepsilon = \kappa$. Accordingly, the threshold damage level, at which ω starts to increase is given by $\kappa_0 = f_t/E$. Beyond this threshold level of damage, i.e. for $\kappa > \kappa_0$, the uniaxial stress-strain relation is

$$\sigma = f_t\left(1 - \frac{\varepsilon - \kappa_0}{\kappa_u - \kappa_0}\right). \tag{4.47}$$

Generally, we also have

$$\sigma = (1 - \omega)E\varepsilon, \tag{4.48}$$

as the uniaxial specialisation of the triaxial relation. We now equate both right-hand sides

$$(1 - \omega)E\varepsilon = f_t\left(1 - \frac{\varepsilon - \kappa_0}{\kappa_u - \kappa_0}\right) \tag{4.49}$$

and consider that for the uniaxial case $\varepsilon = \kappa$, and that $\kappa_0 = f_t/E$. This gives

$$(1 - \omega)\kappa = \kappa_0\left(1 - \frac{\kappa - \kappa_0}{\kappa_u - \kappa_0}\right). \tag{4.50}$$

Rearranging then leads to the following evolution formula for ω (valid for $\kappa_0 < \kappa < \kappa_u$)

$$\omega(\kappa) = \frac{\kappa_u(\kappa - \kappa_0)}{\kappa(\kappa_u - \kappa_0)}. \tag{4.51}$$

Obviously, at $\kappa \leq \kappa_0$ we have $\omega = 0$ and for $\kappa \geq \kappa_u$ we have $\omega = 1$.

More generally, one can state that if the experimentally obtained uniaxial relation is given by $\sigma = f(\kappa - \kappa_0)$, then the damage evolution law is obtained by equating f to $(1 - \omega)E\varepsilon$, or generally, to $(1 - \omega)E\kappa$. The result is then

$$\omega(\kappa) = 1 - \frac{f(\kappa - \kappa_0)}{E\kappa}. \tag{4.52}$$

Eq.(4.51) is in fact a special case of this relation. The formalism of eq.(4.52) can also be used to simulate ideal plasticity. In this case $f = f_t$, the tensile strength. In view of the identity $\kappa_0 = f_t/E$ we then have (for $\kappa > \kappa_0$)

$$\omega(\kappa) = 1 - \frac{\kappa_0}{\kappa}. \tag{4.53}$$

It is noted that the damage formulation still differs from real ideal plasticity in the sense that in the damage formalism unloading is along the secant branch instead of in a linear elastic fashion.

REMARK:

There are (at least) two essential differences between plasticity and damage.

1. Considering the yield function f in plasticity we observe that this function is formulated in terms of *stresses* whereas in the damage formulation the loading function f is dependent upon the *strains*. At least for the virgin state, a simple transfor-

mation from stresses to strains (via the elastic properties) can make both formulations coincide. In fact, there also exists the *strain-space plasticity theory*, which is formulated entirely in terms of strains, i.e. also the yield function is a function of the strains.

2. While strain-space plasticity can (initially) produce exactly the same effect as damage formulations, this is definitely not so for the unloading behaviour. There elasticity-based damage formulations give a secant unloading (to the origin) while strain-space plasticity results in elastic unloading, like conventional stress-space plasticity.

While in the preceding the three-dimensional formulation for damage has been derived from a relatively straightforward generalization of the one-dimensional case, damage theories can also be set up starting from thermodynamics considerations. In fact, one should always verify whether a model is admissible from thermodynamics viewpoints. Not seldom, the thermodynamic requirements impose certain restrictions on the range of material parameter that can be adopted (think of the restriction $-1 \le \nu \le 1/2$ in isotropic, linear elasticity). On the other hand, while a derivation that starts from thermodynamic considerations ensures that thermodynamics principles are not violated, it is incorrect to believe that such derivations are more 'scientific'. They also need assumptions on the material behaviour. It is impossible to derive constitutive formulations directly from thermodynamics, without any additional assumption.

We start our derivation by postulating the following dependence for the free energy per unit mass of material: $\psi = \psi(\varepsilon_{ij}, \omega)$, where index notation has been used as is common for such derivations. It is noted that this dependence restricts attention to isothermal loading conditions. We can now write

$$\dot{\psi} = \frac{\partial \psi}{\partial \varepsilon_{ij}} \dot{\varepsilon}_{ij} + \frac{\partial \psi}{\partial \omega} \dot{\omega} . \tag{4.54}$$

We next introduce the dissipation rate $\dot{\phi}$ which is the difference between the total (internal) variation of mechanical energy for an arbitrary strain rate and the elastic (recoverable) energy rate

$$\dot{\phi} = \sigma_{ij} \dot{\varepsilon}_{ij} - \rho \dot{\psi} , \tag{4.55}$$

with ρ the mass density. Substitution of the expression for the rate of the free energy $\dot{\psi}$ into eq.(4.55) gives

$$\dot{\phi} = \left(\sigma_{ij} - \rho \frac{\partial \psi}{\partial \varepsilon_{ij}} \right) \dot{\varepsilon}_{ij} - \rho \frac{\partial \psi}{\partial \omega} \dot{\omega} . \tag{4.56}$$

Since

$$\sigma_{ij} = \rho \frac{\partial \psi}{\partial \varepsilon_{ij}} , \qquad (4.57)$$

we arrive at the following expression for the energy dissipation rate

$$\dot{\phi} = -\rho \frac{\partial \psi}{\partial \omega} \dot{\omega} = Y \dot{\omega} \qquad (4.58)$$

with

$$Y = -\rho \frac{\partial \psi}{\partial \omega} \qquad (4.59)$$

the thermodynamic 'force' conjugate to the internal (damage) variable ω, is the energy released per unit rate of damage.

Now, we proceed to the restriction which the second law of thermodynamics imposes. Under isothermal conditions we have the requirement that the dissipation rate must be non-negative

$$\dot{\phi} \geq 0 . \qquad (4.60)$$

We now *postulate* the following relation for the free energy ψ

$$\rho \psi = \tfrac{1}{2}(1 - \omega) \varepsilon_{ij} D^e_{ijkl} \varepsilon_{kl} . \qquad (4.61)$$

When we substitute this expression into eq.(4.57), we retrieve the constitutive relation for an elastic damaging solid derived before

$$\sigma_{ij} = (1 - \omega) D^e_{ijkl} \varepsilon_{kl} . \qquad (4.62)$$

Furthermore, the dissipation rate becomes, cf. eq.(4.58)

$$\dot{\phi} = \tfrac{1}{2} \varepsilon_{ij} D^e_{ijkl} \varepsilon_{kl} \dot{\omega} . \qquad (4.63)$$

Since $\tfrac{1}{2} \varepsilon_{ij} D^e_{ijkl} \varepsilon_{kl}$ is a quadratic form it is always positive, and the requirement of a non-negative dissipation rate ($\dot{\phi} \geq 0$) is tantamount to requiring that

$$\dot{\omega} \geq 0 , \qquad (4.64)$$

which implies that damage can only grow.

The algorithm for elasticity-based damage is summarised in Box 4.2. The equation under 5. in Box 4.2 is precisely the relation that has to be linearised to obtain the tangential stiffness operator. One finds directly that

$$\frac{\partial \sigma}{\partial \varepsilon} = (1 - \omega) \mathbf{D}_e - \frac{\partial \omega}{\partial \kappa} \mathbf{D}_e \varepsilon \frac{\partial \kappa}{\partial \varepsilon} . \qquad (4.65)$$

If no damage growth has occurred in the previous iteration, $\partial \omega / \partial \kappa = 0$, and the second term in eq.(4.65) vanishes. What is left is the first term, so that the tangential stiffness is equal to the secant stiffness. For the case that there has been damage growth in the previous iteration, the both terms in eq.(4.65) contribute to the material tangential stiffness matrix.

Box 4.2 Stress update for damage model.

FOR EACH INTEGRATION POINT :

1. ε is given, calculate equivalent strain $\tilde{\varepsilon}$, for instance via eq.(4.43) or (4.44)

2. Evaluate damage loading function : $f(\tilde{\varepsilon}, \kappa^j) = \tilde{\varepsilon} - \kappa^j$

   (κ^j the history parameter computed at the end of the previous load increment)

3. If $f > 0$, update κ such that : $\kappa^{j+1} = \tilde{\varepsilon}$

   If $f \leq 0$, leave κ unchanged : $\kappa^{j+1} = \kappa^j$

4. Update damage variable : $\omega^{j+1} = \omega(\kappa^{j+1})$

5. Compute total stresses : $\boldsymbol{\sigma} = (1 - \omega^{j+1})\mathbf{D}_e \boldsymbol{\varepsilon}$

4.3 Computational fracture

For the continuum approach of fracture a crack model can be defined in the framework of the fixed smeared crack concept (de Borst and Nauta 1985, Rots 1988). In this concept a cracked zone is conceived to be a continuum which permits a description in terms of stress-strain relations. We apply a decomposition of total strain rate into the elastic strain rate $\dot{\boldsymbol{\varepsilon}}_e$ and the crack strain rate $\dot{\boldsymbol{\varepsilon}}_{cr}$

$$\dot{\boldsymbol{\varepsilon}} = \dot{\boldsymbol{\varepsilon}}_e + \dot{\boldsymbol{\varepsilon}}_{cr} \ . \tag{4.66}$$

When incorporating crack stress - crack strain laws it is convenient to set up a local n, t-coordinate system in a two-dimensional configuration, which is aligned with the crack. This necessitates a transformation between the crack strain rate $\dot{\boldsymbol{\varepsilon}}_{cr}$ in the global x, y, z-coordinates and the crack strain rate $\dot{\mathbf{e}}_{cr}$ in the local coordinates. The crack strain rate in the local coordinate system is defined as

$$\dot{\mathbf{e}}_{cr} = [\dot{e}_{nn}, 2\dot{e}_{nt}]^T \ , \tag{4.67}$$

where $\dot{e}_{nn}$ is the mode-I crack normal strain rate and $\dot{e}_{nt}$ is the mode-II crack shear strain rate. The relation between local and global strain rates reads

$$\dot{\boldsymbol{\varepsilon}}_{cr} = \mathbf{N}\dot{\mathbf{e}}_{cr} \ , \tag{4.68}$$

where **N** is the transformation matrix given by

$$\mathbf{N} = \begin{bmatrix} \cos^2 \alpha & -\sin \alpha \cos \alpha \\ \sin^2 \alpha & \sin \alpha \cos \alpha \\ 2\sin \alpha \cos \alpha & \cos^2 \alpha - \sin^2 \alpha \end{bmatrix}, \quad (4.69)$$

with α the inclination angle of the normal of the crack n with the x-axis. The angle is determined by the principal stress direction at the onset of cracking. An essential feature of the model is that $\mathbf{N}$ is fixed upon crack formation so that the concept belongs to the class of fixed crack concepts.

In a similar way we can define a crack stress rate vector

$$\dot{\mathbf{t}}_{cr} = [\,\dot{t}_{nn}, \dot{t}_{nt}\,]^{\mathrm{T}}, \quad (4.70)$$

in which $\dot{t}_{nn}$ is the mode-I normal crack stress rate and $\dot{t}_{nt}$ is the mode-II shear crack stress rate. The relation between the stress rate in the global coordinate system and the local stress rate can be derived to be

$$\dot{\mathbf{t}}_{cr} = \mathbf{N}^{\mathrm{T}} \dot{\boldsymbol{\sigma}}. \quad (4.71)$$

To complete the system of equations we need a constitutive model for the intact concrete and for the smeared cracks. For the concrete between the cracks it is assumed that

$$\dot{\boldsymbol{\sigma}} = \mathbf{D}_e \dot{\boldsymbol{\varepsilon}}_e. \quad (4.72)$$

The relation between the local crack strain rate and the local crack stress rate is

$$\dot{\mathbf{t}}_{cr} = \mathbf{D}_{cr} \dot{\mathbf{e}}_{cr}, \quad (4.73)$$

with $\mathbf{D}_{cr} = diag[h, \beta_s \mu]$, in which h is the mode-I softening modulus ($h<0$), which has been assumed to be a constant for the sake of simplicity. The shear stiffness in the crack is obtained by a multiplication of the elastic shear stiffness μ with a shear reduction factor β_s. Coupling effects between the two modes are not considered. In this model fracture is assumed to be initiated in mode-I and mode-II effects enter upon rotation of the principal stresses.

Now, the overall stress-strain relation of the model with respect to the global coordinate system can be developed. Combining eqs.(4.72) and (4.66) and subsequent substitution of eq.(4.68) yields

$$\dot{\boldsymbol{\sigma}} = \mathbf{D}_e [\,\dot{\boldsymbol{\varepsilon}} - \mathbf{N}\dot{\mathbf{e}}_{cr}\,]. \quad (4.74)$$

Premultiplying this equation by $\mathbf{N}^{\mathrm{T}}$ and substituting eqs.(4.71) and (4.73) yields the relation between the local crack strain rate and the global strain rate

$$\dot{\mathbf{e}}_{cr} = [\,\mathbf{D}_{cr} + \mathbf{N}^{\mathrm{T}} \mathbf{D}_e \mathbf{N}\,]^{-1} \mathbf{N}^{\mathrm{T}} \mathbf{D}_e \dot{\boldsymbol{\varepsilon}}. \quad (4.75)$$

The overall relation between global stress rate and global strain rate is obtained by substituting eq.(4.75) into (4.74)

$$\dot{\boldsymbol{\sigma}} = [\,\mathbf{D}_e - \mathbf{D}_e \mathbf{N}\,[\,\mathbf{D}_{cr} + \mathbf{N}^{\mathrm{T}} \mathbf{D}_e \mathbf{N}\,]^{-1} \mathbf{N}^{\mathrm{T}} \mathbf{D}_e\,] \dot{\boldsymbol{\varepsilon}}. \quad (4.76)$$

In this derivation only one crack is considered, but it is possible that due to the rotation of principal stresses new cracks arise. The crack strain is then decomposed into separate contributions from the multi-directional cracks (de Borst and Nauta 1985, Rots 1988).

The integration of eq.(4.76) can be done by a one-step forward scheme which is exact if the matrices $\mathbf{D}_e$ and $\mathbf{D}_{cr}$ remain constant during the time step. When, for instance, $\mathbf{D}_{cr}$ is non-constant a predictor-corrector method can be used in a inner iteration loop to determine the incremental stresses (Rots 1988).

5. LOCALISATION OF DEFORMATION

An essential problem when plasticity, damage or crack evolution laws, or more general, strain softening type constitutive relations, are used is the inherent mesh dependence that is introduced by it. The conventional approach with these so-called standard continuum models from Chapter 4 as well as novel features of recently proposed, enhanced continuum models in Chapter 6 are best demonstrated by the example of a simple bar loaded in uniaxial tension of Figure 5.1 (de Borst 1986, Sluys 1992). Let the bar be divided into m elements. Prior to reaching the tensile strength f_t a linear relation is assumed between the normal stress σ and the normal strain ε:

$$\sigma = E\varepsilon , \tag{5.1}$$

with E Young's modulus. After reaching the peak strength a descending slope is defined in this diagram through an affine transformation from the measured load-displacement curve. The result has been given in Figure 5.2, where κ_u marks the point where the load-carrying capacity is totally exhausted. In the post-peak regime the constitutive model can thus be written as:

$$\sigma = f_t + h(\varepsilon - \kappa_0) . \tag{5.2}$$

In case of degrading materials $h < 0$ and h may be termed a softening modulus. For linear softening we have

$$h = -\frac{f_t}{\kappa_u - \kappa_0} . \tag{5.3}$$

Now suppose that one element has a tensile strength that is marginally below that of the other $m-1$ elements. Upon reaching the tensile strength of this element failure will occur. In the other, neighbouring elements the tensile strength is not exceeded and they will unload elastically. Beyond the peak strength the average strain in the bar is then given by:

$$\bar{\varepsilon} = \frac{\sigma}{E} + \frac{E-h}{Eh}\frac{\sigma - f_t}{m} . \tag{5.4}$$

Substitution of expression (5.3) for the softening modulus h and introduction of n as the ratio between the strain κ_u at which the residual load-carrying capacity is exhausted and the threshold damage level κ_0, $n = \kappa_u/\kappa_0$ and $h = -E/(n-1)$, so that,

$$\bar{\varepsilon} = \frac{\sigma}{E} + \frac{n(f_t - \sigma)}{mE} . \tag{5.5}$$

The slope in the post-peak regime is then given by:

$$\frac{\dot{\bar{\varepsilon}}}{\dot{\sigma}} = \frac{1}{E} - \frac{n}{mE} . \tag{5.6}$$

The result is plotted in Figure 5.2 for different discretisations of the bar. We observe that there is a tremendous scatter in the results depending on the number of elements that is used. For $m = 1$ the input stress-strain diagram of Figure 4.2 is reproduced, but for $m = n$

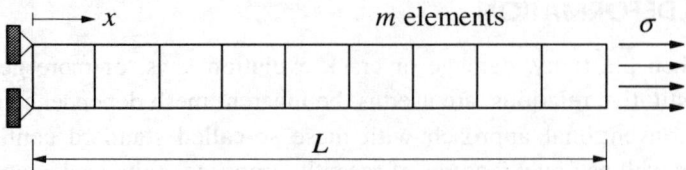

Figure 5.1   Strain-softening bar subject to uniaxial static loading.

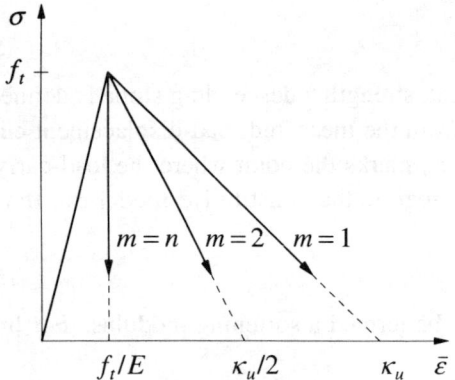

Figure 5.2   Response of imperfect bar in terms of a stress-average strain curve.

the stress drops vertically after exceeding the tensile strength. For $m > n$ the average strain actually decreases after reaching the peak stress. This so-called snap-back behaviour implies that under quasi-static loadings not only the load, but also the displacement of the right end of the bar decreases. Experiments can no longer be kept stable under displacement control. Owing to the fact that the localisation zone cannot absorb the elastic energy released in the unloading remaining parts of the bar, the observed failure of the specimen is of a highly explosive character.

The fact that a structure displays snap-back behaviour can be physical. However, the observation that eqs.(5.5) and (5.6) predict that for an infinite number of elements ($m \to \infty$) the post-peak curve doubles back on the original loading curve is suspect, since this implies that failure occurs without energy dissipation. From a physical point of view this is unacceptable and we must therefore either rephrase our constitutive model in terms of force-displacement relations, which implies the use of special interface elements (Rots 1988), or enrich the continuum description by adding higher-order terms which can accommodate narrow zones of highly localised deformation quite similar to descriptions for boundary layers in fluids (see Chapter 6).

Remarks:

- For multidimensional stress states non-symmetry of the tangent stiffness relation between stress rate and strain rate can generate a descending branch in the load-displacement diagram even though the stress-strain diagram continues to harden monotonically. Prominent examples are non-associated flow rules in soil plasticity and crack-dilatancy relations for rough cracks in concrete and rocks.
- Softening and localisation occur for a large variety of materials, such as concrete, rocks, ceramics, dense sands, overconsolidated clays, structural steels (beyond the onset of necking), alloys, polymers, either reinforced or not.

The mesh-dependence problem can not only be demonstrated by means of this simple static problem but also from an analysis of dynamics. If the initial value problem is considered in one spatial direction the governing equations for motion (cf.eq.(2.1)) and continuity (cf. eq.(2.2)) can be stated in a rate format as

$$\frac{\partial \dot{\sigma}}{\partial x} = \rho \frac{\partial^2 v}{\partial t^2} \tag{5.7}$$

and

$$\dot{\varepsilon} = \frac{\partial v}{\partial x}, \tag{5.8}$$

in which velocity $v = \dot{u}$. In addition to these equations the constitutive equations must be specified. A classical strain-softening model with strain decomposition of the total strain ε in an elastic strain ε_e and an inelastic strain ε_i, as used in the strain-softening models in Chapter 4, is of a general form

$$\sigma = f(\varepsilon_i), \tag{5.9}$$

or in a rate form

$$\dot{\sigma} = f' \dot{\varepsilon}_i. \tag{5.10}$$

Softening occurs if $f' < 0$, in which the superimposed prime denote differentiation with respect to the inelastic strain ε_i. Combination of eq.(5.10) and (5.8), taking $\dot{\varepsilon}_i = \dot{\varepsilon} - \dot{\varepsilon}_e$ and $\dot{\varepsilon}_e = \dot{\sigma}/E$ and differentiation of the result with respect to x yields

$$\frac{\partial \dot{\sigma}}{\partial x} = \frac{f' E}{E + f'} \frac{\partial^2 v}{\partial x^2}. \tag{5.11}$$

If we substitute eq.(5.11) in (5.7) we obtain the wave equation for a one-dimensional strain-softening element

$$\frac{E + f'}{c_e^2} \frac{\partial^2 v}{\partial t^2} - f' \frac{\partial^2 v}{\partial x^2} = 0, \tag{5.12}$$

in which $c_e = \sqrt{E/\rho}$ is the linear elastic, longitudinal wave velocity (so-called bar wave velocity). This second-order partial differential equation is linear if f' is constant (linear strain softening) and quasi-linear if f' is a function of ε_i (nonlinear strain softening). The type of solution of eq.(5.12) can be investigated by means of the characteristics. Characteristics represent the directions along which the solution develops. In case of the wave equation these directions lie in the $x-t$ plane. For a linear or a quasi-linear differential equation we can calculate the characteristics as follows. We consider the variation of the first derivatives of velocity v with respect to t and x

$$d\left(\frac{\partial v}{\partial t}\right) = \frac{\partial^2 v}{\partial t^2} dt + \frac{\partial^2 v}{\partial x \partial t} dx \tag{5.13}$$

$$d\left(\frac{\partial v}{\partial x}\right) = \frac{\partial^2 v}{\partial x \partial t} dt + \frac{\partial^2 v}{\partial x^2} dx . \tag{5.14}$$

Combination of eq.(5.13) and (5.14) with the wave equation for the classical strain-softening bar (eq.(5.12)) yields a system of three second-order differential equations with the characteristic determinant

$$D = \begin{bmatrix} (E+f')/c_e^2 & 0 & -f' \\ dt & dx & 0 \\ 0 & dt & dx \end{bmatrix} = (E+f')/c_e^2\, dx^2 - f'\, dt^2 . \tag{5.15}$$

If $D \neq 0$ a unique solution in the $u-x-t$ space can be determined. However, if $D = 0$ the system of equations is dependent and a curve in the $u-x-t$ plane coincides with the characteristic directions

$$\frac{dx}{dt} = \pm c_e \sqrt{\frac{f'}{E+f'}} . \tag{5.16}$$

For a wave equation the characteristics ($\pm dx/dt$) coincide with the wave speeds ($\pm c$). If we have softening ($f' < 0$) and consider the case of snap-through ($f' > -E$) the characteristics and therefore the wave speeds will be imaginary. So, the wave equation loses hyperbolicity and becomes elliptic whenever strain softening is introduced. In fact, a domain is split up into an elliptic part, in which waves do not have the ability to propagate (standing waves), and into an hyperbolic part with propagating waves. Spatial interaction between the two domains is impossible. The loss of hyperbolicity means that the problem becomes ill-posed as an initial value problem. An ill-posed problem can no longer be a successful description of the underlying physical problem.

It can be determined that a strain-softening medium in non-dispersive. Dispersion is the observation that harmonic waves, with a different frequency, propagate with different velocities. Because a travelling wave is composed of harmonic waves the shape of a travelling wave is altered when the components have mutually different wave speeds. The ability to transform the shape of waves seems a necessary condition for continua to capture localisation phenomena. Waves that propagate through a classical strain-softening medium are not dispersive, i.e. the continuum is not able to transform waves into stationary localisa-

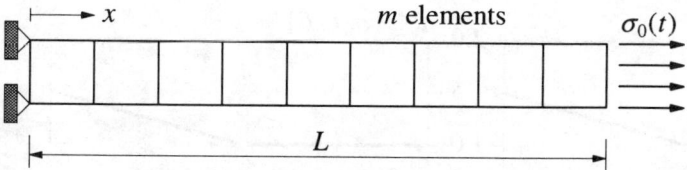

Figure 5.3 Strain-softening bar subject to uniaxial dynamic loading.

tion waves. For a dispersion investigation we assume a single linear harmonic wave propagating through a one-dimensional continuum with a velocity field of the form

$$v(x,t) = Ae^{i(kx-\omega t)}, \tag{5.17}$$

in which ω is the angular frequency and k is the wave number counting the number of wave lengths λ in the bar over 2π

$$k = \frac{2\pi}{\lambda}. \tag{5.18}$$

A dispersion relation can be obtained if eq.(5.17) is substituted in the wave equation (eq.(5.12)), which yields

$$\omega = c_e \sqrt{\frac{f'}{E+f'}} k. \tag{5.19}$$

Waves are called dispersive if the phase velocity $c_f = \omega/k$ is a function of wave number k. For the classical strain-softening bar from eq.(5.12) it becomes clear that c_f is independent of k and therefore waves are non-dispersive (Whitham 1974). For this reason a classical strain-softening bar is not able to change the shape of an arbitrary loading wave into a stationary wave representing the localisation zone.

To investigate the consequences in a mechanical sense of the mathematical statements of ill-posedness and imaginary wave speeds, we will derive the analytical solution of a bar which is fixed at one side and is loaded by a dynamic tensile force at the other side This longitudinal wave propagation problem is given in Figure 5.13 and will be investigated numerically next. We use a linear strain-softening model and a step load, which means that the load remains constant after initiation. The transient wave propagates through the bar and reflects at the left boundary. If $\frac{1}{2} f_t < \sigma_0 < f_t$, with f_t the tensile strength, the tensile strength is exceeded after reflection and a localised softening zone w emerges. In Sluys (1992) an analytical solution for this problem has been derived. The solution for the displacements u, the strains ε, the stresses σ and the energy consumption U in the bar is

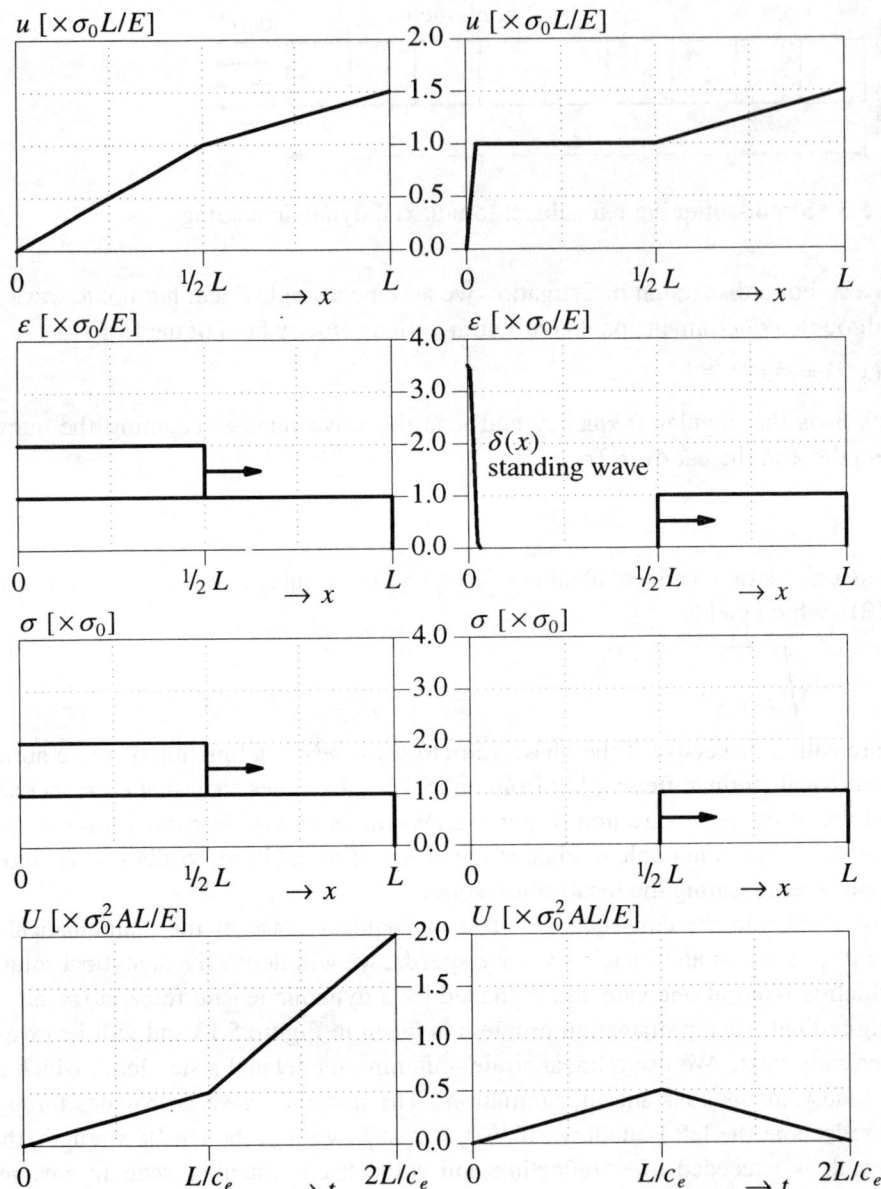

Figure 5.4 Analytical solution for elastic bar (left) and strain-softening bar (right) for displacements, strains, stresses (at $t = 3/2\, L/c_e$) and consumption of energy.

$$u(x,t) = \frac{H(t-(L-x)/c_e)}{\rho c_e} \sigma_0(t-(L-x)/c_e) + \qquad (5.20)$$

$$\frac{H(t-(L+x)/c_e)}{\rho c_e} \sigma_0(t-(L+x)/c_e),$$

$$\varepsilon(x,t) = \frac{\sigma_0}{E}(H(t-(L-x)/c_e) - H(t-(L+x)/c_e)) + \frac{2\sigma_0}{\rho c_e}(t-L/c_e)\delta(x), \qquad (5.21)$$

$$\sigma(x,t) = \sigma_0(H(t-(L-x)/c_e) - H(t-(L+x)/c_e)), \qquad (5.22)$$

$$U(t) = A\int_0^L \sigma\varepsilon\,dx = \frac{\sigma_0^2 A c_e}{2E}(t-2(t-L/c_e)H(t-L/c_e)) \text{ for } 0<t<2L/c_e, \qquad (5.23)$$

respectively, in which A is the cross-sectional area of the bar, H is the Heaviside step function and $\delta(x)$ is the Dirac delta function. The results of the analytical solution of the strain-softening bar have been plotted in Figure 5.4, in which a comparison is made with a purely elastic bar. The spurious character of the solution is obvious, the strain reaches infinity after reflection in a localisation zone of zero length. In fact, the solution of the elliptic equation is a standing wave, described by a Dirac delta function, which does not have the ability to extend. The stress drops to zero instantly and the wave reflects on the softening zone as on a free boundary. As already stated in the previous section spatial interaction between the elliptic and the hyperbolic system is no longer possible. The tensile wave returns as a pressure wave instead of a superposition of tensile waves which is usual after reflection on a fixed boundary. Furthermore, it is obvious that from $t = L/c_e$ after reflection the bar is unable to consume a further amount of energy.

The bar is also analysed numerically. A one-dimensional strain softening law in plasticity, damage or crack format can be used. The bar problem is sketched in Figure 5.3 and the parameter set that is used is as follows : length $L = 100$ mm, cross-section $A = 1$ mm^2, loading $\sigma_0 = 0.75 f_t$, Young's modulus $E = 20000$ N/mm^2, density $\rho = 2 \cdot 10^{-8}$ Ns^2/mm^4, tensile strength $f_t = 2$ N/mm^2 and softening modulus $h = -1000$ N/mm^2. Use of these parameters yields a linear elastic wave speed $c_e = 1000$ m/s. The bar is divided into 10, 20, 40 and 80 elements, respectively. Use has been made of eight-noded elements with a nine-point Gauss integration scheme. The response of the bar is linearly elastic until the loading wave reaches the left boundary. The doubling in stress ($2\sigma_0 = 3/2\,f_t$) due to reflection of the tensile wave causes the initiation of cracking. The material enters the softening regime and a localisation zone of intense straining emerges.

In Figure 5.5 the displacements, the strains and the stresses for the different meshes are plotted at $t = 3/2\, L/c_e = 0.15 \cdot 10^{-3}$ s, that is when the wave has reflected at the left boundary and has returned to $x = 1/2\, L$. Mesh sensitivity is obvious : strain localisation and, consequently, the jump in displacement occur in only one vertical row of three integration points which is the smallest possible zone. Hence, the width of the localisation zone w decreases when more elements are used. After reflection the strain localises in one vertical row of integration points at the left boundary and extension of the zone is not pos-

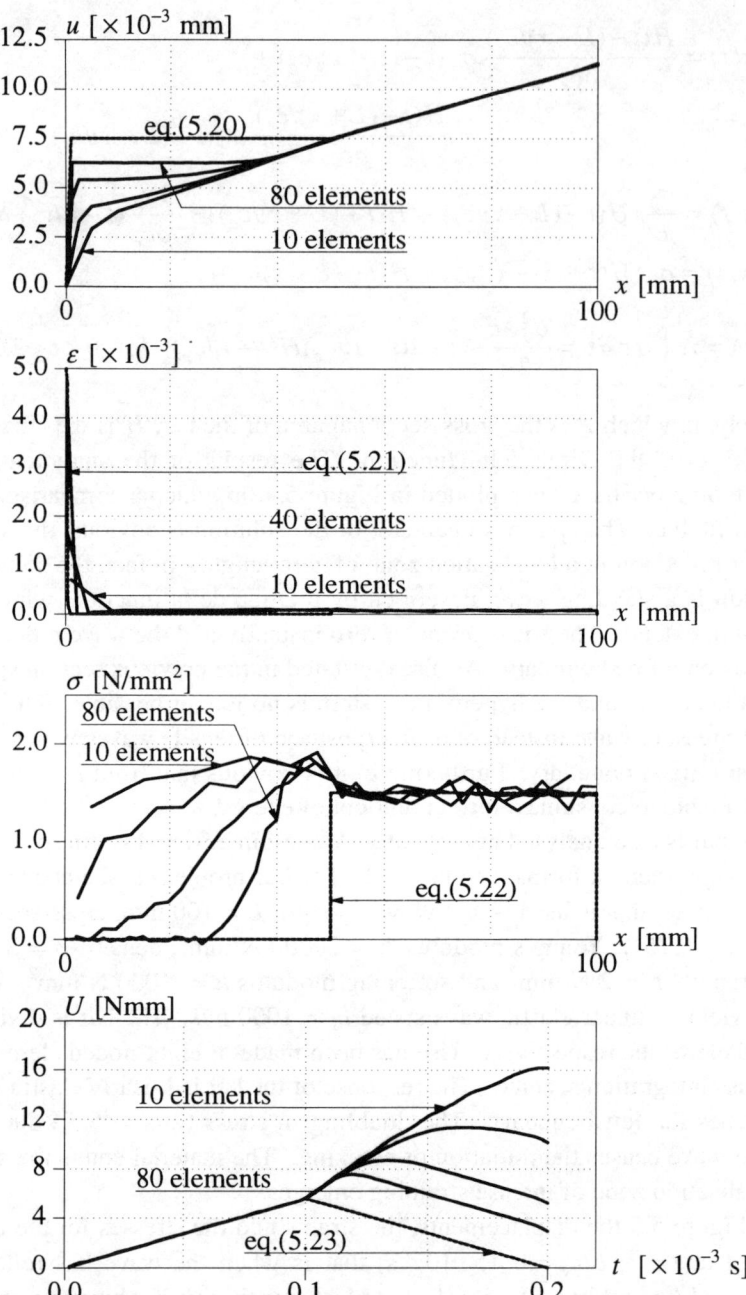

Figure 5.5 Mesh-dependent results with classical strain-softening model: Displacement, strains and stresses at $t = {}^3\!/_2\, L/c_e$ and consumption of energy (bottom).

sible. This is in agreement with the analytical findings previously mentioned. It is obvious that the analytical solution is approached when the mesh is refined. The results for the discretisation with 80 elements has not been plotted because at $t = 0.15 \cdot 10^{-3}$ s the bar has already failed. The stress profiles after reflection show that the amount of wave reflection depends on the mesh : for more elements there is a larger reduction in stress of the reflected wave. When the mesh is finer the stress in the softening zone drops to zero more rapidly which determines the amount of reflection. As soon as the stress has become zero one vertical row of integration points start to act as a free boundary on which the tensile wave reflects as a pressure wave. Summation of a tensile wave propagating to the left and a pressure wave propagating to the right yields a zero stress situation. Moreover, the development of the consumption of energy U in the bar depends on the number of elements in the mesh as can be seen from Figure 5.5. The increase of the sum of elastic and inelastic energy after reflection ($t > L/c_e$) is smaller when the localisation zone is smaller. In the limiting analytical case failure occurs at $t = L/c_e$ without any further energy consumption in the strain-softening zone of the bar. The stress drops to zero instantly and the wave reflects as a pressure wave. The elastic energy gradually vanishes in the bar with the returning pressure wave.

It turns out that under mesh refinement the spurious analytical solution is recovered. The strain in the localisation zone approaches an infinite value, the jump in displacements at the boundary becomes discontinuous, the stress drops to zero in an infinitesimal time interval and the consumption of energy becomes zero because the volume of the strain-softening zone becomes zero.

6. ENHANCED PLASTICITY THEORIES

For proper failure analyses the above observations are unacceptable and we must therefore rephrase our continuum description of the softening solid such that it is able to capture zones of highly localised deformation. For a proper mathematical modelling of the softening solid extra or higher-order derivative terms are necessary in the continuum description. Such enhanced, or higher-order continua do not necessarily lose ellipticity in statics or hyperbolicity in dynamics at the onset of strain softening. In this Chapter we suggest two so-called regularisation techniques to solve mesh sensitivity, namely (i) the addition of viscous, or higher-order time derivatives and (ii) the addition of higher-order deformation gradients. A proper modelling of the softening solid is obtained if the numerical results converge to a finite size of the localisation zone upon mesh refinement with unique properties with respect to energy consumption and wave reflection. Below a certain discretisation level at which accuracy of the results is obtained a refinement of the mesh as well as a change of orientation of the mesh lines should not affect the numerical outcome.

6.1 Viscoplasticity

The inclusion of strain rate dependence in the constitutive modelling of plastic materials seems natural under transient loading conditions. Even for quasi-static problems, final failure is commonly accompanied by high strain rates due to the onset of damage or frictional effects. High strain rates mobilise viscous effects in a material which carry a part of the load, so that, from a physical point of view the extension of the plasticity relations with viscous terms (viscoplasticity) is evident. The inclusion of rate effects in the constitutive equations results in an implicit introduction of an internal length scale and in dispersive wave propagation in the rate-dependent medium. In the viscoplastic theory an important distinction from the inviscid plasticity theory stems from the fact that stress states outside the yield surface are not illegal. If the external loading remains constant the stresses return to the yield surface as a function of time. Because of this feature viscoplastic theories are commonly called overstress laws. The models according to Perzyna and Duvaut-Lions belong to this category. A third model that is treated here is the consistency model in which the yield function is made a function of the strain rate. Consequently, the yield surface may shrink or expand upon variation of the strain rate. Similar to the rate-independent theory the strain rate is decomposed into an elastic and a viscoplastic strain rate

$$\dot{\varepsilon} = \dot{\varepsilon}_e + \dot{\varepsilon}_{vp} . \tag{6.1}$$

Eq.(4.5) then becomes

$$\dot{\sigma} = \mathbf{D}_e (\dot{\varepsilon} - \dot{\varepsilon}_{vp}) . \tag{6.2}$$

The three different approaches in viscoplasticity will be treated subsequently.

6.1.1 Perzyna model

In the viscoplastic model according to Perzyna (Perzyna, 1966) the viscoplastic strain rate is defined similar to the plastic strain rate in standard plasticity with a modified definition of the plastic multiplier

$$\dot{\varepsilon}_{vp} = \dot{\lambda}\,\mathbf{m}, \tag{6.3}$$

$$\dot{\lambda} = \gamma <\phi(f)>, \tag{6.4}$$

in which γ is a fluidity parameter and $\mathbf{m}$ is defined as the gradient to the viscoplastic potential function g_{vp}. The notation $<\,>$ refers to $<y> = y\,H(y)$, where H is the Heaviside step function. $\phi(f)$ is an arbitrary function of the yield function for which we can take a power law expression according to (see Sluys 1992)

$$\phi(f) = \left[\,f/\bar{\sigma}_0\,\right]^N, \tag{6.5}$$

with N a constant and $\bar{\sigma}_0$ the initial yield stress. The choice for $\phi(f)$ determines the regularising effect of the viscoplastic model.

The incremental elastic response is given by

$$\Delta\boldsymbol{\sigma} = \mathbf{D}_e\,(\Delta\boldsymbol{\varepsilon} - \Delta\boldsymbol{\varepsilon}_{vp}), \tag{6.6}$$

and the incremental viscoplastic strain is estimated with the generalised Euler method

$$\Delta\boldsymbol{\varepsilon}_{vp} = [\,(1-\theta)\,\dot{\boldsymbol{\varepsilon}}_{vp}^{t} + \theta\dot{\boldsymbol{\varepsilon}}_{vp}^{t+\Delta t}\,]\,\Delta t, \tag{6.7}$$

where θ is the interpolation parameter for which $0 \le \theta \le 1$. In the one-step Euler integration scheme, the viscoplastic strain rate at the end of the time interval is expressed in a limited Taylor series expansion as

$$\dot{\boldsymbol{\varepsilon}}_{vp}^{t+\Delta t} = \dot{\boldsymbol{\varepsilon}}_{vp}^{t} + \left[\frac{\partial\dot{\boldsymbol{\varepsilon}}_{vp}}{\partial\boldsymbol{\sigma}}\right]^{t}\Delta\boldsymbol{\sigma} + \left[\frac{\partial\dot{\boldsymbol{\varepsilon}}_{vp}}{\partial\kappa}\right]^{t}\Delta\kappa = \dot{\boldsymbol{\varepsilon}}_{vp}^{t} + \mathbf{G}^{t}\Delta\boldsymbol{\sigma} + \mathbf{h}^{t}\Delta\kappa \tag{6.8}$$

where

$$\mathbf{G}^{t} = \gamma\left[\frac{\partial\phi}{\partial\boldsymbol{\sigma}}\mathbf{m}^{T} + \phi\,\frac{\partial\mathbf{m}}{\partial\boldsymbol{\sigma}}\right]^{t},\quad \mathbf{h}^{t} = \gamma\left[\frac{\partial\phi}{\partial\kappa}\mathbf{m} + \phi\,\frac{\partial\mathbf{m}}{\partial\kappa}\right]^{t} \tag{6.9}$$

and eqs.(6.3) and (6.4) have been applied. For $\Delta\kappa$ the value at the previous time step or at the previous global iteration can be used. Substitution of eq.(6.8) into eq.(6.7) yields

$$\Delta\boldsymbol{\varepsilon}_{vp} = (\dot{\boldsymbol{\varepsilon}}_{vp}^{t} + \theta\,\mathbf{G}^{t}\Delta\boldsymbol{\sigma} + \theta\,\mathbf{h}^{t}\Delta\kappa)\,\Delta t. \tag{6.10}$$

This equation can be substituted in the incremental stress-strain relation (6.6), which leads to

$$\Delta\boldsymbol{\sigma} = \mathbf{D}_{vp}\,\Delta\boldsymbol{\varepsilon} - \Delta\mathbf{q}, \tag{6.11}$$

where

$$\mathbf{D}_{vp} = \left[(\mathbf{D}_e)^{-1} + \theta\,\Delta t\,\mathbf{G}^{t}\right]^{-1} \tag{6.12}$$

is the tangent stiffness matrix and

Box 6.1   One step stress update for Perzyna model.

FOR EACH INTEGRATION POINT :

1. $\Delta\varepsilon = \mathbf{B}\Delta\mathbf{a}$, $\sigma = \sigma^t$, $\kappa = \kappa^t$

2. Compute trial stress state : $\sigma^{trial} = \sigma^t + \mathbf{D}_e \Delta\varepsilon$

3. Plasticity ? YES, if $f(\sigma^{trial}, \kappa^t) \geq 0$ ELSE, $\sigma = \sigma^{trial}$ and go to 10.

4. Compute : $\mathbf{G}^t = \gamma \left[\dfrac{\partial \phi}{\partial \sigma} \mathbf{m}^T + \phi \dfrac{\partial \mathbf{m}}{\partial \sigma} \right]^t$

5. Tangent stiffness matrix : $\mathbf{D}_{vp} = [(\mathbf{D}_e)^{-1} + \theta \Delta t \, \mathbf{G}^t]^{-1}$

6. Compute : $\mathbf{h}^t = \gamma \left[\dfrac{\partial \phi}{\partial \kappa} \mathbf{m} + \phi \dfrac{\partial \mathbf{m}}{\partial \kappa} \right]^t$

7. Pseudo-load : $\Delta\mathbf{q} = \mathbf{D}_{vp} (\dot{\varepsilon}^t_{vp} \Delta t + \theta \Delta t \mathbf{h}^t \Delta\kappa^0)$

8. Update of stress : $\sigma^{t+\Delta t} = \sigma^t + \mathbf{D}_{vp} \Delta\varepsilon - \Delta\mathbf{q}$

9. Update eq. plastic strain : $f^{t+\Delta t} = f(\sigma^{t+\Delta t}, \kappa^t)$, $\dot{\varepsilon}^{t+\Delta t}_{vp} = \gamma \phi(\sigma^{t+\Delta t}, \kappa^t) \mathbf{m}^{t+\Delta t}$,

    $\dot{\kappa}^{t+\Delta t} = \sqrt{(\dot{\varepsilon}^{t+\Delta t}_{vp})^T \mathbf{A} \, \dot{\varepsilon}^{t+\Delta t}_{vp}}$, $\Delta\kappa^{t+\Delta t} = [(1-\theta)\dot{\kappa}^t + \theta \dot{\kappa}^{t+\Delta t}] \Delta t$

10. Next integration point

$$\Delta\mathbf{q} = \mathbf{D}_{vp} (\dot{\varepsilon}^t_{vp} \Delta t + \theta \Delta t \, \mathbf{h}^t \Delta\kappa) , \tag{6.13}$$

gives the contribution of an extra pseudo-nodal force in the equation of motion. In Box 6.1 the Euler stress-update algorithm for Perzyna viscoplasticity is outlined, where the matrix $\mathbf{A} = diag \, [2/3 \; 2/3 \; 2/3 \; 1/3 \; 1/3 \; 1/3]$ and the update of the equivalent plastic strain is explained.

In the one-step Euler integration scheme, the current viscoplastic flow is estimated by a limited Taylor series expansion which depends on the gradient of the yield surface at time t. In a fully implicit integration scheme, the viscoplastic flow is determined by the gradient of the yield surface at time $t + \Delta t$. The incremental viscoplastic strain now reads

$$\Delta\varepsilon_{vp} = \Delta\lambda \, \mathbf{m}^{t+\Delta t} \tag{6.14}$$

where a plastic multiplier $\Delta\lambda$ is introduced under the constraint that the residual [cf.eq.(6.4)]

Box 6.2   An iterative implicit stress update for Perzyna model.

FOR EACH INTEGRATION POINT :

1. $\Delta \boldsymbol{\varepsilon} = \mathbf{B}\Delta \boldsymbol{a}$, iteration $k = 0$, $\Delta\lambda^0 = 0$

2. Compute trial stress state : $\boldsymbol{\sigma}^{trial} = \boldsymbol{\sigma}^t + \mathbf{D}_e \Delta \boldsymbol{\varepsilon}$

3. Plasticity ? YES, if $f(\boldsymbol{\sigma}^{trial}, \kappa^t) \geq 0$ ELSE, $\boldsymbol{\sigma} = \boldsymbol{\sigma}^{trial}$ and go to 10.

4. Estimates for zero-th iteration : $\boldsymbol{\sigma}^0 = \boldsymbol{\sigma}^t + \mathbf{D}_e [\Delta \boldsymbol{\varepsilon} - \Delta\lambda^0 \mathbf{m}^{trial}]$

    $r^0 = \phi(\boldsymbol{\sigma}^0, \kappa^t) - \Delta\lambda^0 (\gamma \Delta t)^{-1}$

5. Compute : $\mathbf{P}^k = \left[(\mathbf{D}_e)^{-1} + \Delta\lambda^k \dfrac{\partial \mathbf{m}^k}{\partial \boldsymbol{\sigma}^k} \right]^{-1}$

    $\alpha = \left[\dfrac{\partial \phi}{\partial \boldsymbol{\sigma}} \right]^{kT} \mathbf{P}^k \left[\Delta\lambda \dfrac{\partial \mathbf{m}}{\partial \lambda} + \mathbf{m} \right]^k + \dfrac{1}{\gamma \Delta t} - \left[\dfrac{\partial \phi}{\partial \lambda} \right]^k$

6. Update plastic multiplier : $\Delta\lambda^{k+1} = \Delta\lambda^k + r^k \alpha^{-1}$

7. Update of stress : $\boldsymbol{\sigma}^{k+1} = \boldsymbol{\sigma}^t + \mathbf{D}_e [\Delta \boldsymbol{\varepsilon} - \Delta\lambda^{k+1} \mathbf{m}^k]$

8. Calculate : $r^{k+1} = \phi(\boldsymbol{\sigma}^{k+1}, \lambda^0 + \Delta\lambda^{k+1}) - \Delta\lambda^{k+1} (\gamma \Delta t)^{-1}$

9. Check convergence : $|r^{k+1}| <$ tolerance

    if YES, go to 10.

    if NO, $k = k + 1$ go to 5.

10. Next integration point

$$r = \phi(\boldsymbol{\sigma}^{t+\Delta t}, \lambda^{t+\Delta t}) - \Delta\lambda/(\gamma \Delta t) \qquad (6.15)$$

goes to zero during the local iterative procedure (see also van den Boogaard (1995) and Wang (1997)). Substituting eq.(6.14) into eq.(6.6) yields

$$\Delta \boldsymbol{\sigma} = \mathbf{D}_e \left[\Delta \boldsymbol{\varepsilon} - \Delta\lambda \, \mathbf{m}^{t+\Delta t} \right]. \qquad (6.16)$$

To compute $\Delta\lambda$, a local Newton-Raphson iteration process is applied. The k-th iterative improvements of $\Delta\boldsymbol{\sigma}^k$, $\Delta\boldsymbol{\varepsilon}^k$ and $\Delta\lambda^k$ are denoted by $\delta\boldsymbol{\sigma}^k$, $\delta\boldsymbol{\varepsilon}^k$ and $\delta\lambda^k$, respectively. Differentiation of eq.(6.16) gives the variation of $\Delta\sigma^k$ as a function of the variation of $\Delta\varepsilon^k$ and $\Delta\lambda^k$ during iteration k

$$\delta\sigma = \mathbf{P}\delta\varepsilon - \mathbf{P}\left[\mathbf{m} + \Delta\lambda\frac{\partial \mathbf{m}}{\partial\lambda}\right]\delta\lambda \qquad (6.17)$$

with a pseudo-elastic material stiffness matrix

$$\mathbf{P} = \left[(\mathbf{D}_e)^{-1} + \Delta\lambda\frac{\partial \mathbf{m}}{\partial\sigma}\right]^{-1} \qquad (6.18)$$

where the superscript k has been dropped, and all the variables without specification hereafter, refer to the k-th local iteration during the global iteration at the current time step $t + \Delta t$. Furthermore, by differentiation of eq.(6.15) the correction of the k-th Newton-Raphson process reads

$$\left[\frac{\partial\phi}{\partial\sigma}\right]^T\delta\sigma + \left[\frac{\partial\phi}{\partial\lambda} - \frac{1}{\gamma\Delta t}\right]\delta\lambda = \delta r . \qquad (6.19)$$

Substitution of eq.(6.17) into eq.(6.19) yields

$$\delta\lambda = \frac{1}{\alpha}\left[\left(\frac{\partial\phi}{\partial\sigma}\right)^T\mathbf{P}\delta\varepsilon - \delta r\right] \qquad (6.20)$$

with

$$\alpha = \left[\frac{\partial\phi}{\partial\sigma}\right]^T\mathbf{P}\left[\mathbf{m} + \Delta\lambda\frac{\partial \mathbf{m}}{\partial\lambda}\right] + \frac{1}{\gamma\Delta t} - \frac{\partial\phi}{\partial\lambda} . \qquad (6.21)$$

Substitution of eq.(6.20) into eq.(6.17) subsequently leads to the tangent stiffness matrix

$$\mathbf{D}_{vp} = \mathbf{P} - \frac{1}{\alpha}\mathbf{P}\left[\mathbf{m} + \Delta\lambda\frac{\partial \mathbf{m}}{\partial\lambda}\right]\left[\frac{\partial\phi}{\partial\sigma}\right]^T\mathbf{P} . \qquad (6.22)$$

If we use local iterations during a global iteration, the iterative strain increment $\delta\varepsilon$ vanishes from eq.(6.20) due to a fixed total strain increment. In Box 6.2 the iterative implicit stress-update algorithm for Perzyna viscoplasticity is outlined.

6.1.2 Duvaut-Lions model

A different viscoplastic model, which in its elaboration more closely connects to rate-independent plasticity, has been proposed by Duvaut and Lions (Duvaut & Lions 1972, Simo et al. 1988, Loret & Prevost 1990). The theory is based on the difference in response between the rate-independent material and the viscoplastic material. This is in contrast with the Perzyna model in which the value of the yield surface determines the viscoplastic strain rate. The Duvaut-Lions model has the advantage that it can be combined with a yield surface which has an apex (Drucker-Prager, Mohr-Coulomb) or which is non-smooth (Mohr-Coulomb, Tresca). In the Duvaut-Lions model the viscoplastic strain rate and the hardening law are defined as

$$\dot{\varepsilon}_{vp} = \frac{1}{\eta} (\mathbf{D}_e)^{-1} [\sigma - \bar{\sigma}] \tag{6.23}$$

and

$$\dot{\kappa} = -\frac{1}{\eta} (\kappa - \bar{\kappa}) , \tag{6.24}$$

where $\bar{\sigma}$ is the contribution of the rate-independent material (we use a bar to denote the variable of the inviscid plastic model or back-bone model) which can be viewed as a projection of the current stress to the yield surface. η is a viscous parameter which represents the relaxation time of the material and, in general, is strain and strain-rate dependent.

In the Duvaut-Lions model, the stress update is carried out in two steps. First the inviscid back-bone stress $\bar{\sigma}$ is updated. An Euler backward algorithm is used for the return mapping of the stress to the yield surface. The tangential stiffness relation yields

$$\Delta \bar{\sigma} = \bar{\mathbf{D}}_p \Delta \varepsilon \tag{6.25}$$

with the elasto-plastic stiffness tensor

$$\bar{\mathbf{D}}_p = \mathbf{D}_e - \frac{\mathbf{D}_e \bar{\mathbf{m}} \bar{\mathbf{n}}^T \mathbf{D}_e}{h + \bar{\mathbf{n}}^T \mathbf{D}_e \bar{\mathbf{m}}} , \quad \bar{\mathbf{n}} = \frac{\partial f}{\partial \bar{\sigma}} . \tag{6.26}$$

Next, the viscoplastic response is determined. The current viscoplastic strain rate $\dot{\varepsilon}_{vp}^{t+\Delta t}$ can be expressed as

$$\dot{\varepsilon}_{vp}^{t+\Delta t} = \frac{1}{\eta} (\mathbf{D}_e)^{-1} [\sigma^{t+\Delta t} - \bar{\sigma}^{t+\Delta t}] = \frac{1}{\eta} (\mathbf{D}_e)^{-1} [\Delta \sigma - \Delta \bar{\sigma} + \sigma_{vp}^t] \tag{6.27}$$

where $\sigma_{vp}^t = \sigma^t - \bar{\sigma}^t$ is the viscous stress at the beginning of the time step. Now the viscoplastic strain increment $\Delta \varepsilon_{vp}$ in eq.(6.7) becomes

$$\Delta \varepsilon_{vp} = \left[(1 - \theta) \dot{\varepsilon}_{vp}^t + \frac{\theta}{\eta} (\mathbf{D}_e)^{-1} (\Delta \sigma - \Delta \bar{\sigma} + \sigma_{vp}^t) \right] \Delta t . \tag{6.28}$$

Substitution of eq.(6.28) into eq.(6.6) yields

$$\Delta \sigma = \mathbf{D}_{vp} \Delta \varepsilon - \Delta \mathbf{q} , \tag{6.29}$$

in which

$$\mathbf{D}_{vp} = \frac{\eta}{\eta + \theta \Delta t} \left[\mathbf{D}_e + \frac{\theta \Delta t}{\eta} \bar{\mathbf{D}}_p \right] \tag{6.30}$$

and

Box 6.3 One step implicit stress update for Duvaut-Lions model.

FOR EACH INTEGRATION POINT :

1. $\Delta\varepsilon = \mathbf{B}\Delta\mathbf{a}$, $\sigma = \sigma^t$, $\kappa = \kappa^t$

2. Compute trial stress state : $\sigma^{trial} = \sigma^t + \mathbf{D}_e \Delta\varepsilon$

3. Plasticity ? YES, if $f(\sigma^{trial}, \kappa^t) \geq 0$ ELSE, $\sigma = \sigma^{trial}$ and go to 10.

4. Update back-bone stress : $\bar{\sigma}^{t+\Delta t} = \bar{\sigma}^t + \bar{\mathbf{D}}_p \Delta\varepsilon$

5. Update eq. plastic strain : $\bar{\kappa}^{t+\Delta t} = \bar{\kappa}^t + \dot{\bar{\kappa}}^{t+\Delta t} \Delta t$

6. Tangent stiffness matrix : $\mathbf{D}_{vp} = \dfrac{\eta}{\eta + \theta\Delta t}\left[\mathbf{D}_e + \dfrac{\theta\Delta t}{\eta}\bar{\mathbf{D}}_p\right]$

7. Compute pseudo-load : $\Delta\mathbf{q} = \dfrac{\eta\Delta t}{\eta + \theta\Delta t}\left[(1-\theta)\mathbf{D}_e \dot{\varepsilon}^t_{vp} + \dfrac{\theta}{\eta}\sigma^t_{vp}\right]$

8. Update stress : $\sigma^{t+\Delta t} = \sigma^t + \mathbf{D}_{vp}\Delta\varepsilon - \Delta\mathbf{q}$

9. Update eq. plastic strain : $\dot{\varepsilon}^{t+\Delta t} = \dfrac{1}{\eta}(\mathbf{D}_e)^{-1}(\sigma^{t+\Delta t} - \bar{\sigma}^{t+\Delta t})$

   $\dot{\kappa}^{t+\Delta t} = \sqrt{(\dot{\varepsilon}^{t+\Delta t}_{vp})^T \mathbf{A}\, \dot{\varepsilon}^{t+\Delta t}_{vp}}$, $\kappa^{t+\Delta t} = \bar{\kappa}^{t+\Delta t} - \eta\dot{\kappa}^{t+\Delta t}$

10. Next integration point

$$\Delta\mathbf{q} = \dfrac{\eta\Delta t}{\eta + \theta\Delta t}\left[(1-\theta)\mathbf{D}_e \dot{\varepsilon}^t_{vp} + \dfrac{\theta}{\eta}\sigma^t_{vp}\right]. \tag{6.31}$$

Note that with $\theta = 1$ the tangent $\mathbf{D}_c$ for the backward Euler algorithm as in Ju (1990) is recovered. The algorithm is summarised in Box 6.3.

6.1.3 Consistency model

In the Perzyna viscoplasticity theory the current stress state can be outside the yield surface and the Kuhn-Tucker conditions are not valid. The evolution of the viscoplastic flow is constrained by the plastic relaxation equations and directly defined in the stress space. We now consider another model in which the strain-rate contribution (viscosity) is incorporated through a rate-dependent yield surface (see also Wang (1997)). The viscoplastic strain rate is defined similar to the plastic strain rate in standard plasticity

$$\dot{\varepsilon}^{vp} = \dot{\lambda}\,\mathbf{m}\,,\tag{6.32}$$

$$\mathbf{m} = \partial g_{vp}/\partial \boldsymbol{\sigma}\,.\tag{6.33}$$

A rate-dependent yield function is proposed by introducing the equivalent plastic strain rate

$$f(\boldsymbol{\sigma},\kappa,\dot{\kappa})\tag{6.34}$$

Accordingly, the evolution of the viscoplastic flow is determined by the consistency condition

$$\dot{f}(\boldsymbol{\sigma},\kappa,\dot{\kappa}) = \mathbf{n}^T\dot{\boldsymbol{\sigma}} - h\,\dot{\lambda} - s\,\ddot{\lambda} = 0\,,\tag{6.35}$$

where the gradient vector $\mathbf{n}$, the hardening modulus h and the viscosity parameter s are defined as

$$\mathbf{n} = \frac{\partial f}{\partial \boldsymbol{\sigma}}\,,\quad h = -\frac{\partial f}{\partial \kappa}\frac{\dot{\kappa}}{\dot{\lambda}}\,,\quad s = -\frac{\partial f}{\partial \dot{\kappa}}\frac{\ddot{\kappa}}{\ddot{\lambda}}\,.\tag{6.36}$$

The algorithmic aspects of the consistency model will be given now. The evolution of the plastic flow is determined by the discretised consistency condition of the rate-dependent yield surface

$$f(\boldsymbol{\sigma}^{k+1},\lambda^{k+1},\dot{\lambda}^{k+1}) \approx f^k + \mathbf{n}^T\delta\boldsymbol{\sigma} + h\,\delta\lambda + s\,\delta\dot{\lambda} = 0\tag{6.37}$$

where $f^k = f(\boldsymbol{\sigma}^k,\lambda^k,\dot{\lambda}^k)$ is the k-th residual of the yield function. For algorithmic convenience, we assume that the rate of the history parameter can directly be expressed as a function of the plastic multiplier via $\dot{\kappa} = \alpha\dot{\lambda}$. This assumption seems a limitation of the model but a large class of hardening/softening hypotheses and yield functions satisfy this condition (Pamin 1994). We again use a local Newton-Raphson iteration process to compute the plastic multiplier $\Delta\lambda$. Differentiation of eqs.(6.6) and (6.7) and the estimate for the plastic multiplier

$$\Delta\lambda = [(1-\theta)\dot{\lambda}^t + \theta\dot{\lambda}^{t+\Delta t}]\Delta t\tag{6.38}$$

leads to

$$\delta\boldsymbol{\sigma} = \mathbf{D}_e\,\delta\boldsymbol{\varepsilon} - \mathbf{D}_e\left[\dot{\lambda}\left(\frac{\partial \mathbf{m}}{\partial \boldsymbol{\sigma}}\delta\boldsymbol{\sigma} + \frac{\partial \mathbf{m}}{\partial \lambda}\delta\lambda + \frac{\partial \mathbf{m}}{\partial \dot{\lambda}}\delta\dot{\lambda}\right) + \mathbf{m}\,\delta\dot{\lambda}\right]\theta\,\Delta t\tag{6.39}$$

$$\delta\lambda = \theta\,\Delta t\,\delta\dot{\lambda}\tag{6.40}$$

Substitution of eq.(6.40) into eq.(6.39) yields

$$\delta\boldsymbol{\sigma} = \mathbf{P}\,\delta\boldsymbol{\varepsilon} - \mathbf{P}\,\bar{\mathbf{n}}\,\delta\lambda\tag{6.41}$$

with a pseudo-elastic material stiffness matrix $\mathbf{P}$ and a pseudo plastic flow direction $\bar{\mathbf{n}}$

$$\mathbf{P} = \left[(\mathbf{D}_e)^{-1} + \theta\dot{\lambda}\Delta t\,\frac{\partial \mathbf{m}}{\partial \boldsymbol{\sigma}}\right]^{-1},\tag{6.42}$$

Box 6.4 An iterative implicit stress update for consistency model.

FOR EACH INTEGRATION POINT :

1. $\Delta\boldsymbol{\varepsilon} = \mathbf{B}\Delta\boldsymbol{a}$, iteration $k = 0$

2. Compute trial stress state : $\boldsymbol{\sigma}^{trial} = \boldsymbol{\sigma}^t + \mathbf{D}_e\Delta\boldsymbol{\varepsilon}$

3. Plasticity ? YES, if $f(\boldsymbol{\sigma}^{trial}, \lambda^t, \dot{\lambda}^t) \geq 0$ ELSE, $\boldsymbol{\sigma} = \boldsymbol{\sigma}^{trial}$ and go to 11.

4. In zero-th iteration : $\Delta\lambda^0 = 0$, $\dot{\lambda}^0 = 0$, $\boldsymbol{\sigma}^0 = \boldsymbol{\sigma}^{trial}$, $f^0 = f^{trial}$

5. Compute : $\mathbf{P}^k = \left[(\mathbf{D}_e)^{-1} + \theta \Delta t \, \dot{\lambda}^k \left(\frac{\partial \mathbf{m}}{\partial \boldsymbol{\sigma}}\right)^k \right]^{-1}$

   $\bar{\mathbf{n}}^k = \left[\mathbf{m} + \theta \, \dot{\lambda} \, \Delta t \, \frac{\partial \mathbf{m}}{\partial \lambda} + \dot{\lambda} \, \frac{\partial \mathbf{m}}{\partial \dot{\lambda}} \right]^k$

6. Update plastic multiplier : $\Delta\lambda^{k+1} = \Delta\lambda^k + \dfrac{f^k}{\mathbf{n}^T \mathbf{P} \bar{\mathbf{n}} + h + s\,(\theta\Delta t)^{-1}}$

   $\dot{\lambda}^{k+1} = \dfrac{\Delta\lambda^{k+1}}{\theta \Delta t} - \dfrac{1-\theta}{\theta} \dot{\lambda}^t$

7. Compute : $\Delta\boldsymbol{\varepsilon}_{vp} = \left[(1-\theta)\dot{\boldsymbol{\varepsilon}}^t_{vp} + \theta \, \dot{\lambda}^{k+1} \mathbf{m}^k \right] \Delta t$

8. Update of stress : $\boldsymbol{\sigma}^{k+1} = \boldsymbol{\sigma}^t + \mathbf{D}_e [\Delta\boldsymbol{\varepsilon} - \Delta\boldsymbol{\varepsilon}_{vp}]$

9. Compute : $f^{k+1} = f(\boldsymbol{\sigma}^{k+1}, \lambda^t + \Delta\lambda^{k+1}, \dot{\lambda}^{k+1})$

10. Check convergence : $|f^{k+1}| <$ tolerance

    if YES, go to 11.

    if NO, $k = k + 1$ go to 5.

11. Next integration point

$$\bar{\mathbf{n}} = \mathbf{m} + \theta\dot{\lambda}\Delta t \frac{\partial \mathbf{m}}{\partial \lambda} + \dot{\lambda}\frac{\partial \mathbf{m}}{\partial \dot{\lambda}} \qquad (6.43)$$

respectively. Elimination of $\delta\boldsymbol{\sigma}$ by substitution of eq.(6.41) into eq.(6.37) yields

$$\delta\lambda = \frac{\mathbf{n}^T \mathbf{P} \delta\varepsilon + f}{\mathbf{n}^T \mathbf{P} \bar{\mathbf{n}} + h + s(\theta\Delta t)^{-1}} \tag{6.44}$$

Substitution of eq.(6.44) into eq.(6.41) gives the tangential stiffness matrix

$$\mathbf{D}_{vp} = \mathbf{P} - \frac{\mathbf{P} \bar{\mathbf{n}} \mathbf{n}^T \mathbf{P}}{\mathbf{n}^T \mathbf{P} \bar{\mathbf{n}} + h + s(\theta\Delta t)^{-1}} \, . \tag{6.45}$$

Again, if we use local iterations during a global iteration, the iterative strain increment $\delta\varepsilon$ vanishes from eq.(6.44). In Box 6.4 the stress-update algorithm for the consistency model is outlined.

6.2 Gradient plasticity

From Chapter 5 it appeared that micro-structural modifications, which occur in a localisation zone, cause discontinuous deformation processes which cannot be described with classical continuum models. Therefore, enrichment of the continuum has been proposed to avoid a spurious solution for the localisation zone. In section 6.1 a higher-order time derivative term was included in the material description to overcome the problems. Another type of regularisation methods is based on the inclusion of higher-order spatial derivatives in the constitutive equations (Schreyer and Chen 1986, Lasry and Belytschko 1988, Zbib and Aifantis 1988, Mühlhaus and Aifantis 1991, de Borst and Mühlhaus 1991 and 1992) or the averaging of strains, the so-called non-local models (Bazant et al. 1984, Pijaudier-Cabot and Bazant 1987). In the approach followed in this chapter we assume the presence of second-order strain gradient terms in the stress-strain law. Mühlhaus and Aifantis (1991) and de Borst and Mühlhaus (1991) have shown that such a gradient-dependent model can be derived from non-local models. Gradient models as well as non-local models reflect the fact that the interaction between micro-structural deformations in the localisation zone is non-local.

The use of a higher-order gradient model can result in a well-posed set of partial differential equations. Furthermore, the gradient model explicitly incorporates an internal length scale (see Sluys (1992)). From a dispersion analysis it becomes clear that the wave speeds remain real under strain-softening conditions and that the continuum is capable of transforming a travelling wave into a stationary localisation wave. As a consequence of the well-posedness of the mathematical problem numerical results do not suffer from the pathological mesh dependence.

In the gradient-plasticity model the yield strength does not only depend upon the equivalent plastic strain κ, but also upon the Laplacian thereof. So, we consider the following yield condition

$$f(\sigma, \kappa, \nabla^2 \kappa) = 0 \, . \tag{6.46}$$

As in standard plasticity (see section 4.1) the stress point must remain on the yield surface during plastic deformation, which yields for the consistency condition for continuing plastic flow

$$\frac{\partial f^{\mathrm{T}}}{\partial \boldsymbol{\sigma}} \dot{\boldsymbol{\sigma}} + \frac{\partial f}{\partial \kappa} \dot{\kappa} + \frac{\partial f}{\partial \nabla^2 \kappa} \nabla^2 \dot{\kappa} = 0 .\qquad(6.47)$$

If we use the gradient to the yield surface **n** and the hardening/softening modulus h as defined in the eqs.(4.9) and (4.10) and assume the dependence of f upon $\nabla^2 \kappa$ to be a constant

$$\bar{c} = \frac{\partial f}{\partial \nabla^2 \kappa} ,\qquad(6.48)$$

the consistency equation can be rewritten in

$$\mathbf{n}^{\mathrm{T}} \dot{\boldsymbol{\sigma}} - h \dot{\lambda} + \bar{c} \nabla^2 \dot{\kappa} = 0 .\qquad(6.49)$$

A further assumption is made that the relationship between the plastic multiplier $\dot{\lambda}$ and the hardening/softening rate $\dot{\kappa}$ can be established of the form

$$\dot{\kappa} = \eta \dot{\lambda} ,\qquad(6.50)$$

with η a constant. This assumption seems a limitation of the model but large classes of hardening/softening hypotheses and yield functions satisfy eq.(6.50). In these lecture notes we apply the strain-hardening/softening hypothesis (eq.(4.14)). In combination with the used yield criteria we obtain

$$\eta = 1 \qquad \text{for von Mises and Rankine}\qquad(6.51)$$

and

$$\eta = \sqrt{1 + {}^2\!/_9\, \alpha^2} \qquad \text{for Drucker} - \text{Prager} ,\qquad(6.52)$$

if we assume the cohesion parameter c dependent on the hardening/softening parameter in the Drucker-Prager yield function. If we use $\bar{c}_\eta = \eta \bar{c}$ eq.(6.49) can be written as

$$\mathbf{n}^{\mathrm{T}} \dot{\boldsymbol{\sigma}} - h \dot{\lambda} + \bar{c}_\eta \nabla^2 \dot{\lambda} = 0 .\qquad(6.53)$$

So, for gradient-dependent plasticity the consistency condition results in a differential equation for $\dot{\lambda}$ and an explicit expression for $\dot{\lambda}$ at a local (integration point) level (cf.eq.(4.11)) cannot be obtained.

The discretisation of the gradient model can be carried ou as follows. In Chapter 2 we have derived a weak form of the initial value problem at $t + \Delta t$ according to

$$\int_V \delta \mathbf{u}^{\mathrm{T}} [\mathbf{R}\ddot{\mathbf{u}}^{t+\Delta t}] \, dV + \int_V \delta \boldsymbol{\varepsilon}^{\mathrm{T}} \int_t^{t+\Delta t} \dot{\boldsymbol{\sigma}} \, d\tau \, dV = \int_S \delta \mathbf{u}^{\mathrm{T}} \mathbf{t}^{t+\Delta t} \, dS - \int_V \delta \boldsymbol{\varepsilon}^{\mathrm{T}} \boldsymbol{\sigma}^t \, dV ,\qquad(6.54)$$

in which the body forces **p** have been neglected. Because we cannot satisfy the yield function at a local level we assume satisfaction in a distributed sense. A weak form of eq.(6.46) is given by

$$\int_V \delta \lambda \, f(\boldsymbol{\sigma}^{t+\Delta t}, \kappa^{t+\Delta t}, \nabla^2 \kappa^{t+\Delta t}) \, dV = 0 .\qquad(6.55)$$

The yield function at $t + \Delta t$ can be written as

$$f(\boldsymbol{\sigma}^{t+\Delta t}, \kappa^{t+\Delta t}, \nabla^2 \kappa^{t+\Delta t}) = f(\boldsymbol{\sigma}^t, \kappa^t, \nabla^2 \kappa^t) + \int_t^{t+\Delta t} \dot{f}(\boldsymbol{\sigma}, \kappa, \nabla^2 \kappa) \, d\tau \,. \tag{6.56}$$

With the consistency equation (6.53), eq.(6.55) can now be modified to

$$\int_V \delta\lambda \int_t^{t+\Delta t} [\mathbf{n}^T \dot{\boldsymbol{\sigma}} - h\dot{\lambda} + \bar{c}_\eta \nabla^2 \dot{\lambda}] \, d\tau \, dV = -\int_V \delta\lambda \, f(\boldsymbol{\sigma}^t, \kappa^t, \nabla^2 \kappa^t) \, dV \,. \tag{6.57}$$

Substitution of $\boldsymbol{\varepsilon}_p = \dot{\lambda} \mathbf{m}$ into the stress rate-strain rate equation leads to

$$\dot{\boldsymbol{\sigma}} = \mathbf{D}_e (\dot{\boldsymbol{\varepsilon}} - \dot{\lambda} \mathbf{m}) \tag{6.58}$$

If we substitute this relationship in the eqs.(6.54) and (6.57) we obtain

$$\int_V \delta \mathbf{u}^T [\mathbf{R}\ddot{\mathbf{u}}^{t+\Delta t}] \, dV + \int_V \delta \boldsymbol{\varepsilon}^T \int_t^{t+\Delta t} \mathbf{D}_e (\dot{\boldsymbol{\varepsilon}} - \dot{\lambda} \mathbf{m}) \, d\tau \, dV = \tag{6.59}$$

$$\int_S \delta \mathbf{u}^T \mathbf{t}^{t+\Delta t} \, dS - \int_V \delta \boldsymbol{\varepsilon}^T \boldsymbol{\sigma}^t \, dV = 0 \,,$$

and

$$\int_V \delta\lambda \int_t^{t+\Delta t} [\mathbf{n}^T \mathbf{D}_e \dot{\boldsymbol{\varepsilon}} - (h + \mathbf{n}^T \mathbf{D}_e \mathbf{m})\dot{\lambda} + \bar{c}_\eta \nabla^2 \dot{\lambda}] \, d\tau \, dV = -\int_V \delta\lambda \, f(\boldsymbol{\sigma}^t, \kappa^t, \nabla^2 \kappa^t) \, dV \,. \tag{6.60}$$

It is emphasized that in contrast to the conventional approach in computational plasticity, the plastic multiplier is taken as an independent variable. While this approach is, in principle, also possible in gradient-independent plasticity it does not seem to entail major advantages when compared with the return-mapping algorithms (Ortiz and Simo 1986, de Borst and Feenstra 1990). For gradient plasticity, i.e. when $\bar{c}_\eta \neq 0$ in eq.(6.60), however the discretisation of λ seems natural and automatically gives satisfaction of the consistency condition in a distributed sense (Simo 1989). Other alternatives of dealing with the extra partial differential equation have been discussed by de Borst and Mühlhaus (1991).

The discretisation of strain and displacement field has already been discussed in Chapter 2. We can discretise the plastic multiplier λ in a similar fashion by

$$\lambda = \mathbf{h}^T \boldsymbol{\Lambda} \,, \tag{6.61}$$

with vector $\mathbf{h}$ containing the shape functions $h_1, \ldots\ldots\ldots, h_n$ for the interpolation of the plastic multiplier and $\boldsymbol{\Lambda}$ denoting the vector of additional nodal degrees-of-freedom. Eq.(6.61) in a rate form yields

$$\dot{\lambda} = \mathbf{h}^T \dot{\boldsymbol{\Lambda}} \,. \tag{6.62}$$

Since the Laplacian operator of $\dot{\lambda}$ must also be computed, the differential operator vector $\mathbf{p}$ is introduced

$$\nabla^2 \dot{\lambda} = \mathbf{p}^T \dot{\boldsymbol{\Lambda}} \,, \tag{6.63}$$

in which **p** is defined by

$$\mathbf{p} = [\nabla^2 h_1, \ldots, \nabla^2 h_n]^T .\tag{6.64}$$

Generally, the vector **h** will not contain the same interpolation polynomials as **H**. While the interpolation of the displacement degrees-of-freedom requires only C^0-continuity, the fact that second derivatives of λ enter the weak form of the consistency condition makes it necessary to select C^1-continuous shape functions for the interpolation of λ. For instance, for the one-dimensional examples that will be treated in a subsequent section, a Hermitian interpolation is employed for λ and linear interpolation is used for u_x, quite similar to beam-column elements where the interpolation of the transverse displacements is usually also achieved through Hermitian shape functions and the axial displacements are interpolated linearly.

Next, we substitute eqs.(2.16), (2.17), (2.19), (2.20), (6.62) and (6.63) in eqs.(6.59) and (6.60). The result is

$$\delta \mathbf{a}^T \int_V \mathbf{H}^T \mathbf{R} \mathbf{H} \ddot{\mathbf{a}}^{t+\Delta t} + \delta \mathbf{a}^T \int_V \int_t^{t+\Delta t} [\mathbf{B}^T \mathbf{D}_e \mathbf{B} \dot{\mathbf{a}} - \mathbf{B}^T \mathbf{D}_e \mathbf{m} \mathbf{h}^T \dot{\mathbf{\Lambda}}] \, d\tau \, dV =$$

$$\delta \mathbf{a}^T \int_S \mathbf{H}^T \mathbf{t}^{t+\Delta t} \, dS - \delta \mathbf{a}^T \int_V \mathbf{B}^T \boldsymbol{\sigma}^t \, dV , \tag{6.65}$$

and

$$\delta \mathbf{\Lambda}^T \int_V \int_t^{t+\Delta t} [-\mathbf{h} \mathbf{n}^T \mathbf{D}_e \mathbf{B} \dot{\mathbf{a}} + (h + \mathbf{n}^T \mathbf{D}_e \mathbf{m}) \mathbf{h} \mathbf{h}^T \dot{\mathbf{\Lambda}} - \bar{c}_\eta \mathbf{h} \mathbf{p}^T \dot{\mathbf{\Lambda}}] \, d\tau \, dV =$$

$$\delta \mathbf{\Lambda}^T \int_V f(\sigma^t, \kappa^t, \nabla^2 \kappa^t) \mathbf{h} \, dV . \tag{6.66}$$

As in Chapter 2 we assume Euler forward predictions for the time integrals in eqs.(6.65) and (6.66). Therefore in the zero-th iteration of a Newton-Raphson scheme we take **n**, **m** and h at time t. Furthermore, similar to the definition of $\Delta \mathbf{a}$ in Chapter 2 we define

$$\Delta \mathbf{\Lambda} = \int_t^{t+\Delta t} \dot{\mathbf{\Lambda}} \, d\tau . \tag{6.67}$$

Since the identities (6.65) and (6.66) must hold for any admissible $\delta \mathbf{a}$ and $\delta \mathbf{\Lambda}$ the following set of algebraic equations ensues

$$\int_V \mathbf{H}^T \mathbf{R} \mathbf{H} \ddot{\mathbf{a}}^{t+\Delta t} dV + \int_V [\mathbf{B}^T \mathbf{D}_e \mathbf{B} \Delta \mathbf{a} - \mathbf{B}^T \mathbf{D}_e \mathbf{m}^t \mathbf{h}^T \Delta \mathbf{\Lambda}] dV = \int_S \mathbf{H}^T \mathbf{t}^{t+\Delta t} dS - \int_V \mathbf{B}^T \boldsymbol{\sigma}^t dV , \tag{6.68}$$

and

$$\int_V [-\mathbf{h}(\mathbf{n}^t)^T \mathbf{D}_e \mathbf{B} \Delta \mathbf{a} + (h^t + (\mathbf{n}^t)^T \mathbf{D}_e \mathbf{m}^t) \mathbf{h} \mathbf{h}^T \Delta \mathbf{\Lambda} - \bar{c}_\eta \mathbf{h} \mathbf{p}^T \Delta \mathbf{\Lambda}] \, dV = \tag{6.69}$$

$$\int_V f(\sigma^t, \kappa^t, \nabla^2\kappa^t)\mathbf{h}\, dV.$$

Eqs.(6.68) and (6.69) can be written in a compact fashion as

$$\begin{bmatrix} \mathbf{M}_{aa} & 0 \\ 0 & 0 \end{bmatrix} \begin{bmatrix} \ddot{\mathbf{a}}^{t+\Delta t} \\ \ddot{\Lambda}^{t+\Delta t} \end{bmatrix} + \begin{bmatrix} \mathbf{K}_{aa} & \mathbf{K}_{a\lambda} \\ \mathbf{K}_{\lambda a} & \mathbf{K}_{\lambda\lambda} \end{bmatrix} \begin{bmatrix} \Delta \mathbf{a} \\ \Delta \Lambda \end{bmatrix} = \begin{bmatrix} \mathbf{f}_e^{t+\Delta t} - \mathbf{f}_i^t \\ \mathbf{f}_\lambda^t \end{bmatrix}, \qquad (6.70)$$

where

$$\mathbf{M}_{aa} = \int_V \mathbf{H}^T \mathbf{R} \mathbf{H}\, dV, \qquad (6.71)$$

$$\mathbf{K}_{aa} = \int_V \mathbf{B}^T \mathbf{D}_e \mathbf{B}\, dV, \qquad (6.72)$$

$$\mathbf{K}_{a\lambda} = -\int_V \mathbf{B}^T \mathbf{D}_e \mathbf{m}^t \mathbf{h}^T\, dV, \qquad (6.73)$$

$$\mathbf{K}_{\lambda a} = -\int_V h(\mathbf{n}^t)^T \mathbf{D}_e \mathbf{B}\, dV, \qquad (6.74)$$

$$\mathbf{K}_{\lambda\lambda} = \int_V [(h^t + (\mathbf{n}^t)^T \mathbf{D}_e \mathbf{m}^t) \mathbf{h} \mathbf{h}^T - \bar{c}_\eta \mathbf{h} \mathbf{p}^T]\, dV, \qquad (6.75)$$

and

$$\mathbf{f}_\lambda^t = \int_V f(\sigma^t, \kappa^t, \nabla^2\kappa^t)\mathbf{h}\, dV, \qquad (6.76)$$

while $\mathbf{f}_e^{t+\Delta t}$ and $\mathbf{f}_i^t$ satisfy the definitions in Chapter 2. In the Newton-Raphson iteration scheme, which is used to solve the nonlinear set of algebraic equations, the values at time t are replaced by the values at time $t + \Delta t$ for the last iteration (see Chapter 3). Evidently the tangent stiffness matrix as defined in eqs.(6.72)-(6.75) is non-symmetric because of the gradient terms and the use of a non-associative flow rule. If we assume associative plasticity ($\mathbf{m} = \mathbf{n}$ and thus $\mathbf{K}_{a\lambda} = \mathbf{K}_{\lambda a}$) the non-symmetry in eq.(6.75) disappears if $\bar{c}_\eta = 0$. We retrieve a symmetric operator as one would expect in classical plasticity with an associated flow rule.

7. REFERENCES

[1] Aifantis, E.C. (1984). On the microstructural origin of certain inelastic models. *J. Engng. Mater. Technol.* **106,** 326-334.

[2] Bazant, Z.P., Belytschko, T.B. and Chang, T.-P. (1984). Continuum theory for strain-softening, *ASCE J. Eng. Mech.,* **110,** 1666-1692.

[3] Boogaard, A.H. van den (1995). Implicit integration of the Perzyna viscoplastic material model, TNO-report 95-NMR711, The Netherlands.

[4] Borst, R. de and Nauta, P. (1985). Non-orthogonal cracks in a smeared finite element model, *Eng. Comp.,* **2(1),** 35-46.

[5] Borst, R. de (1986). Non-linear analysis of frictional materials, Dissertation, Delft University of Technology, Delft.

[6] Borst, R. de and Feenstra, P.H. (1990). Studies in anisotropic plasticity with reference to the Hill criterion, *Int. J. Num. Meth. Eng.,* **29,** 315-336.

[7] Borst, R. de and Mühlhaus, H.-B. (1992). Gradient-dependent plasticity : Formulation and algorithmic aspects, *Int. J. Num. Meth. Eng.,* **35,** 521-539.

[8] Borst, R. de, Sluys, L.J., Mühlhaus, H.-B. and Pamin, J. (1993). Fundamental issues in finite element analyses of localization of deformation. *Engng Comput.,* **10,** 99-121

[9] Crisfield, M.A. (1991). Nonlinear finite element analysis of solids and structures, Wiley & Sons, Chichester, U.K.

[10] Duvaut, G. and Lions, J.L. (1972). Les inequations en Mechanique et en Physique, Dunod, Paris.

[11] Hughes, T.J.R. (1987). The finite element method - Linear static and dynamic finite element analysis, Prentice-Hall, New Jersey.

[12] Lasry, D. and Belytschko, T. (1988). Localization limiters in transient problems. *Int. J. Solids Structures* **24,** 581-597.

[13] Loret, B. and Prevost, J.H. (1990). Dynamic strain localization in elasto-(visco-)plastic solids, Part 1. General formulation and one-dimensional examples, *Comp. Meth. Appl. Mech. Eng.,* **83,** 247-273.

[14] Mühlhaus, H.-B. and Aifantis, E.C. (1991). A variational principle for gradient plasticity. *Int. J. Solids Structures* **28,** 845-858.

[15] Ortiz, M. and Simo, J. (1986). An analysis of a new class of integration algorithms for elastoplastic constitutive equations, *Int. J. Num. Meth. Eng.,* **23,** 353-366.

[16] Pamin J. (1994). Gradient-dependent plasticity in numerical simulation of localization phenomena.Dissertation, Delft University of Technology, Delft.

[17] Perzyna, P. (1966). Fundamental problems in viscoplasticity, Recent Advances in Applied Mechanics, Academic Press, New York, **9,** 243-377.

[18] Pijaudier-Cabot, G. and Bazant, Z.P. (1987). Nonlocal damage theory, *ASCE J. Eng. Mech.,* **113,** 1512-1533.

[19] Riks, E. (1979). An incremental approach to the solution of snapping and buckling

problems, *Int. J. Solids Structures,* **15,** 529-551.
[20] Rots, J.G. (1988). Computational modeling of concrete fracture, Dissertation, Delft University of Technology, Delft.
[21] Schreyer, H.L. and Chen, Z. (1986). One-dimensional softening with localization, *J. Appl. Mech.,* **53,** 791-979.
[22] Simo, J.C., Kennedy, J.G. and Govindjee, S. (1988). Non-smooth multisurface plasticity and visco-plasticity - Loading/unloading conditions and numerical algorithms. *Int. J. Num. Meth. Eng.,* **26,** 2161-2185.
[23] Simo, J. (1989). Strain-softening and dissipation : a unification of approaches, Cracking and Damage: Strain Localization and Size Effect, Eds. J. Mazars and Z.P. Bazant, Elsevier, London-New York, 440-461.
[24] Sluys, L.J., Mühlhaus, H.-B. and Borst, R. de (1992). Wave propagation, localization and dispersion in a gradient-dependent medium. *Int. J. Solids Structures* **30,** 1153-1171.
[25] Sluys, L.J. (1992). Wave propagation, localisation and dispersion in softening solids. Dissertation, Delft University of Technology, Delft.
[26] Wang, W.M. (1997). Stationary and propagative instabilities in metals. A computational point of view. Dissertation, Delft University of Technology, Delft.
[27] Whitham, G.B. (1974). Linear and nonlinear waves, John Wiley and Sons, New York-London-Sydney-Toronto.
[28] Zbib, H.M. and Aifantis, E.C. (1988). On the localization and postlocalization behavior of plastic deformation, I,II,III, *Res Mechanica,* **23,** 261-277, 279-292, 293-305.

NUMERICAL SOLUTIONS OF INITIAL-BOUNDARY-VALUE PROBLEMS WITH SHEAR STRAIN LOCALIZATION

R.C. Batra

Virginia Polytechnic Institute and State University, Blacksburg, VA, USA

This work is dedicated with deep respect to my parents,
Amir Chand and Dewki Bai Batra,
who passed away in 1974 and 1980 respectively.

Abstract

The article summarizes some of the results obtained numerically since 1984 by the author and his colleagues on one (simple shearing), two (plane strain and axisymmetric) and three (twisting of a thin-walled and a thick-walled tube) dimensional adiabatic shear banding problems in thermoviscoplastic materials. The material models considered account for strain and strain rate hardening, thermal softening, dipolar effects with the second order spatial gradients of the velocity field taken as independent kinematic variables and the corresponding higher order stresses as kinetic variables, and the nucleation, growth and coalescence of voids. The effect of phase transformation and the consequent change in the material properties has also been accounted for. Different shapes and types of defects considered include geometric such as the variation in the thickness, rigid inclusions, voids, nonuniform initial conditions, and weak elements. Some of the problems studied where no a priori defect is introduced to nucleate a shear band include the plane strain compression of a FCC single crystal, the Taylor impact test, and the penetration of a tungsten heavy alloy or a depleted uranium rod into a steel target.

Introduction

Perzyna's article in this volume provides an extensive review of the experimental work on adiabatic shear banding, and the development of constitutive relations for a microporous thermoviscoplastic single and polycrystalline material. No attempt is made here to provide either an exhaustive survey of all of the works dealing with adiabatic shear bands or even those dealing with the investigation of the phenomenon numerically. The interested reader is referred to two recent books (Dodd and Bai (1987), Bai and Dodd (1992)), and the review articles by Needleman and Tvergaard (1984) and by Tomita (1994). Suffice it to say that the phenomenon was discovered by Tresca in 1878, by Massey in 1921 and then by Zener and Hollomon in 1944. Kalthoff (1988, 1990) has observed shear bands in a prenotched plate impacted on the notched side by a projectile made of the same material as the plate. For low impact speeds, a crack initiated from the notch tip but at high impact speeds a shear band initiated; the transition in the failure mode also depended upon the notch-tip radius. This problem has been studied numerically by Needleman and Tvergaard (1995), Zhou et al. (1996) and Batra and Nechitailo (1997). However, this work is not discussed here.

Shear bands usually precede shear fractures in ductile materials. Whereas their initiation and growth needs to be delayed in most structures, their occurrence in a high kinetic energy penetrator is desired since it is believed to enhance the penetration of the rod into a target by continuously making the penetrator nose shape conical.

Readers interested in the mechanics of adiabatic shear banding and the results for various problems may skip Section 1. Also, all of the work described here has been previously published by the author and his colleagues in refereed journals which should be consulted for additional results and/or details. Because of the page limitations, not every result is included here.

1 A Brief Overview of the Finite Element Method

Equations, in referential description, governing thermomechanical deformations of a body are:

Balance of mass: $\quad\quad\quad\quad\quad\quad \rho J = \rho_0,\quad\quad\quad\quad\quad$ in Ω, $\quad\quad$ (1.1)

Balance of linear momentum: $\rho_0 \dot{v}_i = T_{i\alpha,\alpha} + \rho_0 b_i,\quad\quad$ in Ω, $\quad\quad$ (1.2)

Balance of internal energy: $\quad \rho_0 \dot{e} = -Q_{\alpha,\alpha} + T_{i\alpha}\dot{F}_{i\alpha} + \rho_0 r,\quad$ in Ω. $\quad\quad$ (1.3)

Here ρ is the present mass density of a material particle whose mass density in the reference configuration is ρ_0, $J = \det[F_{i\alpha}]$ is the jacobian, $F_{i\alpha} = \partial x_i/\partial X_\alpha$, x_i is the present position of the material particle that occupied place X_α in the reference configuration, $T_{i\alpha}$ is the first Piola-Kirchhoff stress tensor, v_i is the velocity, a superimposed dot indicates the material time derivative, b_i is the body force per unit mass, a comma followed by α denotes the partial derivative with respect to X_α, e is the specific internal energy, Q_α is the heat flux per unit reference area, r is the supply of internal energy per unit mass, and a repeated index implies summation over the range of the index. Equations (1.1)-(1.3) are written in rectangular Cartesian coordinates, and Ω is the region occupied by the body in the reference configuration. These equations are to be supplemented by constitutive relations which characterize the material of the body and the following side conditions.

$$x_i(X_\alpha, t) = \overline{x}_i, \quad X_\alpha \in \partial_1 \Omega, \tag{1.4}$$

$$\theta(X_\alpha, t) = \overline{\theta}, \quad X_\alpha \in \partial_3 \Omega, \tag{1.5}$$

$$T_{i\alpha}(X_\beta, t) N_\alpha = f_i, \quad X_\beta \in \partial_2 \Omega, \tag{1.6}$$

$$-Q_\alpha(X_\beta, t) N_\alpha = h, \quad X_\beta \in \partial_4 \Omega, \tag{1.7}$$

$$\partial_1 \Omega \cap \partial_2 \Omega = \phi, \quad \partial_1 \overline{\Omega} \cup \partial_2 \overline{\Omega} = \partial \Omega,$$

$$\partial_3 \Omega \cap \partial_4 \Omega = \phi, \quad \partial_3 \overline{\Omega} \cup \partial_4 \overline{\Omega} = \partial \Omega,$$

$$x_i(X_\alpha, 0) = x_i^\circ(X_\alpha), \quad \dot{x}_i(X_\alpha, 0) = \dot{x}_i^\circ(X_\alpha), \quad \theta(X_\alpha, 0) = \theta^0(X_\alpha). \tag{1.8}$$

Boundary conditions (1.4) and (1.5) are usually called essential, and (1.6) and (1.7) natural. In them, N_α is an outward unit normal to the boundary $\partial \Omega$ of Ω in the reference configuration, and functions $x_i^\circ, \theta^\circ, \dot{x}_i^\circ, \overline{x}_i, \overline{\theta}, f_i$ and h are presumed to be given. We will postulate precise constitutive relations later, and merely state for the time being that $T_{i\alpha}$, Q_α and e are functions of $F_{i\alpha}, \dot{F}_{i\alpha}, \theta$ and $\theta_{,\alpha}$ where θ is the present absolute temperature of a material particle. Equation (1.3) can alternatively be written as

$$\rho_0 c \dot{\theta} = -Q_{\alpha,\alpha} + T_{i\alpha}^{ne}\dot{F}_{i\alpha} + \rho_0 r \tag{1.9}$$

where $T_{i\alpha}^{ne}$ is the non-equilibrium part of the first Piola-Kirchhoff stress tensor:

$$T_{i\alpha}^{ne} = T_{i\alpha}(\mathbf{F}, \dot{\mathbf{F}}, \theta, \theta_{,\alpha}) - T_{i\alpha}(\mathbf{F}, 0, \theta, 0), \tag{1.10}$$

and c the specific heat which may depend upon $\mathbf{F}$ and θ. Let for $i = 1, 2, 3$, $\phi_i : \overline{\Omega} \to \mathbb{R}$ be a smooth function such that $\phi_i = 0$ on $\partial_1 \Omega$. Taking the inner product of (1.2) with

ϕ_i, integrating the resulting equation over Ω, using the divergence theorem on the first term on the right-hand side, and the boundary condition (1.6), we arrive at the following.

$$\int_\Omega \rho_0 \dot{v}_i \phi_i d\Omega = \int_{\partial_2 \Omega} f_i \phi_i d\Gamma - \int_\Omega T_{i\alpha} \phi_{i,\alpha} d\Omega + \int_\Omega \rho_0 b_i \phi_i d\Omega . \tag{1.11}$$

Let $\eta : \overline{\Omega} \to \mathbb{R}$ be a smooth function such that $\eta = 0$ on $\partial_3 \Omega$. Multiplying both sides of (1.9) with η, integrating the resulting equation over Ω, using the divergence theorem on the first term on the right-hand side, and the boundary condition (1.7), we obtain

$$\int_\Omega \rho_0 c \dot{\theta} \eta d\Omega = \int_{\partial_4 \Omega} h \eta d\Gamma + \int_\Omega Q_\alpha \eta_{,\alpha} d\Omega + \int_\Omega (T^{ne}_{i\alpha} \dot{F}_{i\alpha} + \rho_0 r) \eta d\Omega. \tag{1.12}$$

Let

$$H = \left\{ \phi | \phi : \overline{\Omega} \to \mathbb{R}, \int_\Omega \phi_{,\alpha} \phi_{,\alpha} d\Omega < \infty \right\}, \tag{1.13}$$

$$\mathcal{T}_m = \{\phi_i | \phi_i \in H, \phi_i = 0 \text{ on } \partial_1 \Omega\},$$
$$\mathcal{S}_m = \{\phi_i | \phi_i : \overline{\Omega} \times (0, \overline{t}) \to \mathbb{R}, \phi_i(\cdot, t) \in H, \phi_i(X_\alpha, t) = \overline{x}_i(X_\alpha, t) \text{ on } \partial_1 \Omega\},$$
$$\mathcal{T}_\theta = \{\phi | \phi \in H, \phi = 0 \text{ on } \partial_3 \Omega\},$$
$$\mathcal{S}_\theta = \{\phi | \phi : \overline{\Omega} \times (0, \overline{t}) \to \mathbb{R}, \phi(\cdot, t) \in H, \phi(X_{\alpha,t}) = \overline{\theta}(X_{\alpha,t}) \text{ on } \partial_3 \Omega\},$$
$$\mathcal{S} = \mathcal{S}_m \otimes \mathcal{S}_\theta, \mathcal{T} = \mathcal{T}_m \otimes \mathcal{T}_\theta. \tag{1.14}$$

Then a weak formulation of the given problem can be stated as follows. Find $(x_i, \theta) \in \mathcal{S}$ such that equations (1.11) and (1.12) hold for every $(\phi_i, \eta) \in \mathcal{T}$. One can think of $(\phi_i, \eta) \in \mathcal{T}$ as representing virtual displacements and virtual temperature fields. Functions in $\mathcal{S}$ and $\mathcal{T}$ have the same smoothness requirements and differ only in the boundary conditions on $\partial_1 \Omega$ and $\partial_3 \Omega$. When $\mathcal{S} \neq \mathcal{T}$, the weak formulation is called Petrov-Galerkin. Let $g \in \mathcal{S}$ be a fixed function. Then every function $u \in \mathcal{S}$ can be written as $u = g + v$ for some $v \in \mathcal{T}$. Equivalently, $\mathcal{S} = \mathcal{T} \oplus \{g\}$. Regarding g as a known function, the problem reduces to finding a $v \in \mathcal{T}$ such that (1.11) and (1.12) hold for $\forall (\phi_i, \eta) \in \mathcal{T}$, and the weak formulation is called Bubnov-Galerkin, and simply Galerkin in many books.

Let $\mathcal{S}^n \subset \mathcal{S}$ and $\mathcal{T}^n \subset \mathcal{T}$ be finite-dimensional sets. Then an approximate solution of the given problem is functions $(x_i^n, \theta^n) \in \mathcal{S}^n$ such that equations (1.11) and (1.12) hold for every $(\phi_i^n, \eta^n) \in \mathcal{T}^n$ with $T_{i\alpha}$, Q_α and $T^{ne}_{i\alpha}$ now evaluated from (x_i^n, θ^n) rather than (x_i, θ).

Let $\psi_1, \psi_2, \ldots, \psi_n$ be bases functions in $\mathcal{T}_\theta^n$. Then

$$\begin{aligned}
\eta^n(X_\alpha) &= \psi_A(X_\alpha) \eta_A, A = 1, 2, \ldots, n, \\
\phi_i^n(X_\alpha) &= \psi_A(X_\alpha) \phi_{iA}, i = 1, 2, 3, \\
x_i^n(X_\alpha, t) &= \psi_A(X_\alpha) d_{iA}(t), \\
\theta^n(X_\alpha, t) &= \psi_A(X_\alpha) \theta_A(t).
\end{aligned} \tag{1.15}$$

Recall that a repeated index implies summation over the range of the index. Substitution from (1.15) into (1.11) and (1.12) and using the fact that these equations must hold for all choices of η_A and ϕ_{iA}, we arrive at the following set of coupled ordinary differential equations in time.

$$M_{AB}\ddot{d}_{iB} = F_{Ai}^m, \tag{1.16}$$

$$H_{AB}\dot{\theta}_B = F_A^\theta, \tag{1.17}$$

$$F_{Ai}^m = \int_{\partial_2 \Omega} f_i \psi_A d\Gamma - \int_\Omega T_{i\alpha} \psi_{A,\alpha} d\Omega + \int_\Omega \rho_0 b_i \psi_A d\Omega, \tag{1.18_1}$$

$$F_A^\theta = \int_{\partial_4 \Omega} h \psi_A d\Gamma + \int_\Omega Q_\alpha \psi_{A,\alpha} d\Omega + \int_\Omega G \psi_A d\Omega, \tag{1.18_2}$$

$$G = T_{i\alpha}^{ne} \dot{F}_{i\alpha} + \rho_0 r, \tag{1.18_3}$$

$$M_{AB} = \int_\Omega \rho_0 \psi_A \psi_B d\Omega, \tag{1.18_4}$$

$$H_{AB} = \int_\Omega \rho_0 c \psi_A \psi_B d\Omega. \tag{1.18_5}$$

In these equations $T_{i\alpha}$, Q_α and $T_{i\alpha}^{ne}$ are functions of unknown fields $x_i^n(X_\alpha, t)$ and $\theta^n(X_\alpha, t)$ or equivalently of $d_{iA}(t)$ and $\theta_B(t)$.

From initial conditions (1.8) we can similarly derive

$$\begin{aligned} M_{AB} d_{iB}(0) &= F_{iA}^\circ, \\ M_{AB} \dot{d}_{iB}(0) &= \dot{F}_{iA}^\circ, \\ H_{AB} \theta_B(0) &= F_A^{\circ \theta}, \end{aligned} \tag{1.19}$$

and thus evaluate initial conditions for integrating (1.16) and (1.17).

Recall that the natural boundary conditions have been embedded into the weak formulation of the problem, and the essential boundary conditions are satisfied by either selecting a function g or by another equivalent method.

Because of the referential description of the balance laws used, a weak formulation of the balance of mass is not needed. The mass density in the deformed configuration can be determined from the computed displacement field.

Let Ω be divided into the union of disjoint subdomains Ω_e satisfying

$$\Omega = \bigcup_{e=1}^m \Omega_e; \quad \Omega_e \cap \Omega_f = \phi, \ e \neq f; \ \overline{\Omega}_e \cap \overline{\Omega}_f \neq \phi,$$

when Ω_e and Ω_f are adjacent subdomains. That is, adjacent subdomains share atmost a common boundary. These subdomains are called finite elements, and generally have straight boundaries (e.g. Ω_e is a triangle or a quadrilateral in a plane, and a cube or a tetrahedron in a 3-dimensional space). The points of intersection of the boundaries of all adjacent elements are called nodes. The collection of elements and nodes is called a finite element mesh or simply a mesh.

In the finite element work, the number of bases functions $\{\psi_A\}$ equals the number of nodes in the mesh. The bases functions are simple polynomials and the basis function ψ_A corresponding to node A equals 1 at node A, and equals zero on all elements that do not meet at node A. Thus ψ_A equals zero at all nodes except node A, has a compact support, and (1.15$_3$) and (1.15$_4$) imply that $d_{iA}(t)$ and $\theta_A(t)$ equal, respectively, x_i and θ at node A. The restrictions of bases functions to an element are called shape functions. The number of shape functions for an element equals the number of nodes on the element, and the shape function corresponding to node A of element Ω_e equals 1 at node A, zero at all other nodes and also vanishes on all sides of the element Ω_e that do not pass through node A. Generally, shape functions and bases functions are simple polynomials; the former are defined on an element and the latter on the entire domain Ω.

The integrals appearing in (1.18) are evaluated numerically. To illustrate the procedure, we consider one such integral:

$$I = \int_\Omega f d\Omega = \sum_e \int_{\Omega_e} f d\Omega \equiv \sum_e I^e, \tag{1.20}$$

$$I^e = \int_{\Omega_e} f d\Omega. \tag{1.21}$$

Let

$$T^e : \Omega_M \to \Omega_e; \quad \Omega_M = [-1, 1] \times [-1, 1] \times [-1, 1], \tag{1.22}$$

be a one-to-one, continuously differentiable and invertible map with continuously differentiable inverse. Here we have tacitly assumed that Ω_e is a "brick" or cubic element. We express T^e as

$$X_\alpha = X_\alpha(\xi_j) \tag{1.23}$$

and require that

$$\det\left[\frac{\partial X_\alpha}{\partial \xi_j}\right] > 0 \,\forall \xi_i \in \Omega_M. \tag{1.24}$$

With

$$J = \det\left[\frac{\partial X_\alpha}{\partial \xi_j}\right], \tag{1.25}$$

the integral (1.21) can be written as

$$I^e = \int_{\Omega_e} f dX_1 dX_2 dX_3 = \int_{\Omega_M} f J d\xi_1 d\xi_2 d\xi_3 = \int_{-1}^1 d\xi_3 \int_{-1}^1 d\xi_2 \int_{-1}^1 f J d\xi_1. \tag{1.26}$$

Recall that, for a polynomial of degree $2n + 1$,

$$\int_{-1}^1 f ds = \sum_{i=1}^{n+1} W_i f(s_i). \tag{1.27}$$

Here $s_1, s_2, \ldots, s_{n+1}$ are the sampling points or quadrature points and $W_1, W_2, \ldots, W_{n+1}$ the corresponding weights. The sampling points are the roots of

$$P_{n+1}(s) = 0$$

and

$$W_i = \int_{-1}^{1} L_i(s) ds \tag{1.28}$$

where P_{n+1} is a Legender polynomial of degree $n+1$ and L_i is a Lagrange polynomial of degree n defined by

$$L_i(s) = \prod_{\substack{j=1 \\ j \neq i}}^{n+1} (s - s_j) / \prod_{\substack{j=1 \\ j \neq i}}^{n+1} (s_i - s_j). \tag{1.29}$$

Thus when the integrand in (1.26) is a polynomial, I^e can be evaluated exactly by using an appropriate quadrature rule in each coordinate direction. However, in general, the integrand is not a polynomial and hence all of the integrals in (1.18) are evaluated approximately by employing a quadrature rule.

We note that the mapping T^e is generally taken to be

$$X_\alpha = \sum_{a=1}^{8} N_a(\xi_1, \xi_2, \xi_3) X_\alpha^a,$$

$$N_a(\xi_1, \xi_2, \xi_3) = \frac{1}{8}(1 + \xi_1^a \xi_1)(1 + \xi_2^a \xi_2)(1 + \xi_3^a \xi_3), \tag{1.30}$$

where $(\xi_1^a, \xi_2^a, \xi_3^a)$ are coordinates $(\pm 1, \pm 1, \pm 1)$ of node a of the element Ω_M in the local coordinate system, and X_α^a coordinates of the corresponding node on Ω_M in the global system.

A field variable, e.g., temperature θ defined on Ω_e can be expressed as a function of ξ_1, ξ_2 and ξ_3 through the transformation (1.23). In order to evaluate stresses and other constitutive quantities, we will need to determine gradients of field variables with respsect to X_α. Since it may be a nontrivial task to solve (1.30_1) for ξ_i, we proceed as follows. Note that

$$dX_\alpha = \frac{\partial N_a}{\partial \xi_i} d\xi_i X_\alpha^a \qquad \text{(summed on } a \text{ and } i\text{)}$$

$$= \frac{\partial N_a}{\partial \xi_i} X_\alpha^a \frac{\partial \xi_i}{\partial X_\beta} dX_\beta,$$

therefore

$$\left[\frac{\partial \xi_i}{\partial X_\beta}\right] = \left[\frac{\partial N_a}{\partial \xi_i} X_\beta^a\right]^{-1}.$$

Thus

$$\frac{\partial \theta}{\partial X_\alpha} = \frac{\partial \theta}{\partial \xi_i} \frac{\partial \xi_i}{\partial X_\beta} \qquad (1.31)$$

can be computed.

Lumping of the Mass Matrix

The mass matrix computed from (1.18_4) is a banded matrix with the band width depending upon the node numbering scheme; it is called consistent mass matrix. In order to reduce the computational effort, we generally approximate it by a lumped (or diagonal) mass matrix; the same is also done for the heat capacity matrix given by (1.18_5). The use of node points as quadrature points to numerically evaluate (1.18_4) will yield a diagonal mass matrix but will result in zero entries for axisymmetric problems. Zero or negative entries in a diagonal mass matrix can have disastrous consequences. The following two techniques are commonly used to obtain a lumped mass matrix.

- a) Row-Sum Technique: The elements in each row of the consistent mass matrix obtained from (1.18_4) are summed and lumped on the diagonal. It can sometimes produce negative masses.

- b) Special-Lumping Technique: The entries of the lumped mass matrix are set proportional to the diagonal entries of the consistent mass matrix and the constant of proportionality is selected to conserve the total mass. By virtue of the positive-definiteness of the mass matrix, the diagonal entries of the consistent mass matrix are necessarily positive.

Numerical Integration of Ordinary Differential Equations

Equations (1.16) and (1.17) are nonlinear, coupled ordinary differential equations in d_{iB} and θ_B since their right-hand sides are also functions of d_{iB} and θ_B. In shear localization problems, one also has constitutive relations expressed in the rate form, e.g., the Jaumann rate of Cauchy stress tensor is expressed as a linear function of the strain-rate tensor for a hypoelastic material. Invariably, one postulates evolution laws for internal variables describing the microstructural changes in the body. The time scales for these processes may differ by several orders of magnitude which makes the system of equations very stiff.

An example of stiff equations is given by Gear. Equations

$$\begin{aligned}\dot{u} &= 998u + 1998v, \\ \dot{v} &= -999u - 1999v,\end{aligned} \qquad (1.32)$$

with initial conditions

$$u(0) = 1, \ v(0) = 1, \qquad (1.33)$$

have the solution

$$u(t) = 2e^{-t} - e^{-1000t},$$
$$v(t) = -e^{-t} + e^{-1000t}. \qquad (1.34)$$

The presence of the e^{-1000t} term requires a time stepsize of $\leq 1/1000$ for the method to be stable. Of course, the term e^{-1000t} is completely negligible in determining the values of u and v away from $t = 0$. The generic problem with stiff equations is that we are required to follow the variation in the solution on the shortest time scale to maintain the stability of the integration scheme.

The simplest cure is to resort to implicit differencing, where the time derivative is evaluated by using a backward difference scheme. For example, consider

$$\dot{y} = -cy. \qquad (1.35)$$

Then

$$y_{n+1} = \begin{cases} y_n/(1 + c\Delta t) & \text{backward-difference} \\ y_n(1 - c\Delta t) & \text{forward-difference} \end{cases} \qquad (1.36)$$

where $y_n \simeq y(t_n)$. Clearly the forward-difference method is unstable if $\Delta t > 2/c$ but the backward-difference method is unconditionally stable (i.e. is stable for all choices of Δt). We give up accuracy in following the evolution towards equilibrium if we use large timesteps, but we maintain stability.

For a system of linear equations with constant coefficients

$$\dot{\mathbf{y}} = -\mathbf{C}\mathbf{y} \qquad (1.37)$$

where $\mathbf{C}$ is a positive definite matrix, the explicit method is stable only if

$$\Delta t < \frac{2}{\lambda_{\max}} \qquad (1.38)$$

where $\lambda_{\max}$ is the largest eigenvalue of $\mathbf{C}$. The implicit differencing gives

$$\mathbf{y}_{n+1} = (\mathbf{1} + \mathbf{C}\Delta t)^{-1}\mathbf{y}_n \qquad (1.39)$$

and is stable for all values of Δt. For a system of nonlinear equations

$$\dot{\mathbf{y}} = \mathbf{f}(t, \mathbf{y}), \qquad (1.40)$$

the implicit differencing gives

$$\mathbf{y}_{n+1} = \mathbf{y}_n + \Delta t \mathbf{f}(t_{n+1}, \mathbf{y}_{n+1}),$$
$$\simeq \mathbf{y}_n + \Delta t \left[\mathbf{f}(t_{n+1}, \mathbf{y}_n) + \frac{\partial \mathbf{f}}{\partial \mathbf{y}} \bigg|_{\mathbf{y}_n} \cdot (\mathbf{y}_{n+1} - \mathbf{y}_n) \right], \qquad (1.41)$$

and requires the inversion of $\left(1 - \Delta t \dfrac{\partial \mathbf{f}}{\partial \mathbf{y}}\right)$. This procedure is called a "semi-implicit" method and is not guaranteed to be stable, but it usually is. By averaging the explicit and implicit first-order methods, we can construct a second-order method:

$$\mathbf{y}_{n+1} = \mathbf{y}_n + \frac{\Delta t}{2}\left[\mathbf{f}(t_{n+1}, \mathbf{y}_n) + \left.\frac{\partial \mathbf{f}}{\partial \mathbf{y}}\right|_{\mathbf{y}_n} \cdot (\mathbf{y}_{n+1} - \mathbf{y}_n) + \mathbf{f}(t_n, \mathbf{y}_n)\right]. \qquad (1.42)$$

One can download solvers like LSODE (Livermore Solver for Ordinary Differential Equations) or VODEPK from the network. These solvers adjust the timestep size adaptively to compute a solution of the set of coupled nonlinear ordinary differential equations within a prescribed tolerance. In strain localization problems the timestep size drops dramatically once the deformation begins to localize indicating that the system of equations is getting stiffer.

Satisfaction of Essential Boundary Conditions

Instead of using the function g, the system of equations (1.16) and (1.17) is assembeled for all nodes and is then modified as follows to satisfy the essential boundary conditions prescribed at node A. From the given time history of the displacement at node A, one can compute its acceleration at any time t. For a lumped mass matrix, the displacement boundary condition at node A is satisfied when F_{Ai} is replaced by $M_{AA}(\ddot{d}_{Ai})$, (no sum on A). One follows this procedure for every node where an essential boundary condition is given. For a consistent mass matrix, one can use the penalty method in which M_{AA} is replaced by $M_{AA} + \lambda$ and F_{Ai} by $\lambda \ddot{d}_{Ai}$ where λ is typically much larger than the magnitude of any entry in the mass matrix.

Interpretation of the Finite Element Solution

Nodal values of displacements and temperatures are reasonably accurate. However, stresses, heat flux and other quantities involving derivatives of displacements and temperatures are evaluated at quadrature points which are generally in the interior of Ω_e. They can then be extrapolated or interpolated to other points of interest, e.g. see Zienkiewicz and Zhu (1992).

Factors Affecting the Quality of the Approximate Solution

The following list is not exhaustive.

1. Dimensionality of the finite-dimensional space: the quality of the approximate solution generally improves with an increase in the dimensionality of the finite dimensional space.

2. Choice of bases functions: for localization problems, lowest order bases functions are recommended.

3. Design of the finite element mesh: adaptively refined meshes with element size inversely proportional to a measure of deformation within the element are recommended. Care should be taken to properly grade the mesh to avoid a very large element abutting an extremely small one.

4. Number of quadrature points used to evaluate various integrals: the CPU time increases with an increase in the number of quadrature points. However, too few quadrature points may result in hour-glass or spurious modes.

5. Consistent/lumped mass matrix: consistent mass matrices generally require smaller time step size for stability.

6. Explicit/implicit time integration scheme: numerical integration schemes generally introduce time period errors and numerical dissipation. For the one-dimensional linear elastic problem, the lumped mass matrix and the explicit (central-difference) method with $\Delta t = h/c$ gives exact values of nodal displacements. Here h is the element size and c the speed of the wave in the bar.

7. Tolerances in solving nonlinear algebraic equations if an implicit time integration scheme is used.

8. Time step size. Note that as $\Delta t \to 0$, the solution of (1.16) and (1.17) will converge to the analytical solution of the discretized problem and not of the original continuous problem defined by (1.1)-(1.3).

9. Points where stresses/fluxes are computed for postprocessing of results. It is recommended that fluxes be computed at quadrature points.

10. The time integration scheme should maintain the objectivity of the rate form of constitutive relations.

Material for Additional Reading

There are several books on the finite element method, e.g., by Becker et al. (1981), Hughes (1987), and Belytschko and Hughes (eds.) (1986). Gear (1971) and Press et al. (1986), amongst others, have discussed schemes for integrating a set of coupled ordinary differential equations.

2 Analysis of 1-dimensional Shear Band Problems

Marchand and Duffy (1988) analysed the initiation and growth of shear bands by twisting thin-walled tubes. Because of machining defects, the thickness of the tube varied. As a first approximation, the torsional deformations of a thin-walled tube can be regarded as equivalent to simple shearing deformations of a block situated near the center of the tube. The work presented below follows closely that of Batra and Kim (1990, 1991); some of the other similar works are due to Wright and Batra (1985), Wright and Walter (1989), Shawki and Clifton (1989) and Batra (1987). This list is representative rather than complete. The goal here is to exhibit the effects of different viscoplastic constitutive equations, thermal conductivity, inertia forces, and defect size.

We choose a fixed set of rectangular Cartesian coordinate axes with origin at the lower surface of the block and the direction of shearing along the x-axis. The governing equations are

$$\rho w \dot{v} = (ws)_{,y}, \quad 0 < y < H, \tag{2.1}$$

$$\rho c(\theta) w \dot{\theta} = [wk(\theta)\theta_{,y}]_{,y} + ws\dot{\gamma}_p, \quad 0 < y < H, \tag{2.2}$$

$$\dot{s} = \mu(\theta)(v_{,y} - \dot{\gamma}_p), \tag{2.3}$$

$$\dot{\gamma}_p = g(s, \gamma_p, \theta). \tag{2.4}$$

Here v, θ, s, γ_p and w represent, respectively, the velocity of a material particle in the direction of shearing, temperature rise, shear stress, plastic strain, and thickness of the block. Furthermore, k is the thermal conductivity, ρ the mass density, μ the shear modulus and c the specific heat. All of the plastic working is assumed to be converted into heating.

We non-dimensionalize the variables as follows.

$$\begin{aligned}
&\bar{y} = y/H, \quad \bar{t} = v_0 t/H, \quad \bar{w} = w/H, \quad \bar{\theta} = \theta/\theta_0, \quad \theta_0 = \sigma_0/\rho c_R, \\
&\bar{s} = s/\sigma_0, \quad \bar{\rho} = \rho v_0^2/\sigma_0, \quad \bar{k} = k/(\rho v_0 c_R H), \\
&\bar{c} = c(\theta)/c_R, \quad \bar{\mu} = \mu/\sigma_0, \quad c_R = c(\theta_R), \quad \theta_R = 300K.
\end{aligned} \tag{2.5}$$

Here v_0 is the steady value of the shearing velocity applied to the top surface of the block when its lower surface is kept fixed. In terms of non-dimensional variables, (2.1)-(2.4) remain unchanged except that dimensional quantities are replaced by non-dimensional ones and the domain of the problem becomes $0 < \bar{y} < 1$. Henceforth in this section only nondimensional variables are used and the superimposed bars have been dropped.

For the initial and boundary conditions we take

$$\begin{aligned}
&\theta(y, 0) = 0, \quad v(y, 0) = 0, \quad s(y, 0) = 0, \quad \gamma_p(y, 0) = 0, \\
&\theta_{,y}(0, t) = 0, \quad \theta_{,y}(1, t) = 0, \quad v(0, t) = 0, \\
&v(1, t) = t/0.01, \quad 0 \leq t \leq 0.01, \\
&\quad\quad\quad\;\; = 1, \quad\quad\quad\; t \geq 0.01.
\end{aligned} \tag{2.6}$$

That is, the block is initially stress free, is undeformed, is at rest, and has a uniform temperature, normalized to be zero. The overall deformations of the block are taken to be adiabatic and the lower surface is at rest, whereas the upper surface is assigned a velocity that increases from 0 to 1 in a non-dimensional time of 0.01 and then stays at 1.0. The block is taken to be thinnest at the center, $y = 1/2$, and thickest at the boundary surfaces, $y = 0, 1$, with the thickness variation given by

$$w(y) = w_0 \left[1 + \frac{\delta}{2} \sin\left(\frac{1}{2} + 2y\right)\pi\right]. \tag{2.7}$$

Marchand and Duffy (1988) reported nearly 10% variation in the thickness of the steel tubes they tested in torsion. Our choice of locating the thinnest section at the center is for convenience only and should not affect the computed results.

2.1 Viscoplastic Flow Rules

2.1.1 Litonski's law

Wright and Batra (1985) modified the Litonski law (1977) to account for elastic unloading of a material point, and postulated that

$$\dot{\gamma}_p = \Lambda s, \tag{2.8}$$

$$\Lambda = \max\left[0, \left(\left(\frac{s}{(1-\nu\theta)\left(1+\frac{\psi}{\psi_0}\right)^n}\right)^{1/m} - 1\right)/bs\right], \tag{2.9}$$

$$\dot{\psi} = s\dot{\gamma}_p/(1+\psi/\psi_0)^n. \tag{2.10}$$

Here ψ may be viewed as an internal variable that describes the work hardening of the material. Equation (2.10) implies that the rate of growth of ψ is proportional to the plastic working. In (2.9), $(1-\nu\theta)$ describes the softening of the material as a result of its being heated up, b and m characterize its strain-rate sensitivity, and ψ_0 and n its work hardening. Equations (2.8) and (2.9) imply that

$$\dot{\gamma}_p = 0 \text{ if } s \leq (1-\nu\theta)(1+\psi/\psi_0)^n. \tag{2.11}$$

Thus $s = (1-\nu\theta)(1+\psi/\psi_0)^n$ describes a loading surface, and if the local state given by (s, ψ, θ) lies inside or on this surface, the plastic strain-rate is zero and the material then is deforming elastically. Besides σ_0, which has been used to non-dimensionalize stress-like quantities, five material parameters, ν, b, m, ψ_0, and n are needed to specify the viscoplastic response of the material.

2.1.2 Bodner-Partom law

Bodner and Partom (1975) assumed that there is no loading surface and that plastic strain-rate $\dot{\gamma}_p$, albeit very small at low values of s, is always non-zero. Their constitutive relation can be written as

$$\dot{\gamma}_p = D_0 \exp\left[-\frac{1}{2}\left(\frac{z^2}{3s^2}\right)^n\right], \quad n = \frac{a}{T} + b, \tag{2.12}$$

$$z = z_1 - (z_1 - z_0)\exp(-mW_p), \tag{2.13}$$

$$\dot{W}_p = s\dot{\gamma}_p. \tag{2.14}$$

Here T is the absolute temperature of a material particle, W_p is the plastic work done, z may be regarded as an internal variable, and D_0 is the limiting value of the plastic strain-rate, usually taken as $10^8 s^{-1}$. Besides D_0, we need to specify a, z_1, z_0, m, and b to characterize the material.

2.1.3 Johnson-Cook law

Johnson and Cook (1983) tested 12 materials in simple shear and compression at different strain-rates and found that

$$\dot{\gamma}_p = \dot{\gamma}_0 \exp\left[\left(\frac{s}{(A + B\gamma_p^n)(1 - \overline{T}^m)} - 1.0\right)/C\right], \qquad (2.15)$$

$$\overline{T} = (\theta - \theta_0)/(\theta_m - \theta_0), \qquad (2.16)$$

describe well the test data. For θ_m equal to the melting temperature of the material, θ_0 equal to the ambient temperature and $\dot{\gamma}_0 = 1/s$, they tabulated values of A, B, n, m, and C for 12 materials. It should be noted that there is no loading surface assumed in this case, too.

2.2 Results

2.2.1 Computational considerations

The governing equations (2.1)-(2.4) with the function g given by one of the flow rules described above are highly nonlinear, and difficult to solve analytically under the side conditions (2.6); their approximate solution has been computed numerically by using the finite element method. These partial differential equations are first reduced to a set of coupled nonlinear ordinary differential equations by using the Galerkin approximation resulting in

$$\mathbf{M}\dot{\mathbf{d}} = \mathbf{F} \qquad (2.17)$$

where $\mathbf{M}$ is the generalized mass matrix, $\mathbf{F}$ the generalized force vector and $\mathbf{d}$ the vector of nodal values of v, θ, s and ψ. The stiff nonlinear and coupled ordinary differential equations (2.17) are integrated with respect to time by using the subroutine LSODE included in the package ODEPACK developed by Hindmarsh (1983). The subroutine adjusts the time increment adaptively until a solution of the ordinary differential equations has been computed to the desired accuracy.

In the computation of results given below, the following values of various material parameters were used: $\rho = 7860$ kg/m^3, $\sigma_0 = 405$ MPa, and $c = 473$ J/kg°C.

(a) Litonski's law: $\nu = 6 \times 10^{-4}$/K, $\psi_0 = 0.012$, $m = 0.01872$, $n = 0.054$, and $b = 10^4$s.

(b) Bodner-Partom law: $D_0 = 1000/s$, $z_1 = 3.778$, $z_2 = 3.185$, $m = 2.5$, $a = 1800$K, and $b = 0$.

(c) Johnson-Cook law: $\dot{\gamma}_0 = 1/s$, $A = 0.275$, $B = 1.433$, $C = 0.36$, $n = 0.054$, $m = 0.8$, $\theta_m = 1800$K and $\theta_0 = 300$K.

The values of geometric parameters used are $H = 2.5$ mm, $w_0 = 0.38$ mm, and $\delta = 0.05$. The values of material parameters given above are such that for $k = 50$ W/m°C and average strain-rate of 3300s^{-1}, the average shear stress, s_a, versus the average shear strain, γ_{avg}, curve approximated well the experimental stress-strain curve for HY-100 steel given by Marchand and Duffy (1988). The average shear stress s_a is defined as

$$s_a = \int_0^1 s(y, t) dy.$$

For $\dot{\gamma}_{\text{avg}} = 3300\text{s}^{-1}$, the inertia effects do not play a noticeable role, and the shear stress depends upon y mainly because of the dependence of w upon y. Subsequently, the values of material parameters and the average strain-rate were kept fixed, and results were computed for $k = 0$, 5, 50, 500, and 5000 W/m°C. These results are identified below as follows.

Curve type	$k(\text{W/m}°\text{C})$
- - - - -	0
————	5
- - - -	50
- - - - - -	500
— - - — - - — - -	5000

For the Litonski law, and for $k = 0$ and 5 W/m°C, results could not be computed satisfactorily once the shear stress began to drop precipitously.

2.2.2 Numerical results

Figure 2.1 depicts the average shear stress, s_a, versus the average shear strain, γ_{avg}, curves for the three constitutive models and the five values of the thermal conductivity k. For each constitutive relation used, the $s_a - \gamma_{\text{avg}}$ curves for $k = 0$ and 5 W/m°C are essentially identical with each other. The values of γ_{avg} at which s_a begins to drop increases a little with an increase in the value of the thermal conductivity. However, the rate of stress drop decreases dramatically as the value of k is increased from 50 to 500 W/m°C as compared with that when k is increased from 5 to 50 W/m°C. For each value of k considered, the value of γ_{avg} when the average shear stress s_a becomes maximum is the least for the Johnson-Cook law. The $s_a - \gamma_{\text{avg}}$ curves look alike for the Litonski law and the Bodner-Partom law, except that the rate of stress drop is a little less for the Bodner-Partom law than that for the Litonski law.

Figure 2.2 depicts the evolution of the homologous temperature θ_H, defined as the ratio of the absolute temperature of a material point to the melting temperature of the material, at the center of the specimen. Because of the non-dimensional variables being used, the horizontal scale representing the average strain can also be interpreted as the time elapsed. For each of the three constitutive relations used, the rate of temperature rise is largest for $k = 0$ and decreases as the value of k is increased. For $k = 0$ and 5 W/m°C, the Johnson-Cook law gives the steepest rise in the temperature at the specimen center. It should be recalled that the shear stress is greatest at the specimen center because the thickness there is the least. For $k = 50$ W/m°C, the Litonski law gives the most rapid rate of temperature increase at the center of the specimen. The value of γ_{avg} when the temperature at the specimen center begins to rise sharply is

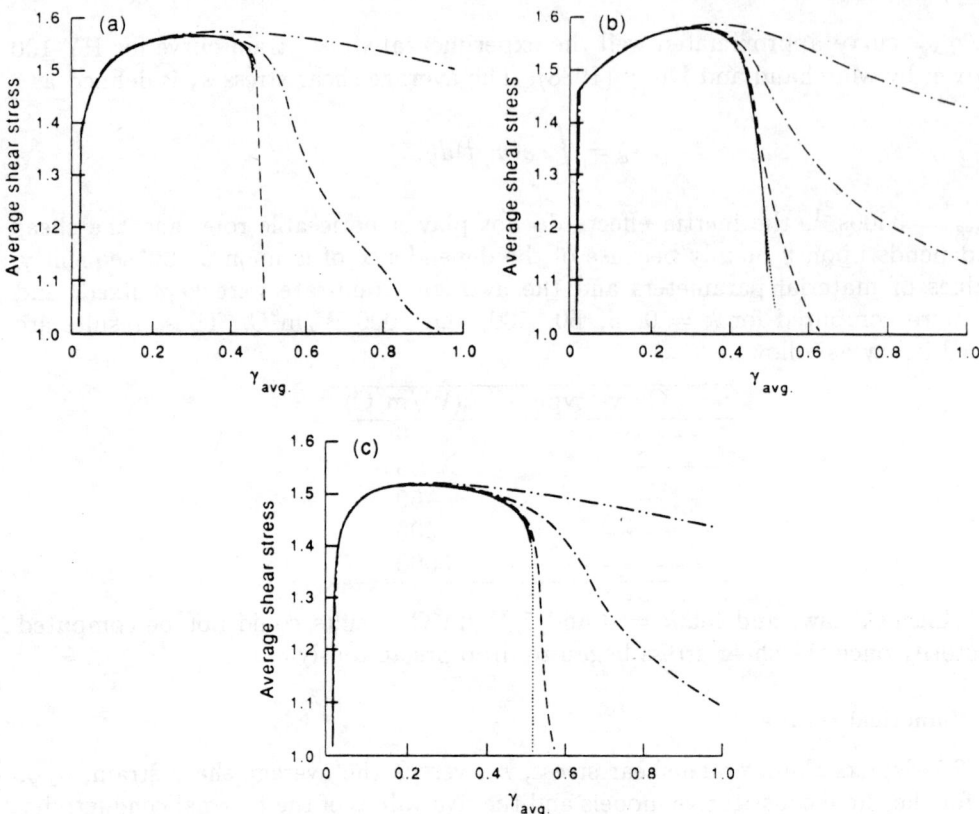

Figure 2.1: Average shear stress vs average shear strain for the three constitutive relations and the five values of the thermal conductivity. (a) Litonski, (b) Bodner-Partom, (c) Johnson-Cook.

different for the three constitutive relations. For $k = 5000$ W/m°C and for $0 < \gamma_{avg} < 1$, the temperature at the specimen center increases nearly linearly for each of the three constitutive relations used, except that for the Bodner-Partom law the slope of the θ_H vs. γ_{avg} curve increases at $\gamma_{avg} \simeq 0.4$. As the value of k increases, the heat conducted away from the central hotter region to the outer parts of the specimen increases and the rate of temperature rise at the specimen center decreases. Because of the adiabatic boundary conditions assumed, the temperature everywhere in the specimen increases.

2.2.3 Comparison of numerical results with experimental findings

Results presented in this section were computed by setting $k = 49.2$W/m°C. Initally the thickness at the specimen center was varied so as to obtain the rapid drop in the shear stress at about the same value of the average strain with each constitutive relation. Subsequently the thickness variation was kept fixed. The curves plotted in Fig. 2.3 vividly reveal that until the time the shear stress begins to drop rapidly, all of the flow rules considered predict material behavior in reasonable agreement with the

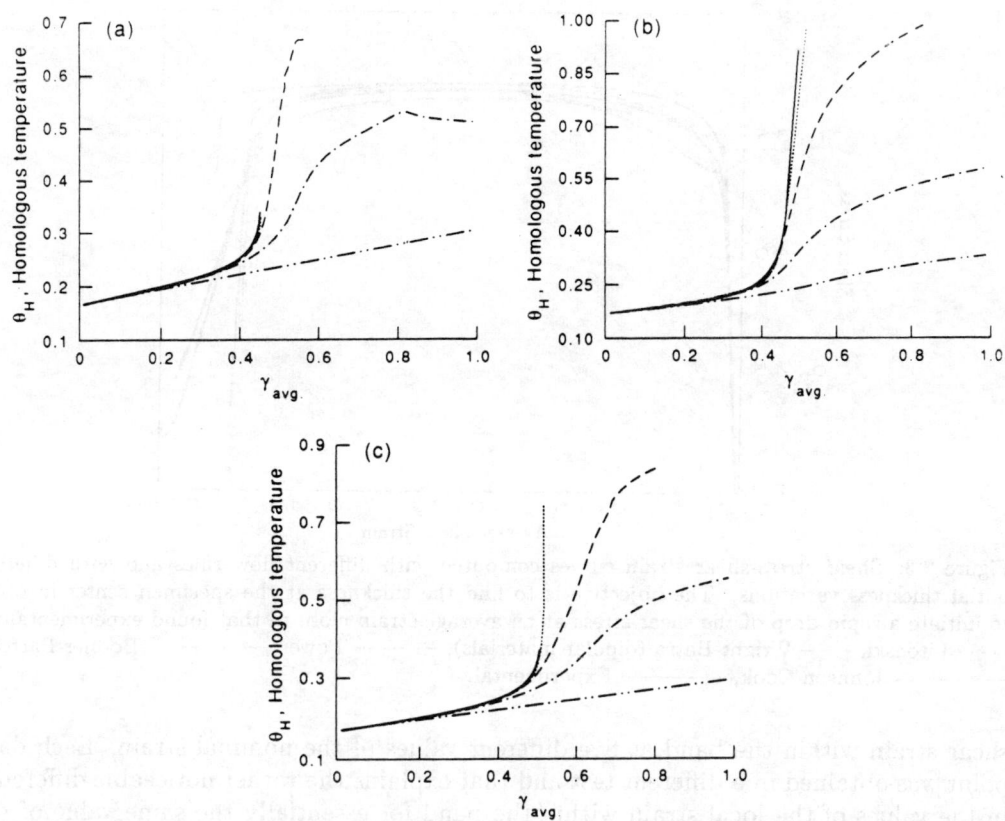

Figure 2.2: Evolution of the homologous temperature at the center of the specimen for the three constitutive relations and five values of the thermal conductivity. (a) Litonski, (b) Bodner-Partom, (c) Johnson-Cook.

experimental observations. Litonski's law, the power law (e.g. see Shawki and Clifton (1989)) and the Johnson-Cook law give essentially a catastrophic drop in the shear stress with virtually no increase in the nominal shear strain; this does not agree with the experimental data since Marchand and Duffy observed that during the drop of the shear stress, the nominal strain increases by approximately 5%. The Wright-Batra (1987) law for dipolar materials[2] and the Bodner-Partom law for nonpolar materials do predict the gradual drop in the shear stress which agrees with the experimental data. However, for the Bodner-Partom law the shear stress does not drop as much as it does during the tests, and the computed value of the shear stress reaches a plateau. Since curves plotted in Fig. 2.3 were for calibration purposes, these remarks should be regarded as general observations rather than a test of the validity of any of the flow rules.

For a nominal strain-rate of 1600 s^{-1}, Marchand and Duffy also gave values of the

---

[2]Dipolar materials are discussed in Section 5.

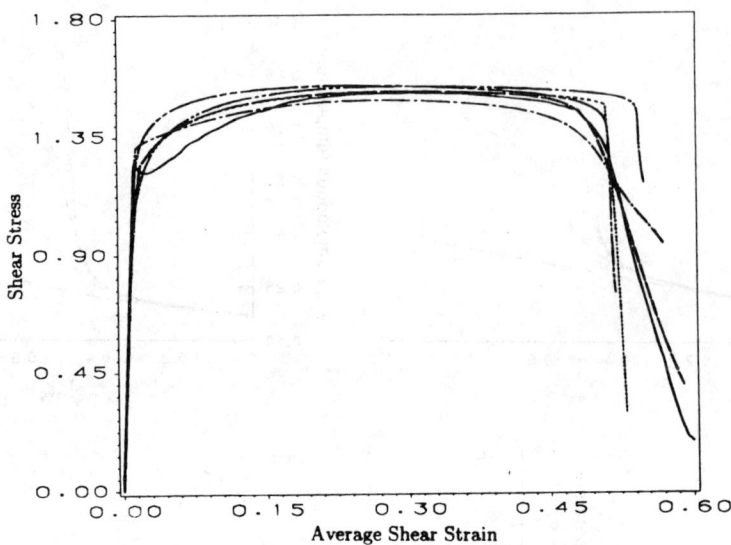

Figure 2.3: Shear stress-shear strain curves computed with different flow rules and with different initial thickness variations. The objective is to find the thickness at the specimen center in order to initiate a rapid drop of the shear stress at an average strain close to that found experimentally. ····· Litonski, − − − Wright-Batra (dipolar materials), — — — Power, — ·· — ·· Bodner-Partom, — ·· — ·· Johnson-Cook, ———— Experimental.

shear strain within the band at five different values of the nominal strain. Each data point was obtained in a different test and that explains the rather noticeable difference in the values of the local strain within the band for essentially the same value of the nominal strain for the last two data points. These and the corresponding numerically computed results with the different flow rules are plotted in Fig. 2.4. Whereas the Litonski law, the power law and the Johnson-Cook law give a rapid increase in the local strain once a shear band initiates, the Bodner-Partom law and the Wright-Batra law for dipolar materials give general trends which agree with the experimental data. We should add that the values of the material parameters and the initial thickness variation were those used to plot results depicted in Fig. 2.3. Also, the computed local strain equals the strain at the center.

With the power law and the Johnson-Cook law, the plastic strain started to oscillate during the time the shear stress was dropping. This was earlier pointed out by Batra and Kim (1990) and has also been noticed by Walter (1992). Possible explanations for this are the interplay between the material hardening due to the strain and strain-rate effects and the thermal softening, and the conversion of elastic energy into plastic deformations. This explains the discontinuities in the curves computed with these two flow rules.

The experimental data points plotted in Fig. 2.5 were taken from the data given in Table 4 of Marchand and Duffy's paper. Each data point represents a different test performed at an average strain-rate of approximately 1600/s. Since the nominal strain,

Initial-Boundary-Value Problems with Shear Strain Localization

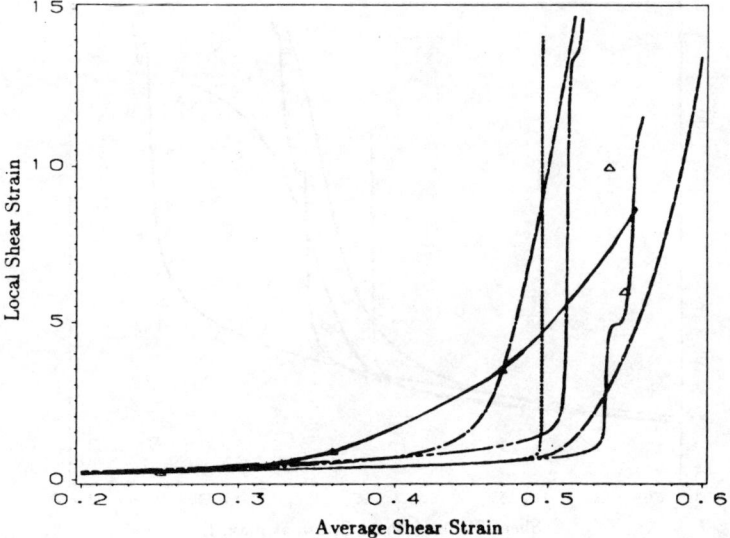

Figure 2.4: Growth of the local shear strain within the band as the specimen deforms. See Fig. 2.3 for the description of various curves. The experimental data points are denoted by a △.

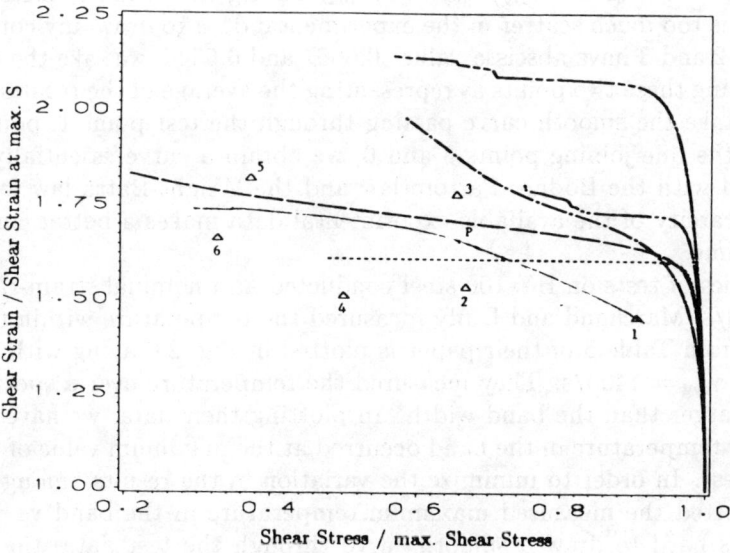

Figure 2.5: Normalized shear strain vs the normalized shear stress during the time shear stress is dropping with increasing strain. See Fig. 2.3 for the description of various curves.

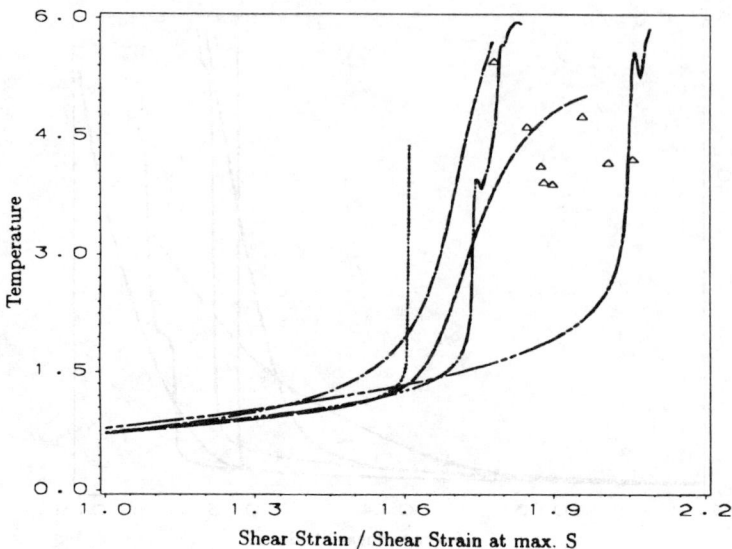

Figure 2.6: Temperature at the band center vs the normalized shear strain. The experimental data points are denoted by a Δ. See Fig. 2.3 for the description of various curves.

γ^*_{avg}, at which the shear stress attained the maximum value $s_{\max}$ is different in each test we have plotted in Fig. 2.5 $\gamma_{\text{avg}}/\gamma^*_{\text{avg}}$ vs $s/s_{\max}$ during the time the shear stress is dropping. There is too much scatter in the experimental data to draw any conclusions. Since test points 2 and 3 have abscissa values 0.6667 and 0.6546, we take the midpoint P on the line joining these two points as representing the average of the results for these two tests. If we take the smooth curve passing through the test point 1, point P and the midpoint of the line joining points 5 and 6, we obtain a curve essentially parallel to that computed with the Bodner-Partom law and the Wright-Batra law for dipolar materials. The scarcity of the available experimental data makes a better comparison difficult at this time.

In another series of tests on HY-100 steel conducted at a nominal strain-rate of approximately 1400/s, Marchand and Duffy measured the temperature within the band. The data taken from Table 5 of their paper is plotted in Fig. 2.6 along with the computed results for $\dot{\gamma}_{\text{avg}} = 1400$/s. They measured the temperature over a spot width of 35 μm which is larger than the band width. In plotting their data, we have assumed that the reported temperature in the band occurred at the maximum value of the nominal strain in a test. In order to minimize the variation in the results among different tests we have plotted the measured maximum temperature in the band vs $\gamma_{\text{avg}}/\gamma^*_{\text{avg}}$. Even though it is hard to draw a smooth curve through the test data, the detector output plotted in Fig. 19 of Marchand and Duffy's paper reveals that the temperature rises during the last stage of the localization process when the shear stress is dropping and that the increase in the average strain during the time temperature rises is about 8%. This observation is in closer agreement with the results computed with the

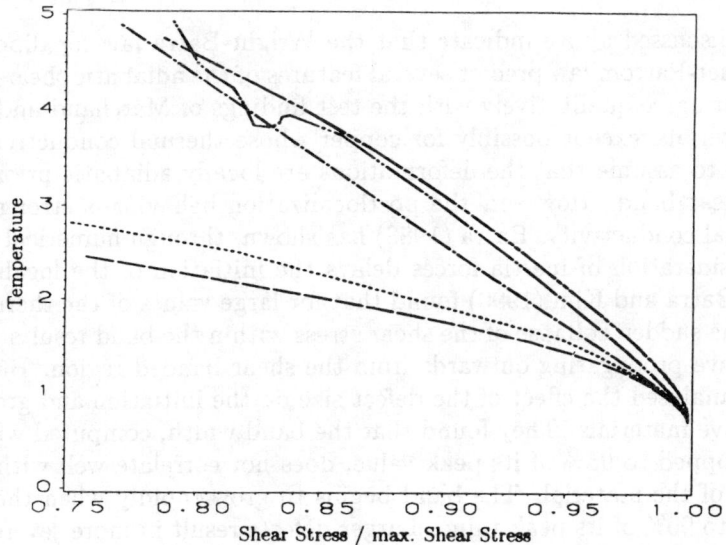

Figure 2.7: The evolution of the temperature at the center vs the normalized shear strain during the time the shear stress is dropping. See Fig. 2.3 for the description of various curves.

Wright-Batra law for dipolar materials. Also, the computed temperature rise of 539°C with this flow rule when $\gamma_{\text{avg}}/\gamma^*_{\text{avg}} = 1.91$ agrees well with the average value of 475°C found in the eight tests. We should note that the computed temperature within the band of 50 μm width came out to be nearly uniform. Marchand and Duffy estimated that the maximum temperature in the band reached a little over 900°C. Since we do not have any failure criterion included in our work, it is hard to decide when to stop the computations and thus estimate the maximum temperature rise.

Figure 2.7 shows how the temperature at the center increases after the peak in the shear stress has been attained. It is interesting to note that the temperature, when the shear stress attains the maximum value, is essentially the same for all flow rules. However, the rate of rise of temperature with the drop in the shear stress for the Johnson-Cook law, the power law and the Bodner-Partom law is nearly the same but differs significantly from that for the Litonski law and the Wright-Batra law for dipolar materials. The transition in the slope of the curves near $s/s_{\max} = 1.0$ indicates the point when the rapid drop in the shear stress occurs and the plastic strain rate rises sharply. Thus, the computed temperature rise will depend upon the point when the material is taken to have failed. As pointed out by Marchand and Duffy, once the shear stress begins to collapse, the load carrying capacity of the member is drastically reduced and the material has failed.

The band width was computed to be 2.56, 34.3, 10.3, 10.3 and 4.5 μm for the Litonski, Wright-Batra (dipolar materials, $\ell = 0.01$), Bodner-Partom, power, and Johnson-Cook relations respectively; the experimental value was approximately 20 μm.

2.3 Remarks

The results discussed above indicate that the Wright-Batra law for dipolar materials and the Bodner-Partom law predict several features of the adiabatic shear banding phenomenon that agree qualitatively with the test findings of Marchand and Duffy. Also, for most materials, except possibly for copper whose thermal conductivity is high, it is reasonable to assume that the deformations are locally adiabatic prior to the initiation of a shear band. However, the postlocalization behavior is strongly influenced by the thermal conductivity. Batra (1988) has shown, through numerical experiments, that the consideration of inertia forces delays the initiation of the localization of deformations. Batra and Kim (1990) found that for large values of the thermal softening coefficient, the sudden collapse of the shear stress within the band results in an unloading elastic wave propagating outwards from the shear banded region. Batra and Kim (1992) have analysed the effect of the defect size on the initiation and growth of shear bands in twelve materials. They found that the band width, computed when the shear stress has dropped to 95% of its peak value, does not correlate well with the thermal conductivity of the material. The band begins to grow rapidly when the shear stress has dropped to 90% of its peak value. Larger defects result in more severe localization of the deformation for the same percentage drop in the shear stress. However, the defect size influences very little the homologous temperature when either the shear stress has attained its maximum value or it has dropped to 85% of its maximum value.

Kim and Batra (1992) have accounted for the dependence of material properties upon the temperature, and found that the band width and the average strain at which a shear band forms decrease with a decrease in the initial temperature of the specimen. This partially explains why large deformation processes are performed on preheated specimens.

3 Adaptive Mesh Refinement for 2-dimensional Problems

Because of large deformations of the material within a shear band, the finite element mesh is severely distorted and an interior angle of an element may become either too small or too large. Thus the deforming region should be frequently remeshed. Several studies (e.g. see Needleman (1988)) have indicated that the results for localization problems are mesh dependent. One way to overcome this is to adaptively refine the mesh. Here we discuss two such techniques for 2-dimensional problems; the h-method in which the element size is varied (Batra and Ko (1992)) and the r-method wherein the mesh topology (number of elements and nodes connecting various elements) is kept fixed but the locations of nodes are varied (Batra and Hwang (1993)).

3.1 The h-method

We first select a coarse mesh and find a solution of the problem. This mesh is refined so that

$$a_e = \int_{\Omega_e} I\, d\Omega, \quad e = 1, 2, \ldots, n_{el}, \tag{3.1}$$

is nearly the same for each element Ω_e. In (3.1), I is the second invariant of the deviatoric strain-rate tensor, n_{el} equals the number of elements in the coarse mesh and Ω_e is one of the elements. Since points where the solution will exhibit sharp gradients are unknown a priori, the initial coarse mesh may be chosen as uniform. The motivation behind making a_e the same over each element Ω_e is that within the region of localization of the deformation values of I are very high as compared to those in the remaining region. Other variables such as the temperature rise, the maximum principal strain, and the equivalent strain which are also quite large within the band will be suitable replacements for I in (3.1). The topology of the refined mesh will depend upon the variable used in (3.1). In order to refine the mesh, we find

$$\bar{a} = \frac{1}{n_{el}} \sum_{e=1}^{n_{el}} a_e, \quad \varepsilon_e = \frac{a_e}{\bar{a}}, \quad h_e = \frac{\bar{h}_e}{\varepsilon_e}, \quad \text{and} \quad H_n = \frac{1}{N_e} \sum_{e=1}^{N_e} h_e, \quad n = 1, 2, \ldots n_{od}. \tag{3.2}$$

Here, $\bar{h}_e$ is the size of the element Ω_e in the coarse mesh, N_e equals the number of elements meeting at node n, and n_{od} equals the number of nodes in the coarse mesh. We refer to H_n as the nodal element size at node n.

In order to generate the new mesh, we first discretize the boundary by following the procedure given by Cescotto and Zhou (1989). Let AB be a segment of the contour to be discretized, s the arc length measured from point A, and H_A and H_B be nodal element sizes for nodes located at points A and B, respectively. From a knowledge of the values of H at discrete points, corresponding to the nodes in the coarse mesh, we define on AB a piecewise linear continuous function $H(s)$ that takes the previously computed values at the node points. In order to discretize AB for the new mesh, we start from point A if $H_A < H_B$; otherwise we start from B. Let A be the starting point. We first find temporary positions of nodes on segment AB by using the following recursive procedure. Assume that points $1, 2, \ldots, k$ have been found. Then the temporary location of point $(k+1)$ is given by

$$s_{k+1} = s_k + \tfrac{1}{2}[H(s_k) + H(s^*_{k+1})], \tag{3.3}$$

where

$$s^*_{k+1} = s_k + H(s_k). \tag{3.4}$$

Referring to Fig. 3.1, the above procedure will give rise to the following four alternatives: $a = b = 0$, $a < b$, $a > b$, $a = b \neq 0$. If $a = b = 0$, then the temporary locations of node points are their final positions. Depending upon whether $a < b$ or $b \leq a$, node points 2 to p or 2 to $p+1$ are moved, the displacement of a node being proportional to the value of H there, so that either node p or node $(p+1)$ coincides with B. This determines the final positions of nodes on segment AB.

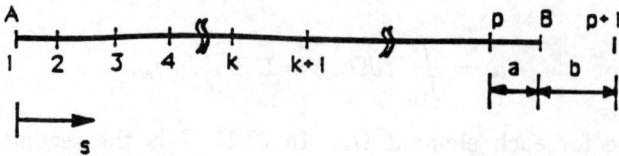

Figure 3.1: Discretization of a boundary segment for mesh refinement

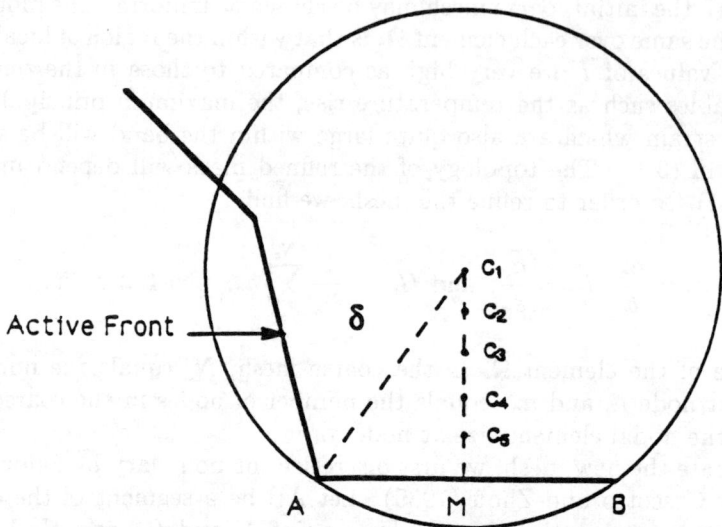

Figure 3.2: Advancing front and new element generation

Having discretized the boundary, we use the concept of advancing front(e.g., see Lo (1985); Peraire et al. (1987), Habraken and Cescotto (1990)) to generate the elements. An advancing front consists of straight line segments which are available to form a side of an element. Thus, to start with, it consists of the discretized boundary. We choose the smallest line segment (say side AB) connecting two adjoining nodes, and determine the nodal element size $H_M \equiv H(s_M) = (H_A + H_B)/2$ at the midpoint M of AB. We set

$$\delta = \begin{cases} 0.8\overline{AB} & \text{if } H_M < 0.8\overline{AB}, \\ H_M & \text{if } 0.8\overline{AB} \leq H_M \leq 1.4\overline{AB}, \\ 1.4\overline{AB} & \text{if } 1.4\overline{AB} < H_M, \end{cases} \qquad (3.5)$$

and find point C_1 at a distance δ from A and B (cf. Fig. 3.2). Here $\overline{AB}$ equals the length of segment AB, and numbers 0.8 and 1.4 can be changed to generate different size elements. We search for all nodes on the active front that lie inside the circle with center at C_1 and radius δ, and order them according to their distance from C_1 with the first node in the list being closest to C_1. At the end of this list are added points C_1, C_2, C_3, C_4, and C_5, which lie on C_1M and divide it into five equal parts. We next determine the first point C in the list that satisfies the following three conditions.

(i) Area of triangle ABC > 0.

(ii) Sides AC and BC do not cut any of the existing sides in the front.

(iii) If any of the points $C_1, C_2, \ldots, C_5$ is chosen, that point is not too close to the front.

The triangle ABC is an element in the new mesh. If C is one of the points $C_1, C_2, \ldots, C_5$, then a new node is also created. The advancing front is updated by removing the line segment AB from it, and adding line segments AC and CB to it. The element generation process ceases when there is no side left in the active front.

We determine values of solution variables at a newly created node by first finding out to which element in the coarse mesh this node belongs, and then finding values of solution variables at this node by interpolation. This process and that of searching for line segments and points in the aforestated element generation technique consume a considerable amount of CPU time. These operations are optimized to some extent by using the heap list algorithm (e.g., see Löhner (1988)) for deleting and inserting new line segments, and quadtree structures and linked lists for searching line segments and points and also for the interpolation of solution variables at the newly created nodes.

3.2 The r-method

The goal here is to reposition the nodes so that a_e defined by (3.1) with I replaced by the temperature rise θ is nearly the same for each element Ω_e. Having solved the problem on an initial mesh we refine it as follows. We begin with either the horizontal boundary or the vertical one and relocate nodes on it according to the following criterion. To be definite, let us begin with the left vertical edge. After having repositioned nodes on it we do the same on the almost vertical curve that passes through nodes next to the left vertical side, and continue the process till we reach the right vertical edge. The procedure is then repeated beginning with the top or bottom horizontal edge and going to the other end.

Referring to Fig. 3.3, let AB be the curve on which nodes are to be relocated. We plot the temperature distribution on AB with abscissa as the distance of a point from A measured along AB and ordinate as the temperature at that point. Values of temperature at numerous points on AB are obtained by linear interpolation from the values at node points. If S equals the total area under the curve, the approximate location s_n^a of the nth node on AB is given by

$$\int_{s_{n-1}}^{s_n^a} \theta ds = \frac{S}{N_{es}}, \qquad (3.6)$$

where N_{es} equals the number of elements on AB. We reposition the node to the interpolation point immediately to the left of its approximate location determined from (3.6). In Fig. 3.3c, the position of a node as found from (3.6) is shown by a superimposed prime, and its relocation in Fig. 3.3d by superimposed two primes. Since the end

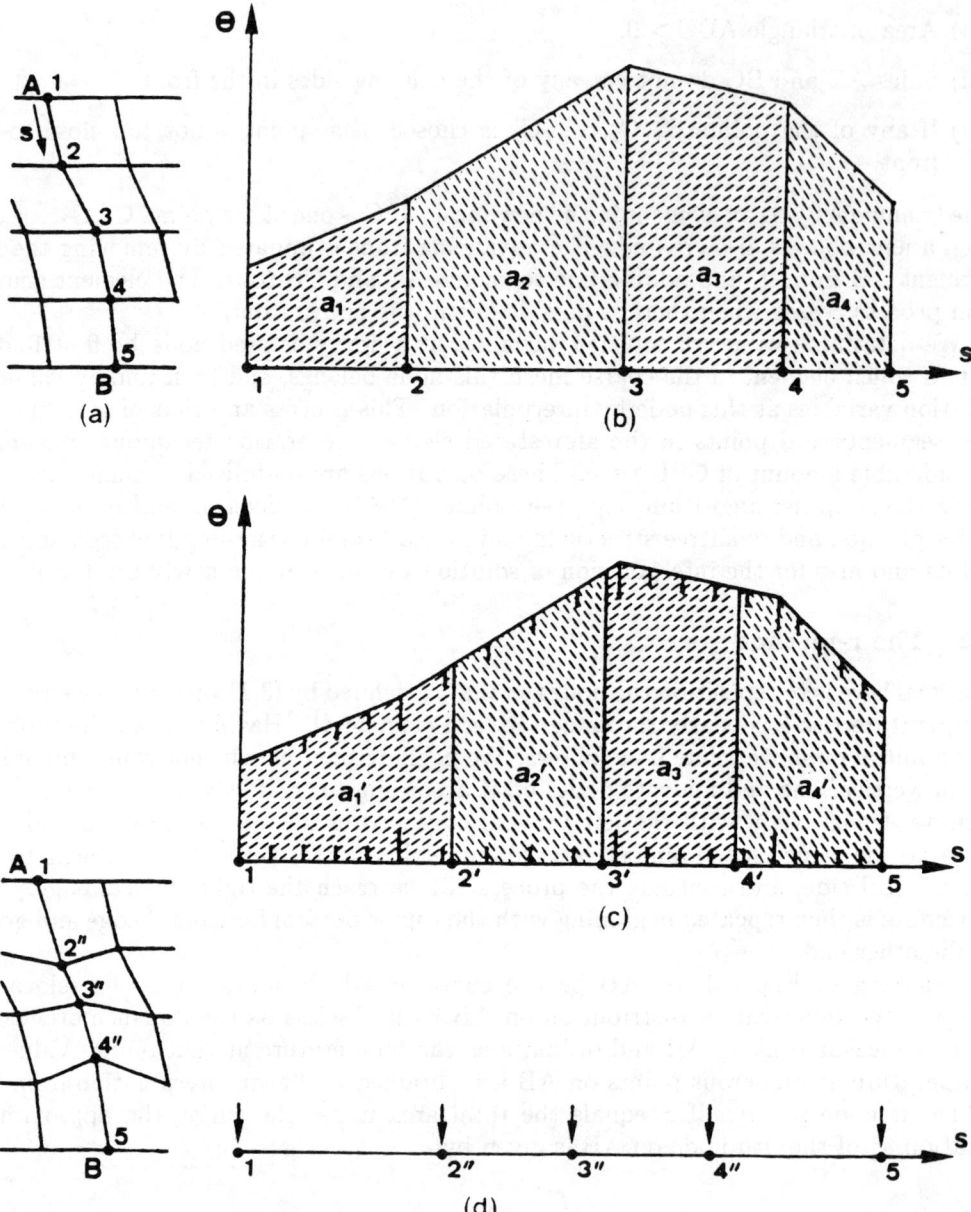

Figure 3.3: **a** Curve AB on which nodes are to be relocated. **b** Temperature distribution on curve AB on which nodes are to be repositioned. **c** Temporary position on curve AB of relocated nodes. **d** Repositioned nodes on curve AB.

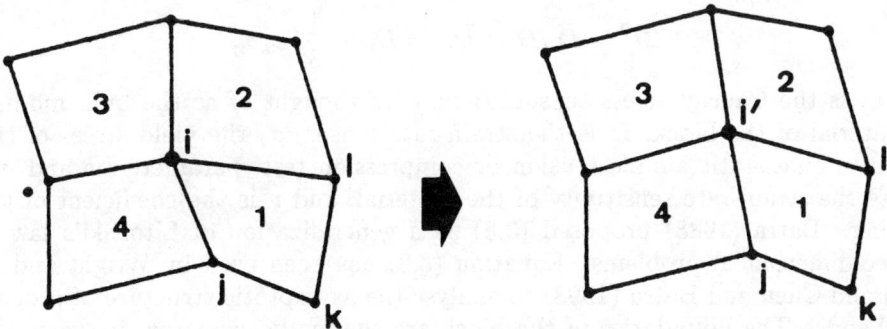

Figure 3.4: Relocation of an interior node to smoothen out the generated mesh

points on AB are kept fixed, the aforestated procedure can be employed by starting from either A or B. Note that when nodes on an approximate horizontal curve are relocated, positions of nodes A and B will change.

The quadrilateral elements produced by the aforestated simple technique are not always well shaped in the sense that one of the interior angles may be either too small or too large. It usually happens in regions where the element size varies noticeably. We use the mesh smoothing method of Zhu et al. (1991) to improve upon the shapes of quadrilateral elements. Each internal node is repositioned to the centroid of the polygon formed by all of the elements meeting at the node. As illustrated in Fig. 3.4, the internal node i is moved to i' with coordinates given by

$$x'_i = \frac{1}{4M} \sum_{a=1}^{M} (x_j + 2x_k + x_l)_a, \quad y'_i = \frac{1}{4M} \sum_{a=1}^{M} (y_j + 2y_k + y_l)_a, \qquad (3.7)$$

where M is the number of elements sharing node i. After having relocated all of the internal nodes, the element shapes are checked to see if all interior angles of every element are between 20° and 160°; these limiting values of interior angles are arbitrarily chosen. If not, the nodes are repositioned according to (3.7) till such is the case. Because of the smoothening of the mesh, the value of a_e defined by (3.1) is only approximately same for all elements in the mesh.

3.3 Numerical Results

The two aforestated adaptive mesh refinement techniques have been used to delineate the initiation and development of shear bands in plane strain compression of a square block. Because of the presumed symmetry of deformations about the horizontal and vertical centroidal axes, deformations of only a quarter of the block are investigated. The constitutive relation for the material is taken to be

$$\sigma_{ij} = -B\left(\frac{\rho}{\rho_0} - 1\right)\delta_{ij} + \frac{\sigma_0}{\sqrt{3I}}(1 + bI)^m(1 - \nu\theta)D_{ij}, \qquad (3.8)$$

$$2I^2 = \overline{D}_{ij}\overline{D}_{ij}, \ \overline{D}_{ij} = D_{ij} - \frac{1}{3}D_{kk}\delta_{ij}. \tag{3.9}$$

Here σ_{ij} is the Cauchy stress tensor, B may be thought of as the bulk modulus for the material of the block, $\mathbf{D}$ is the strain-rate tensor, σ_0 the yield stress of the material in a quasistatic simple tension or compression test, parameters b and m characterize the strain-rate sensitivity of the material, and ν is the coefficient of thermal softening. Batra (1988) proposed (3.8) as a generalization of Litonski's law (1977) to three-dimensional problems. Equation (3.8) has been used by Wright and Walter (1996) and Chen and Batra (1998) to analyse the asymptotic structure of propagating shear bands. The boundaries of the block are thermally insulated, its vertical edges traction free, and upper and lower smooth horizontal surfaces are moved vertically so as to induce a nominal strain-rate of 5000/s.

3.3.1 The h-refinement

Figure 3.5 depicts the initial coarse mesh at time $t = 0$, and the generated refined meshes at an average strain or non-dimensional time $t = 0.025$, 0.040, and 0.047. In the solution of the problem, the mesh was also adaptively refined at $t = 0.015$, 0.030, and 0.035; however, these are not shown for the sake of brevity. The times at which the mesh is refined were selected manually, and are arbitrary. A possible criterion could be to refine the mesh when the second invariant of the strain-rate tensor or the temperature at the center has risen by a certain amount. The meshes shown in Fig. 3.5 vividly reveal that the refinement technique outlined in Sec. 3 gives rise to nonuniform meshes with finer mesh in the severely deforming region and coarse mesh elsewhere. We did not impose any restriction on the number of new nodes that can be introduced when the mesh is refined. Practical considerations such as the core storage available may restrict the number of nodes. The distribution of the velocity field in the deforming region at $t = 0.047$, shown in Fig. 3.6a, supports Tresca's (1878) and Massey's (1921) assertions that the tangential velocity is discontinuous across a shear band. In our work the velocity field is forced to be continuous throughout the domain. The sharp jumps in v_1 and v_2 across the shear band lend credence to the discontinuity of the tangential velocity across the shear band. The plot of the effective stress s_e, defined as

$$s_e = \sqrt{\frac{2}{3}(1-\nu\theta)(1+bI)^m},$$

in Fig. 3.6b reveals that s_e drops considerably within the shear band. These observations qualitatively agree with Batra and Liu's (1989) results, except that the present results are sharper in the sense that the region of localized deformation is significantly narrower.

We now investigate the change, if any, in the approximate solution caused by refining the mesh. Since the analytical solution of the problem is unknown, we compare the approximate solution with a higher-order approximate solution (Hinton and Campbell

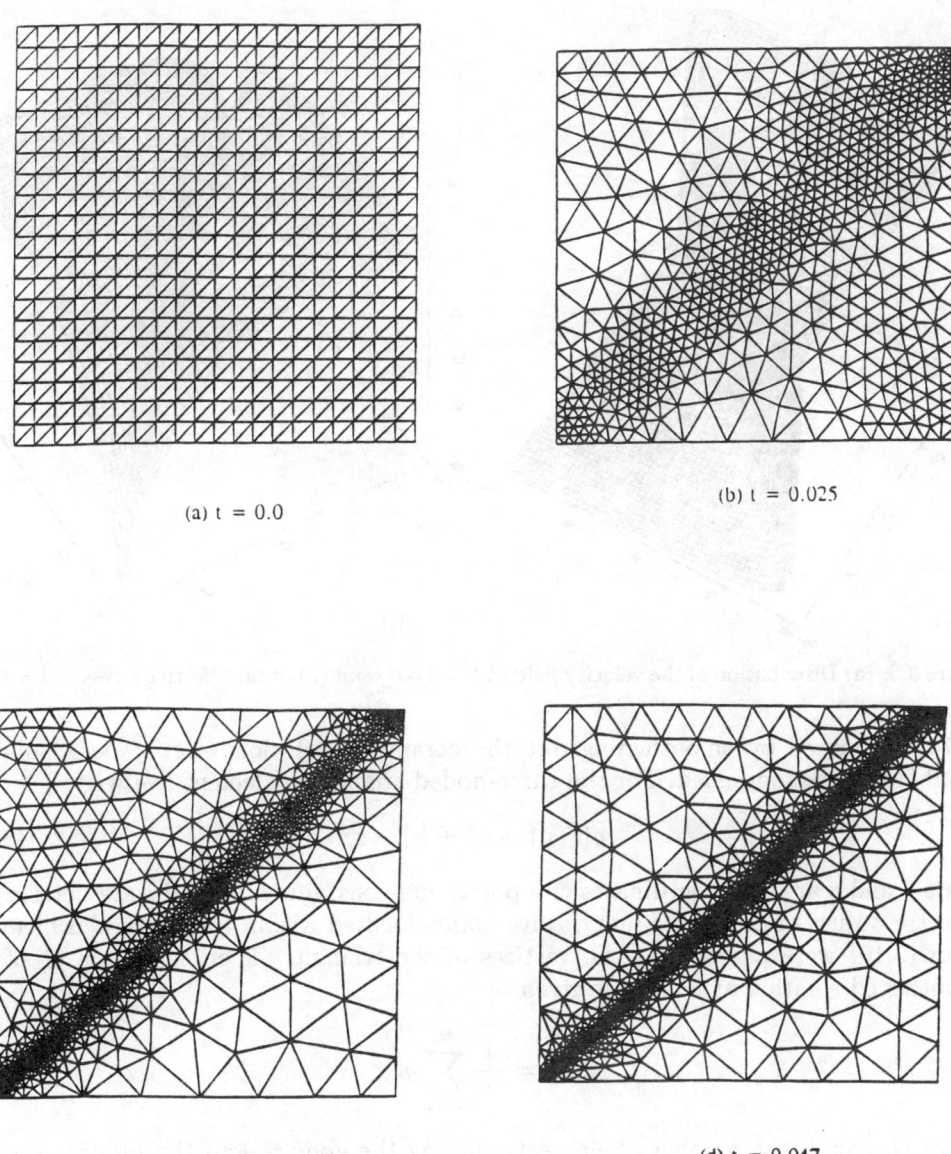

Figure 3.5: Finite element meshes at **a** $t = 0.0$, **b** $t = 0.025$, **c** $t = 0.040$, and **d** $t = 0.047$

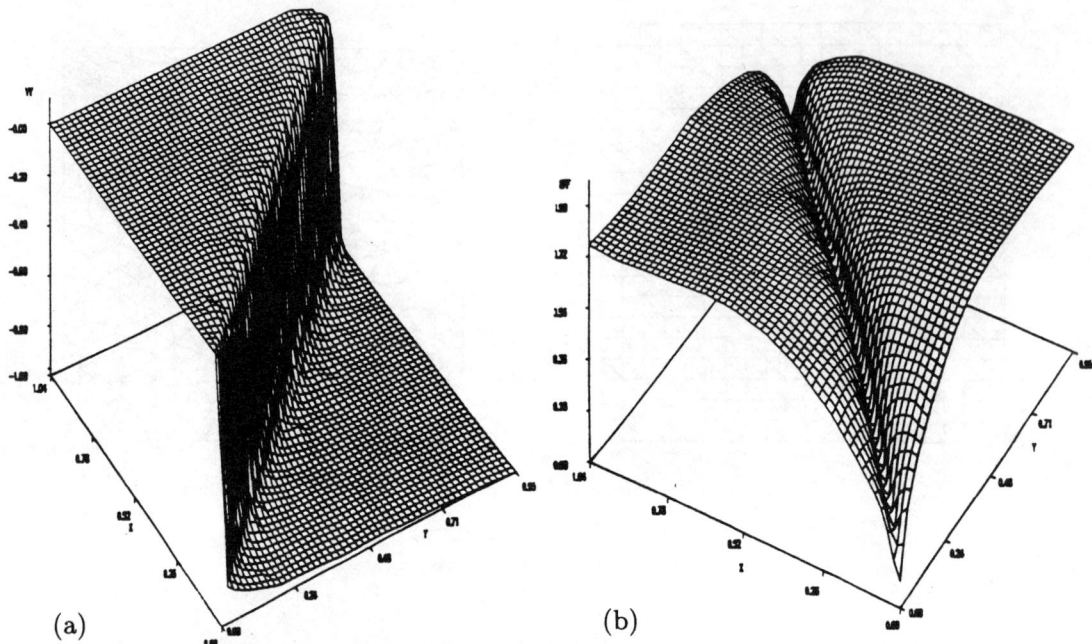

Figure 3.6: (a) Distribution of the velocity field at $t = 0.047$, and (b) of the effective stress at $t = 0.047$

(1974)) obtained by smoothening out the computed solution. Let g be a solution variable to be smoothened. For the three-noded triangular element, we write

$$g(\xi, \eta) = a\xi + b\eta + c, \qquad (3.10)$$

where ξ and η are area coordinates of a point, and constants a, b, and c are determined from the values of g at three quadrature points located within the triangular element. From (3.10) we evaluate g at the vertices of the triangle. Then the value g_n^* of the smoothened solution at node n is given by

$$g_n^* = \frac{1}{N_e} \sum_{n=1}^{N_e} g_n, \qquad (3.11)$$

where N_e equals the number of elements sharing the node n, and the summation sign on the right-hand side implies the sum of the values of g at node n evaluated for each element meeting at that node. Knowing g^* at each node, we interpolate its value at any other point by using the finite element bases functions. We define the percentage error η in the deviatoric strain-rate tensor $\overline{\mathbf{D}}$ by the relation

$$\eta = \left(\frac{\|\mathbf{e}\|_0^2}{\|\mathbf{e}\|_0^2 + \|\overline{\mathbf{D}}\|_0^2} \right)^{1/2} \times 100, \qquad (3.12)$$

Initial-Boundary-Value Problems with Shear Strain Localization

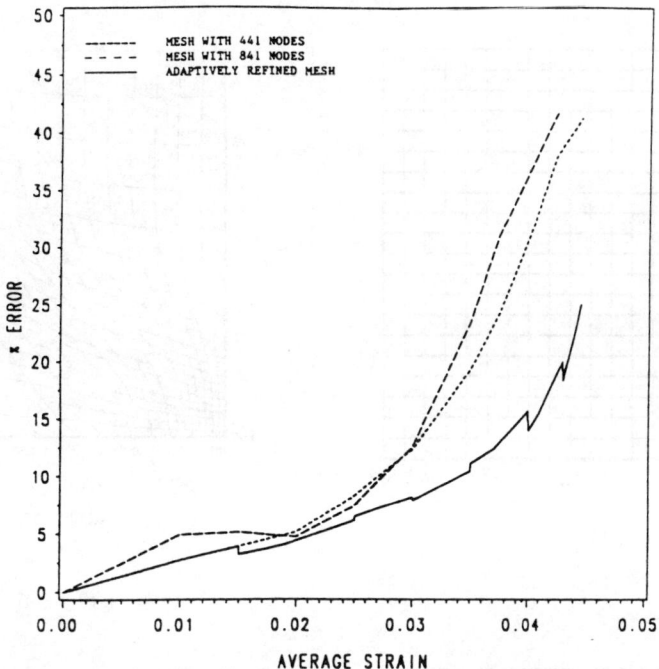

Figure 3.7: Comparison of the error in the computed approximate solution and the higher-order approximate solution for three different meshes; fixed mesh with 441 nodes, fixed mesh with 841 nodes, and an adaptively refined mesh.

where

$$\mathbf{e} = \overline{\mathbf{D}} - \overline{\mathbf{D}}^*, \quad \|\mathbf{e}\|_0^2 = \sum_{e=1}^{N_{el}} \int_{\Omega_e} \mathbf{e}^T \mathbf{e} d\Omega, \qquad (3.13)$$

and N_{el} equals the number of elements in the mesh. The plot of the percentage error η in Fig. 3.7 for the three meshes shows that the error is lower for the approximate solution obtained by using the adaptively refined mesh as compared to that for the other two meshes. That the error measure is rather crude is indicated by the slightly larger errors obtained with a fixed mesh of 841 nodes as compared to that with 441 nodes. It could be due to the larger errors caused by smoothening out of the approximate solution with 841 nodes since the band in this case is more intense than that for the mesh with 441 nodes.

3.3.2 The r-refinement

Figure 3.8a shows the initial mesh consisting of 400 uniform elements, and the generated refined meshes when the nondimensional temperature θ at the centroid equaled 0.25, 0.35, and 0.45 are plotted in Figs. 3.8b, 3.8c, and 3.8d, respectively; the mesh was

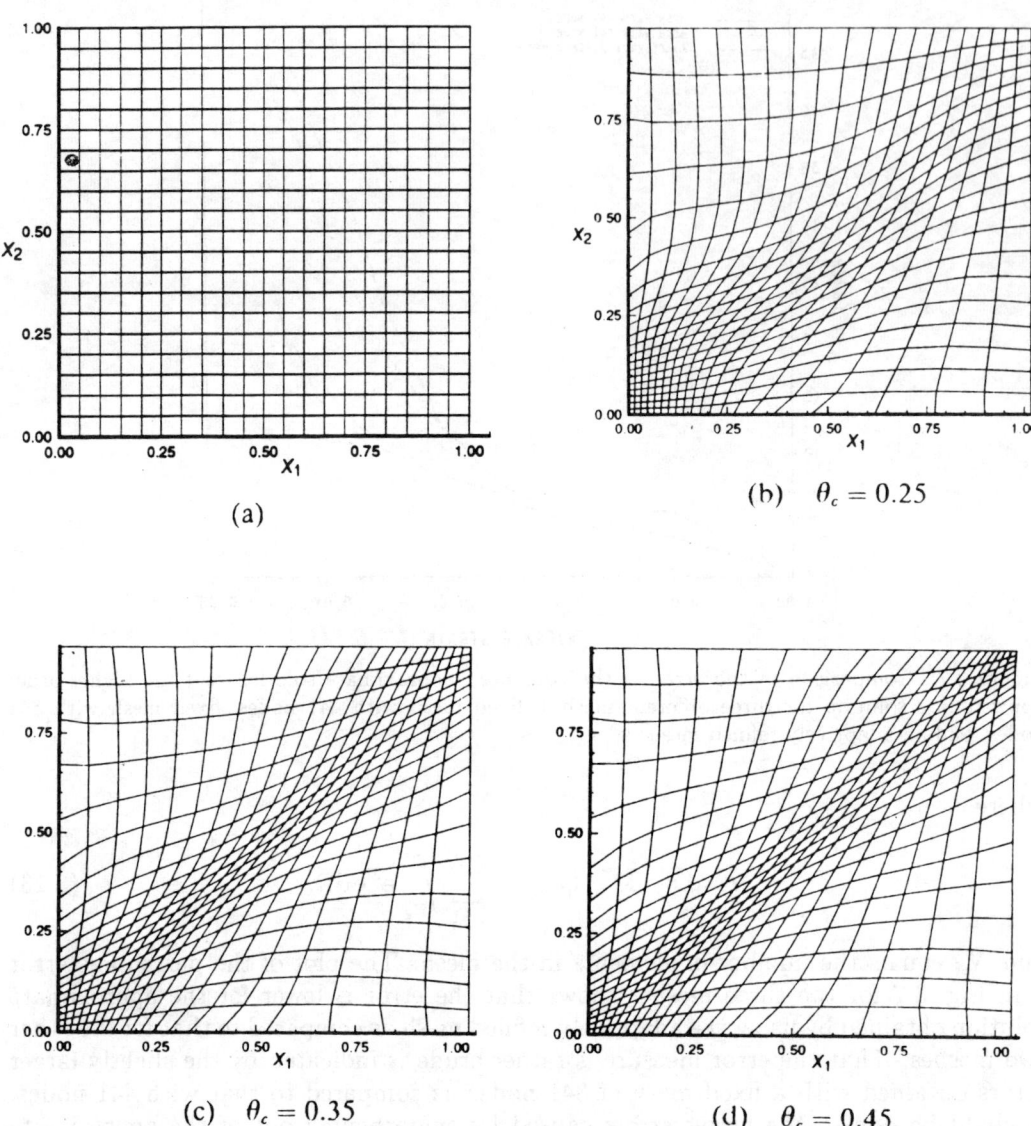

Figure 3.8: **a** Initial uniform mesh of 400 elements. **b-d** Finite element meshes generated by using the mesh refinement technique when the temperature at the center of the specimen reached 0.25, 0.35, and 0.45.

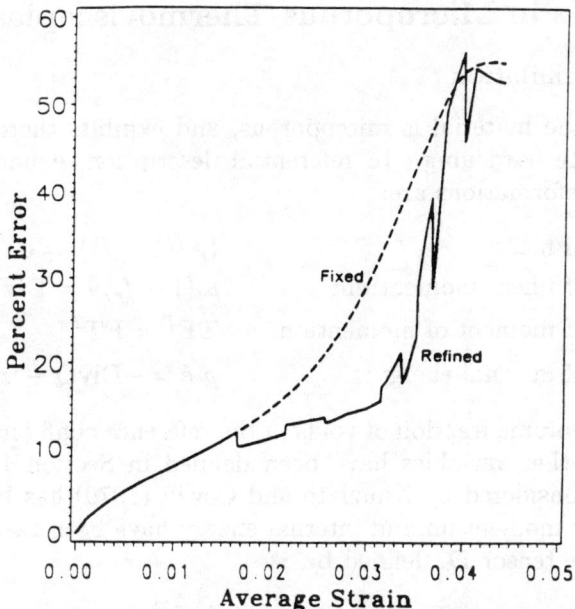

Figure 3.9: Comparison of the error in the approximate solution computed with a fixed mesh and an adaptively refined mesh. In each case, the error is determined by comparing the computed solution with a higher-order approximate solution.

also refined when θ at the centroid reached 0.30 and 0.40, but these are not depicted. We chose to refine the mesh for equal increments of the temperature rise. However, other criteria such as the second invariant I of the strain-rate tensor attaining certain values, would be equally good. The meshes shown in Fig. 3.8 vividly reveal that the r-refinement technique generated a nonuniform mesh with finer elements in the severely deformed region and coarse elements elsewhere. The mesh smoothening criterion (3.7) had to be applied atmost three times to satisfy the requirement that the interior angles of every quadrilateral element be between 20° and 160°. We note that the mesh generation scheme does not impose any restriction on the ratio of the area of the largest to that of the smallest element in the mesh. Even though this did not cause an unduly skewed mesh to be generated for the present problem, in other situations such a restriction may be necessary. One could avoid this either by having more elements in the initial mesh or by adding more elements at a few intermediate stages. The latter would necessitate the creation of a new mesh topology.

The percentage error in the deviatoric strain-rate tensor $\overline{\mathbf{D}}$ is plotted in Fig. 3.9. The error for the solution obtained by using the adaptively refined mesh is lower upto an average strain of 0.037 and it then suddenly increases and becomes larger than that obtained for the solution computed by using a fixed mesh. It probably is due to large errors caused by smoothening out the approximate solution in the late stages of the band development when the deformations within the band are very intense.

4 Shear Bands in Microporous Thermoviscoplastic Solids

4.1 Problem Formulation

It is assumed that the material is microporous, and exhibits thermal softening, and strain and strain-rate hardening. In referential description, equations governing its thermomechanical deformations are:

Balance of mass: $\quad (\rho J(1-f))^{\cdot} = 0,$ (4.1)

Balance of linear momentum: $\quad \rho_r(1-f_r)\dot{\mathbf{v}} = \text{Div}\mathbf{T},$ (4.2)

Balance of moment of momentum: $\quad \mathbf{TF}^T = \mathbf{FT}^T,$ (4.3)

Balance of internal energy: $\quad \rho_r \dot{e} = -\text{Div}\mathbf{Q} + \text{tr}(\mathbf{T}\dot{\mathbf{F}}^T),$ (4.4)

where f_r equals the volume fraction of voids in the reference configuration, f in present configuration, and other variables have been defined in Section 1. The balance of equilibrated forces considered by Nunziato and Cowin (1979) has been neglected and the supplies of linear momentum and internal energy have been taken to be zero.

Let the strain-rate tensor $\mathbf{D}$, defined by

$$2\mathbf{D} = \text{grad } \mathbf{v} + (\text{grad } \mathbf{v})^T, \qquad (4.5)$$

with grad denoting the gradient operator applied to a field quantity defined as a function of $\mathbf{x}$ and t, have the additive decomposition into elastic $\mathbf{D}^e$, plastic $\mathbf{D}^p$, and thermal parts:

$$\mathbf{D} = \mathbf{D}^e + \mathbf{D}^p + \alpha\dot{\theta}\mathbf{1}. \qquad (4.6)$$

The following constitutive assumptions are made.

$$\overset{\triangledown}{\boldsymbol{\sigma}} \equiv \dot{\boldsymbol{\sigma}} + \boldsymbol{\sigma}\mathbf{W} - \mathbf{W}\boldsymbol{\sigma} = \frac{E(1-f)}{1+\nu}\mathbf{D}^e + \frac{E\nu(1-f)}{(1+\nu)(1-2\nu)}\text{tr}(\mathbf{D}^e)\mathbf{1}, \qquad (4.7)$$

$$\frac{3}{2}\frac{\text{tr}(\mathbf{ss}^T)}{\sigma_m^2} + 2f^*\beta_1 \cosh\left(\frac{\beta_2 \text{tr}\boldsymbol{\sigma}}{2\sigma_m}\right) - 1 - \beta_1^2 f^{*2} = 0, \qquad (4.8)$$

$$\mathbf{D}^p = \frac{(1-f)\sigma_m \dot{\epsilon}_m^p}{\text{tr}(\boldsymbol{\sigma}\mathbf{N}^T)}\mathbf{N}, \quad \mathbf{s} = \boldsymbol{\sigma} - \frac{1}{3}(\text{tr }\boldsymbol{\sigma})\mathbf{1}, \qquad (4.9)$$

$$\mathbf{N} = \frac{3\mathbf{s}}{\sigma_m^2} + \frac{f^* q_1 q_2}{\sigma_m}\left[\sinh\left(\frac{q_2 \text{tr}\boldsymbol{\sigma}}{2\sigma_m}\right)\right]\mathbf{1}, \qquad (4.10)$$

$$f^* = \begin{cases} f & \text{if } f \leq f_c, \\ f_c + \left(\dfrac{f_u^* - f_c}{f_f - f_c}\right)(f - f_c), & \text{otherwise,} \end{cases} \qquad (4.11)$$

$$\sigma_m = \sigma_0(1 + b\dot{\epsilon}_m^p)^m \left(1 + \frac{\epsilon_m^p}{\epsilon_y}\right)^n (1 - \beta\theta), \qquad (4.12)$$

$$\dot{f} = (1-f)\mathrm{tr}\mathbf{D}^p + \frac{f_2 \dot{\epsilon}_m^p}{s_2\sqrt{2\pi}}\exp\left(-\frac{1}{2}\left(\frac{\bar{\epsilon}_m^p - e_N}{s_2}\right)^2\right), \tag{4.13}$$

$$\mathbf{q} = -k\left(1 - \frac{3}{2}f\right)\mathrm{grad}\theta, \tag{4.14}$$

$$\dot{e} = (1 - f_r)c\dot{\theta} + \frac{1}{\rho}\,\mathrm{tr}(\boldsymbol{\sigma}(\mathbf{D} - \mathbf{D}^p)), \tag{4.15}$$

where

$$\mathbf{T} = J\boldsymbol{\sigma}(\mathbf{F}^{-1})^T, \quad \mathbf{Q} = J\mathbf{F}^{-1}\mathbf{q}, \tag{4.16}$$

$$2\mathbf{W} = \mathrm{grad}\mathbf{v} - (\mathrm{grad}\mathbf{v})^T. \tag{4.17}$$

Equation (4.7) is the constitutive relation for a hypoelastic material with the left-hand side equal to the Jaumann derivative of the Cauchy stress tensor $\boldsymbol{\sigma}$, $\mathbf{W}$ defined by (4.17) is the skew-symmetric part of the velocity gradient, E, ν, and α, respectively, are the Young's modulus, Poisson's ratio, and the coefficient of thermal expansion for the matrix material, and $\mathbf{1}$ is the unit tensor. Note that we have the factor $(1-f)$ appearing on the right-hand side of (4.7) to account for the material damage caused by the porosity of the material; this factor has also been considered by Passman and Batra (1984) and Kobayashi and Dodd (1989); MacKenzie (1950) proposed a different factor. Budiansky (1990) has given the dependence of the material parameters upon f for a macroscopically isotropic composite of a random dispersion of roughly spherical holes in a matrix material. These relations are more involved than the simple reduction of E by the factor $(1-f)$ used in (4.7).

Equations (4.8) and (4.9) follow from the plastic yield function Φ proposed by Gurson (1977) and subsequently modified by Tvergaard (1981), and the assumptions that $\mathbf{D}^p$ is directed along the outward normal to Φ and the plastic working $\mathrm{tr}(\boldsymbol{\sigma}\mathbf{D}^p)$ equals $(1-f)\sigma_m \dot{\epsilon}_m^p$ with σ_m and $\dot{\epsilon}_m^p$ denoting the effective stress and the equivalent plastic strain rate in the matrix material. Expressions (4.11) for f^* were given by Tvergaard and Needleman (1984) so that the computed results matched well with the test findings for the cup-cone fracture in a round tensile bar. They suggest the values $f_c \approx 0.15$ and $f_f \approx 0.25$. As $f \to f_f$, $f^* \to f_u^*$, and the material loses all stress-carrying capacity.

Equation (4.12), relating the effective stress σ_m in the matrix to the equivalent plastic strain, equivalent plastic strain-rate, and the temperature, is a generalization due to Batra (1988) to the three-dimensional state of deformation of that proposed by Litonski (1977) for the simple shearing problem. In it, σ_0 equals the yield stress of the matrix material in a quasistatic simple compression test, the parameters b and m characterize the strain-rate sensitivity of the material, ϵ_y and n the strain-hardening, and β the thermal softening of the matrix material. The first term on the right-hand side of (4.13) describes the growth of voids due to plastic dilatation, and the second term describes the plastic strain-controlled nucleation of voids. Chu and Needleman (1980) suggested this form by assuming that void nucleation follows a normal distribution about some mean critical plastic strain. In (4.13), s_2 is the standard deviation of the normal distribution, f_2 equals the volume fraction of voids that would be nucleated if the deformation continued indefinitely, and e_N equals the strain at which the void nucleation rate achieves a maximum. The experimental studies of LeRoy (1978) and Fisher (1980) on spheroidized carbon steel indicate that, at least in these materials, a void perfusion strain can be identified at which the rate of nucleation is maximal. The void perfusion strain can be taken as e_N. Here the stress-controlled nucleation of voids has not been considered. Equation (4.14) is the Fourier law of heat conduction with **q** the heat flux per unit deformed area and k the thermal conductivity of the matrix. Budiansky (1990) has proposed that the thermal conductivity of the porous material equal $(1 - 3/2 \cdot f)$ times that of the matrix material. In the constitutive relation (4.15) for the rate of change of internal energy, c is the specific heat. Nearly all of the thermophysical material parameters depend upon the temperature; however, such dependencies have been neglected for the sake of simplicity.

The foregoing formulation of the problem follows closely that of Pan et al. (1983), and generalizes their work to include thermal softening, and the dependence of material parameters on the porosity. Perzyna's article in this volume gives an extensive discussion and development of a theory of microporous thermoviscoplastic solids. In order to solve a problem, equations (4.1) through (4.17) need to be supplemented with side conditions such as the initial and boundary conditions, and also various material parameters assigned values.

We analyse plane strain deformations of a prismatic body of square cross-section with all of its sides thermally insulated, the vertical edges traction free, and the lower and upper horizontal surfaces pulled apart at a uniform speed which increases from zero to the steady value v_0 in time t_r giving an eventual nominal strain-rate of v_0/H.

Initially the block is assumed to be at rest, stress-free and at a uniform temperature θ_0. However, the initial porosity is taken to be non-uniform and given by

$$f(X_1, X_2, 0) = f_0 + \varepsilon(1 - r^2)^9 e^{-5r^2},$$
$$r^2 = (X_1^2 + X_2^2)/H^2. \tag{4.18}$$

The second term on the right-hand side of (4.18_1) models a material defect or inhomogeneity, and the value of ε is a measure of the strength of the defect.

The aforestated coupled and nonlinear partial differential equations (4.1) through (4.14) under the prescribed initial and boundary conditions are solved numerically.

4.2 Computational Considerations

We rewrite (4.12) as

$$\dot{\varepsilon}_m^p = \max\left[0, \frac{1}{b}\left(\left(\frac{\sigma_m}{\sigma_0\left(\left(1 + \frac{\varepsilon_m^p}{\varepsilon_y}\right)^n (1 - \beta\theta)\right)}\right)^{\frac{1}{m}} - 1\right)\right]. \tag{4.19}$$

Thus the equivalent plastic strain-rate is positive only when

$$\sigma_m > \sigma_0 \left(1 + \frac{\varepsilon_m^p}{\varepsilon_y}\right)^n (1 - \beta\theta), \tag{4.20}$$

otherwise it equals zero implying thereby that all components of the plastic strain-rate tensor at the material point under consideration and at that instant vanish identically. The value of σ_m is computed from the yield function (4.8) once $\boldsymbol{\sigma}$ or $\boldsymbol{s}$ has been found.

Substitution from the constitutive relations into the balance laws yields evolution equations for ρ, $\mathbf{v}$ and θ which, when combined with (4.7), (4.13) and (4.19), give a system of equations for the determination of ρ, $\mathbf{v}$, θ, $\boldsymbol{\sigma}$, f and ε_m^p at a material point and at any instant of time. In the numerical solution of the problem, we employed the lumped mass matrix obtained by using the row sum technique, and evaluated various integrals over an element by using the 3-point quadrature rule. At each node point in the mesh there are 10 unknowns, namely, $\rho, \mathbf{v}$, θ, $\boldsymbol{\sigma}$, f and ε_m^p. Thus the number of ordinary differential equations equals 10 times the number of nodes. These are integrated with respect to time t by using the subroutine LSODE with variables ATOL and RTOL that control the absolute and relative errors in the solution vector each set equal to 10^{-6}. The initial discretization of the domain consisted of uniform 3-noded triangular elements, but subsequent meshes were refined adaptively by using the r-method discussed in Section 3 with $I = \dot{\varepsilon}_m^p$. The finite element mesh was refined whenever the porosity at the block centroid increased by a preassigned amount, and the computations were stopped when the porosity at any point in the domain reached the critical value f_f. Depending upon the initial distribution of f, atmost six mesh refinements had to be performed.

4.3 Numerical Results and Discussion

Numerical results have been computed with the following values of various geometric and material parameters which are for a typical steel.

$$\sigma_0 = 333 \text{ MPa}, \quad E = 210 \text{ GPa}, \quad \nu = 0.27, \quad \beta = 6.67 \times 10^{-4}/°\text{C},$$
$$\rho_r = 7800 \text{ kg/m}^3, \quad k = 49.2 \text{ W/m}°\text{C}, \quad c = 473 \text{ J/kg}°\text{C},$$
$$b = 10000 \text{ s}, \quad m = 0.025, \quad n = 0.02, \quad f_2 = 0.04, \quad s_2 = 0.1, \tag{4.21}$$
$$e_N = 0.5, \quad \varepsilon_y = 0.017, \quad \beta_1 = 1.5, \quad \beta_2 = 1.0, \quad f_c = 0.15,$$
$$f_f = 0.35, \quad f_u^* = 2/3, \quad H = 5 \text{ mm}, \quad t_r = 0.005H/v_0.$$

The value of β equals the reciprocal of the melting temperature, taken here to be 1500°C. For $v_0 = 25$ m/s, the nominal strain-rate equals 5000/s and the rise time for the axial speed at the top surface to reach its steady value equals one μs.

4.3.1 Effect of nominal strain-rate and initial porosity distribution

Figure 4.1 depicts the history of the axial load required to pull the bar at a speed of

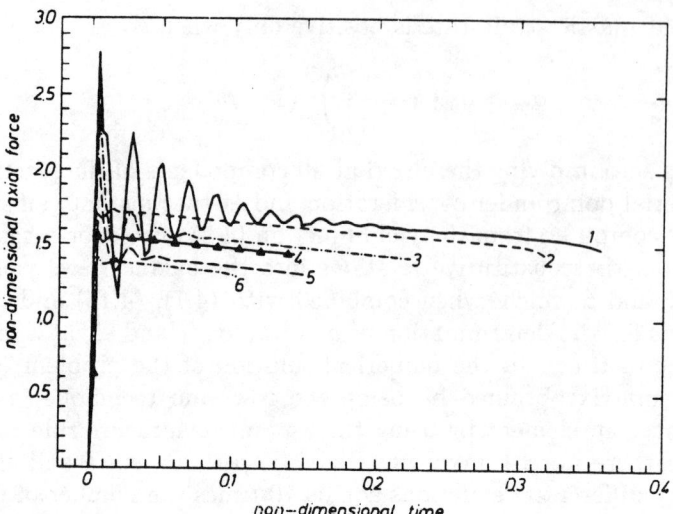

Figure 4.1: Axial load vs. time for three distributions of the initial porosity and at nominal strain-rates of $500s^{-1}$ and $5000s^{-1}$.

1— $f = 0.025(1 - r^2)^9\exp(-5r^2)$, $\quad v_0 = 25$ m/s,
2— $f = 0.025(1 - r^2)^9\exp(-5r^2)$, $\quad v_0 = 5$ m/s,
3— $f = 0.05 + 0.025(1 - r^2)^9\exp(-5r^2)$, $\quad v_0 = 25$ m/s,
4— $f = 0.05 + 0.025(1 - r^2)^9\exp(-5r^2)$, $\quad v_0 = 5$ m/s,
5— $f = 0.1 + 0.025(1 - r^2)^9\exp(-5r^2)$, $\quad v_0 = 25$ m/s,
6— $f = 0.1 + 0.025(1 - r^2)^9\exp(-5r^2)$, $\quad v_0 = 5$ m/s.

either 5 m/s or 25 m/s and for three values of the initial porosity. In each case the initial porosity f is nonuniform with its highest value occurring at the block centroid, and it quickly decreases to the uniform value. The oscillations of higher amplitude and longer duration at $v_0 = 25$ m/s are attributed to the predominance of inertia effects. At the lower value 5 m/s of v_0, after the initial rise in the load because of the increase in the axial speed, the load decreases gradually for each one of the three values of the initial porosity distribution. For a nonporous thermoviscoplastic material Batra (1988) pointed out that inertia forces start playing a noticeable role at a nominal strain-rate of 5000/s. Just when the axial load began to drop suddenly, the value of f at a node point adjacent to the block centroid reached f_f and the computations were stopped. Because of this we do not see in plots of Fig. 4.1 the precipitous drop in the load usually associated with the initiation of the localization of the deformation. As will be shown below, the deformation does localize into a narrow band.

Figure 4.2 shows the evolution of porosity at the block centroid for the three different distributions of the initial porosity and two values of the axial speed. The porosity evolves gradually first, and its rate of increase picks up substantially once the deformation has begun to localize. For the same value of $f(\mathbf{x}, 0)$, the increase in the nominal

Initial-Boundary-Value Problems with Shear Strain Localization

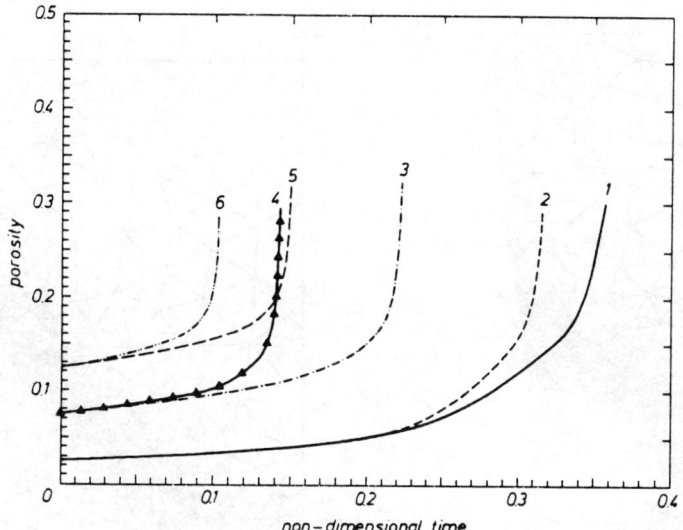

Figure 4.2: Evolution at the block centroid of the porosity for three distributions of the initial porosity and at nominal strain-rates of 500s^{-1} and 5000s^{-1}.

$$
\begin{array}{lll}
1- & f = 0.025(1-r^2)^9\exp(-5r^2), & v_0 = 25 \text{ m/s}, \\
2- & f = 0.025(1-r^2)^9\exp(-5r^2), & v_0 = 5 \text{ m/s}, \\
3- & f = 0.05 + 0.025(1-r^2)^9\exp(-5r^2), & v_0 = 25 \text{ m/s}, \\
4- & f = 0.05 + 0.025(1-r^2)^9\exp(-5r^2), & v_0 = 5 \text{ m/s}, \\
5- & f = 0.1 + 0.025(1-r^2)^9\exp(-5r^2), & v_0 = 25 \text{ m/s}, \\
6- & f = 0.1 + 0.025(1-r^2)^9\exp(-5r^2), & v_0 = 5 \text{ m/s}.
\end{array}
$$

strain-rate from 500/s to 5000/s delays the initiation of the localization of the deformation, primarily due to the effect of inertia forces. For a given nominal strain-rate of 500/s or 5000/s, an increase in the value of $f(\mathbf{x},0)$ causes the localization of the deformation to occur earlier. This is because a more porous material is plastically deformed at a lower value of σ_m, thus facilitating the growth and nucleation of voids and also causing the material to heat up sooner, both of which further enhance its plastic deformations. Hence it is a self-feeding mechanism.

In order to illustrate the localization of deformation and to depict how finite element meshes adapt to the deformations, results are shown below for

$$f(\mathbf{x},0) = 0.025(1-r^2)^9\exp(-5r^2), \qquad v_0 = 25 \text{ m/s}. \qquad (4.22)$$

Figures 4.3a and 4.3b show the finite element meshes generated at non-dimensional times $t = 0.339$ and 0.357 when the initial mesh consisted of 1600 uniform triangular elements. The finite element mesh was refined whenever the porosity f at the block centroid increased by 0.025, a criterion chosen arbitrarily. It is clear that a narrow region of material is deforming severely at times $t = 0.339$ and 0.357. As stated earlier, computations were stopped when the porosity at any point in the deforming region reached the critical value f_f. This implies failure of the material at a point

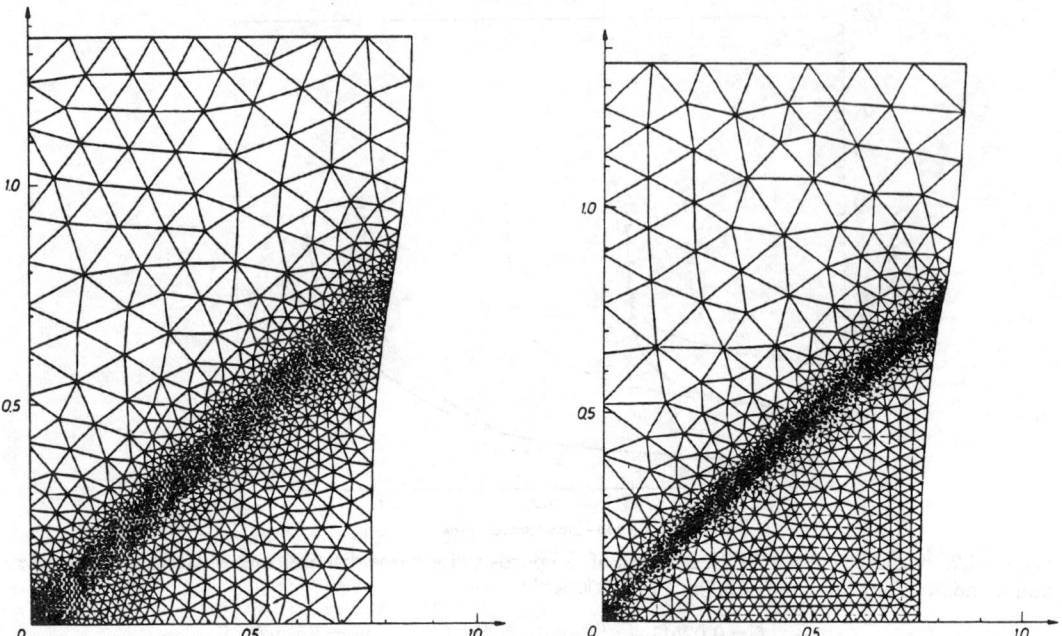

Figure 4.3: Adaptively refined meshes at non-dimensional times (a) $t = 0.339$ and (b) $t = 0.357$.

which does not necessarily result in the instantaneous failure of the block. Figure 4.4 evinces the distribution of the velocity within the deforming region at non-dimensional time $t = 0.357$. Initially, because of the lateral motion of the block, the velocity component in the horizontal direction has a significant value everywhere. However, once the deformation has started to localize, the body is essentially divided into three regions. In the region above the shear band, the material particles are moving nearly vertically, and those below the band - horizontally. Within the band, the velocity changes direction sharply. Contours of the porosity within the deforming region at time $t = 0.357$ are shown in Fig. 4.5.

4.3.2 Effect of strain-induced void nucleation

In order to delineate the effect of the softening caused by the strain-controlled nucleation of voids, we have plotted in Fig. 4.6 the evolution of the temperature and porosity at the block centroid for the two cases: (i) $f_2 = 0$, and (ii) $f_2 = 0.04$. It is apparent that the consideration of strain-induced void nucleation enhances the onset of the localization of the deformation, as shown by the sharp rise in the rate of increase of the temperature and the porosity at the block centroid.

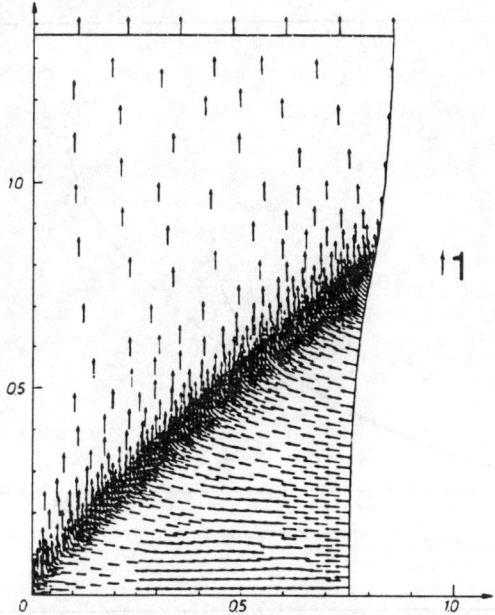

Figure 4.4: Velocity field at nondimensional time $t = 0.357$

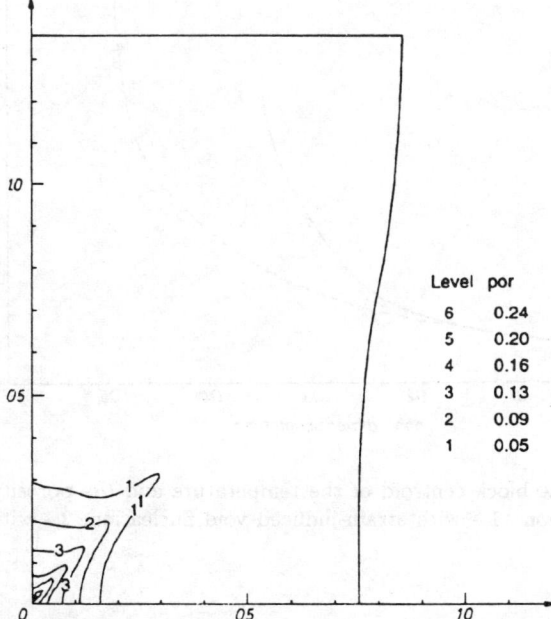

Figure 4.5: Contours of the porosity at non-dimensional time $t = 0.357$.

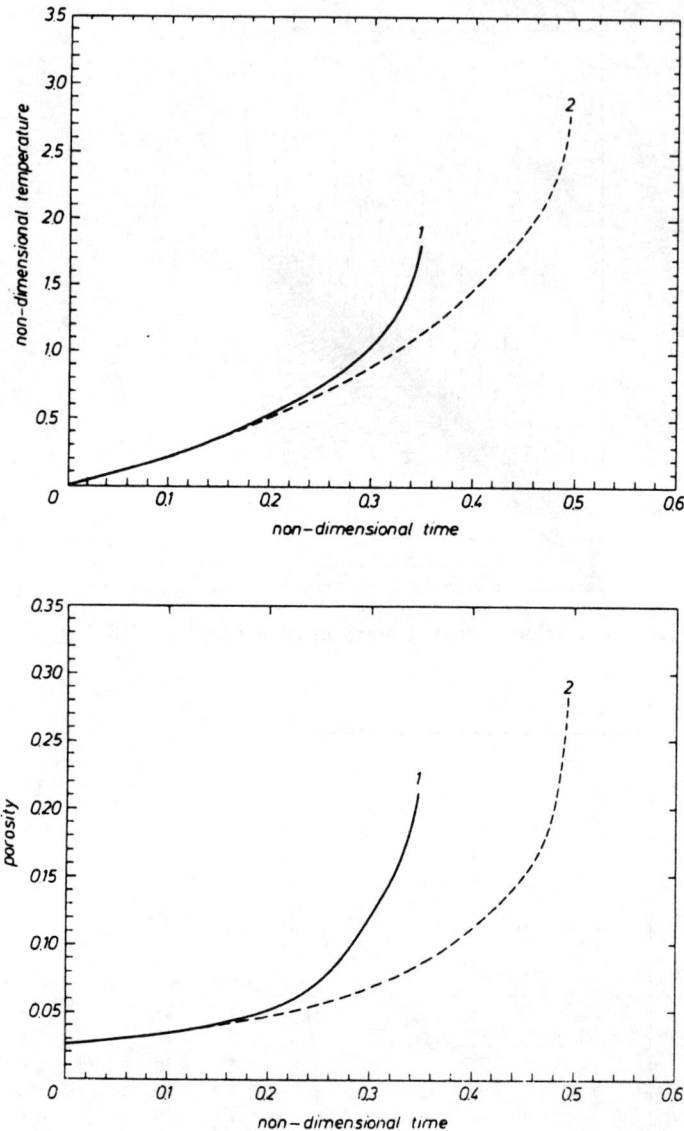

Figure 4.6: Evolution at the block centroid of the temperature and the porosity with and without strain-induced void nucleation. 1 - with strain-induced void nucleation, 2 - without strain-induced void nucleation.

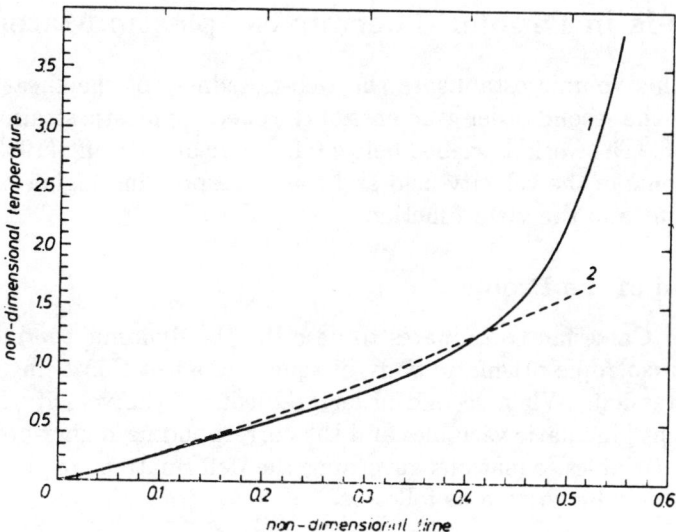

Figure 4.7: Variation of the temperature at the block centroid vs. time, 1- no thermal softening, 2- thermal softening only

4.3.3 Thermal softening vs. softening due to porosity change

For the choice (4.21) of parameters we assessed the effect of softening caused by the rise in temperature versus that induced due to the increase in porosity by performing two sets of calculations, one with $\beta = 0$ and the other with $f(\mathbf{x}, t) \equiv 0$. The material defect for these computations was modelled by assuming that the yield stress $\hat{\sigma}_0$ in a quasistatic simple compression test was given by

$$\hat{\sigma}_0(\mathbf{X}) = \begin{cases} \sigma_0 \left(1 - 0.1(1 - r^2)^9 e^{-5r^2}\right), & r^2 \equiv (X_1^2 + X_2^2)/H^2 \leq 1, \\ \sigma_0, & r \geq 1. \end{cases}$$

Figure 4.7 exhibits the evolution of the temperature rise at the block centroid for the two cases with $v_0 = 25$ m/s. It is clear that for the parameters considered herein, the softening due to the increase in porosity is considerably higher than that caused by the rise in the temperature. Whereas a shear band initiates, as indicated by the rise in the rate of increase of the temperature at the block centroid at non-dimensional time $t \simeq 0.5$ when softening is caused by the change in porosity, no shear band forms till a non-dimensional time of 0.5 when the softening is induced by the temperature rise, since both the effective plastic strain and the temperature at the block centroid increase essentially linearly with time. It is possible that a band will initiate at a later time. Thus for values of material parameters chosen for this study, softening caused by the growth and nucleation of voids is stronger than that induced by the temperature rise.

5 Shear Bands in Dipolar Thermoviscoplastic Materials

Sluys's article in this volume establishes the well-posedness of the shear localization problem when only the second order gradients of the effective plastic strain are included in the yield function. The work described below (cf. Batra and Hwang (1993)) considers second order gradients of the velocity field and the corresponding higher-order stresses in the field equations and the yield function.

5.1 Formulation of the Problem

We use rectangular Cartesian coordinates to describe the dynamic thermomechanical deformations of an isotropic prismatic body of square cross-section being deformed in plane strain compression. When second order gradients of the velocity field are also taken as independent kinematic variables and the corresponding higher-order (dipolar) stresses as kinetic variables, equations governing the deformations of the body in the spatial description may be written as follows.

Balance of mass: $v_{i,i} = 0$, (5.1)

Balance of linear momentum: $\rho \dot{v}_i = \sigma_{ij,j} - \tau_{ijk,jk}$, (5.2)

Balance of internal energy: $\rho \dot{e} = -q_{i,i} + \sigma_{ij} v_{i,j} + \tau_{ijk} v_{i,jk}$. (5.3)

Here τ is the dipolar stress tensor, and ϵ_{ijk} is the permutation symbol with

$$\epsilon_{ijk} = \begin{cases} 1, & \text{according as } i, j, k \text{ form an even permutation of } 1, 2, \text{ and } 3, \\ -1, & \text{according as } i, j, k \text{ form an odd permutation of } 1, 2, \text{ and } 3, \\ 0, & \text{if any two of the three indices are equal.} \end{cases} \quad (5.4)$$

We refer the reader to Mindlin (1965), Toupin (1962), Green et al. (1968), and Dillon and Kratochvil (1970) for motivation and derivation of these equations. Here we have followed an approach similar to that of Mindlin, and have assumed that the deformations of the body are isochoric, and the supplies of linear momentum and internal energy are null. Since our interest is in studying the intense plastic deformations, we neglect elastic deformations. We postulate that

$$\tau_{i[jk]} \equiv \tfrac{1}{2}(\tau_{ijk} - \tau_{ikj}) = 0 \quad (5.5)$$

and note that $\tau_{i[jk]}$ contributes nothing to the balance of linear momentum and the balance of internal energy. Green et al. (1968) made a similar assumption; our reason for assuming (5.5) is to avoid finding constitutive relations for $\tau_{i[jk]}$ and keeping the formulation of the problem simple. A closer look at (5.2) reveals that $(\sigma_{ij} - \tau_{ijk,k})$ equals the flux of linear momentum.

For the one-dimensional theory of dipolar materials, Wright and Batra (1987) provided some motivation for the assumption that the plastic strain-rate and plastic part of the dipolar strain-rate equal the corresponding stresses multiplied by the same plastic multiplier. Here we make a similar postulate, viz.,

$$v_{i,j} = \Lambda s_{ij}, \tag{5.6}$$

$$A_{ijk} = A_{ikj} = v_{i,jk} = \frac{\Lambda}{l^2}\tau_{ijk}, \tag{5.7}$$

where

$$s_{ij} = \sigma_{ij} + p\delta_{ij}, \tag{5.8}$$

p being the hydrostatic pressure not determined by the deformation of the body, and l is a material characteristic length. As in classical plasticity, a scalar yield or loading function f is assumed to exist such that

$$f(\mathbf{s}, \boldsymbol{\tau}, \theta, \mathbf{D}, \mathbf{A}) = \kappa \tag{5.9}$$

where κ describes the work hardening of the material. In (5.9), θ is the temperature rise of a material particle. We assume that plastic flow occurs for every value of $\mathbf{s}, \boldsymbol{\tau}$, and θ, and find Λ from

$$f(\mathbf{s}, \boldsymbol{\tau}, \theta, \Lambda\mathbf{s}, \Lambda\boldsymbol{\tau}) = \kappa. \tag{5.10}$$

The loading function f is such that the derivative f_Λ is negative for all values of other arguments. It ensures that (5.10) will have a unique solution with $\Lambda > 0$. Motivated by the von Mises yield criterion in classical plasticity and the one-dimensional dipolar theory of Wright and Batra (1987), we select f and κ as follows.

$$f = \frac{s_e}{(1 + b\Lambda s_e)^m(1 - \alpha(\theta - \theta_o))}, \tag{5.11}$$

$$\kappa(\psi) = \sigma_0(1 + \frac{\psi}{\psi_0})^n, \tag{5.12}$$

$$\dot{\psi} = \frac{\Lambda s_e^2}{\sigma_0(1 + \frac{\psi}{\psi_0})^n}, \quad s_e^2 = s_{ij}s_{ij} + \frac{1}{\ell^2}\tau_{i(jk)}\tau_{i(jk)}. \tag{5.13}$$

We presume that the initial values of θ and ψ are symmetric and those of v_1 and v_2 are antisymmetric in x_1 and x_2, and seek solutions with the same symmetries. Thus, the problem is studied over the spatial domain $[0, b] \times [0, h]$ under the boundary conditions

$$\begin{aligned} &v_1(0, x_2, t) = 0, & &\theta_{,1}(0, x_2, t) = 0, & &(\sigma_{21} - \tau_{21k,k})|_{(0,x_2,t)} = 0, \\ &v_2(x_1, 0, t) = 0, & &\theta_{,2}(x_1, 0, t) = 0, & &(\sigma_{12} - \tau_{12k,k})|_{(x_1,0,t)} = 0, \\ &v_2(x_1, h, t) = v_0, & &\theta_{,2}(x_1, h, t) = 0, & &(\sigma_{12} - \tau_{12k,k})|_{(x_1,h,t)} = 0, \\ &n_j\theta_{,j}(b, x_2, t) = 0, & &n_j(\sigma_{ij} - \tau_{ijk,k})|_{(b,x_2,t)} = 0. \end{aligned} \tag{5.14}$$

For the initial conditions, we take

$$\begin{aligned} &v_1(x_1, x_2, 0) = x_1, \quad v_2(x_1, x_2, 0) = -x_2, \\ &\theta(x_1, x_2, 0) = \theta_0 + \epsilon(1 - r^2)^9 e^{-5r^2}, r^2 = x_1^2 + x_2^2, \quad r \leq 1, \\ &\quad = \theta_0, \ r > 1. \end{aligned} \tag{5.15}$$

Here b and h equal, in the present configuration, the length of the base and the height of the quarter of the block, and **n** is an outward unit normal. The boundary conditions (5.14) imply that the boundaries of the block are thermally insulated, the right surface is free of flux of linear momentum, there is no flux of linear momentum in the tangential direction on the other three bounding surfaces, and the normal component of velocity on the left and bottom surfaces vanishes. The boundary conditions on the left and bottom surfaces follow from the assumed symmetry of the deformation field. The initial conditions on the velocity field represent the situation when the transients have died out. This assumption is justified because it significantly reduces the CPU effort required to solve the problem and does not affect the qualitative nature of computed results. The initial temperature distribution given by $(5.15)_2$ models a material inhomogeneity; the amplitude ϵ of the perturbation can be thought of as representing the strength of the singularity.

5.2 Computational Considerations

Since second order spatial derivatives of the dipolar stress τ appear in (5.2) we introduce auxiliary variables **f** by

$$f_{ij} = \tau_{ijk,k} , \qquad (5.16)$$

and use the Galerkin approximation to derive the weak form of (5.1)-(5.3). The auxiliary variables f are eliminated at the element level, and nodal values of **s** and τ are expressed in terms of the nodal values of **D** and **A** by using (5.6) and (5.7). By using $(5.6)_1$ and the intermediate variable $g_{ij} \equiv v_{i,j}$, we express nodal values of **D** and **A** in terms of nodal values of **v**. The reason for introducing auxiliary or intermediate variables is to have, atmost, first order derivatives of various field quantities in the weak formulation. It enables us to select test functions and trial solutions from the space H^1 of functions, which includes functions defined on the domain of interest and whose first order derivatives are square integrable. The disadvantage of having auxiliary and/or intermediate variables is that the number of unknowns becomes very large. It is obviated somewhat by eliminating these auxiliary variables before integrating the ordinary differential equations with respect to time t. We use four-noded quadrilateral elements, compute the lumped mass matrix by using the special lumping technique, regard the pressure to be constant within each element, and employ 2×2 Gauss quadrature rule to evaluate various integrals over an element. This results in a set of coupled highly nonlinear ordinary differential equations (ODEs) for nodal values of the two components of the velocity, temperature, and the internal variable ψ, and a set of algebraic equations for values of the hydrostatic pressure at the element centroids. Thus, the number of ODEs equals four times the number of nodes, and the number of algebraic equations equals the number of elements in the mesh.

The ODEs are integrated with respect to time t by using the trapezoidal rule (Hughes (1987)), which is a member of the Newmark family of methods. For linear problems, the method is implicit, second order accurate, and unconditionally stable. For the nonlinear problem studied herein, the time step had to be controlled to achieve

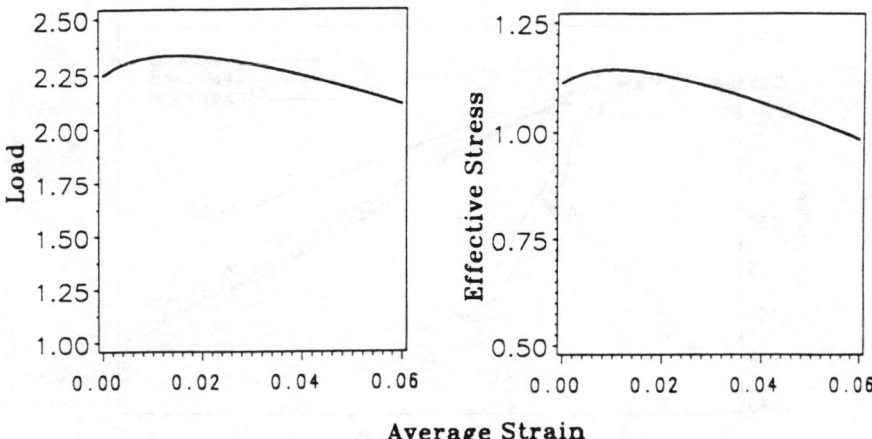

Figure 5.1: Vertical load and effective stress versus the average strain in the vertical direction for homogeneous deformations of the block.

stability. The application of the trapezoidal rule results in a system of coupled nonlinear algebraic equations, which are solved iteratively. The iterative process is stopped when, at each node point, either

$$\frac{|\Delta v_1|}{|v_1|} + \frac{|\Delta v_2|}{|v_2|} + \frac{|\Delta \theta|}{\theta} + \frac{|\Delta \psi|}{\psi} \leq \epsilon_1, \text{ or} \tag{5.17}$$

$$|\Delta v_1| + |\Delta v_2| + |\Delta \theta| + |\Delta \psi| \leq \epsilon_2, \tag{5.18}$$

where ϵ_1 and ϵ_2 are preassigned small numbers, and $\Delta \theta$ denotes the difference in the nodal values of θ during two successive iterations within the same time increment. A reason for applying either criterion (5.17) or (5.18) is that, at nodes on the boundary where essential boundary conditions are prescribed to be zero, criterion (5.17) is meaningless.

A few trial runs indicated that non-dimensional $\Delta t = 5 \times 10^{-5}$ was a good starting step size. The time step size was reduced by a factor of 0.7 chosen by trial and error, every time the convergence criterion failed. In a typical run, Δt had to be reduced ten times.

For the one-dimensional problem, the effects of material elasticity are included and the approach followed is similar to that for nonpolar materials described in Section 2. Also, the ODES are integrated by using LSODE.

5.3 Discussion of Results

5.3.1 Plane strain problems

Figure 5.1 depicts the vertical load and the effective stress s_e versus the average strain in the vertical direction when the block is deformed homogeneously, i.e., with no initial temperature perturbation introduced. The vertical load P on the top surface is given

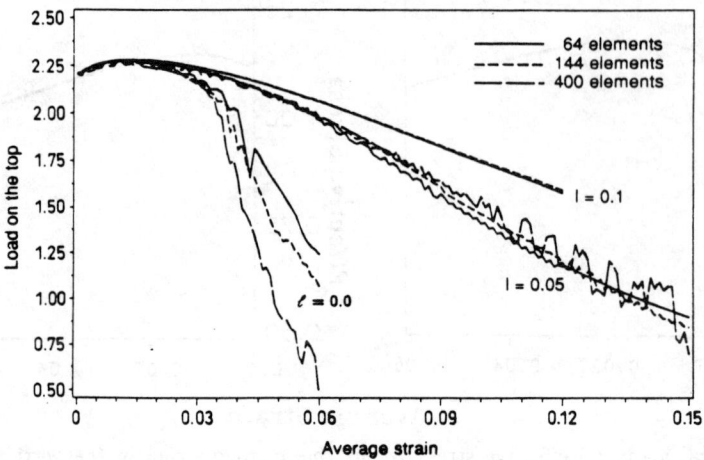

Figure 5.2: Vertical load versus the average strain in the vertical direction for non-dimensional $l = 0$, 0.05, and 0.1, and three different finite element meshes of 64, 144 and 400 uniform elements in the reference configuration.

by

$$P = -\int_0^1 (\sigma_{22}(x_1, h, t) - \tau_{22k,k}(x_1, h, t))dx_1 \qquad (5.19)$$

where the negative sign in front of the integral is to get a positive value of P. For homogeneous deformations of the block, the dipolar stress vanishes identically. In order to evaluate P, we need to find the value of the hydrostatic pressure p at points on the top surface. Since p is assumed to be constant within each element, its values at the node points are computed by using the following smoothening technique.

$$\sum_\beta \left(\int_\Omega \Phi_\alpha \Phi_\beta d\Omega \right) p_\beta = \sum_e \int_{\Omega_e} \Phi_\alpha p d\Omega. \qquad (5.20)$$

Here $\{\Phi_\alpha, \alpha = 1, 2, \ldots\}$ is the set of piecewise linear finite element bases functions defined on Ω, $p_1, p_2, \ldots$ are nodal values of the hydrostatic pressure, and p on the right hand side of (5.20) is the piecewise constant pressure field computed as a solution of the problem. It is obvious from the plot of Fig. 5.1 that the peak in the load occurs at an average strain of 0.012, and beyond this value of the average strain, the softening caused by the heating of the material exceeds the hardening due to the strain and strain-rate effects. The difference between the magnitude of the vertical load and the effective stress is due to the hydrostatic pressure; our definition of the effective stress differs from the usual one by a constant factor.

In Fig. 5.2, we have plotted the vertical load versus the average strain curves for nondimensional $l = 0, 0.05$, and 0.1 when there is a temperature perturbation introduced. The results were computed with three initial meshes having eight, twelve, or

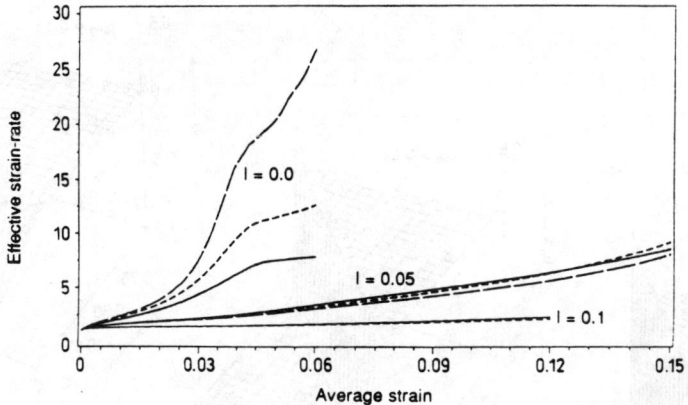

Figure 5.3: The evolution of the effective strain-rate at the block center for three different meshes and three different values of l; ——— 64 elements, - - - - - 144 elements, —— - 400 elements.

twenty uniform elements in both the horizontal and vertical directions. The coordinates of the node points are updated after each time increment so that once the block begins to deform nonhomogeneously, the finite element mesh becomes nonuniform. The plotted results reveal that, for nonpolar materials with $l = 0.0$, the load drops severely soon after its peak occurs. At an average strain of 0.06, the load has dropped to nearly half of its peak value for the 64-element mesh and to 22% of the peak value for the 400-element mesh. The results computed with the 144-element and 400-element meshes are smoother and the drop in the load is more than that obtained by using the 64-element mesh. For $l = 0.05$ and $l = 0.1$, the drop in the load is less rapid and the rate of the load drop with increasing average strain decreases with an increase in the value of l. Also, the dependence of the computed value of the load upon the mesh used decreases with an increase in the value of l, the results for the three meshes used being essentially identical for $l = 0.1$. At an average strain of 0.06 and for the 144 element mesh, the vertical load has dropped to 1.06, 2.0 and 2.05 from a peak of 2.275 for $l = 0.0, 0.05,$ and 0.1, respectively.

The evolution of the effective strain-rate I_e at the block center is depicted in Fig. 5.3. The finer meshes of 144 and 400 elements give sharper results than those obtained with 64 elements in the sense that the effective strain-rate increases more rapidly, the temperature rise is more, and the drop in the effective stress is more severe. For $l = 0.05$ and 0.1, the differences in the solution variables obtained with different meshes become minuscule enough to conclude that the results are independent of the mesh used for $l = 0.1$. The consideration of dipolar effects does not alter the qualitative nature of computed results, except that the temperature rise, the drop in the effective stress, and the evolution of the effective strain-rate at the block center for a fixed value of the average strain are higher for nonpolar materials than that for dipolar materials.

The distribution of v_1 and v_2 within the domain at $\gamma_{\text{avg}} = 0.06$ for $l = 0$ and 0.01, and at $\gamma_{\text{avg}} = 0.15$ for $l = 0.05$ plotted in Fig. 5.4, wherein the distribution of only v_2 is shown, indicates that the deforming region is divided into two parts essentially

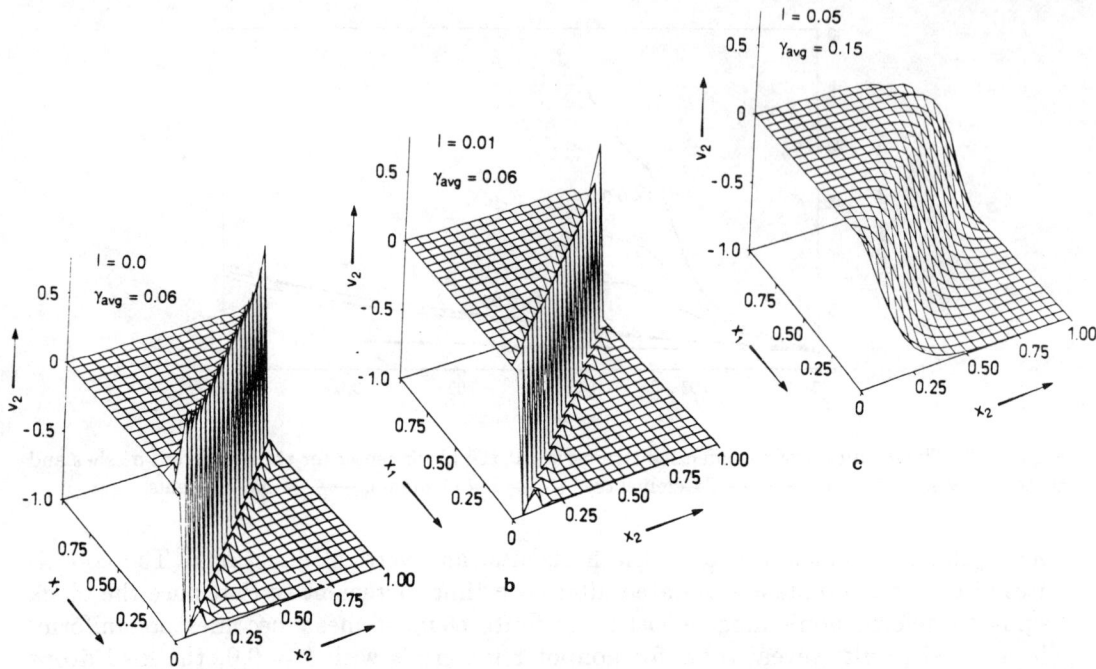

Figure 5.4: Distribution of v_2 within the deforming region at $\gamma_{\text{avg}} = 0.06$ for $l = 0$ and 0.01, and at $\gamma_{\text{avg}} = 0.15$ for $l = 0.05$.

separated along the diagonal passing through the block center. Each region is moving as a rigid body with all of the deformations concentrated in the narrow region separating the two parts. The plotted velocity field supports Massey's (1921) assertion that the tangential velocity field is discontinuous across the shear band. In our computations, the velocity field is assumed to be continuous. However, the sharp jumps in the values of v_1 and v_2 across the narrow region lend credence to Massey's proposal.

5.3.2 Simple shearing problems

For the one-dimensional (simple shearing) problem (e.g. see Section 2) sharper results can be obtained; results reported here are from Batra and Kim (1990, 1988). Computations were stopped when the shear stress, s, at any point became zero. Note that the flux of linear momentum, due to the contribution from the dipolar stress, need not vanish there. Values of average strain, $\gamma_{avg}, y_s = y-$ coordinate of the point where s first becomes zero, $\dot{\gamma}_{pmax}=$ maximum plastic strain-rate, $\theta_{max} =$ maximum temperature, $|\tau|_{max} =$ maximum value of the dipolar stress, $y_\tau =$ location of the point of the maximum value of the dipolar stress, and band width = thickness of the severely deforming region defined as being equal to twice the distance of the point from the center beyond which the effective plastic strain-rate is one-thousandth of its peak value, are given in the Table for different values of l nondimensionalized by the height of the

Initial-Boundary-Value Problems with Shear Strain Localization

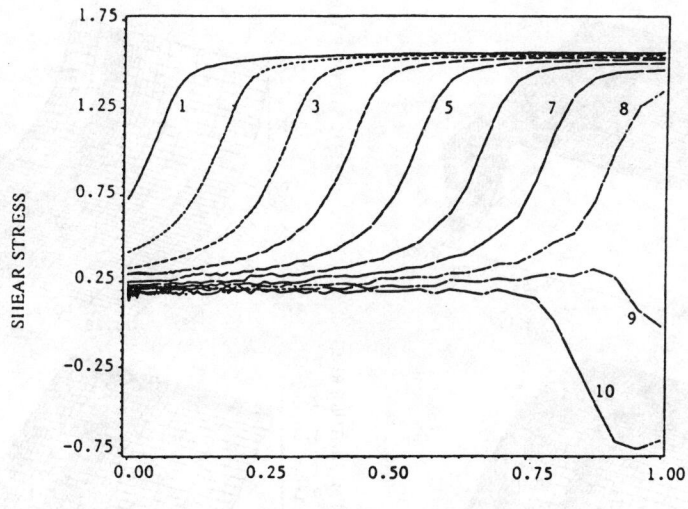

Figure 5.5: Distribution of the shear stress within the specimen at different times during the localization of the deformation for nonpolar materials. These curves are plotted at intervals of 0.1 μs with curve 1 at $t = 64.0$ μs, curve 2 at $t = 64.1$ μs, ..., and curve 10 at $t = 64.9$ μs.

Table: Dependence of Solution Variables Upon ℓ

ℓ	γ_{avg} (%)	y_s (μm)	$\dot{\gamma}_{pmax}$	$\alpha(\theta_{max})$	$|\tau|_{max}$	y_τ (μm)	$d(\mu m)$
0.01	16.42	41.35	139	0.952	0.989	153.5	66.4
0.005	13.50	22.54	347	0.960	0.953	80.6	29.6
0.001	10.40	2.22	19,516	0.983	0.931	16.1	3.9
0.0005	10.23	0.59	99,606	0.994	0.899	8.5	1.0

block. It is clear that the band width and $|\tau|_{max}$ strongly depend upon l, and decrease rapidly with a decrease in the value of l, but $\dot{\gamma}_{pmax}$ increases sharply.

Figure 5.5 taken from Batra and Kim (1990) depicts for nonpolar materials (i.e. $l = 0$) the distribution of the shear stress within a steel specimen at intervals of 0.1 μs starting with the time when the deformation begins to localize. It is clear that an unloading elastic shear wave emanates outwards from the region of severe deformation. The emanation of the elastic unloading wave is probably associated with the sudden collapse of the shear stress within the band. The computed speed, 3178 m/s, of the wave essentially equals $(\mu/\rho)^{1/2} = 3,190$ m/s. It takes 0.807 μs for the shear wave to reach the outer boundary from which it is reflected back with a negative value of the shear stress. The numerical calculations were not pursued any further.

In Fig. 5.6 is plotted for the same steel specimen but the steel modeled as a dipolar material, with $l = 0.01$ the evolution of the shear stress s, the dipolar stress σ, the temperature change θ and the plastic strain-rate $\dot{\gamma}_p$. Now the shear stress drops gradually rather than suddenly, no unloading wave emanates out of the localization region, and the plastic strain rate does not attain the enormously high values it achieved for

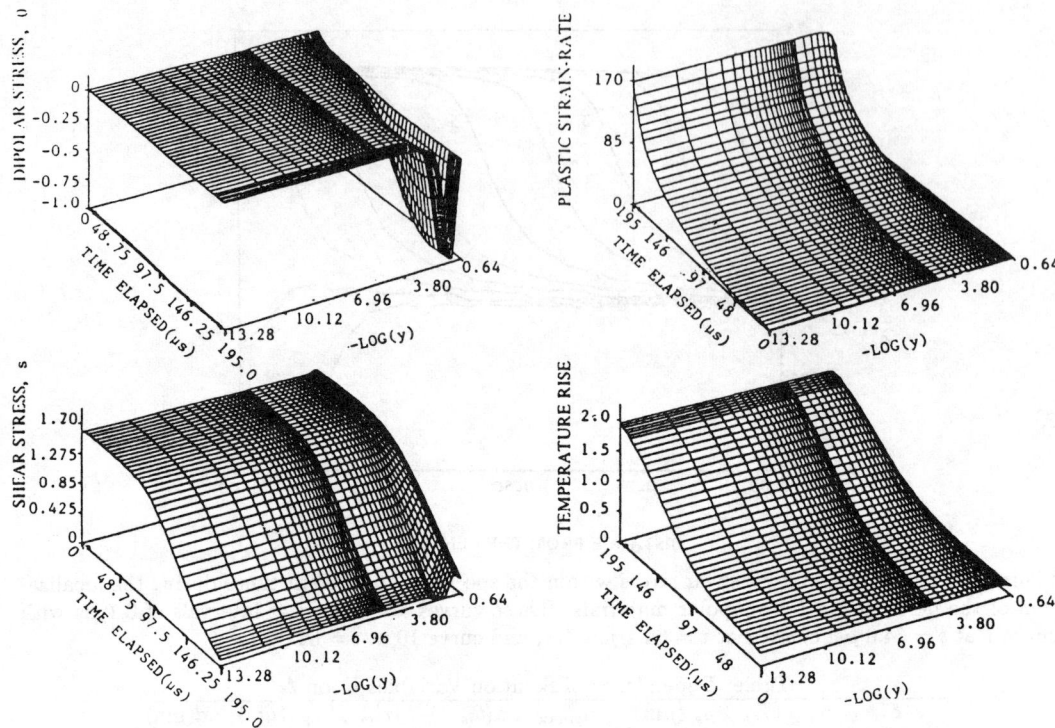

Figure 5.6: Evolution of the shear stress, the dipolar stress, temperature rise and plastic strain-rate at points near the center of the specimen for dipolar materials with $l = 0.01$.

nonpolar materials. Also the localization of the deformation is delayed considerably as compared to that for nonpolar materials. At points where the magnitude of the gradient of the dipolar stress is maximum, the shear stress attains minimum values. Additional results are given in Batra and Kim (1990).

6 Adiabatic Shear Bands in Axisymmetric Impact and Penetration Problems

Batra and Stevens (1997) have studied axisymmetric deformations of depleted uranium (DU) and tungsten heavy alloy (WHA) rods impacting at normal incidence either a flat rigid or a thick steel target with the objective of ascertaining when and where a shear band forms in each material. The Taylor (1948) impact simulations show that shear bands form earlier in WHA than in DU for values of material parameters used here. In the penetration simulations, shear bands form continuously in the ejecta of the DU penetrator but only one shear band forms in the WHA ejecta followed by its more uniform deformations. Some of Batra and Stevens's results are discussed below.

6.1 Formulation of the Problem

The Lagrangian or referential description of motion is used to describe dynamic, thermomechanical and axisymmetric deformations of a cylindrical rod impacting at normal incidence a target; the deformations are governed by (1.1)-(1.3). Here the effect of heat conduction is neglected and thus deformations are assumed to be locally adiabatic. This assumption facilitates the computation of the temperature rise from the incremental plastic work done without numerical integration of the balance of internal energy. Equations (1.1)-(1.3) are supplemented with the following constitutive relations.

$$\mathbf{T} = J\boldsymbol{\sigma}(\mathbf{F}^{-1})^T, \quad \boldsymbol{\sigma} = -p\mathbf{1} + \mathbf{S}, \quad p = K(\rho/\rho_0 - 1), \tag{6.1}$$

$$\overset{\nabla}{\mathbf{S}} = 2\mu(\overline{\mathbf{D}} - \overline{\mathbf{D}}^p), \tag{6.2}$$

$$\overline{\mathbf{D}} = \mathbf{D} - \frac{1}{3}(tr\,\mathbf{D})\mathbf{1}, \quad tr\,\mathbf{D}^p = 0, \quad \mathbf{D}^p = \Lambda\mathbf{S}, \quad S_e^2 \equiv \frac{3}{2}tr(\mathbf{SS}^T), \tag{6.3}$$

$$\dot{e} = c\dot{\theta} + tr(\boldsymbol{\sigma}\mathbf{D}^e), \quad \sigma_y = (A + B(\epsilon^p)^n)(1 + C\,ln(\dot{\epsilon}^p/\dot{\epsilon}_0))(1 - T^m), \tag{6.4}$$

$$T = (\theta - \theta_0)/(\theta_m - \theta_0), \quad (\dot{\epsilon}^p)^2 = \frac{2}{3}tr(\mathbf{D}^p\mathbf{D}^p). \tag{6.5}$$

Here $\mathbf{S}$ is the deviatoric part of the Cauchy stress tensor $\boldsymbol{\sigma}$, p the hydrostatic pressure taken to be positive in compression, K the bulk modulus, $\overline{\mathbf{D}}$ the deviatoric strain-rate tensor, S_e the effective stress, and ϵ^p the effective plastic strain; other variables have been defined before. Equation (6.1)$_3$ implies that the volumetric response of the material is elastic, and (6.2) is the constitutive relation in terms of deviatoric stresses for a linear, isotropic hypoelastic material, (6.3)$_4$ signifies the von Mises yield criterion with isotropic hardening, and (6.4)$_2$ is the Johnson-Cook relation. Truesdell and Noll (1965) have pointed out that (6.2) is not invariant with respect to the choice of different objective (or material frame indifferent) time derivatives of the stress tensor. Equation (6.3)$_3$ signifies that the plastic strain-rate is along the normal to the yield surface (6.3)$_4$, and the factor of proportionality Λ is given by

$$\Lambda = 0 \text{ when either } S_e < \sigma_y, \text{ or } S_e = \sigma_y \text{ and } tr(\mathbf{S}\overset{\nabla}{\mathbf{S}}) < 0; \tag{6.6}$$

otherwise it is a solution of

$$S_e = (A + B(\epsilon^p)^n)\left(1 + C\,ln\left(\frac{2}{3}\Lambda S_e/\dot{\epsilon}_0\right)\right)(1 - T^m). \tag{6.7}$$

Once $\theta = \theta_m$ at a material point, its flow stress, σ_y, is set equal to zero, and it behaves like a compressible, nonviscous fluid. In physical experiments, fracture in the form of a crack will ensue from the point much before it is heated up to its melting temperature. Here we have not incorporated any fracture criterion into the problem formulation. Because a Lagrangian formulation is used and a perfect fluid can not support shear stresses, once a material point melts, the mesh will be distorted quickly and the computations must be stopped.

Zhou et al. (1994) found that the thermal softening of the WHA they tested is better described by $(1 - \beta(-1 + (\theta/\theta_0)^\alpha))$ where α and β are material parameters and for their WHA, $\beta = 2.4$ and $\alpha = 0.2$. The effect of replacing $(1 - T^m)$ in $(6.4)_2$ by Zhou et al.'s expression is investigated.

Initially, the cylindrical rod is stress free, at room temperature θ_0 and is moving with a uniform speed V_0 in a direction normal to the plane surface of the target, and strikes it at time $t = 0$. All bounding surfaces of the rod except that contacting the target are taken to be traction free. Because of the assumption of locally adiabatic deformations, no thermal boundary conditions are needed.

In the Taylor impact test, the target is taken as rigid and stationary, and the contact surfaces smooth. In the penetration problem, the deformable steel target is initially at rest, at room temperature θ_0 and is stress free. It is unconstrained and all of its bounding surfaces except the one contacting the penetrator are taken to be traction free. On the target/penetrator interface, the normal component of relative velocity between the target and penetrator vanishes, and

$$\mathbf{f}_t = -|f_n|(\mu_s + (\mu_k - \mu_s)e^{-\gamma \bar{v}})\bar{\mathbf{v}}/|\bar{\mathbf{v}}|, \tag{6.8}$$

where, $\mathbf{f}_t$ is the tangential traction, $\bar{\mathbf{v}}$ the relative velocity between the target and the penetrator, f_n the normal traction at a point on the target/penetrator interface, μ_s and μ_k the static and kinetic coefficients of friction respectively, and γ describes the dependence of the coefficient of friction upon the relative speed of sliding between the two contacting surfaces.

6.2 Computation and Discussion of Results

An approximate solution of the problem is obtained by the finite element large scale explicit code DYNA2D. In the weak formulation of the problem, the bases functions for the test functions equal those for the trial solution divided by the radial coordinate. The code uses 4-noded quadrilateral elements, lumped mass matrix, one-point integration rule, an hour-glass control to suppress the spurious or zero energy modes, and the central-difference method to integrate the coupled ordinary differential equations. The time step is adjusted adaptively and equals a fraction of the time taken for an elastic wave to travel through the smallest element in the mesh. As the bodies deform, elements near the target/penetrator interface become severely distorted and the time step size drops drastically. The time step is also affected by the restoring force to be applied to the interpenetrating nodes at the target/penetrator interface. In the symmetric penalty method used to enforce the continuity of velocity components normal to the interface, this force is proportional to the depth of interpenetration, the bulk modulus of the penetrated element, that element's dimensions and a user defined scale factor. Once the time step size has become too small, the deforming region needs to be rezoned for computations to continue at a reasonable pace.

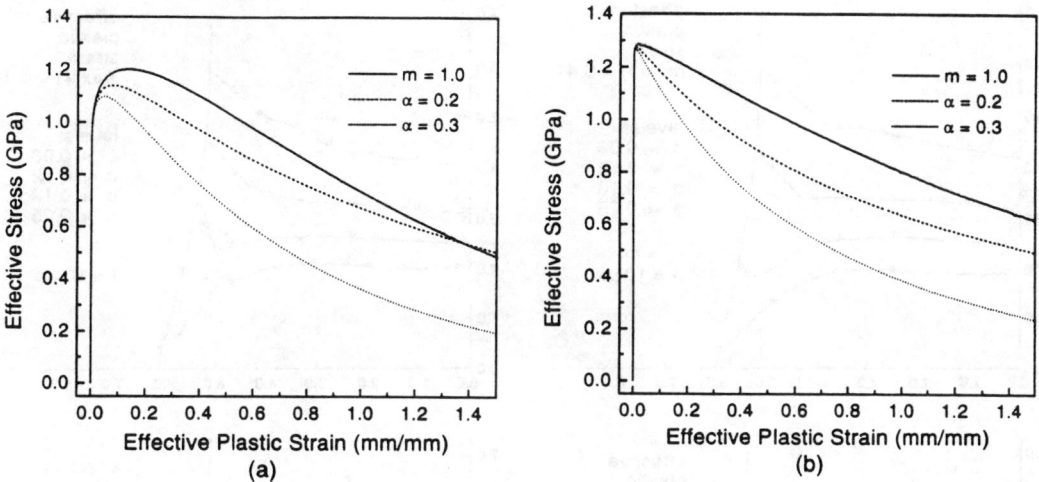

Figure 6.1: Effective stress vs. effective strain during axisymmetric compression of homogeneous WHA and DU cylinders deformed at a nonimal strain-rate of 5000/s.

6.2.1 Results for the Taylor Impact test

6.2.1.1 Results for the Johnson-Cook thermal softening function

Various material and geometric parameters were assigned the following values.

Rod length = 60 mm, Rod diameter = 10 mm, $V_0 = 150$ m/s, $\theta_0 = 293$ K.

Depleted Uranium (DU):

$A = 1,079$ MPa, $B = 1,120$ MPa, $C = 0.007$, $n = 0.25$, $m = 1.0$, $\dot{\epsilon}_0 = 1$/s,
$\rho = 18,600$ kg/m^3, $\mu = 58$ GPa, $K = 119$ GPa, $c = 117$ J/kg°C, $\theta_m = 1473$ K.
(6.9)

Tungsten Heavy Alloy (WHA):

$A = 1,506$ MPa, $B = 177$ MPa, $C = 0.016$, $n = 0.12$, $m = 1.0$, $\dot{\epsilon}_0 = 1$/s,
$\rho = 18,600$ kg/m^3, $\mu = 160$ GPa, $K = 328$ GPa, $c = 134$ J/kg°C, $\theta_m = 1723$K.

The values of A, B, C, θ_m, m, $\dot{\epsilon}_0$ and n for both materials are taken from Rajendran's report (1992); those of other material parameters are taken from a handbook. Figure 6.1 depicts the effective stress vs. the effective strain curves for axisymmetric compression of a homogeneous cylinder deformed at a nominal strain-rate of 5,000/s for both the Johnson-Cook and the Zhou et al. thermal softening functions. It is clear that for each material, the Zhou et al. expression exhibits enhanced softening.

The finite element mesh had 40 uniform elements in the radial direction and 250 elements in the axial direction as described below. The portion $0 \leq z \leq 10$ mm was divided into 80 uniform elements and segments $10 \leq z \leq 25$ mm, and $25 \leq z \leq$

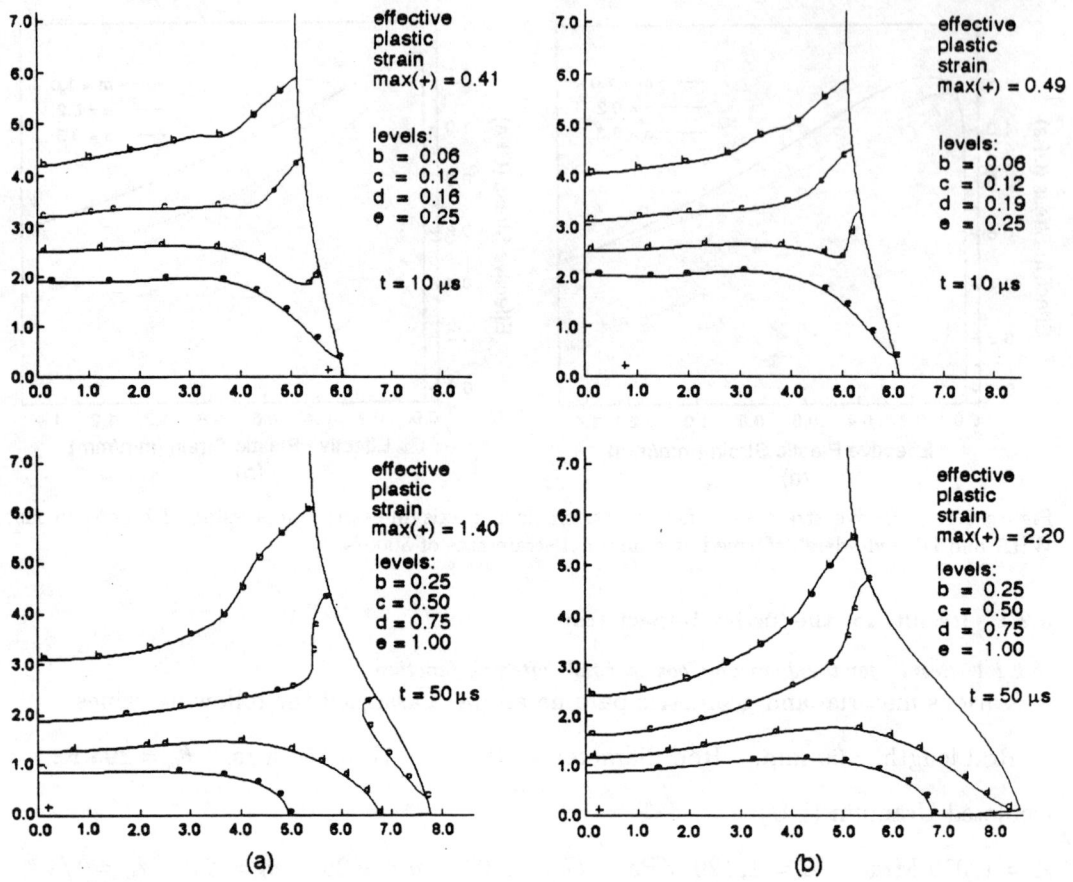

Figure 6.2: Contours of the effective plastic strain at $t = 10$ μs and $t = 50$ μs in (a) the DU rod; (b) the WHA rod.

60 mm were each divided into 85 nonuniform elements whose height increased as one moved away from the impact face. The mesh is fine near the impact surface and should sufficiently capture intense deformations of the rod in the mushroomed region. Computed results for a trial problem indicated that the height of the mushroomed region was approximately 5mm. The contact condition is satisfied by nullifying the axial velocity of nodes about to penetrate the target.

The contours of the effective plastic strain exhibited in Fig. 6.2 for the DU and WHA rods at $t = 10$ and 50 μs suggest that rod particles outside of the mushroomed region are not noticeably deformed. At $t = 10$ μs the maximum effective plastic strain equals 0.41 and 0.49 in the DU and WHA rods and occurs at a point near the periphery of the impact face in DU but near the centroidal axis in WHA. This point where the maximumn effective plastic strain occurs gradually moves towards the stagnation point.

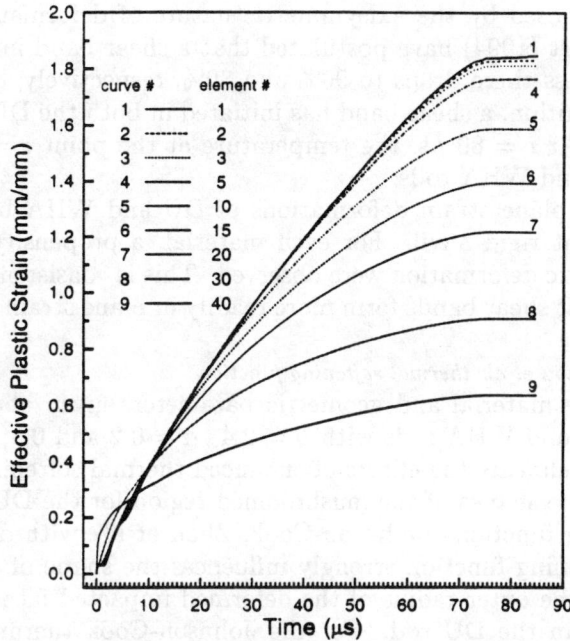

Figure 6.3: Time history of the effective plastic strain at the centroids of several elements on a radial line in the impact face of the DU rod; elements are uniform and element 1 abuts the centroidal axis and 40 the mantle of the rod.

At $t = 50$ μs, the maximum effective plastic strain equals 1.4 in DU and 2.2 in WHA and, in each rod, occurs at a point of the impact face that is close to the centroidal axis. The severely deformed region with the effective plastic strain exceeding 1.0 extends, on the impact face to nearly 6.8 mm for the WHA rod and 5.00 mm for the DU rod; its thickness equals approximately 0.8 mm for both materials.

In Fig. 6.3 is plotted the time history of the effective plastic strain at several points on the impact face of the DU rod; similar results were obtained for the WHA rod and also for points on the centroidal axial line and the mantle of each rod. It is evident that the effective plastic strain increases monotonically at these points and the material at points close to the centroidal axial line is intensely deformed. The plots of the effective stress vs. effective plastic strain at these points, not included herein, reveal that the effective stress drops essentially monotonically and gradually except when the gap occurs between the impact face and the anvil. At a point near the center of the DU rod, it drops from 1.2 GPa at $t = 10$ μs to 0.45 GPa at $t = 60$ μs and its rate of drop slowly decreases. At numerous points on the impact face, centroidal axial line and the mantle of the rods where the time histories of the effective plastic strain and the effective stress were plotted, neither a catastrophic drop in the effective stress nor a dramatic rise in the rate of growth of the effective plastic strain was seen; these two phenomena were observed at the initiation of a shear band in a thermoviscoplastic body deformed either in simple shear or plane strain compression. This could be due to the constraints imposed by the axisymmetric nature of deformations. Batra and Kim (1992) and Deltort (1994) have postulated that a shear band initiates at a point when the effective stress there drops to 90% and 80%, respectively, of its peak value. According to this definition, a shear band has initiated in both the DU and WHA rods at numerous points. At $t = 80$ μs, the temperature at the point $r = z = 0$ equalled 1250 K for both DU and WHA rods.

We also simulated plane strain deformations of DU and WHA blocks striking at normal incidence a flat rigid anvil. For each material, a propensity of thin narrow regions of intense plastic deformation were observed. This is consistent with Batra and Ko's (1993) results that shear bands form more readily in plane strain simulations than axisymmetric ones.

6.2.1.2 Results for the Zhou et al. thermal softening function

Using values of the material and geometric parameters given above, results were computed for the DU and WHA rods with $\beta = 2.4$, $\alpha = 0.2$ and 0.3; the higher value of α is considered to delineate the effects of enhanced thermal softening. Figures 6.4a, 6.4b and 6.4c depict the shapes of the mushroomed region for the DU and WHA rods for the three softening functions (Johnson-Cook, Zhou et al. with $\alpha = 0.2$ and 0.3). As expected, the softening function strongly influences the shape of the mushroomed region. In each case, the outer radius of the deformed impacted face is greater in the WHA rod than that in the DU rod. For the Johnson-Cook thermal softening, the inflection point in the shape of the mantle is closer to the impacted end for the DU rod as compared to that for the WHA rod. For $\alpha = 0.3$ in the Zhou et al. thermal softening, the curve describing the mantle has a cusp for each material signifying severe deformations there. The height of the mushroomed region is smaller for $\alpha = 0.3$ and the mushroomed region forms sooner in each rod than that for $\alpha = 0.2$. The shapes of the mushroomed regions suggest that severe deformations and hence a shear band may initiate at one or more of the following three locations: adjacent to the stagnation point, near the periphery of the impacted end, and close to the inflection point in the curve describing the mantle of the deformed rod; points selected in these areas for further study of deformations are marked in Figs. 6.4a-c as P, Q and R respectively. For comparison purposes, we have also investigated histories of various deformation measures at the point S close to the centroidal axis and about 2.5 mm above the impact face. Time histories of the effective plastic strain at these four points for the DU rod are plotted in Fig. 6.5; similar results were obtained for the WHA rod. Results for the Johnson-Cook model, discussed in the previous section, are shown for comparison purposes. It is clear that for both DU and WHA rods and the three thermal softening functions considered, the effective plastic strain at point S is much lower than that at the other three points. We recall that during plane strain compression and simple shearing deformations of a thermoviscoplastic block, the initiation of a shear band is indicated by a rapid growth of the effective plastic strain accompanied by a sharp drop in the effective stress. For the DU and WHA rods, these two conditions are fulfilled at points P and R for $\alpha = 0.3$, thus shear bands initiate there and propagate into the mushroomed head of the rod. For $\alpha = 0.2$ and 0.3 the effective plastic strain at each one of the four points considered grows less rapidly in the DU rod than that in

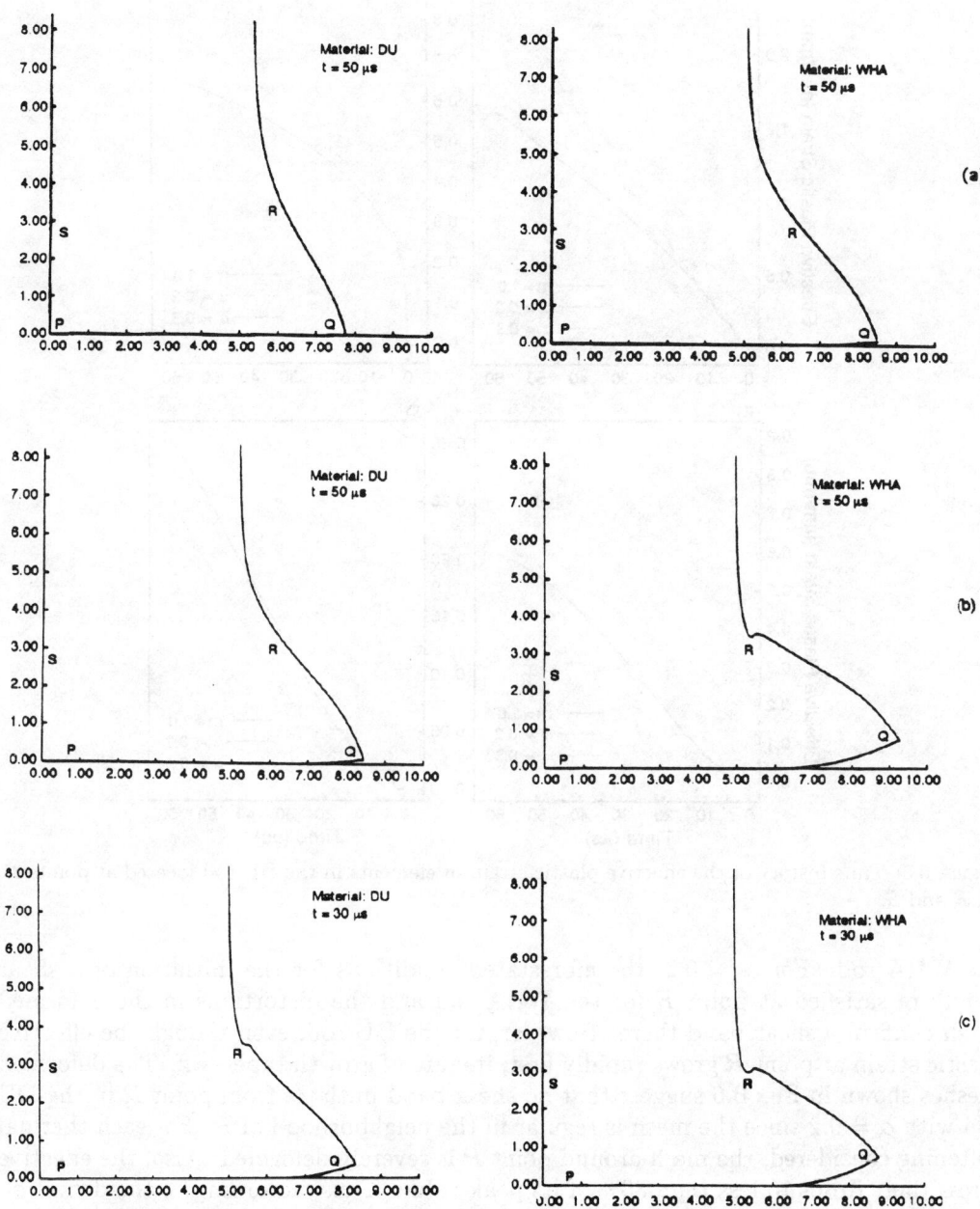

Figure 6.4: Deformed shape of the mushroomed regions for the DU and WHA rods; (a) Johnson-Cook thermal softening, $m = 1.0$; (b) Zhou et al. thermal softening, $\alpha = 0.2$; (c) Zhou et al. thermal softening, $\alpha = 0.3$.

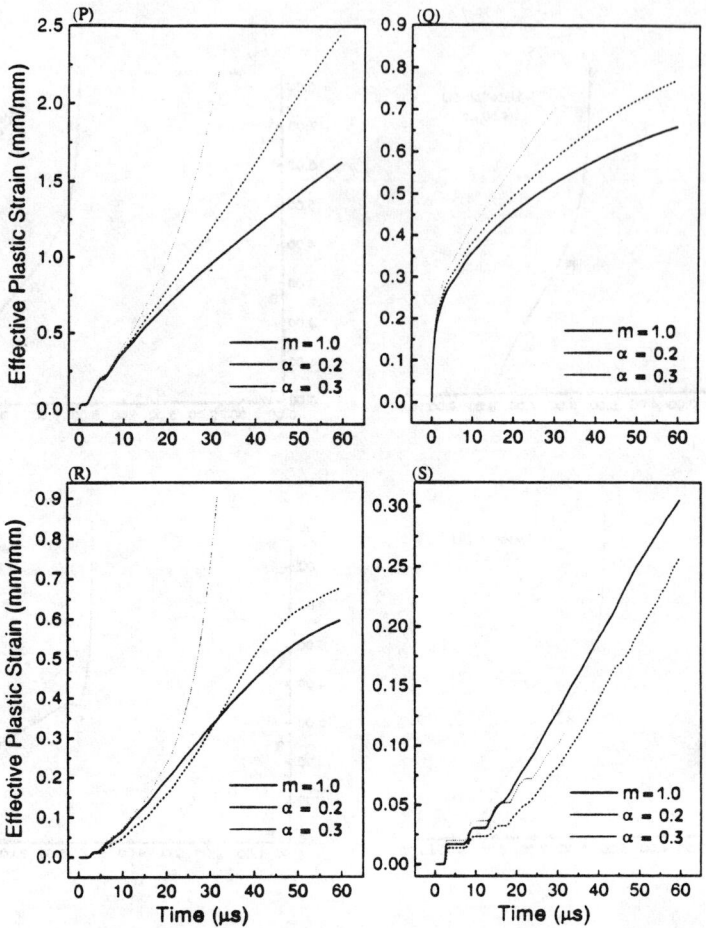

Figure 6.5: Time history of the effective plastic strain in elements in the DU rod located at points P, Q, R and S.

the WHA rod. For $\alpha = 0.2$, the aforestated conditions for the initiation of a shear band are satisfied at point R for the WHA rod and the distortions in the deformed mesh confirm a shear band there. However, for the DU rod, even though the effective plastic strain at point R grows rapidly first, its rate of growth tapers off. The deformed meshes shown in Fig. 6.6 suggest that no shear band initiates from point R in the DU rod with $\alpha = 0.2$ since the mesh is regular in the neighborhood of R. For each thermal softening considered, the mesh around point P is severely deformed. Also, the effective stress there drops to less than 80% of its peak value. Thus, according to the definition of the drop in the effective stress, a shear band initiates at point P for each thermal softening studied. Recall that Dick et al. (1991) observed a shear band passing through the transition region in their reverse ballistic tests on the WHA rod, and a sharp cusp developed where the band intersected the mantle. The mesh used herein is not fine

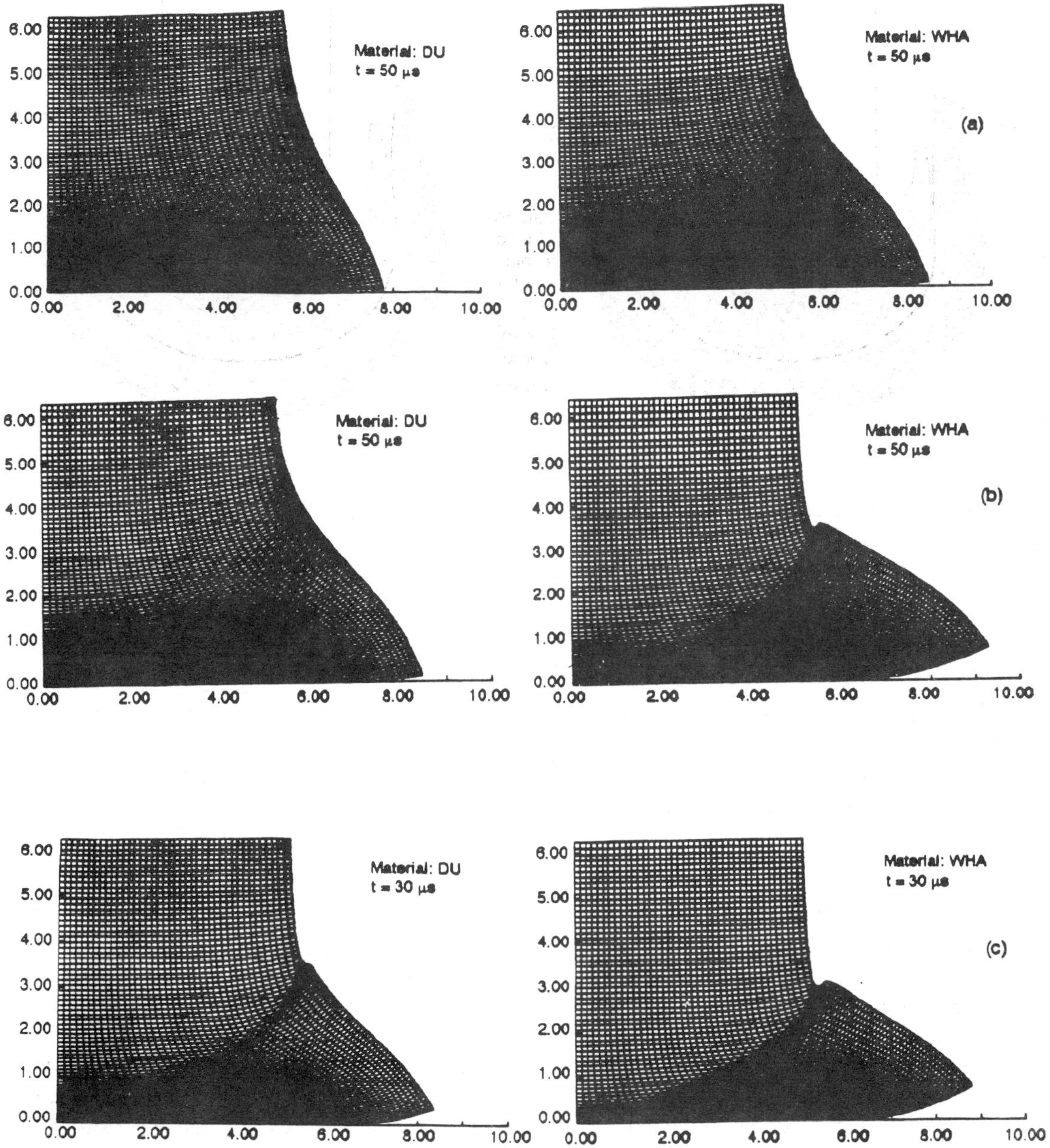

Initial-Boundary-Value Problems with Shear Strain Localization

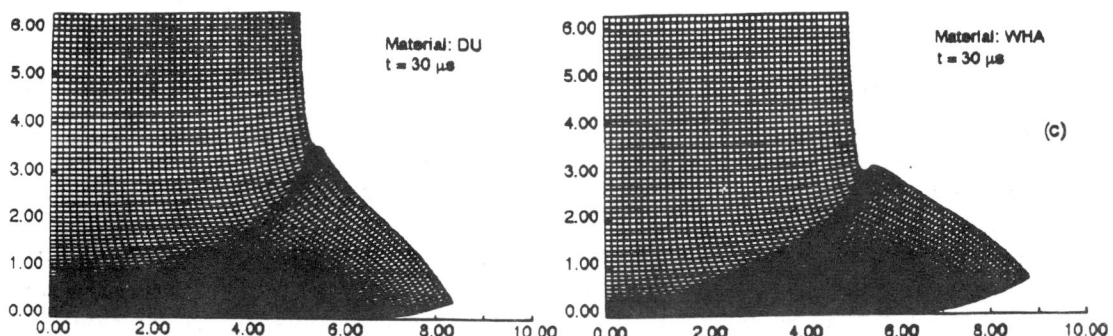

Figure 6.6: Deformed meshes near the impacted ends of the DU and WHA rods: (a) Johnson-Cook thermal softening, $m = 1.0$; (b) the Zhou et al. thermal softening, $a = 0.2$; (c) the Zhou et al. thermal softening, $\alpha = 0.3$.

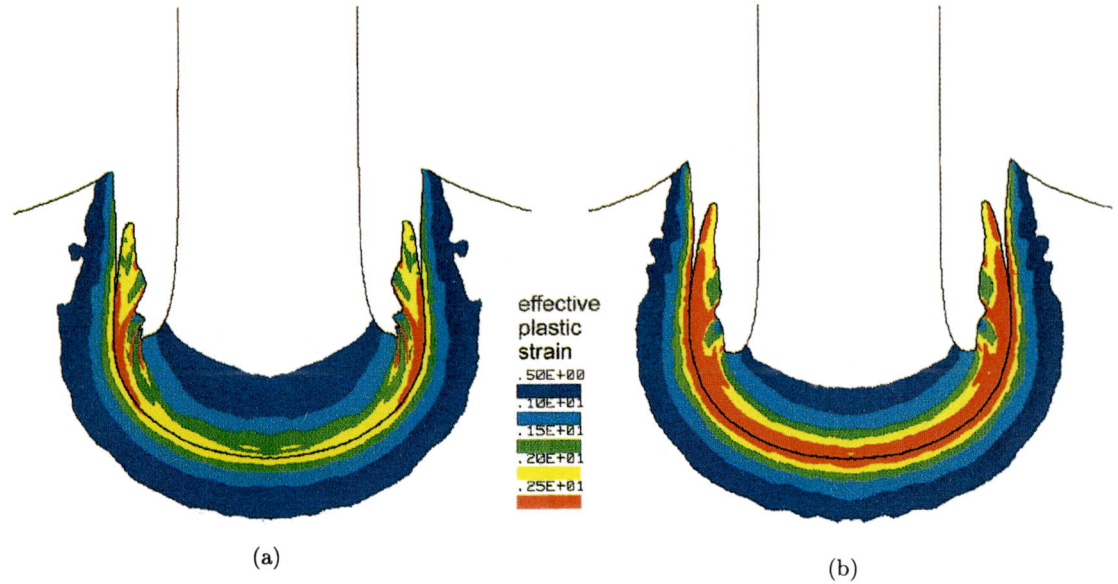

Figure 6.7: Fringes of the effective plastic strain at $t = 25\mu s$ for (a) DU, (b) WHA penetrators.

enough to capture the dimensions of this cusp or of the band.

6.2.2 Results for the penetration problem

The normal impact of 7.69-mm diameter and 76.9-mm long DU and WHA rods moving at 843 m/s on a 60-mm diameter and 70-mm deep steel target has been analysed. Each material was modeled by the Johnson-Cook relation and for the steel we took

$$A = 792.2 \text{ MPa}, \quad B = 509.5 \text{ MPa}, \quad C = 0.014, \quad n = 0.26, \quad m = 1.03, \quad T_m = 1793 \text{ } K,$$
$$\rho = 7,840 \text{ kg/m}^3, \quad c = 477 \text{ J/kg}°\text{C}, \quad \mu = 76 \text{ GPa}, \quad K = 147 \text{ GPa}$$

Parameters in the friction law (6.8) at the target/penetrator interface were assigned the values $\mu_s = 0.78$, $\mu_k = 0.06$, $\gamma = 0.0055$. The penetrator region was divided into uniform 0.16 mm × 0.171 mm elements near the impact face with properly graded larger elements away from it. The mesh in the target region was also graded with 0.16 mm × 0.171 mm elements in the region adjoining the impact face and larger elements elsewhere. We used the automatic contact option for the first 18.5 μs for DU and 15 μs for WHA penetrator. Subsequently, the noninterpenetration conditions were satisfied by using the slideline type 4 (kinetic with sliding, separation and friction).

Figure 6.7 depicts at $t = 25$ fringes of the effective plastic strain in the deformed region adjoining the target/penetrator interface. Similar plots and a movie (available at the URL www.sv.vt.edu/research/batra-stevens/pent.html) showing the distribution of the effective plastic strain in the deformed regions indicate that a propensity of narrow regions of intense plastic deformation form in the DU penetrator in the region abutting the target/penetrator interface and also in the ejecta where the penetrator particles turn to flow backwards. Fewer of these narrow regions occur in the WHA rod and they are smeared out resulting in a rather uniformly deformed zone of intense plastic deformation. Also, the ejecta consists of intensely deformed material, separated by shear banded material for the DU rod but continuously deformed material for the WHA rod. The deformed meshes indicated the occurrence of narrow severely deformed regions in the DU penetrator passing through the points where the flow reverses but none in the WHA penetrator. The deforming region was remeshed at $t = 14$ μs and subsequently at every instant when the time step required to integrate the governing equations decreased to one-hundredth of the starting value, or when the mesh distorted so severely as to cause kinks in the boundary. Every attempt was made to remesh identically the deforming penetrator and target regions in the two cases. The fluid-like material at the head of the ejecta had to be removed in order for the computations to continue at a reasonable pace; such deletions were kept to a minimum and no fluid-like material was removed till $t = 15$ μs. Values of velocities, mass density, temperature and other variables at the newly created nodes were computed from the previous solution so that the total linear momentum and energy were conserved. Such mappings smoothen out the fields and may retard the growth of shear bands.

Figure 6.8a illustrates the deformed mesh at $t = 22.5$ μs for the DU penetrator in the region where the flow is turning around, and Fig. 6.8b exhibits the distribution of the effective plastic strain on line AB shown in Fig. 6.8a. It is clear that three shear bands have formed on this line. The deformed mesh for the WHA penetrator at $t = 22.5$ μs is illustrated in Fig. 6.9; it does not exhibit any narrow intensely deformed regions. For each penetrator, the effective stress within a thin layer adjoining the target/penetrator interface is reduced to almost zero at $t = 30$ μs indicating that the material has essentially lost its strength and behaves like an ideal compressible fluid. The time step has now become too small to continue the computations at a desirable pace.

The time histories of the axial velocity of the tail end of the two penetrators and also of the axial point on the penetrator/target interface were very close to each other. This suggests that differences in the occurrence of shear bands in the ejectas of the two penetrators can not be discerned by examining either the penetration speed or the speed of the tail end of the rod.

The time histories of the effective plastic strain, effective stress, the temperature and the stress-strain curve for a penetrator particle initially located near the stagnation point are exhibited in Fig. 6.10. Even though the effective plastic strain has nearly the same value for the DU and WHA rods, the temperature rise is more for the DU because of its greater strain hardening. The initial rate of drop of the effective stress is same for the two materials, and the effective stress begins to drop a little earlier for the WHA. This suggests that, during the early stages of impact, the WHA particle shear bands sooner than the corresponding DU particle; a trend consistent with that observed in simulations of simple shear (Batra et al., 1995), plane strain compression (Batra and Peng, 1995) and the Taylor impact test. However, as the ejecta grows, the

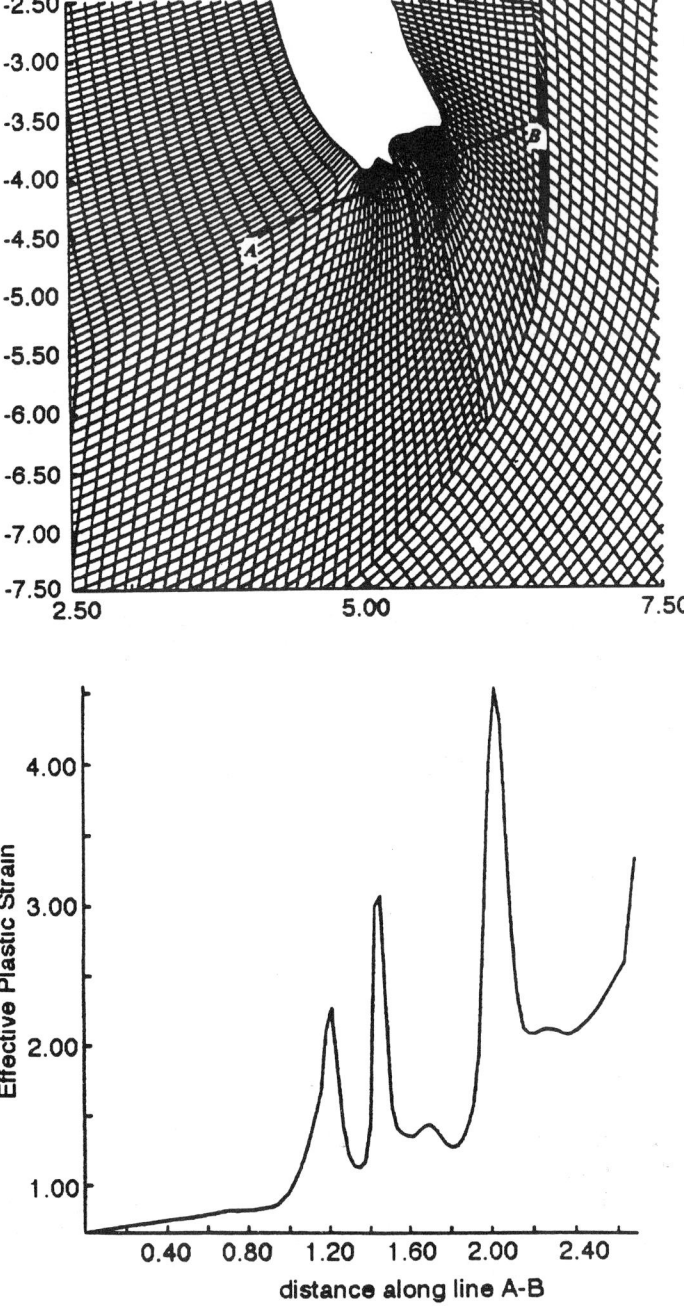

Figure 6.8: (a) Deformed mesh of the ejecta of the DU rod at $t = 22.5$ μs, (b) Distribution of the effective plastic strain on line AB shown in Fig. 6.8a

Initial-Boundary-Value Problems with Shear Strain Localization

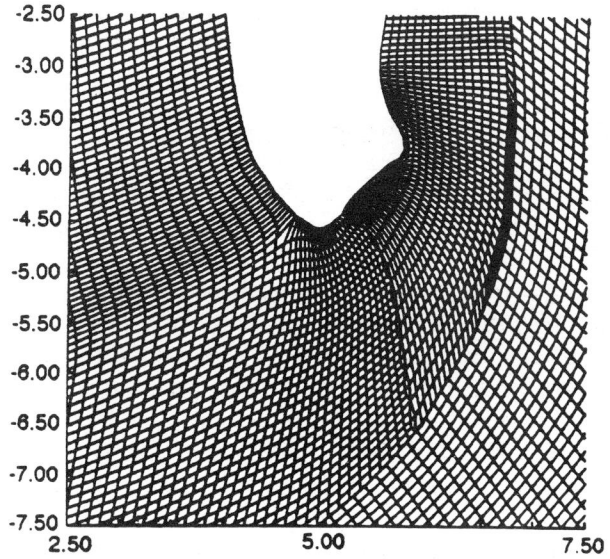

Figure 6.9: Deformed mesh of the ejecta of the WHA rod at $t = 22.5$ μs.

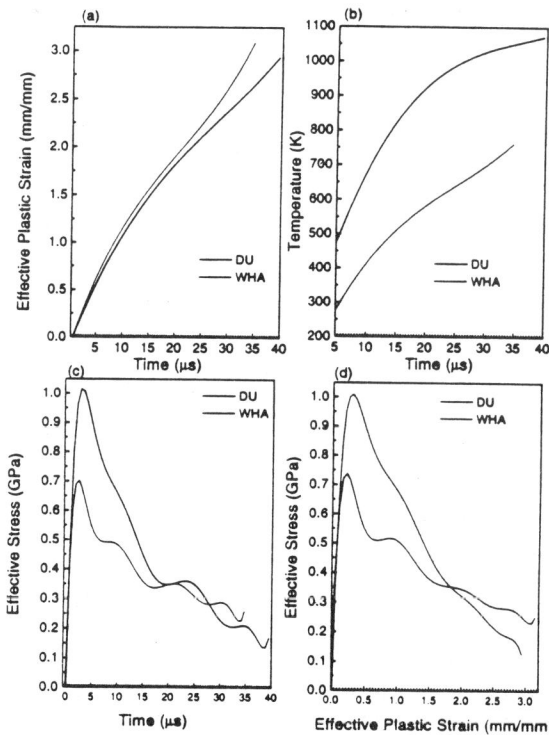

Figure 6.10: (a)-(c) Time history, for an element near the stagnation point, of the effective plastic strain, temperature and effective stress; (d) effective stress vs. effective plastic strain for the element.

deformation gradients in the WHA penetrator get smeared out while the DU ejecta experiences periodic localizations.

7 Shear Bands in a FCC Single Crystal

One way to understand the micromechanics of shear band formation in polycrystalline materials is to study their initiation and growth in a single crystal. Several investigators, e.g. Sawkill and Honeycombe (1954), Price and Kelley (1964), Saimoto et al. (1965), and Chang and Asaro (1981), have observed regions of localized shearing in fcc single crystals deformed quasistatically. Zikry and Nemat-Nasser (1990) have recently studied numerically the phenomenon of shear banding in a fcc single crystal undergoing plane-strain tensile deformations at high strain rates. They used the double cross-slip model proposed by Koehler (1952) and later by Orowan (1954) during the entire loading history. Zhu and Batra (1993) and Batra and Zhu (1995) studied plane strain compression of a single crystal and assumed that all 12 slip systems are potentially active at any instant of loading. With the axis of compression aligned along the crystallographic direction [010], the plane of deformation was taken to be either parallel to the plane (001) or (10$\bar{1}$) of the single crystal. Also studied was the case when the crystal is loaded along the crystallographic direction [380] with the plane of deformation parallel to the plane (001). Some results from Zhu and Batra (1993) and Batra and Zhu (1995) are described below. The constitutive model used is a little different from that proposed by Perzyna (see his article in this volume).

7.1 Formulation of the Problem

A fixed set of rectangular Cartesian coordinates is used to study the thermomechanical deformations of a fcc single crystal of square cross-section compressed along the crystallographic direction [010] which is taken to coincide with the x_3-axis, and the x_1-x_2 plane of deformation either parallel to the plane (001) or (10$\bar{1}$) of the single crystal. In each case, the 12 slip systems are aligned symmetrically about the two centroidal axes. Both elastic and plastic deformations of the single crystal are presumed to be symmetric about the two centroidal axes, even after the band has formed, and accordingly deformations of the material in the first quadrant only are studied. In Eulerian description, equations governing the deformations of the single crystal are:

The balance of mass:	$\dot{\rho} + \rho v_{i,i} = 0.$	(7.1)
The balance of linear momentum:	$\rho \dot{v}_i = \sigma_{ij,j}.$	(7.2)
The balance of internal energy:	$\rho c \dot{\theta} = -q_{i,i} + \sigma_{ij} D^p_{ij}.$	(7.3)

The plastic part D^p_{ij} of the strain-rate tensor D_{ij} is determined by the local plastic slip rate of all active slip systems at a material particle. For plane strain deformations in the x_1-x_2 plane, various quantities are functions of x_1, x_2 and time t, and subscripts i, j range over 1 and 2. However, in the second term on the right-hand side of (7.3), indices i and j extend to 3, since in plane strain deformations $\sigma_{33} \neq 0$ in general, and D^p_{33} need not equal zero even though $D_{33} = 0$.

We postulate Fourier's law of heat conduction (5.14). The strain-rate tensor D_{ij} and the spin tensor W_{ij} are assumed to have additive decompositions into elastic, plastic and thermal parts, *viz.*

$$D_{ij} = D^e_{ij} + D^p_{ij} + \alpha \dot\theta \delta_{ij}, \quad W_{ij} = W^e_{ij} + W^p_{ij}. \tag{7.4}$$

The Cauchy stress rate corotational with the elastic distortion of the single crystal is assumed to be related to the elastic distortion rate by

$$\overset{\nabla}{\sigma}{}^e_{ij} = L_{ijkl} D^e_{kl}, \tag{7.5}$$

where

$$\overset{\nabla}{\sigma}{}^e_{ij} = \dot\sigma_{ij} + \sigma_{ik} W^e_{kj} - W^e_{ik} \sigma_{kj}, \tag{7.6}$$

and L_{ijkl} is the fourth order tensor of the elasticities of the single crystal. The crystal lattice is taken to be elastically isotropic and

$$L_{ijkl} = \lambda \delta_{ij} \delta_{kl} + \mu(\delta_{ik}\delta_{jl} + \delta_{il}\delta_{jk}), \tag{7.7}$$

where λ and μ are Lamé's constants for the crystal material. Recall that the Jaumann stress rate $\overset{\nabla}{\sigma}_{ij}$ defined in (4.7) is corotational with the material element.

The Schmid stress or the resolved shear stress $\overset{(\alpha)}{\tau}$ of the αth slip system is assumed to be related to the local Cauchy stress σ_{ij} through

$$\overset{(\alpha)}{\tau} = \overset{(\alpha)}{\nu_{ij}} \sigma_{ij}, \tag{7.8}$$

where the Schmid factor $\overset{(\alpha)}{\nu_{ij}}$ is defined as

$$\overset{(\alpha)}{\nu_{ij}} = \tfrac{1}{2}(\overset{(\alpha)}{b_i}\overset{(\alpha)}{n_j} + \overset{(\alpha)}{b_j}\overset{(\alpha)}{n_i}), \tag{7.9}$$

$\overset{(\alpha)}{\mathbf{b}}$ and $\overset{(\alpha)}{\mathbf{n}}$ being the unit slip direction and the unit normal to the slip-plane of the αth slip system.

For a strain rate dependent material of the single crystal, the slip rate of the αth slip system is assumed to be related to the resolved shear stress by the power law.

$$\overset{(\alpha)}{\dot\gamma}{}^p = \begin{cases} \overset{(\alpha)}{\dot\gamma_0} \left[\dfrac{\overset{(\alpha)}{\tau}}{\overset{(\alpha)}{\tau_c}}\right] \left[\dfrac{|\overset{(\alpha)}{\tau}|}{\overset{(\alpha)}{\tau_c}}\right]^{1/m-1}, & \overset{(\alpha)}{\tau} \geq \overset{(\alpha)}{\tau_c}, \\ 0, & \overset{(\alpha)}{\tau} < \overset{(\alpha)}{\tau_c}, \end{cases} \tag{7.10}$$

where m is the rate sensitivity parameter, and $\overset{(\alpha)}{\dot\gamma_0}$ is a reference shear strain rate such that if the crystal is to be deformed with each $\overset{(\alpha)}{\dot\gamma}{}^p$ set equal to $\overset{(\alpha)}{\dot\gamma_0}$, then $\overset{(\alpha)}{\tau} = \overset{(\alpha)}{\tau_c}$ (Pan and

Rice, 1983). When the resolved shear stress of the αth slip system is below the critical resolved shear stress $\overset{(\alpha)}{\tau_c}$ required to cause plastic deformation on that slip system, the αth slip system will be inactive. The critical resolved shear stress is assumed to be a function of the initial flow stress τ_0, work hardening, and temperature θ through

$$\overset{(\alpha)}{\tau_c} = \left\{ \tau_0 + \sum_\beta [g + (1-g) \cos \overset{(\alpha\beta)}{\psi} \cos \overset{(\alpha\beta)}{\phi}] (\overset{(\beta)}{\gamma}{}^p)^n h \right\} (1 - \nu\theta) \qquad (7.11)$$

where $\overset{(\alpha\beta)}{\psi}$ is the angle between the slip directions of the αth and βth slip systems, $\overset{(\alpha\beta)}{\phi}$ the angle between their slip normals, $\overset{(\beta)}{\gamma}{}^p$ the plastic strain of the βth slip system, h the strength coefficient, n the work hardening exponent, g the degree of isotropy in work hardening, and ν the thermal softening coefficient. The quantity in the square bracket represents the latent hardening coefficient, and the summation index β ranges over all slip systems. Taylor's (1938) isotropic hardening law follows from (7.11) by setting $g = 1$, and $g = 0$ corresponds to kinematic hardening. Equation (7.11) without the thermal softening term was proposed by Weng (1980).

We assume that the plastic slip rates $\overset{(\alpha)}{\dot\gamma}{}^p$ of all active slip systems at a material point contribute linearly to the plastic parts of the strain rate and spin tensors there through the Schmid factor $\overset{(\alpha)}{\nu_{ij}}$ and the antisymmetric part $\overset{(\alpha)}{\omega_{ij}}$ of the dyad **bn**. Thus,

$$D^p_{ij} = \sum_\alpha \overset{(\alpha)}{\nu_{ij}} \overset{(\alpha)}{\dot\gamma}{}^p, \quad W^p_{ij} = \sum_\alpha \overset{(\alpha)}{\omega_{ij}} \overset{(\alpha)}{\dot\gamma}{}^p, \qquad (7.12)$$

where

$$\overset{(\alpha)}{\omega_{ij}} = \tfrac{1}{2}(\overset{(\alpha)}{b_i}\overset{(\alpha)}{n_j} - \overset{(\alpha)}{b_j}\overset{(\alpha)}{n_i}). \qquad (7.13)$$

The slip direction **b** and the unit normal **n** to the slip plane are orthogonal unit vectors, and are assumed to rotate with the elastic spin of the lattice. Thus, their rates of change are given by

$$\dot b_i = W^e_{ij} b_j, \quad \dot n_i = W^e_{ij} n_j. \qquad (7.14)$$

In plane strain deformations of the crystal, the rotation of a slip system can be characterized by the angle change ϕ of the projective direction of the slip vector in the $x_1 - x_2$ plane. Using $(7.4)_2$ and $(7.12)_2$ we obtain

$$\dot\phi = W^e_{21} = W_{21} - \sum_\alpha \overset{(\alpha)}{\omega_{21}} \overset{(\alpha)}{\dot\gamma}{}^p, \qquad (7.15)$$

and rewrite (7.14) as

$$\begin{aligned} b_1 = \sqrt{1 - b_3^2} \cos\phi_b, \quad & b_2 = \sqrt{1 - b_3^2} \sin\phi_b, \\ n_1 = \sqrt{1 - n_3^2} \cos\phi_n, \quad & n_2 = \sqrt{1 - n_3^2} \sin\phi_n, \end{aligned} \qquad (7.16)$$

where ϕ_b and ϕ_n are, respectively, the current angles between the x_1-axis and the projective directions of the slip vector and the slip plane normal to the x_1-x_2 plane. They equal the sum of their initial values and their changes with respect to the rotated lattice.

Scaling stress-like quantities by τ_0, mass density by ρ_0, length by H, time by H/v_0, and the temperature by θ_r, we rewrite the above equations in terms of nondimensional variables and henceforth use nondimensional variables only. Note that $2H$ equals the height of the block, v_0 the steady value of the vertical component of velocity imposed on the top and bottom surfaces, ρ_0 the mass density in the undeformed and unstressed configuration of the single crystal, and $\theta_r = \tau_0/(\rho_0 c)$.

Because of the presumed symmetry of deformations about the horizontal and vertical centroidal axes, boundary conditions that follow from the symmetry of deformations are applied on the left and bottom surfaces. Both the top and the right surfaces are taken to be thermally insulated, the right surface is taken to be traction free, and on the top surface zero tangential tractions and a vertical component v_2 of velocity given by

$$-v_2(t) = \begin{cases} t/0.005, & 0 \le t \le 0.005, \\ 1, & t \ge 0.005, \end{cases} \quad (7.17)$$

are prescribed. For the initial conditions, we take

$$\rho(\mathbf{x},0) = 1.0, \quad \mathbf{v}(\mathbf{x},0) = \mathbf{0}, \quad \boldsymbol{\sigma}(\mathbf{x},0) = \mathbf{0},$$
$$\phi(\mathbf{x},0) = 0, \quad \theta(\mathbf{x},0) = \begin{cases} \epsilon(1-r^2)^9 \exp(-5r^2), & r \le 1, \\ 0, & r > 1, \end{cases} \quad (7.18)$$

where $r^2 = x_1^2 + x_2^2$. The initial nonuniform temperature field represents a possible imperfection in the single crystal and serves as a triggering mechanism for the localization of the deformation.

7.2 Numerical Solution and Results

An approximate solution of the aforestated problem is computed by the finite element method by employing 4-noded quadrilateral elements. The lumped mass matrix is obtained by assigning one-fourth of the mass of an element to each one of its four nodes. At each node, the mass density, two components of the velocity, temperature, three components σ_{11}, σ_{22}, and σ_{12} of the Cauchy stress, and the angle ϕ characterizing the rotation of the slip system are taken as unknowns. The coordinates of nodes are updated after each time increment. Therefore, the spatial domain occupied by the body and the shapes of these elements vary with time. The coupled nonlinear ordinary differential equations are integrated by using the subroutine LSODE with ATOL = 10^{-3}, and RTOL = 10^{-3}. From the computed solution we evaluated $\overset{(\alpha)}{\tau}$, $\overset{(\alpha)}{\dot{\gamma}}{}^p$, $D_{ij}^{(p)}$, and $W_{ij}^{(p)}$ at each quadrature point, and found the plastic slip strain of the active slip system from

$$\overset{(\alpha)}{\gamma}{}^p(t+\Delta t) = \overset{(\alpha)}{\gamma}{}^p(t) + \Delta t[\overset{(\alpha)}{\dot{\gamma}}{}^p(t) + \overset{(\alpha)}{\dot{\gamma}}{}^p(t+\Delta t)]/2. \quad (7.19)$$

Following values were assigned to various material and geometric parameters to compute numerical results.

$$k = 237 \text{W m}^{-1}\,°\text{C}^{-1}, \quad c = 960 \text{ J kg}^{-1}\,°\text{C}^{-1}, \quad \rho_0 = 270 \text{ kg m}^{-3},$$
$$\mu = 27.6 \text{ GPa}, \quad K = 81.48 \text{ GPa}, \quad \tau_0 = 55 \text{ MPa}, \quad n = 0.52,$$
$$h = 11.02 \text{ MPa}, \quad m = 0.02, \quad \nu = 0.0222°\text{C}^{-1}, \quad H = 5 \text{ mm}, \quad (7.20)$$
$$g = 0.28, \quad v_0 = 5 \text{ m s}^{-1}, \quad \epsilon = 1.0,$$

where K is the bulk modulus. Thus, the average applied strain rate equals 1000 s^{-1}, and $\theta_0 = 21.2°\text{C}$. The aforestated values are for a typical single crystal of aluminum, except that a rather large value of the thermal softening coefficient ν is used to reduce the CPU time required to initiate a shear band.

An aluminum single crystal has a face-centered-cubic lattice structure, which is characterized by four octahedral slip planes $\{111\}$ and three slip directions $<110>$ on each plane to give 12 slip systems. Herein all slip systems are assumed to be equally active.

We use the maximum principal logarithmic strain ϵ_p, defined as

$$\epsilon_p = \ln \lambda_1 \cong -\ln \lambda_2 \qquad (7.21)$$

to find the deformation at a point. Here λ_1^2, λ_2^2, and 1 are eigenvalues of the right Cauchy-Green tensor $C_{\alpha\beta} = x_{i,\alpha} x_{i,\beta}$, or the left Cauchy-Green tensor $B_{ij} = x_{i,\alpha} x_{j,\alpha}$, where $x_{i,\alpha} \equiv \partial x_i/\partial X_\alpha$, X_α being the coordinates of a material point in the stress-free undeformed configuration. The second equality in (7.21) holds because plastic deformations of the crystal are isochoric, and within the band elastic deformations are minuscule.

We employed a finite element mesh consisting of 32×32 uniform square elements in the undeformed configuration, and used 2×2 Gaussian quadrature rule to evaluate various integrals numerically.

7.2.1 The plane of deformation is parallel to the plane (001) of the single crystal

Figure 7.1 depicts contours of the maximum principal logarithmic strain ϵ_p for average strain, $\gamma_{\text{avg}} = 0.07755$ and 0.10755. These suggest that a shear band, indicated by higher values of the contours of the maximum principal logarithmic strain near the center, originates at the center, propagates along $\pm 45°$ directions and is reflected back from the top surface, with the angle of reflection being nearly equal to the angle of incidence. The severely deforming region narrows down initially, but then widens, probably because of a change in the locations of the active slip systems. A closer look at the computed results suggests that in the beginning, the block is uniformly deformed elastically and all slip systems are inactive in the entire body. As the block continues to be deformed at a nominal strain rate of 1000/s, the top part of the square cross-section yields first, and the plastic deformation spreads into the body to make the four slip systems $(111)[1\bar{1}0]$, $(11\bar{1})[1\bar{1}0]$, $(1\bar{1}\bar{1})[110]$, and $(1\bar{1}1)[110]$ active. The material

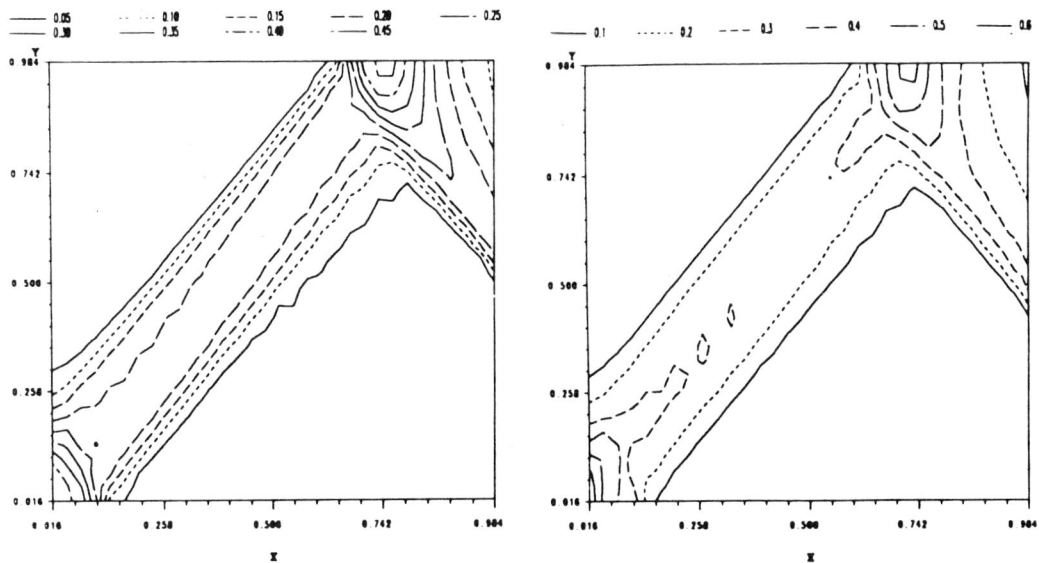

Figure 7.1: Contours of the maximum principal logarithmic strain at $\gamma_{\text{avg}} = 0.07755$ and 0.10755.

surrounding the origin where the temperature perturbation is applied also yields early due to the lower value of the critical shear stress of slip systems at relatively higher temperature. The material adjoining the centroid of the cross-section undergoes more severe plastic deformations than the rest of the material. With further straining of the block, the plastic deformation spreads throughout the body.

The accumulated plastic strain of each active slip system is plotted in Fig. 7.2 at an average strain of 0.10755. It is clear that four primary slip systems $(111)[1\bar{1}0]$, $(11\bar{1})[1\bar{1}0]$, $(1\bar{1}\bar{1})[110]$, and $(1\bar{1}1)[110]$ contribute significantly to plastic deformations, that the maximum slip strain equals 0.4262, and the average slip strain within the band is approximately 0.175. These slip systems are more favorable to plastic deformation than the slip systems $(111)[0\bar{1}1]$, $(11\bar{1})[0\bar{1}\bar{1}]$, $(1\bar{1}\bar{1})[101]$, and $(1\bar{1}1)[10\bar{1}]$ in the central band, and $(\bar{1}11)[0\bar{1}1]$, $(1\bar{1}1)[011]$, $(111)[10\bar{1}]$, and $(11\bar{1})[101]$ in the reflected band. Note that the average slip strain of the four secondary slip systems in the central band equals 0.025, and that of the slip systems in the reflected band equals 0.01. During early stages of the shear band formation, only the primary slip systems are active and contribute to the intense plastic deformation within the band. For simple compression in the crystallographic direction [010] and plane of deformation parallel to the crystallographic plane (001), the four primary slip systems are equally favorable to slip throughout the loading history. However, in a double-slip model for a single crystal employed by Zikry and Nemat-Nasser (1990), the slip system $(111)[\bar{1}01]$, corresponding to $(111)[1\bar{1}0]$ in our coordinate system, is chosen as the primary slip system, and $(\bar{1}11)[011]$ $((1\bar{1}\bar{1})[101]$ in our model) as the conjugate one. These two slip systems are not equally active, with the result that the primary slip system dominates the slip deformation. In our model, all potentially active slip systems are employed, and the slip system becomes active if

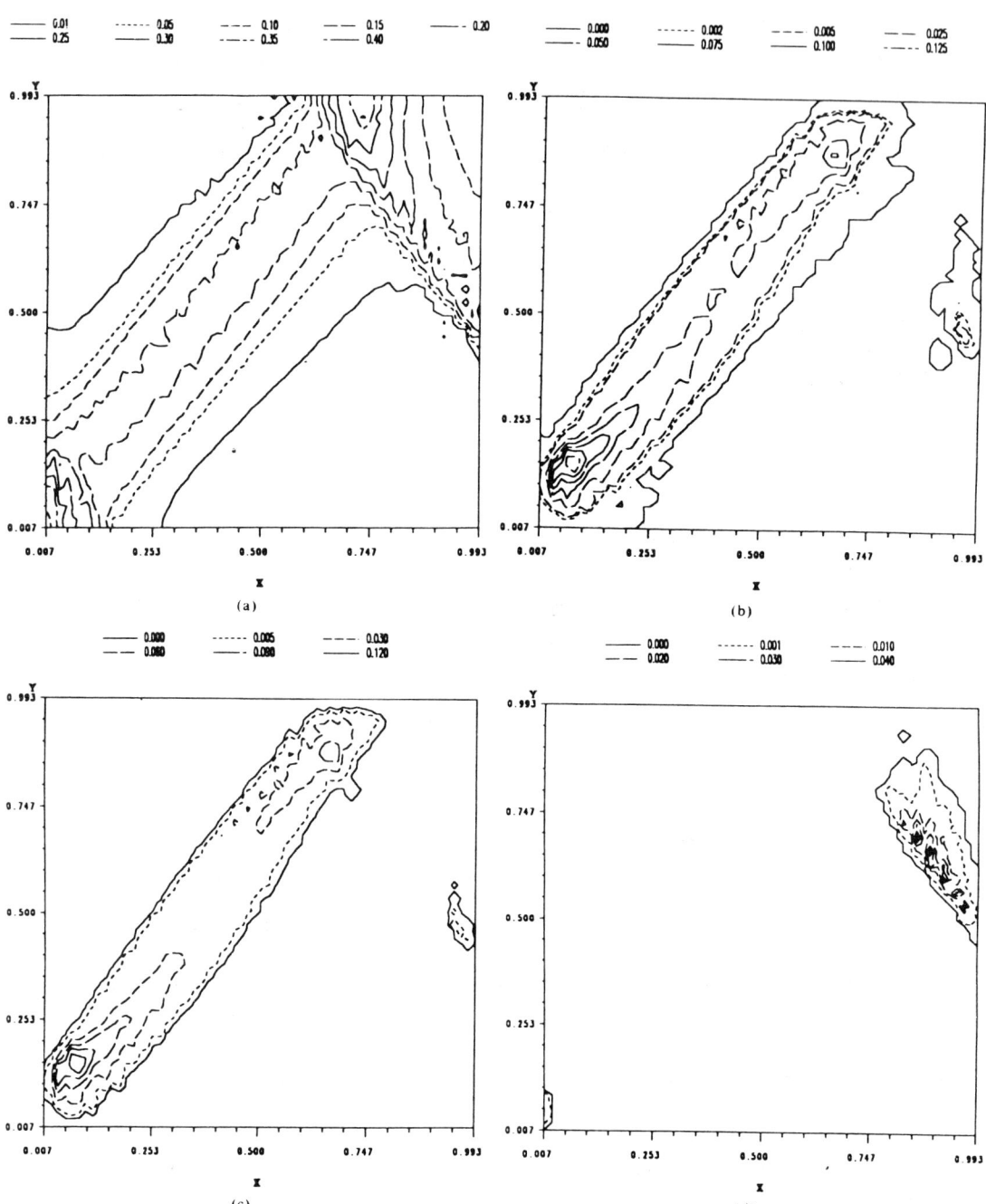

Figure 7.2: Contours of the accumulated plastic slip strains of different slip systems at an average strain of 0.10755. (a) Slip systems: $(111)[1\bar{1}0]$, $(11\bar{1})[1\bar{1}0]$, $(1\bar{1}\bar{1})[110]$, and $(1\bar{1}1)[110]$, (b) Slip systems: $(111)[0\bar{1}1]$ and $(11\bar{1})[0\bar{1}\bar{1}]$, (c) Slip systems: $(1\bar{1}\bar{1})[101]$ and $(1\bar{1}1)[10\bar{1}]$, (d) Slip systems: $(111)[10\bar{1}]$ and $(11\bar{1})[101]$.

its resolved shear stress reaches the critical value. The computed results show that all four primary slip systems, namely, $(111)[1\bar{1}0]$, $(11\bar{1})[1\bar{1}0]$, $(1\bar{1}\bar{1})[110]$, and $(1\bar{1}1)[110]$ are equally active. As the single crystal is deformed and the crystal lattice is reoriented by the deformation, other slip systems become active as conjugate slip systems resulting in multiple gliding. The slip systems $(111)[0\bar{1}1]$, $(11\bar{1})[0\bar{1}\bar{1}]$, $(1\bar{1}\bar{1})[101]$, and $(1\bar{1}1)[10\bar{1}]$ in the central band, and $(111)[10\bar{1}]$, $(11\bar{1})[101]$, $(\bar{1}11)[0\bar{1}1]$, and $(1\bar{1}1)[011]$ in the reflected band are the conjugate slip systems. Zhu and Batra (1993) have given additional results.

The double slip model gives a misorientation between the bands of the primary slip system and the global one due to the heterogeneous slip deformations of the primary and the secondary slip systems. An examination of the slip-rate bands of the primary slip system at four different values of the average strain (cf., Fig. 3 of Zhu and Batra (1993)) suggests that the slip-rate bands broaden as the crystal is deformed. One reason for this widening of the slip-rate bands is that once plastic deformation occurs within the slip bands, the work hardening raises the critical shear stress, and further slip deformation in the center of the band may become more difficult than that in the adjacent regions. This facilitates plastic deformation of the material adjacent to the centerline of the slip-rate band. Another reason is that the lattice of the single crystal is reoriented by the deformation, and the widening of the slip-rate band ensures that the centerline of the global band makes an angle of ±45° with the direction of the compression loading; cf. the article by Korbel in this volume.

The contours of the angle of rotation ϕ of the crystal lattice are given in Fig. 7.3. Within the shear band passing through the block center, the average angle of rotation of slip systems at a nominal strain of 0.10755 is 14.5° counterclockwise, the maximum angle of rotation is 18.54° counterclockwise, and their values in the reflected shear band near the top right corner of the block equal 14.3° and 20.3° clockwise.

7.2.2 The plane of deformation is parallel to the plane $(10\bar{1})$ of the single crystal

Like the previous case when the plane of deformation is parallel to the plane (001), the initial plastic deformations of the block are essentially uniformly distributed throughout the cross-section, except near the center where a temperature perturbation is applied. With continuous compression of the block, a shear band initiates from the center and propagates into the body. This is evidenced by the plots given in Fig. 7.4 of the contours of the maximum principal logarithmic strain at $\gamma_{\text{avg}} = 0.02755$ and 0.1075. At an average strain of 0.02755, a shear band passing through the center and inclined at an angle of approximately 39.5° with the horizontal has developed. A comparison of this with the results plotted in Fig. 7.1 suggests that the direction of the shear band in a single crystal depends upon the orientation of the crystal relative to the axis of loading. The contours of the maximum principal logarithmic strain plotted at average strains of 0.02755 and 0.10755 suggest that the band widens as the single crystal continues to be compressed, and the width of the severely deformed region is more than that in the previous case discussed above.

Figure 7.5 depicts the angle of rotation of the crystal lattice at a nominal strain of

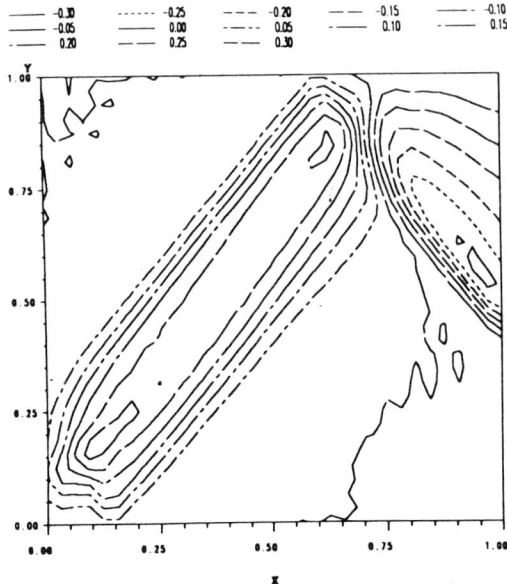

Figure 7.3: Contours of the angle of rotation of the crystal lattice at $\gamma_{\text{avg}} = 0.10755$.

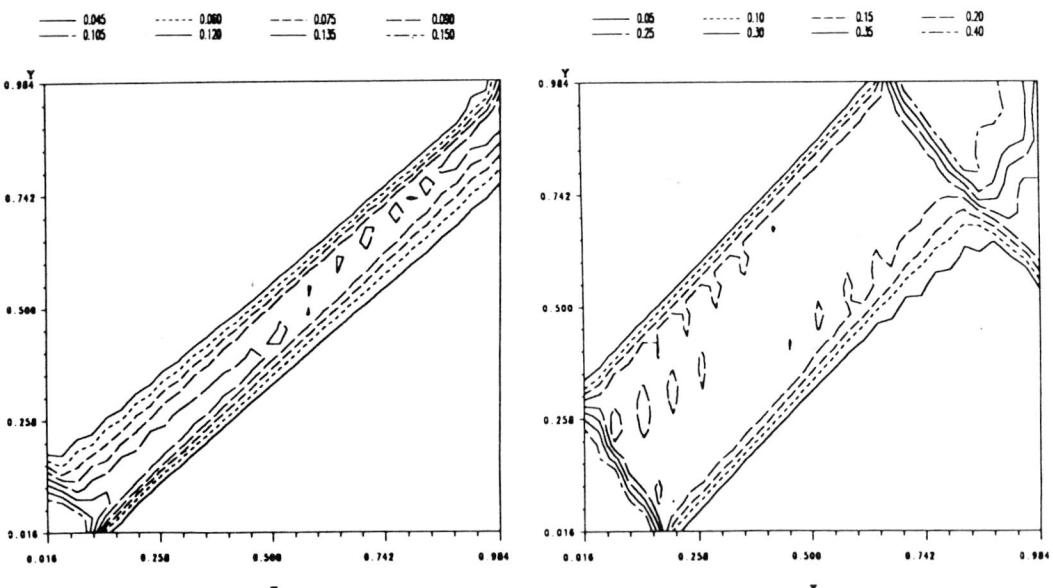

Figure 7.4: Contours of the maximum principal logarithmic strain at $\gamma_{\text{avg}} = 0.02755$ and 0.10759.

Initial-Boundary-Value Problems with Shear Strain Localization

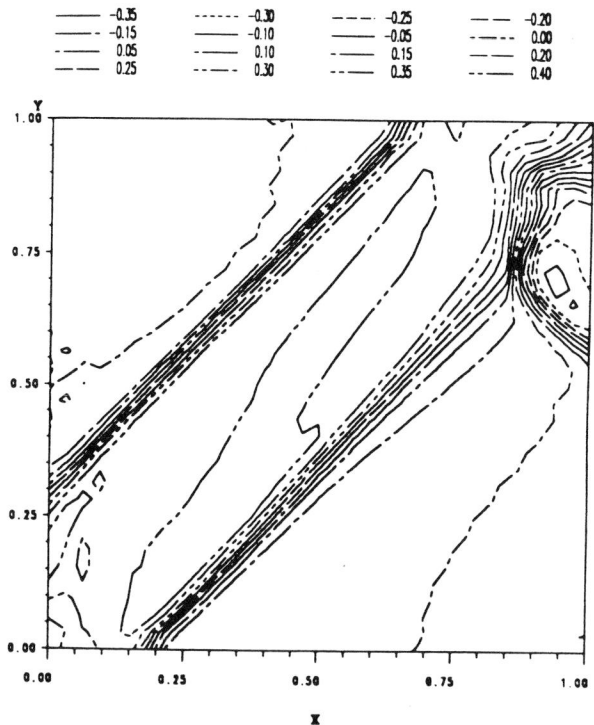

Figure 7.5: Contours of the angle of rotation of the crystal lattice at $\gamma_{\text{avg}} = 0.10755$.

0.10755. It is clear that the average angle of rotation within the central shear band is 22.9° counterclockwise, and in the reflected band is 17.2° clockwise. These values indicate that the crystal lattice undergoes significant rotations within the shear band.

The accumulated slip strains in different slip systems revealed that the slip systems $(111)[10\bar{1}]$ and $(1\bar{1}1)[10\bar{1}]$ remained inactive throughout the entire loading history. Slip systems $(111)[1\bar{1}0]$ and $(111)[0\bar{1}1]$ in both the central and the reflected band, and slip systems $(1\bar{1}1)[110]$ and $(1\bar{1}1)[011]$ in the reflected band were found to be more active than other slip systems. At the nominal strain of 0.10755, the average slip strain of slip systems $(111)[1\bar{1}0]$ and $(111)[0\bar{1}1]$ in the central band equals 0.2 and that in the reflected band is 0.5, and slip systems $(1\bar{1}1)[110]$ and $(1\bar{1}1)[011]$ have an average slip strain of 0.17 in the reflected band. Other active slip systems give very small values of the slip strains within the bands. From contours of slip strain rates we conclude that at an average strain of 0.00255, the four slip systems $(111)[1\bar{1}0]$, $(111)[0\bar{1}1]$, $(1\bar{1}1)[110]$, and $(1\bar{1}1)[011]$ are active everywhere in the block. The narrow region with intensive slip-rate deformation for slip systems $(111)[1\bar{1}0]$ and $(111)[0\bar{1}1]$ differs from that for slip systems $(1\bar{1}1)[110]$ and $(1\bar{1}1)[011]$, in contrast with the case discussed above wherein all four primary slip systems are equally active in the same narrow region. However, the intensity of slip-rates in the two narrow regions seems to be nearly the same. Since, in simple compression, the slip systems $(111)[1\bar{1}0]$ and $(111)[0\bar{1}1]$ are more favorable to slip than the other two primary slip systems, these two slip systems eventually dominate the slip deformation of the single crystal and the slip systems $(1\bar{1}1)[110]$ and $(1\bar{1}1)[011]$ become inactive in the central bands.

7.2.3 Loading along the crystallographic direction [380]

7.2.3.1 Results for no material imperfection

A shear band develops due to the heterogeneity of deformations caused by significantly varying contributions to the overall plastic deformations of the crystal from different slip systems. Two loadings, namely when the crystal is pulled and when it is compressed at an average strain-rate of 1000/s are examined; in each case, the initial temperature is assumed to be uniform, and deformations of the entire cross-section are studied.

7.2.3.1a Tensile loading

A study of the evolution of the accumulated plastic slip strains on different slip systems indicated that the slip systems $(111)[\bar{1}10]$, $(11\bar{1})[\bar{1}10]$, $(\bar{1}11)[110]$ and $(\bar{1}1\bar{1})[110]$ contributed equally and significantly to the plastic deformations of the single crystal. The plastic deformation on these slip systems first ensued at the top right and bottom left corners possibly because of the singularity of the deformations there since the boundary surfaces meeting there have different types of boundary conditions prescribed on them. This plastic deformation propagated into the body, and gradually concentrated into two narrow parallel regions at an angle of approximately 60° with the horizontal axis. Figure 7.6 shows contours of the accumulated plastic strain on one of these four slip systems at an average strain of 0.0575. During subsequent loading of the block, most of the deformations occurred within the two parallel narrow regions. The slip systems $(111)[01\bar{1}]$, $(11\bar{1})[011]$, $(\bar{1}11)[101]$ and $(1\bar{1}1)[\bar{1}01]$ also contributed to the plastic deformation of the body, the severely deformed regions of these slip systems were wider and were aligned along lines almost perpendicular to the centerlines of the narrow regions in which intense plastic deformations of the previous four slip systems were concentrated. Throughout the loading history studied herein, the plastic deformation everywhere in the body stayed minuscule (negligible for all practical purposes) on the remaining four slip systems, viz., $(111)[01\bar{1}]$, $(\bar{1}1\bar{1})[101]$, $(\bar{1}11)[10\bar{1}]$, and $(\bar{1}1\bar{1})[011]$. The value of the maximum plastic strain within the aforestated sets of severely deforming regions were essentially the same and it equalled 0.45 at an average strain of 0.0975.

Figure 7.7 depicts contours of the angle ϕ of rotation of the crystal lattice at $\gamma_{avg} = 0.0975$. It is clear that the crystal lattice undergoes significant rotations in severely deformed regions where the two sets of aforestated slip systems are active, and the maximum value of ϕ in each region equals 16.4° but the directions of rotation are opposite of each other. Contours of slip strain-rates on different slip systems resemble those of slip strains and are, therefore, not exhibited herein.

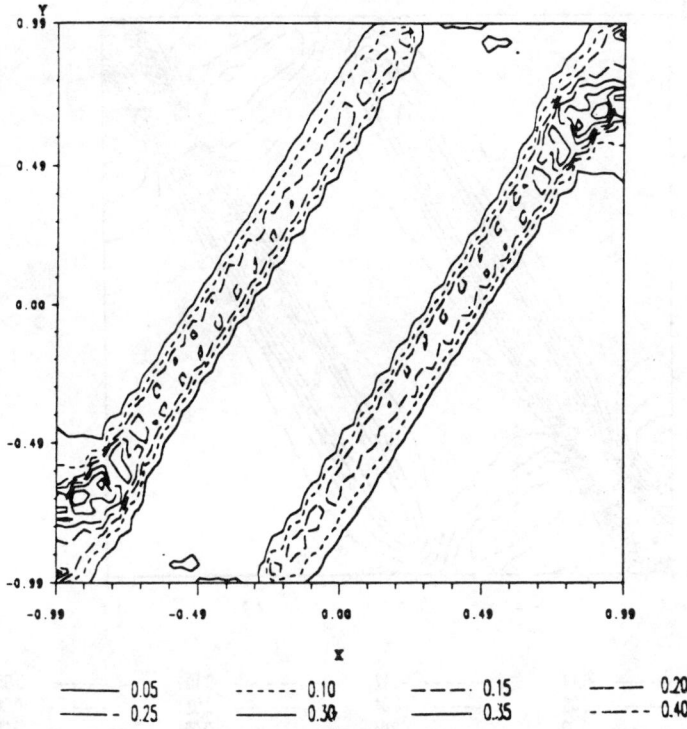

Figure 7.6: Contours of slip strains on any one of the four slip systems $(111)[\bar{1}10]$, $(1,1,\bar{1})[\bar{1},1,0]$, $(\bar{1}11)[110]$, and $(\bar{1}1\bar{1})[110]$ at an average strain of 0.0575.

7.2.3.1b Compressive loading

As for tensile loading, two sets of slip systems namely $(111)[\bar{1}10]$, $(11\bar{1})[\bar{1}10]$, $(\bar{1}11)[110]$, $(\bar{1}1\bar{1})[110]$, and $(111)[01\bar{1}]$, $(11\bar{1})[011]$, $(\bar{1}11)[101]$, $(1\bar{1}1)[\bar{1}01]$ are quite active and contribute significantly to the plastic deformations of the body. Whereas slip strains on the first set of slip systems contribute to the plastic deformation of the region near the top right and bottom left corners, that on the second set of slip systems deform noticeably the central longitudinal region. At an average strain of 0.0975, the maximum value of the slip strain in the first and second sets of slip systems equals 1.076 and 0.381, respectively. The other four slip systems, i.e., $(111)[\bar{1}01]$, $(\bar{1}\bar{1}1)[101]$, $(\bar{1}11)[01\bar{1}]$, and $(\bar{1}1\bar{1})[011]$ stay dormant until the average axial strain reaches 0.0575 at which instant they start making a contribution to the plastic deformations of a very small region near the top right and bottom left corners. At $\gamma_{\text{avg}} = 0.0575$, 0.0775, and 0.0975, the maximum values of the slip strain on a slip system from this set equal 0.048, 0.235, and 0.417, respectively. However, the region over which these slip systems are active is quite small. One reason for these slip systems to begin contributing towards the plastic deformation of the body at $\gamma_{\text{avg}} = 0.0575$ is the noticeable rotation of the lattice structure during its plastic deformation; Fig. 7.8 depicts the contours of the angle of rotation ϕ at $\gamma_{\text{avg}} = 0.0575$. The maximum values of ϕ in the severely deformed regions near the top right and bottom left corners and in the central longitudinal region equal

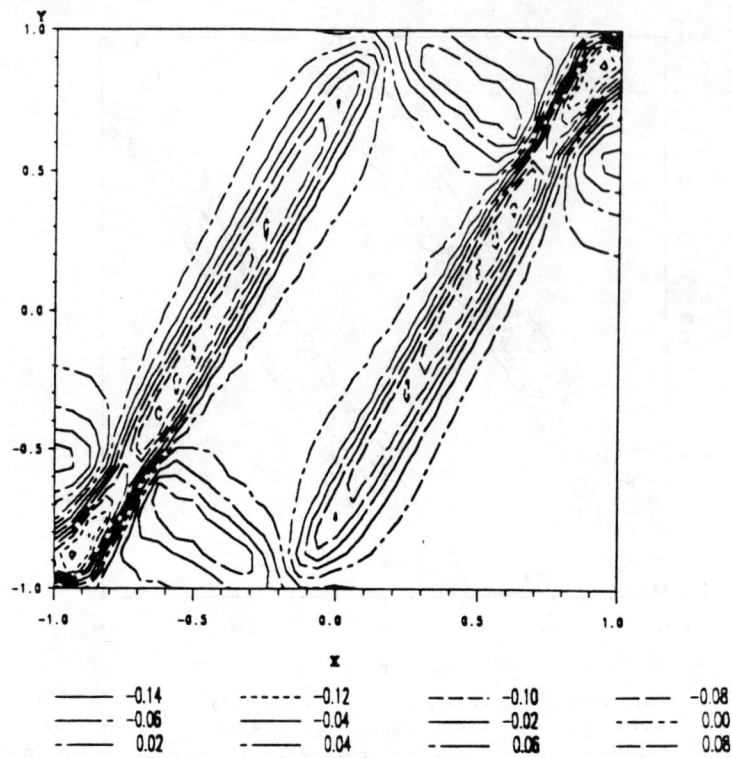

Figure 7.7: Contours of the angle of rotation of the crystal lattice at an average strain of 0.0575.

25.5° counterclockwise and 15.8° clockwise, respectively.

Note that the regions in which the first two sets of slip systems are active are quite different when the crystal is loaded in tension and compression. At an average strain of 0.0975 the maximum value of slip strain on any slip system equalled 1.076 and 0.486 for compression and tension loading, respectively. One reason for this difference is that the more severely deformed region is smaller when the body is compressed as compared to that when it is pulled.

7.2.3.2 Material imperfection modelled by nonuniform initial temperature

We assume that the inital temperature is given by (7.18) with $\epsilon = 1.0$, and the single crystal is compressed along the crystallographic direction [380]. The maximum value of the initial temperature perturbation is intentionally taken to be large so as to reduce the computational time. Contours of slip strains on slip systems $(111)[\bar{1}10]$, $(11\bar{1})[\bar{1}10]$, $(\bar{1}11)[110]$, and $(\bar{1}1\bar{1})[110]$ are nearly identical. As the single crystal continues to be compressed, its plastic deformations ensuing from the center propagate outwards in the form of the letter x, and the material near the top right and bottom left corners also begins to deform plastically. Interestingly enough, plastic deformations of the material in these regions recede rather than intensify with the passage of time, and eventually intense plastic deformations of the material along the line making an angle

Initial-Boundary-Value Problems with Shear Strain Localization

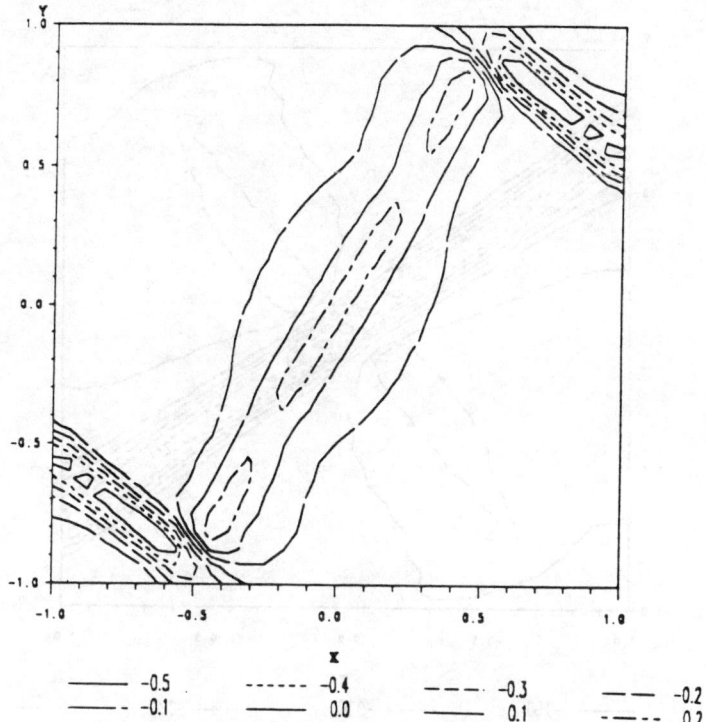

Figure 7.8: Contours of the angle of rotation of the crystal lattice at an average strain of 0.0575.

of 34° clockwise with the horizontal axis persist. It is because in compression the 34° direction is more favorable to plastic deformation than the one perpendicular to it. Contours of slip strains on slip systems $(111)[01\bar{1}]$, $(111)[01\bar{1}]$, $(\bar{1}11)[101]$ and $(1\bar{1}1)[\bar{1}01]$ are almost identical to each other. The material region wherein these slip systems are active looks like a star and the plastic deformation therein continues to intensify and propagate outwards with an increase in the overall deformations of the crystal. No measureable or detectable plastic deformation occurs on the other four slip systems, viz. $(111)[\bar{1}01]$, $(\bar{1}\bar{1}1)[101]$, $(\bar{1}11)[01\bar{1}]$, and $(\bar{1}1\bar{1})[011]$, until the average strain of 0.0275 at which instant these slip systems begin contributing to the plastic deformation of the body. Slip strains on these slip systems are essentially the same, and the narrow intensely deformed region is oriented at an angle of approximately 30° clockwise from the horizontal axis. The maximum values of the slip strain on the three sets of slip systems when $\gamma_{avg} = 0.0575$ equal 0.3986, 0.1209, and 0.3348, respectively. However, these need not occur at the same point.

The contours of the angle ϕ of rotation of the crystal lattice exhibited in Fig. 7.9 at an average strain of 0.0575 vividly illustrate the region where significant values of ϕ occur. At $\gamma_{avg} = 0.0575$, ϕ varies from 2° clockwise to 39.8° counterclockwise, the minimum and maximum values of ϵ_p equal 0 and 0.737 signifying that some of the region has not been plastically deformed at all. The maximum value 19.24 of

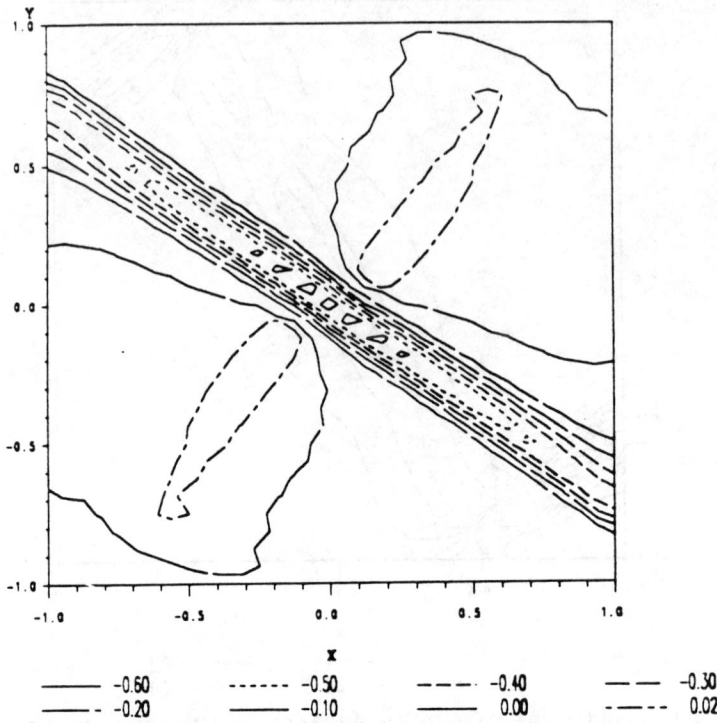

Figure 7.9: Contours of the angle of rotation of the crystal lattice at an average strain of 0.0575.

the second invariant I of the nondimensional strain-rate tensor $\mathbf{D}$ indicates that peak strain-rates equal 1.9×10^4 s^{-1}, and the peak nondimensional temperature of 1.97 implies that the critical shear stress at the point where the peak temperature occurs equals 7.2% of its value in the absence of the thermal softening effect. The deformed mesh at $\gamma_{\text{avg}} = 0.0575$ illustrates that the shear band is inclined at an angle of nearly 30° clockwise with the horizontal axis and does not pass through a corner (cf. Batra and Zhu (1995)).

7.2.3.3 Material imperfection modelled by a misorientation of the crystal lattice

The initial temperature is assumed to be uniform but four elements meeting at the centroid of the cross-section are misoriented by 10°. Thus the deformations of these four elements will be different from that of the rest of the body, and these elements may act as nuclei of shear bands or may not deform much. For the single crystal compressed along the crystallographic direction [380], the contours of the accumulated slip strains on any one of the four slip systems $(111)[\bar{1}10]$, $(11\bar{1})[\bar{1}10]$, $(\bar{1}11)[110]$, and $(\bar{1}1\bar{1})[110]$ at an average strain, γ_{avg}, of 0.0775 vividly demonstrate that intense plastic deformation on these slip systems initiates from the top right and bottom left corners and propagates inwards; cf. Batra and Zhu (1995). These bands do not pass through the center. Since the slip systems within the central four elements are different from those

Initial-Boundary-Value Problems with Shear Strain Localization

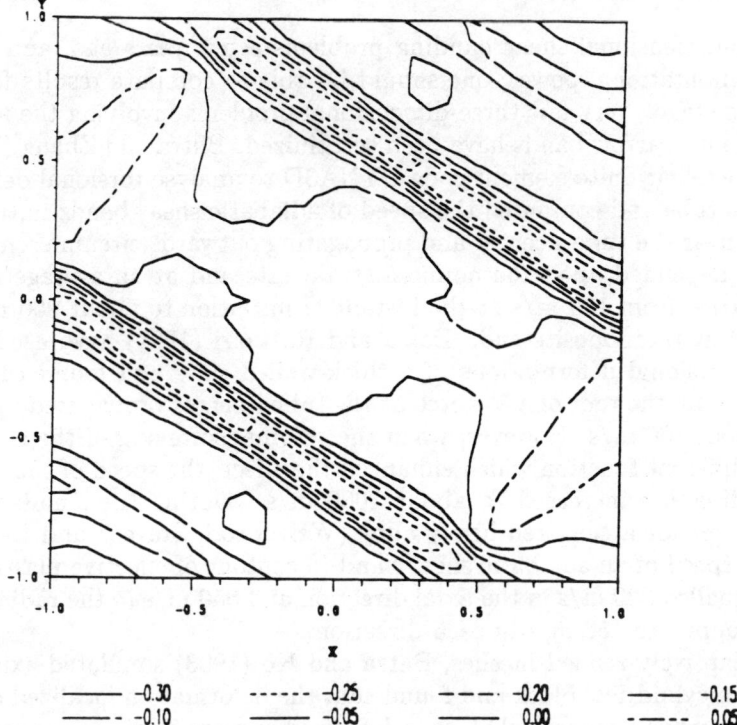

Figure 7.10: Contours of the angle of rotation of the crystal lattice at an average strain of 0.0775.

outside of them, this region is found to be less amenable to severe plastic deformations and resists the propagation of shear bands through it. The contours of the accumulated slip strains on slip systems $(111)[01\bar{1}]$, $(11\bar{1})[011]$, $(\bar{1}11)[101]$, and $(\bar{1}1\bar{1})[011]$ at $\gamma_{avg} = 0.0075$, 0.0475, 0.0775 and 0.0975 reveal that severe plastic deformations on them occur near the boundaries of the central four elements and propagate outwards. The remaining four slip systems $(111)[\bar{1}01]$, $(\bar{1}\bar{1}1)[101]$, $(\bar{1}11)[01\bar{1}]$ and $(\bar{1}1\bar{1})[011]$ stay inactive until the crystal has been compressed to an average strain of 0.057 at which point they become active in a very narrow region. At an average strain of 9.75%, the maximum value of the accumulated plastic strain on these three sets of slip systems equals 37.1%, 16.9%, and 17.4% respectively. Contours of the angle of rotation ϕ of the crystal lattice at $\gamma_{avg} = 0.0775$ are plotted in Fig. 7.10. At an average strain of 0.0975, peak values of ϕ in the two severely deformed regions equal 9° clockwise and 25.4° counterclockwise.

Concluding Remarks

As should be clear from the results discussed above, one-dimensional problems can be analysed with fine enough and properly graded meshes so as to obtain essentially mesh independent results. With the adaptive mesh refinement techniques developed for the analysis of two-dimensional shear banding problems a few years ago, and the recent advances in computational power, one should be able to compute results far into the postlocalization stage. Very few three-dimensional problems involving the localization of deformation into narrow bands have been scrutinized. Batra and Zhang (1994) used the large scale explicit finite element code DYNA3D to analyse torsional deformations of a thin-walled tube and computed the speed of adiabatic shear bands initiating from weak elements near the tube's center and propagating outwards circumferentially. The speed strongly depended upon the nominal strain rate and at an average strain rate of 5000/s, it varied from 180 m/s at the instant of initiation to about 900 m/s by the time it reached at the opposite end. Batra and Rattazzi (1997) have used the same code to study torsional deformations of a thick-walled tube and found that a shear band initiating from the root of a V-notch at the tube's center propagated in the radial direction at about 100 m/s. However, when the thermal softening of the material was modeled by a different function which enhanced the effect, the speed of the shear band in the radial direction increased to about 1000 m/s. During the simulation of the Taylor impact test for a tungsten heavy alloy (WHA) rod, Stevens and Batra (1998) found that the speed of an adiabatic shear band (a contour of effective plastic strain of 1.0) initially equalled 800 m/s in the axial direction and 550 m/s in the radial direction and quickly dropped to 150 m/s in each direction.

By using adaptively refined meshes, Batra and Ko (1993) simulated axisymmetric compression of a cylindrical block and found that the deformation localized much later for this case than when an identical block is deformed in plane strain compression. Also, when the loaded ends were constrained from moving radially, the cylinder barrelled outwards near the center but reverse barrelling effect was observed when the loaded ends were free to move radially. A similar delay in the shear band initiation was computed by Batra and Stevens (1997) during the simulation of the impact of a WHA rod against a rigid smooth target.

Since in real problems, there generally are several defects where shear bands can nucleate, Batra (1987), Batra and Liu (1989), Batra and Hwang (1995) and Kwon and Batra (1988) investigated the effect of perturbations with multiple peaks of not necessarily the same amplitudes. Kwon and Batra found that for nonpolar materials deformed in simple shear, a shear band initiated from each cusp in the inital sinusoidal temperature distribution, but for dipolar materials only one shear band developed at the loaded boundary.

Zhu and Batra (1992) studied the possibility of phase transformation within the shear banded region. In the undeformed state, the specimen was assumed to be fully annealed, isotropic, and its microstructure to be a mixture of coarse ferrite and cementite. A material point was assumed to transform into austenite once its temperature exceeded the transformation temperature with the rate of transformation governed by a simple kinetic equation. Proper account was taken of the latent heat required for the transformation, the associated volume change, and the variation in the thermophysical properties. It was found that the austenite is quenched rapidly enough by the surrounding material for it to be converted into martensite rather than a mixture of pearlite and martensite.

Wang and Batra (1994) accounted for the texture development by using a theory with two internal variables, a scalar to account for the isotropic hardening of the material and a symmetric traceless second-order tensor to account for the kinematic hardening. They found that the consideration of kinematic hardening does not alter the qualitative nature of results.

It is hoped that this article has elucidated upon some aspects of adiabatic shear banding and has raised enough questions in the reader's mind to warrant further enquiry into the subject.

References

1. Y. L. Bai and B. Dodd, *Adiabatic Shear Localization: Occurrence, Theories and Applications*, Pergamon, Oxford 1992.

2. R. C. Batra, The Initiation and Growth of, and the Interaction Among Adiabatic Shear Bands in Simple and Dipolar Materials, *Int. J. Plasticity*, **3** (1987), 75.

3. R. C. Batra, Effect of Nominal Strain-Rate on the Initiation and Growth of Adiabatic Shear Bands in Steels, *J. Appl. Mechs.*, **55** (1988), 229-230.

4. R. C. Batra, Steady State Penetration of Thermoviscoplastic Targets, *Computat. Mech. an Int. J.*, **3** (1988), 1-11.

5. R. C. Batra and C. H. Kim, Effect of Material Characteristic Length on the Initiation, Growth and Band Width of Adiabatic Shear Bands in Dipolar Materials, *J. de Physique*, **49** (1988), C3/41-C3/46.

6. R. C. Batra and De-Shin Liu, Adiabatic Shear Banding in Plane Strain Problems, *J. Appl. Mechs.*, **56** (1989), 527-534.

7. R. C. Batra and C. H. Kim, Adiabatic Shear Banding in Elastic-Viscoplastic Nonpolar and Dipolar Materials, *Int. J. Plasticity*, **6** (1990), 127-141.

8. R. C. Batra and C. H. Kim, The Interaction Among Adiabatic Shear Bands in Simple and Dipolar Materials, *Int. J. Engng Sci.*, **28** (1990), 927-942.

9. R. C. Batra and C. H. Kim, Effect of Viscoplastic Flow Rules on the Initiation and Growth of Shear Bands at High Strain Rates, *J. Mechs. Phys. Solids*, **38** (1990), 859-874.

10. R. C. Batra and C. H. Kim, Effect of Thermal Conductivity on the Initiation, Growth and Band Width of Adiabatic Shear Bands, *Int. J. Engng Sci.*, **29** (1991), 949-960.

11. R. C. Batra and K. I. Ko, An Adaptive Mesh Refinement Technique for the Analysis of Shear Bands in Plane Strain Compression of a Thermoviscoplastic Solid, *Computational Mechs.: an Int. J.*, **10** (1992), 369-379.

12. R. C. Batra and C. H. Kim, Analysis of Shear Bands in Twelve Materials, *Int. J. Plasticity*, **8** (1992), 425-452.

13. R. C. Batra and J. Hwang, An Adaptive Mesh Refinement Technique for Two-Dimensional Shear Band Problems, *Computat'l Mechs.: an Int. J.*, **12** (1993), 255-268.

14. R. C. Batra and K. I. Ko, Analysis of Shear Bands in Dynamic Axisymmetric Compression of a Thermoviscoplastic Cylinder, *Int. J. Engng Sci.*, **31** (1993), 529-547.

15. R. C. Batra and X. S. Jin, Analysis of Dynamic Shear Bands in Porous Thermally Softening Viscoplastic Materials, *Archives of Mechanics*, **41** (1994), 13-36.

16. R. C. Batra and J. Hwang, Dynamic Shear Band Development in Dipolar Thermoviscoplastic Materials, *Computational Mechs.*, **12** (1994), 354-369.

17. R. C. Batra and X. Zhang, On the Propagation of a Shear Band in a Steel Tube, *J. Eng'g Materials & Technology*, **116** (1994), 155-161.

18. R. C. Batra and Z. Peng, Development of Shear Bands in Dynamic Plane Strain Compression of Depleted Uranium and Tungsten Blocks, *Int. J. Impact Eng'g*, **16** (1995), 375-395.

19. R. C. Batra, X. Zhang and T. W. Wright, Critical Strain Ranking of Twelve Materials in Deformations Involving Adiabatic Shear Bands, *J. Appl. Mechs.*, **62** (1995), 252-255.

20. R. C. Batra and Z. G. Zhu, Effect of Loading Direction and Initial Imperfections on the Development of Dynamic Shear Bands in a FCC Single Crystal, *Acta Mechanica*, **113** (1995), 185-203.

21. R. C. Batra and J. B. Stevens, Adiabatic Shear Bands in Axisymmetric Impact and Penetration Problems, *Comp. Meth. Appl. Mechs. & Eng'g* (to appear in 1997).

22. R. C. Batra and N. Nechitailo, Analysis of Failure Modes in Impulsively Loaded Prenotched Steel Plates, *Int. J. Plasticity*, **13** (1997), 291-308.

23. R. C. Batra and D. Rattazzi, Adiabatic Shear Banding in a Thick-Walled Steel Tube, *Comp. Mechs.*, (to appear, 1997).

24. E. B. Becker, G. F. Carey and J. T. Oden, *Finite Elements. An Introduction, Vol. I*, Prentice Hall, Englewood Cliffs, NJ 1981.

25. T. Belytschko and T. J. R. Hughes (Editors), Computational Methods for Transient Analysis, North-Holland, Amsterdam 1986.

26. S. R. Bodner and Y. Partom, Mechanical Properties at High Rate of Strain, *Inst. Phys. Conf.*, Ser. No. 21 (1975), 102.

27. B. Budiansky, Thermal and Thermoelastic Properties of Isotropic Composites, *J. Compos. Mater.*, **4** (1990), 701-744.

28. S. Cescotto and D. W. Zhou, A Variable Density Mesh Generation for Planar Domains, *Comm. Appl. Num. Meth.*, **5** (1989), 473-481.

29. Y. W. Chang and R. J. Asaro, An Experimental Study of Shear-Localization in Aluminum-Copper Single Crystals, *Acta Metall.*, **29** (1981), 241.

30. L. Chen and R. C. Batra, Shear Instability Direction at a Crack Tip in Thermoviscoplastic Materials, *Theoretical and Appl. Fracture Mechs.* (in press).

31. C. C. Chu and A. Needleman, Void Nucleation Effects in Biaxially Stretched Sheets, *J. Engng. Mat. and Technol.*, **102** (1980), 249-256.

32. B. Deltort, Experimental and Numerical Aspects of Adiabatic Shear in a 4340 Steel, *J. de Physique*, C8, **4** (1994), 447-452.

33. R. D. Dick, V. Ramachandran, J. D. Williams, R. W. Armstrong, W. H. Holt and W. Mock, Jr., Dynamic Deformation of W7Ni3Fe Alloy via Reverse-Ballistic Impact, in: A. Crowson and E. S. Chen, eds., *Tungsten and Tungsten Alloys – Recent Advances* (The Minerals, Metals & Materials Society) 1991, 269-276.

34. O. D. Dillon, Jr. and J. Kratochvil, A Strain Gradient Theory of Plasticity, *Int. J. Solids and Structures*, **6** (1970), 1513-1533.

35. B. Dodd and Y. Bai, *Ductile Fracture and Ductility with Applications to Metal Working*, Academic Press, London 1987.

36. J. R. Fisher, Void Nucleation in Spheriodized Steels during Tensile Deformation, Ph.D. Thesis, Brown Univ., 1980.

37. C. W. Gear, *Numerical Initial Value Problems in Ordinary Differential Equations*, Prentice Hall, Englewood Cliffs, NJ 1971.

38. A. E. Green, B. McInnis and P. M. Naghdi, Elastic-Plastic Continua with Simple Force Dipole, *Int. J. Engng. Sci*, **1** (1968), 373-394.

39. A. L. Gurson, Continuum Theory of Ductile Rupture by Void Nucleation and Growth. Part 1. Yield Criteria and Flow Rules for Porous Ductile Media, *J. Eng. Mater. Technol.*, **99** (1977), 2-15.

40. A. M. Habraken and S. Cescotto, An Automatic Remeshing Technique for Finite Element Simulation of Forming Processes, *Int. J. Num. Meth. Eng.*, **30** (1990), 1503-1525.

41. A. C. Hindmarsh, ODEPACK: A Systematized Collection of ODE Solvers. Scientific Computing (R. A. Stepleman *et al.*, eds.), *Scientific Computing*, Amsterdam, North Holland 1983, pp. 55-64.

42. E. Hinton and J. S. Campbell, Local and Global Smoothing of Discontinuous Finite Element Functions Using a Least Squares Method, *Int. J. Num. Meth. Eng.*, **8** (1974), 461-480.

43. E. Hinton, T. Rock and O. C. Zienkiewicz, A Note on Mass Lumping and Related Processes in the Finite Element Method, *Earthquake Engng. and Structural Dynamics*, **4** (1976), 245-249.

44. T. J. R. Hughes, *The Finite Element Method. Linear Static and Dynamic Finite Element Analysis*, Prentice-Hall, Englewood Cliffs, NJ 1987.

45. G. R. Johnson and W. H. Cook, A Constitutive Model and Data for Metals Subjected to Large Strains, High Strain Rates and High Temperatures, *Proc. 7th Int. Symp. Ballistics*, The Hague, The Netherlands 1983, 1-7.

46. J. F. Kalthoff, Shadow Optical Methods of Caustics, in *Handbook on Experimental Mechanics*, (Editor A. S. Kobayashi) Prentice-Hall, Englewood Cliffs, NJ 1987, 430-500.

47. J. F. Kalthoff and S. Winkler, Failure Mode Transition at High Strain Rates of Loading, *Proc. Int. Conf. On Impact Loading and Dynamic Behavior of Materials*, (Editors: C. Y. Chiem, H. D. Kunze and L. W. Meyer), Deutsche Gewsellschaft für Metallkundcle, DGM, Bremen 1988, 185-196.

48. J. F. Kalthoff, Transition in the Failure Behavior of Dynamically Shear Loaded Cracks, *Appl. Mech. Rev.*, **43** (1990), S247.

49. C. H. Kim and R. C. Batra, Effect of Initial Temperature on the Initiation and Growth of Shear Bands in a Plain Carbon Steel, *Int. J. Nonlinear Mechs.*, **27** (1992), 279-291.

50. H. Kobayashi and B. Dodd, A Numerical Analysis of the Formation of Adiabatic Shear Bands Including Void Nucelation and Growth, *Int. J. Impact Engng.*, **8** (1989), 1-13.

51. J. S. Koehler, The Nature of Work-Hardening, *Phys. Review*, **86** (1952), 51.

52. Y. W. Kwon and R. C. Batra, Effect of Multiple Initial Imperfections on the Initiation and Growth of Adiabatic Shear Bands in Nonpolar and Dipolar Materials, *Int. J. Engng Sci.*, **26** (1988), 1177-1187.

53. G. H. LeRoy, Large Scale Plastic Deformation and Fracture for Multiaxial Stress States, Ph.D. Thesis, McMaster University 1978.

54. J. Litoński, Plastic Flow of a Tube under Adiabatic Torsion, *Bull. Acad. Polon. Sci.*, **25** (1977), 7.

55. S. H. Lo, A New Mesh Generation Scheme for Arbitrary Planar Domains, *Int. J. Num. Meth. Eng.*, **21** (1985), 1403-1426.

56. R. Löhner, Some Useful Data Structures for the Generation of Unstructured Grids, *Comm. Appl. Num. Meth.*, **4** (1988), 123-135.

57. J. H. MacKenzie, The Elastic Constants of a Solid Containing Spherical Holes, *Proc. Phys. Soc.*, **63B** (1950), 2-11.

58. A. Marchand and J. Duffy, An Experimental Study of the Formation Process of Adiabatic Shear Bands in a Structural Steel, *J. Mechs. Phys. Solids*, **36** (1988), 251.

59. H. F. Massey, The Flow of Metal During Forging, *Proc. Manchester Assoc. Engrs.* (1921), 21-26.

60. R. D. Mindlin, Second Gradient of Strain and Surface Tension in Linear Elasticity, *Int. J. Solids and Structures*, **1** (1965), 417-438.

61. A. Needleman and V. Tvergaard, *Finite Element Analysis of Localization in Plasticity, in Finite Elements: Special Problems in Solid Mechanics*, (J. T. Oden and G. F. Carey, eds.), Prentice Hall 1984, 94-157.

62. A. Needleman, Material Rate Dependence and Mesh Sensitivity in Localization Problems, *Comp. Meth. Appl. Mech. Engng.*, **67** (1988), 69-85.

63. J. W. Nunziato and S. C. Cowin, A Nonlinear Thoery of Elastic Materials with Voids, *Arch. Rational Mech. Anal.*, **72** (1979), 175-201.

64. E. Orowan, Dislocation in Metals. American Institute of Mining, Metallurgical and Petroleum Engineers, NY (1954), 103.

65. J. Pan, M. Saje and A. Needleman, Localization of Deformation in Rate-Sensitive Porous Plastic Solids, *Int. J. Fract.*, **21** (1983), 261-278.

66. J. Pan and J. R. Rice, Rate Sensitivity of Plastic Flow and Implication for Yield Surface Vertices, *Int. J. Solids and Structures*, **19** (1983), 973-987.

67. S. L. Passman and R. C. Batra, A Thermomechanical Theory for a Porous Anisotropic Elastic Solid with Inclusions, *Arch. Rat. Mech. Anal.*, **87** (1984), 11-33.

68. J. Peraire, M. Vahdati, K. Morgan and O. C. Zienkiewicz, Adaptive Remeshing for Compressible Flow Computations, *J. Comp. Phys.*, **72** (1987), 449-466.

69. W. H. Press, B. P. Flannery, S. A. Teukolsky and W. T. Vetterling, *Numerical Recipes*, Cambridge University Press, Cambridge 1986.

70. R. J. Price and A. Kelly, Deformation of Age-hardened Aluminum Alloy Crystals – II. *Acta Metall.*, **12** (1964), 979.

71. A. M. Rajendran, High Strain Rate Behavior of Metals, Ceramics and Concrete, Report #WL-TR-92-4006, Wright Patterson Air Force Base 1992.

72. S. Saimoto, W. F. Hosford and W. A. Backofen, Ductile Fracture in Copper Single Crystals, *Phil. Mag.*, **12** (1965), 319.

73. M. Saje, J. Pan and A. Needleman, Void Nucleation Effects on Shear Localization in Porous Plastic Solids, *Int. J. Fracture*, **19** (1982), 163-182.

74. J. Sawkill and R. W. K. Honeycombe, Strain-Hardening in Face Centered Cubic Metal Crystals, *Acta Metall.*, **2** (1954), 854.

75. T. J. Shawki and R. J. Clifton, Shear Band Formation in Thermal Viscoplastic Materials, *Mechanics of Materials*, **8** (1989), 13-43.

76. J. B. Stevens and R. C. Batra, Adiabatic Shear Bands in the Taylor Impact Test for a WHA Rod, *Int. J. Plasticity* (to appear in 1998).

77. G. I. Taylor, Plastic Strains in Metals, *J. Inst. Metals*, **62** (1938), 307-324.

78. G. I. Taylor, The Use of Flat-Ended Projectiles for Determining Dynamic Yield Stress 1. Theoretical Considerations, *Proc. Roy. Soc.*, **A194** (1948), 289-299.

79. Y. Tomita, Simulations of Plastic Instabilities in Solid Mechanics, *Appl. Mechs. Revs.*, **47** (1994), 171-205.

80. R. A. Toupin, Elastic Materials with Couple Stresses, *Arch. Rat'l Mechs. Anal.*, **11** (1962), 385-414.

81. H. Tresca, On Further Application of the Flow of Solids, *Proc. Inst. Mech. Engrs.*, **30** (1878), 301-345.

82. C. A. Truesdell and W. Noll, The Nonlinear Field Theories of Mechanics, in: S. Flügge, ed.: *Handbuch der Physik, Vol. III/3*, Springer-Verlag, Berlin 1965.

83. V. Tvergaard, Influence of Voids on Shear Band Instabilities Under Plane Strain Conditions, *Int. J. Fracture*, **17** (1981), 389-407.

84. V. Tvergaard, On Localization in Ductile Materials Containing Spherical Voids, *Int. J. Fracture*, **128** (1982), 237-252.

85. V. Tvergaard and A. Needleman, Analysis of the Cup-Cone Fracture in a Round Tensile Bar, *Acta Metall.*, **32** (1984), 157-196.

86. J. W. Walter, Numerical Experiments on Adiabatic Shear Band Formation in One-Dimension, *Int. J. Plasticity*, **8** (1992), 657-693.

87. Y. Wang and R. C. Batra, Effect of Kinematic Hardening on the Initiation and Growth of Shear Bands in Plane Strain Compression of a Thermoviscoplastic Solid, *Acta Mechanica*, **102** (1994), 217-233.

88. G. J. Weng, Dislocation Theories of Work Hardening and Yield Surfaces of Single Crystal, *Acta Mechanica*, **37** (1980), 217-230.

89. T. W. Wright and R. C. Batra, The Initiation and Growth of Adiabatic Shear Bands, *Int. J. Plasticity*, **1** (1985), 205-212.

90. T. W. Wright and R. C. Batra, Adiabatic Shear Bands in Simple and Dipolar Plastic Materials. In: K. Kawata and J. Shiori, (eds.) *Macro- and Micro Mechanics of High Velocity Deformation and Fracture*, IUTAM Symp. on MMMHVDF, Tokyo, Japan, Springer-Verlag, Berlin, Heidelberg 1987, 189-201.

91. T. W. Wright and J. W. Walter, On Stress Collapse in Adiabatic Shear Bands, *J. Mech. Phys. Solids*, **35** (1987), 701-715.

92. T. W. Wright and J. W. Walter, The Asymptotic Structure of an Adiabatic Shear Band in Antiplane Motion, *J. Mech. Phys. Solids*, **44** (1996), 77-97.

93. C. Zener and J. H. Hollomon, Effect of Strain Rate on Plastic Flow of Steel, *J. Appl. Phys.*, **14** (1944), 22-32.

94. M. Zhou, A. Needleman and R. J. Clifton, Finite Element Simulations of Plate Impact, *J. Mech. Phys. Solids*, **42** (1994), 423-458.

95. M. Zhou, G. Ravichandran and A. J. Rosakis, Dynamically Propagating Shear Bands in Impact-Loaded Prenotched Plates-II. Numerical Simulations, *J. Mechs. Phys. Solids*, **44** (1996) 1007-1032.

96. J. Z. Zhu, O. C. Zienkiewicz, O. Hinton and J. Wu, A New Approach to the Development of Automatic Quadrilateral Mesh Generation, *Int. J. Num. Meth. Eng.*, **32** (1991), 849-886.

97. Z. G. Zhu and R. C. Batra, Consideration of Phase Transformations in the Study of Shear Bands in a Dynamically Loaded Steel Block, *J. Eng'g Mat'l Tech.*, **114** (1992), 368-377.

98. Z. G. Zhu and R. C. Batra, Analysis of Dynamic Shear Bands in a Single Crystal, *Int. J. Plasticity*, **9** (1993), 653-696.

99. O. C. Zienkiewicz and J. Z. Zhu, The Superconvergent Patch Recovery and a Posteriori Error Estimates, Part I, II, *Int. J. Numer. Meth. Engng.*, **33** (1992), 1331-1382.

100. M. A. Zikry and S. Nemat-Nasser, High Strain-Rate Localization and Failure of Crystalline Materials, *Mechs. Materials*, **10** (1990), 215-237.

NUMERICAL SOLUTIONS OF INITIAL BOUNDARY VALUE PROBLEMS FOR METALS AND SOILS

T. Lodygowski
Technical University of Poznan, Poznan, Poland

Abstract

This work considers the numerical aspects of the problems of plastic strain localization for strain softening materials in both soil-like materials and ductile metals. In mathematical formulation the attention should be paid on the well-posedness of the initial boundary value problem which guarantees the uniqueness of the solution in the whole domain of the incremental analysis. Viscoplasticity serves as a mathematical tool of regularization of the system of governing equations. This formulation which is the base of numerics allows to stay free of the effects of the results sensitivity to the finite element mesh density. The numerical examples for the initial boundary value problems of ductile and clay-type materials are presented.

There are also presented some preliminary thoughts on the sensitivity analysis for these ruther complicated problems of plastic strain localization in softening materials.

1. Introduction

The localization of deformations is a physical phenomenon which is observed for a wide range of materials such as ductile metals or polymers and brittle concretes, rocks and soils. The deformations in specimens or elements of structures localize into relatively narrow zones of intense straining.

The results of numerous experiments expressed in the load-displacement space exhibit descending branches after the peak loads have been obtained. This is reported by dozens papers describing the experiments for quasi static as well as for dynamic cases. For ductile materials the damage usually appears as a result of the increasing number of microcracks which finally localize in a particular domain. The directions of propagation of microdamage have a random character and concentrate in the certain domains of the weakest material structure.

In civil engineering practice the localization of deformations is clearly observed for concrete and rocks, some types of soils and the steel. Also the theoretical description which is used to predict the limit load, and which bases on the concept of localized plastic hinges (for beams and frames), widely used by designers to some extend reflects the localization phenomenon.

Below, in figures we report the results of two quasi static experiments for granular and ductile materials. After DESRUES' thesis [19] we cite for a sand the results of testing the compressed specimen. The localization of deformations and the softening behaviour, after the critical intensity of stresses σ_1 has been reached (points 6 and 7), are shown in Fig.1.

The initiation and the development of shear bands for a set of specimens (plane stress to plane strain) were reported by CHAKRABARTI and SPRETNAK [17]. In Fig.2 we observe the appearing and grows of the zones of plastic localization which are the precursors of failure. In Fig.3 there are presented four stages of a tensile deformation process of the specimen of oxygen-free, high conductivity copper: see CURRAN, SEAMAN and SHOCKEY [18]. The nucleation of voids and their grows, which increases the porosity of the material, and finally the localized necking type of damage of the specimen is reported.

The localization phenomenon even stronger accompanies the dynamical loading. Among many experiments reported in the journals let us recall, as the representative one, this presented by MARCHAND and DUFFY [62]. Hot rolled steel (HRS) and cold rolled steel (CRS) were investigated in a fast torsion test of so called Hopkinson bar. The specimen with hexagonal mounting flanges, the effect of localized plastic deformations and the formation of shear bands in 1018 CRS are presented in Fig.4. The softening behaviour in the dynamic cases for ductile materials usually is strongly connected with temperature and heat transfer effects. As a conclusion some engineering materials, including metals, soils, concrete and ice are classified as so-called softening materials. These materials show the reduction of the load-carrying capacity together with the increasing localized deformations after the limit load has been reached.

In many practical engineering boundary value problems the knowledge on post-critical behavior significantly helps to predict the safety of the whole system. The detailed information on both global and local effects seems to be very important.

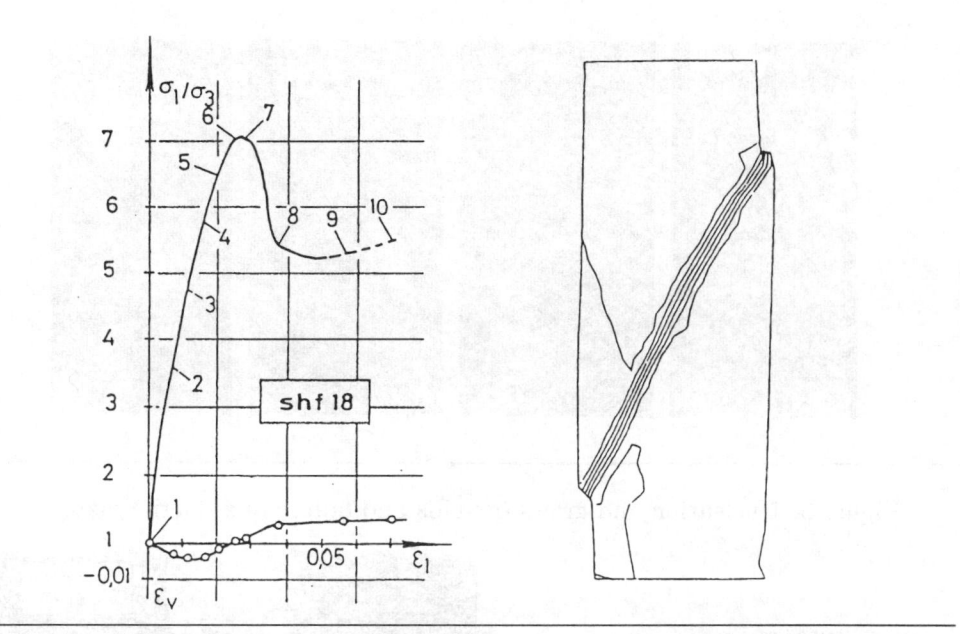

Figure 1: Stress strain curves and the mode of failure for sand specimen

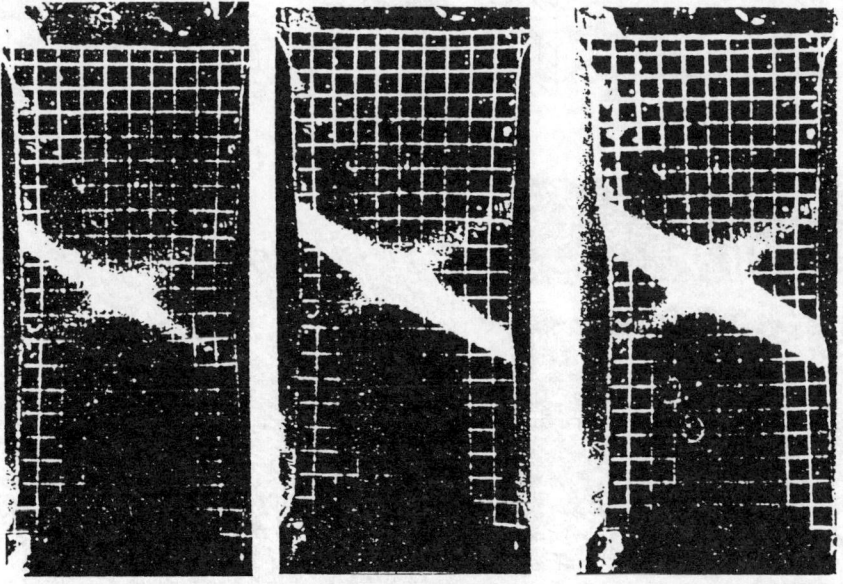

Figure 2: Nucleation and grows of shear bands for ductile material

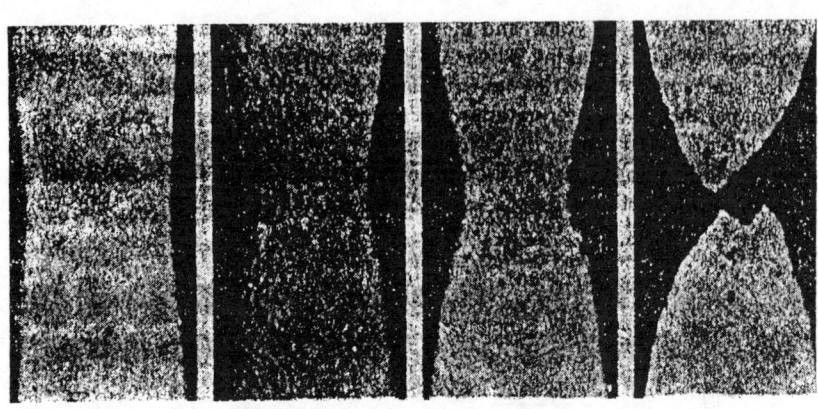

Figure 3: Nucleation and grows of voids and failure of a ductile material

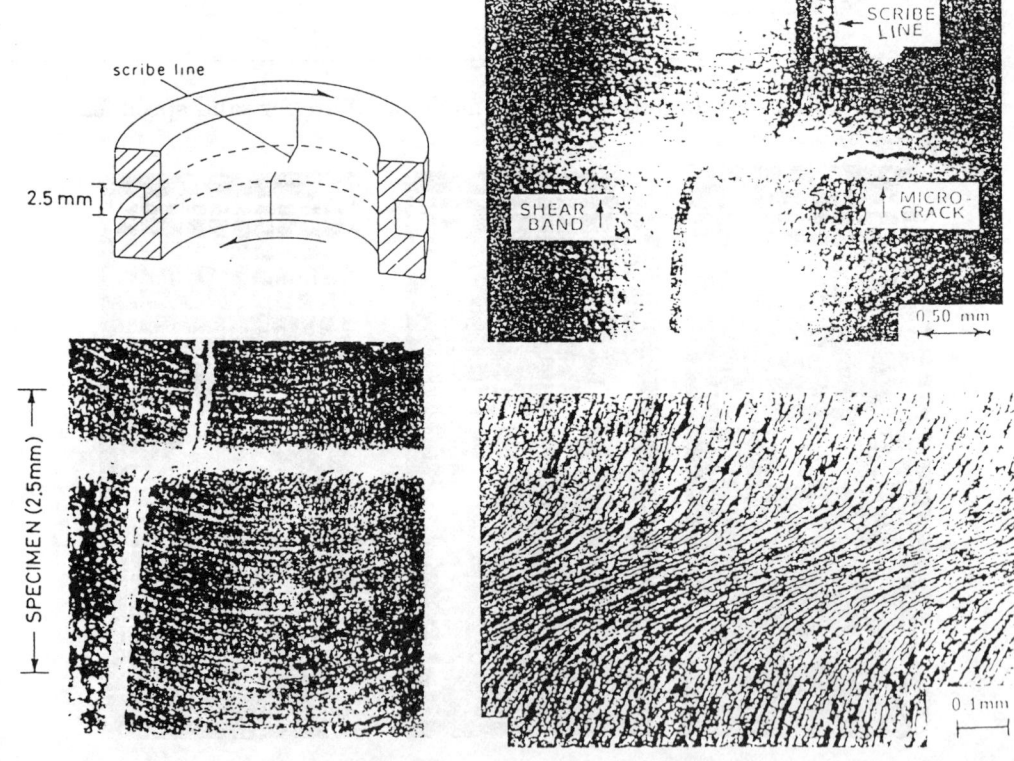

Figure 4: The formation of shear bands in dynamic twisting test of ductile material

2. Aims and motivations

To present the motivation which is based on experimentally observed physical facts and to define the background to this presentation, several levels of treatment of the localization are pointed out in BOX 1. If we concentrate on the continuum mechanics basis, one can notice that the phenomenon of localization which starts from the microlevel is not directly observable from this level of description. It seems to be natural that we have to introduce the additional information that allows us to model the nucleation and growth of microdefects which are macroscopically observable as a localization.

BOX 1

> **Level of Experimental Results**
>
> **Level of Mathematical Description of Experiments**
>
> – Statistic Mechanics
> – Microlevel Description
> – **Level of Continuum Mechanics Description**
>
>> – Continuum Damage Mechanics
>> – Fracture Mechanics
>> – **Level of Regularization of Plasticity with softening**
>>
>>> – Smeared Crack Models
>>> – Higher Gradients Models
>>> – Embedded Models
>>> – Cosserat Models
>>> – **Rate Dependent Models**
>>> – others
>
> **Level of Numerical Approximation**
>
>> – Finite Difference Method
>> – **Finite Element Method**
>> – Boundary Element Method
>> – others
>
> **Level of Comparison (Experiments vs. Results)**

If one restricts the attention to strain-softening rate independent materials, the local Cauchy problem and as a consequence the initial boundary value problems (IBVP) do not become well-posed after the critical loading has been reached. So, also the achieved numerical results that are based on such a formulation become meaningless.

The decision on using the continuum mechanics description for the analysis of strain localization has two important faces. From the first side, it is profitable because it allows to incorporate into the analysis of localization phenomenon, in particular for large, complicated engineering problems, the variety of well known and verified methods. On the other hand, by introducing of the softening (non positive definiteness of constitutive operators) there appear a new source of mathematical and numerical subtle difficulties which have to be overcome.

Two important questions have to be answered when we intend to solve any initial boundary value problem with localization. First, on the level of mathematical modelling one has to be able to predict when, where and in which direction the localization zone will appear and what will be its width. The problem of well-posedness has to be the focus of interest. Secondly, from the viewpoint of numerical simulation one should not avoid the discussion of mesh dependency. The last is particularly important because in some presentations there appeared the notions on high (unusual) mesh sensitivity of the results of calculations to the density of mesh adopted in computations.

The widths and the directions of localized zones, however, are not known a priori; they depend not only on material parameters but also upon the shape of the specimen and the initial and boundary conditions. In the phenomenological approach, which also is used in the present paper and is always used in continuum mechanics, a simple mapping of such experimental data into stress-strain relations provides a negative stiffness in the constitutive model. This fact can create considerable mathematical difficulties which, as a consequence, can result in unusual mesh sensitivity in the numerical calculations.

There is no agreement to the question, whether strain softening is a true material property and can be used in stress-strain relations. In FEM formulation softening is attributed to the material and is incorporated into the constitutive equations. However, from the other side we know, that the descending branch in the stress-deformation curve is the result of non-homogeneous deformation on the micro level which in fact is not visible from the level of continuum mechanics description. When talking about localization the other two facts have to be mentioned which are omitted in our consideration. The first is the question, whether the behaviour of the material in the plastic localization zones can be assumed as isotropic. The second is connected with the description of damage of the material. The development of the constitutive models which take into account the above pointed properties is the natural continuation of the formulations proposed in this presentation but will not be discussed here any more. Therefore special modelling techniques should be used to derive the proper material models with reliable parameters for strain softening.

For isothermal processes and rate–independent materials, this phenomenon has been analyzed as a material instability e.g. RUDNICKI & RICE [83]. This kind of treatment, which does not include length–scale parameters, practically precludes post–localization analysis. Usually, the analysis of eigenvalues and eigenvectors of the acoustic tensor leads only to the formulation of localization criteria (places and directions of singular localization lines). This formulation also simplifies the observed physical fact that in any experiment the width of the intensive straining zone is always finite even for very fast processes e.g. MARCHAND & DUFFY [62]. For such a rate independent description after the critical point has been reached, the problem becomes ill–posed, which prevents reliable analysis and as a result – reliable (useful) computations e.g. PERZYNA [75] and KIBLER et al. [40]. When one would like to continue the post–critical analysis after the plastic localization appears, then the analysis of well–posedness of the initial boundary value problem (IBVP) plays a crucial role [40]. Then, if the condition for well–posedness are fulfilled, one can continue the numerical calculations while avoiding the well-known problem of observed strong mesh sensitivity, which is the result of changing the type of the operator of the governing system of incremental equations.

According to the opinion of LEMAITRE and CHABOCHE [46]: *Tout solide est un fluide qui s'ignore* and many other experimental confirmations the choice of viscoplasticity as a rate dependent constitutive model for the plastic strain localization has a deep physical background for a wide range of materials and types of loading.

2.1 Short review of references

The basic works which are cited here in general reflect the following branches of interests:

- experimental evidence of the plastic strain localization effects that appear in laboratory tests of specimens and in engineering practice,

- mathematical background which describes the properties of the solution of the set of differential governing equations,

- the mechanical assumptions which drive to the creations of constitutive models,

- numerical aspects which, in particular, discuss the sensitivity of the results to the spatial discretization.

The book of RENARDY and ROGERS [81] was used as a classical one and it introduces the fundamental mathematical formulations and collects the definitions in the field of differential equations.

The idea of regularization of differential equations with the application to mechanics was presented in 1972 by DUVAUT and LIONS [23]. The other important achievements which should be mentioned here due to the theory of stability and bifurcation, often discussed and used as tools of treatment of localization are by LYAPUNOV [60]

(English translation in 100 anniversary of his thesis) and by HADAMARD [33]. The elements of linear and nonlinear functional analysis were presented by KURCYUSZ [44] and IONESCU and SOFONEA [38]. The ideas taken from these books mainly helped to proof the stability of the solution of the problem under analysis. The book by PAZY [71] among others introduces the method of semigroups which served as a powerful tool for the discussion of the contractive properties of the solutions of governing equations.

The mathematical method of regularization is also intensively used for the solution of ill-posed problems (e.g. inverse problems) see TIKHONOV and ARSENIN [91].

To introduce a good numerical approximation one has to be precise in the original mathematical description which is the basis of numerical analysis. Therefore the problem of mathematical well-posedness is fundamental to further discussion of the uniqueness of numerical results. Well-posedness has been studied by PAO [69], PAO and VOGT [70], KATO [39], HUGHES, KATO & MARSDEN [37], PERZYNA [75], de BORST et al. [16]. In the papers of LENGNICK et al. [47] and ŁODYGOWSKI et al. [56] the examples of viscoplastic regularization of brittle and ductile types of constitutive models were presented.

The crucial question from the viewpoint of constitutive modelling is whether the formation of strain concentration zones can be preceded by any other kind of nonuniqueness. The answer to this question was studied by HILL, RICE, RUDNICKI, HUTCHINSON, NEEDLEMAN [35, 34, 36, 83, 82] and has been given for particular conditions. HILL has shown [35] that for homogeneously deformed elastoplastic body with associative plasticity the sufficient condition for strain localization is the loss of uniqueness which appears as the vanishing speed of acceleration waves. For the plain strain tension and compression in the presence of large displacements HILL and HUTCHINSON [36] found that shear band localization may occur when the system of governing equations is hyperbolic (for statics). The review of the types of governing equations for different mechanical and thermal problems is discussed for example in the paper of GAWĘCKI and JANIŃSKA [28]. The criteria related to the uniqueness of the solution in elastoplastic associative and non-associative response are summarized in the work of BIGONI and HÜCKEL [12]. The loss of positiveness of the second order work is emphasized as a condition of appearing the strain localization into a narrow band. It is also known, that for statics the loss of ellipticity of the system of governing equations or the loss of positiveness of eigenvalues of the acoustic tensor may occur simultaneously or in the neighborhood of the loss of positiveness of second order work, and appearing the localization phenomenon.

The sufficient criterion for uniqueness of the solution of the boundary value problem appears in the form

$$\int_\Omega \Delta\dot{\sigma}\Delta\dot{\varepsilon}d\Omega > 0, \qquad (1)$$

where $\Delta\dot{\sigma}$ and $\Delta\dot{\varepsilon}$ are the stress and the strain rate fields and Ω represents the whole domain of the body. Δ refers to the difference between alternative fields. If we restrict our attention to the elastic-plastic solids, where stress and strain rates $\dot{\sigma}, \dot{\varepsilon}$ are related through a fourth order **D** tensor, different for loading and unloading, (symbol **D** will

be also used for matrix)
$$\dot{\sigma} = \mathbf{D}\dot{\varepsilon}, \qquad (2)$$
we can arrive at second order work expression:
$$W_2 = \frac{1}{2}\dot{\sigma}\dot{\varepsilon} > 0 \qquad (3)$$
as a restrictive local condition of uniqueness which assures the fulfillment of the global criterion of stability. Using the previous equations leads to the equivalent condition for uniqueness that is the positive definiteness of constitutive rate tensor $\mathbf{D}$. This condition is not so restrictive as positiveness of the second order work, but still sufficient to exclude strain localization if the rate deformation mode defined by a tensor product is introduced. This appears as the requirement of the positive definiteness of so called acoustic tensor $\mathbf{A} = \mathbf{n} \otimes \mathbf{D} \otimes \mathbf{n}$. Mathematically this is the condition of strong ellipticity of the system of differential equations governing the local equilibrium. The strain localization is attained when the system of local governing equations looses its ellipticity and become parabolic
$$det\,(\mathbf{nDn}) = 0, \qquad (4)$$
where $\mathbf{n}$ is the vector normal to the planar band. The persuasive interpretation of relationships between uniqueness criteria, second order work, strong ellipticity of equations and stability is presented in the paper of BIGONI and HÜCKEL [12].

For 1-D cases with a band appearing perpendicularly to the force direction the last criterion is equivalent to
$$det\,\mathbf{K} = 0, \qquad (5)$$
where $\mathbf{K}$ is a stiffness matrix of the system. The last equation can serve as a criterion for detecting the bifurcation and the eigenvectors of $\mathbf{K}$ for choosing the post-bifurcation mode for materials that exhibit softening (see e.g. LODYGOWSKI [49]). In the rate-independent continuous formulation which should always be the base of any computational procedure there appears new difficulties when we start to deal with numerical models which include the softening behaviour. The softening manifests in introducing, on the local level, not positively defined constitutive operators. This in consequence leads to the lost of positive definiteness of acoustic tensor. The attention has been paid to the formulation of numerical procedures for resolving the problems of numerical simulation of onset and grows of the localized deformation in elastic-plastic solids. Before we start the detailed discussion on how the problem can be treated numerically, the following issues should be emphasized: First, it is important to notice that since the incremental boundary value problem can have non-unique solutions (this can be the consequence of theoretical formulation) it is necessary to have some additional criteria to choose the post-bifurcation path. Secondly, it is necessary to obtain 'objective' parameters of the material when dealing with smeared models which are used to represent the discontinuities as in plastically deformed or brittle materials. Thirdly,

the numerical algorithm should capture in post critical states the "anticipated" mode of deformation; should be stable and fast convergent. It should be emphasized that the difficulties which are observed in numerics (choosing of post-bifurcation path, weak convergence and sensitivity of the results to the finite element mesh) arise from the theoretical formulation. For statics this exhibits loosing of ellipticity of governing equations and of course, as a consequence, nonuniqueness of the solutions in post critical states. Some ideas which are discussed below concentrate on a kind of enrichment of the formulation to reach the final set of equations elliptic (or hyperbolic for dynamics) to assure the unique solutions, and also to omit the problems with bifurcation and the choice of the post critical paths which usually requires the using in numerical calculation some kinds of imperfections (physical or geometrical type).

To avoid the strong mesh dependency observed on the level of computations the finite elements with embedded localization zones PIETRUSZCZAK and MRÓZ, BELYTSCHKO et al. [80, 9, 8] or with so called internal zone of localization KLISIŃSKI et al. [42] and also the spectral overlays concept BELYTSCHKO and FISH [7, 25] were developed.

To some extend the classical methods of limit analysis which assume the existance of concentrated plastic deformations (eg. plastic hinges in frames) and its development to the theory of gaps and distortions, like presented by GAWĘCKI [27] describe the localized deformations and its influence on the behaviour of the structure.

Application of rate independent material models does not ensure the well–posedness in post critical states. The classical papers of HILL [35], MANDEL [61], and RICE [82] stated that loss of material stability, and what follows localization, would not occur until at least one eigenvalue of the acoustic tensor is equal to zero. The used here Maxwell's or Hadamard's compatibility conditions define that the localization is associated with a strain rate jump within a planar band and it does not enforce any kinematic incompatibilities with the remaining material. According to rate independent model the further study of the evolution of the localization domain is not possible because of the change of the type of the incremental governing operator. Instead of studying the abstract Cauchy problem to discuss the posedness it is sometimes easier to concentrate the attention only on the acoustic tensor. An interesting study on bifurcation in inelastic materials has been recently published by NEILSEN and SCHREYER [65].

Using rate independent constitutive models, for static problems, if the system of governing equations changes from elliptic to hyperbolic (if one of the eigenvalue $\lambda = 0 \rightarrow$ parabolic), we can not expect the results of the described physical phenomena in real spaces (ill-posedness).

Moreover, the bifurcation analysis mentioned above predicts only the localization line which basically is not to full extend confirmed in laboratory experiments. It appears, that always the localization zones have finite width, which depends upon material properties but also very strongly on boundary and initial conditions.

The standard, rate independent continuum, models when one introduces material instability in the Drucker's sense

$$\dot{\varepsilon}^T \dot{\sigma} < 0 \qquad (6)$$

lose the ellipticity of the governing material operator, and the initial boundary value problem (IBVP) loses well-posedness at the development of the localization zones. Then, for the wide class of commonly acceptable, and from the point of view of engineering applications, useful materials both the analytical and numerical solutions become meaningless.

Furthermore, when the computational results are obtained for these materials one can easily prove that they exhibit so called "pathological mesh dependency". For these reasons one has to be very careful in starting or implementing the results of numerical calculations until some fundamental problems on the level of mathematical modelling are clear.

The classical rate-independent continuum signals the onset of strain localization via bifurcation at the level of constitutive relations (see RUDNICKI and RICE [83]). For example, the velocity gradient may exhibit discontinuity across the singular surface with the specific orientation **n**. This bifurcation coincides with the loss of ellipticity (for statics) or hyperbollicity (for dynamics) of the governing system of equations, and it is reflected by a singularity of the acoustic tensor **A**.

To regularize the system of equations which drives the incremental process in post localization states, one has to introduce a length-scale parameter which leads to the specification of the width of localization.

A comprehensive study was presented by SLUYS [84]. To enrich the softening continuum the length-scale parameter can be introduced into the formulation in a different way; for example, by using nonlocal theories BAZANT & PIJAUDIER–CABOT [6], by the addition of higher-order strain terms (second order) BELYTSCHKO [10], by introducing a gradient model de BORST [13, 14] and PAMIN [68], by the inclusion of micro–polar effects de BORST [16], STEINMANN & WILLAM [87] or by using rate dependent formulation PERZYNA [75], NEEDLEMAN [64], LENGNICK et al. [47], ŁODYGOWSKI et al. [54].

The common feature of the proposed methods of regularization is the effect of dispersion, that is the observation that harmonic waves, with a different frequency propagate with different velocity. The classical strain-softening medium is not dispersive, i.e. the continuum is not able to transform waves into stationary localization waves and as a consequence is not able to reproduce the effect of starin localization.

Finally, an avoiding of mesh dependency in the problems with softening materials is also possible by introducing the width of the localization zone (length-scale parameter) explicitly into the formulation on the level of approximation (finite elements) as it was shown by PIETRUSZCZAK & MRÓZ [80], BELYTSCHKO et al. [11], ORTIZ et al. [66] or ŁODYGOWSKI [50]. But in these cases, in view of the experimental evidence which confirms that the width of localization strongly depends on initial and boundary conditions, one can expect serious difficulties with its explicit definition. Moreover, some of the proposed enrichments, however, computationally efficient, physically are not justified.

The only drawback of the rate-dependent formulation seems to be the necessity of full dynamic analysis of the process under consideration. In the author's opinion the rate-dependent model is physically well–founded especially for ductile metals and thus has a variety of advantages in comparison with the other models. Using such parameters as viscosity, particularly for fast mechanical processes, has a deep physical explanation on the micromechanical level. For these reasons we will use viscoplasticity to describe the physical process as well as a tool for the mathematical regularization of softening behaviour.

In view of the recent achievements [47, 54, 40] for viscoplastic (rate-dependent) formulation, the problem remains well–posed at each interval of time so the unique solution in numerical calculations can be obtained. Using a dynamic formulation for two-dimensional cases, both failure modes (I-mode and II-mode) can be performed, and also because of the wave propagation phenomenon no artificial imperfections are necessary in computations to be superposed to activate the process of localization.

In the work, we are going to simulate numerically the softening behaviour of the structures in the range of large plastic deformations which exhibit localization. The process of numerical modelling of physical phenomena requires several steps which should precede the computations. The consistent way of treating the modelling of physical behaviour starts from the mathematical formulation. In author's opinion, before the computations start the system of governing equations should be studied with respect to the properties of its solution without explicit construction of the solution. After such a discussion one can be sure that the solution obtained by numerical calculations will be unique and free of unexpected mesh sensitivity.

For the society of mathematicians some thoughts which clarify the problems of existence, uniqueness and the stability of the solutions could not be new, however the successful application of quite complicated constitutive models to solve the boundary value problems could be surprising. For mechanicians, who deal with numerical solutions using powerful finite element codes, the attraction of the attention towards the mathematical basis of the formulation could be instructive. The results and the treatment of the problem shown in this work should warn the analysts against *to fast numerical analysis* and the interpretation of the results before the mathematical description of real physics is fully recognized.

Many aspects of the treatment of plastic strain localization in continuum mechanics on the levels of material science [43, 88] theoretical formulation and experimental evidence [78] as well as algorithmic and numerical methods [4, 85] are widly discussed in this volume.

One of the aims of this presentation is to show numerically that attaching the importance to proper mathematical formulation helps to avoid the serious mesh sensitivity. Some aims of this work are as follows:

- to introduce the definitions of two types of mesh dependency,

- to present the IBVP which includes the constitutive equations namely viscoplastic which base on the physically motivated behaviour for ductile and soil type materials,

- to present the numerical results that support the concept of viscoplastic regularization,

- finally to clarify the methodology of treatment the problem of the numerical simulation of plastic strain localization with softening effects.

To achieve the goals the work is organized as follows.

In the following two sections we introduce two constitutive models of the materials (soil, critical state line type model and ductile metal). In Section 3 the effect of regularization of plastic flow localization in a soil material is presented. The effect of *pathological* mesh dependency for classical rate independent model find there its evidence. The viscoplastic regularization and the numerical results that confirm the reduction of spurious mesh sensitivity are presented. Also the numerical problems are discussed in this chapter. At the end, the examples taken from the engineering practice (stability of slopes and breakwater under dynamic loading) are shown.

The using of viscoplastic ductile type material under dynamic load cases is presented in Section 4. Among the other aspects, the attention has been paid to show the influence of boundary conditions on the creation of localization domains.

Section 5 briefly presents the aspects of sensitivity analysis in the problems with plastic strain localization in softening materials. This analysis should not be mixed with the sensitivity in the pure analysis of the results to the discussed mesh density. The problem of the choice of the objective functional seems to be the crucial one.

The concluding remarks are the essence of the last section.

The list of selected references close the work.

3. Soil strain–softenning material

3.1 Notation

Let us introduce the well-known multiplicative decomposition of the deformation gradient $\mathbf{F}$ in the form (see LEE [45])

$$\mathbf{F} = \mathbf{F}^e \mathbf{F}^p, \tag{7}$$

where $\mathbf{F}(\mathbf{X}, t) = \frac{\partial \mathbf{x}(\mathbf{X}, t)}{\partial \mathbf{X}}$ and $\mathbf{F}^e$, $\mathbf{F}^p$ are the so-called elastic and plastic deformation gradients, respectively. The velocity gradient $\mathbf{L}$ can be then decomposed into the elastic and plastic parts

$$\mathbf{L} = \mathbf{L}^e + \mathbf{L}^p, \tag{8}$$

and each of them can be further presented as a sum of its symmetric and nonsymmetric parts; for example, the elastic part is written as

$$\mathbf{L}^e = \mathbf{D}^e + \mathbf{W}^e. \tag{9}$$

The symmetric part of $\mathbf{L}$, namely $\mathbf{D}$, represents the stretching tensor and $\mathbf{W}$ the spin tensor.

Next, let us define the bar form of the selected second order tensors with respect to the group of rotations $\mathbf{Q}^e$ by the following transformations (see GURTIN [32])

$$\bar{\mathbf{T}} = \mathbf{Q}^{eT} \mathbf{T} \mathbf{Q}^e, \qquad \bar{\mathbf{D}} = \mathbf{Q}^{eT} \mathbf{D} \mathbf{Q}^e, \tag{10}$$

where $\mathbf{T}$ represents the Cauchy stress tensor and $\mathbf{Q}^e(t)$ defines the group of time-dependent rotations through the initial value problem

$$\dot{\mathbf{Q}}^e = \mathbf{W}^e \mathbf{Q}^e, \qquad with \qquad \mathbf{Q}^e(0) = \mathbf{1}, \tag{11}$$

where $\mathbf{1}$ is identity tensor and $(\ )^T$ denotes transposition. This rotational-neutralized (pull-back) tensorial quantities which are introduced here (see also GURTIN [32]) will be used to define the convenient framework for integration of the constitutive model. With these definitions the time derivative of the Cauchy stress tensor satisfies the equation

$$\dot{\mathbf{T}} = \mathbf{Q}^{eT} \mathbf{T}^{\nabla e} \mathbf{Q}^e \equiv \bar{\mathbf{T}}^{\nabla e}, \tag{12}$$

and is objective in the sense of material frame-indifference (cf. TRUESDELL and NOLL [92]), i.e. is invariant with respect to any superposed rigid motion.

3.2 Rate independent material and mesh sensitivity

Let us summarize some basic assumptions of an elastic-plastic model of soil-like material with the yield conditions which depend upon the hydrostatic pressure and changes of porosity. This material model belongs to the so-called critical state models (see WOOD [96]) and was originally introduced by PIETRUSZCZAK & MRÓZ in [79] to model granular and rock materials. Let us assume also that the yield condition is a function of stresses and depends on the irreversible part of porosity or density variation η, which is the internal state variable. Hardening and/or softening behaviours are introduced through the evolution of this internal scalar state variable. The model which is presented intends to simulate, contrary to the known Cam-Clays, the constitutive behaviour which is not restricted to the purely cohesionless materials. Similarly to the other Cam-Clay-type models, yielding depends on the hydrostatic pressure and critical state line separates two regions of different behaviours: hardening or softening (for detailed discussion see WOOD [96], LORET [59], DRESCHER [21], ADACHI & OKA [2] et al.). In this model, on the so-called "dry" side the material dilates and, in a consequence softens, while on the so-called "wet" side it hardens what accompanies compaction. Main property of this model is that on the critical state line the material can yield at constant shear stress with no volume changes. The other important property is that the yield is influenced by the mean principal stress. We restrict our attention to the one-phase material model.

The relative bulk density, that is the introduced scalar variable, is defined by

$$\eta = \frac{\rho}{\rho_0}, \tag{13}$$

where ρ denotes the mean bulk density, and ρ_0 the intrinsic material bulk density at a reference, in general, unloaded configuration, respectively. The porosity which could be defined as

$$\beta = \frac{V_v}{V_t}, \tag{14}$$

in which V_v and V_t denote the volume of voids and the total volume of the representative specimen can play also a similar role of an internal state variable. η can be expressed also as

$$\eta = \frac{V_m}{V_t} = \frac{V_t - V_v}{V_t} = 1 - \beta, \tag{15}$$

where additionally V_m describes the material volume. The change of the internal state variable η is given by

$$\dot{\eta} = \frac{\dot{\rho}}{\rho_0}. \tag{16}$$

From the continuity condition

$$\dot{\rho} + \rho \, div \mathbf{v} = 0, \tag{17}$$

where $\mathbf{v}$ denotes the velocity, we arrive at

$$\dot{\rho} = -\rho tr(\bar{\mathbf{D}}), \tag{18}$$

and after the decomposition of $\bar{\mathbf{D}}$ into elastic and plastic parts we obtain

$$\dot{\eta} = \underbrace{-\eta tr(\mathbf{D}^e)}_{\dot{\eta}^e} - \underbrace{\eta tr(\mathbf{D}^p)}_{\dot{\eta}^p}. \tag{19}$$

The second term of the right-hand side of (19) $\dot{\eta}^p$ is the irreversible rate of change of porosity. Finally, the set of equations for the rate-independent material model expressed in transformed form with respect to rotation group $\mathbf{Q}$ is as follows:

- Evolution equation for the Cauchy stresses

$$\dot{\bar{\mathbf{T}}} = \mathbf{C} : (\bar{\mathbf{D}} - \bar{\mathbf{D}}^p), \tag{20}$$

where $\mathbf{C} = 2G\mathbf{I} + (K - \frac{2}{3}G)\mathbf{1} \otimes \mathbf{1}$ is an elastic material tensor; $\bar{\mathbf{D}}^p$ is the tensor of the rate of plastic deformation which has the form

$$\bar{\mathbf{D}}^p = \langle \dot{\lambda} \rangle \bar{\mathbf{N}}, \quad \bar{\mathbf{N}} = \frac{\partial f}{\partial \bar{\mathbf{T}}}, \tag{21}$$

and where

$$\langle \dot\lambda \rangle = \begin{cases} \dot\lambda & : \quad \text{if} \quad f = 0 \quad \text{and} \quad \bar{\mathbf{N}} : \mathbf{C} : \bar{\mathbf{D}} > 0, \\ 0 & : \quad \text{if} \quad f \neq 0 \quad \text{or} \quad f = 0 \quad \text{and} \quad \bar{\mathbf{N}} : \mathbf{C} : \bar{\mathbf{D}} \leq 0. \end{cases}$$

f is the yield function which depends, like in critical state line type models, upon deviatoric stresses, mean pressure and also on the introduced internal state variable η and its evolution (19). The yield function is assumed in the form

$$f = \hat{f}(S, p, c) = (p - c)^2 + \frac{1}{2}\left(\frac{S}{d}\right)^2 - \left(\frac{\mu_c c}{d}\right)^2 \leq 0, \qquad (22)$$

and the following notations are used:

$$S = \sqrt{\bar{\mathbf{S}} : \bar{\mathbf{S}}}, \qquad \bar{\mathbf{S}} = \text{dev}\left(\bar{\mathbf{T}}\right), \qquad p = -\text{tr}\left(\bar{\mathbf{T}}\right)/3. \qquad (23)$$

The geometrical interpretation of this yield function in (S, p) space and its evolution are discussed in many papers, e.g. see DRESCHER [21]. Also following this work we have adopted the material function c in the form

$$c = \hat{c}(\eta^p) = \alpha(\eta_0 + \eta^p) - c_0 \qquad (24)$$

as well as the constant parameters α, η_0, c_0 used in numerical calculations. The system of equations which describes the constitutive relation consists of eqns. (19)–(24). Different values of parameters c, d, μ_c define in $(S, -p)$ space different yield surfaces for which, contrary to the well documented Cam–Clay models, carrying of small tension stresses is also possible. As an alternative of the discussed yield function f, that one which defines the modified Cam–Clay can also be used. For the latter model the yield surface consists of two elliptical segments in the (S, p) space

$$\frac{1}{\psi^2}\left(\frac{p}{a} - 1\right)^2 + \left(\frac{S}{Ma}\right)^2 - 1 = 0, \qquad (25)$$

where ψ is a constant used to modify the shape of the yield surface on the "wet" side of the critical state, M is the slope of the critical state line and defines the position of the yield origin along the axis of mean pressure value p, a is a constant. The response on the "dry" side of the critical state for both above models expressed in the (S, γ) space (γ is the shear angle) exhibits the softening behaviour, while on the "wet" side it tends to some limit value with no negative slope. The detailed discussion of the models can be found in PIETRUSZCZAK & MRÓZ [79], WOOD [96]. Some practical applications as well as the discussion of the used parameters one can find in the works of ADACHI & OKA [2, 3]. For rate-independent formulation one can expect the ill–posed problems and, as a consequence, in numerical computations the significant mesh sensitivity when the yielding enforce the shrinking of the yield surface (softening) (see discussion in the work of PERZYNA [76] and also the study of De BORST [15], SLUYS [86] and

ŁODYGOWSKI [51]).
Application of the viscoplastic model to the cases involving soil materials may be criticized. But one has to agree that all the formulations indicated in BOX 1, and used to describe the localization, have their advantages and drawbacks. We believe, that each soil-clay material exhibits certain amount of viscosity, and in particular for the dynamical load cases applications of viscoplasticity is justified and finds its experimental confirmation; e.g. see the works of TAVENAS et al. [90, 89] or LEROUEIL et al. [48].

For this purpose, the viscoplastic regularization is proposed to overcome the ill–posedness and to assure invariability of the type of the governing operator for the whole process, even in the post-critical states.

3.3 Remarks on mesh–dependence

In many papers the problem of unexpected mesh-dependence of the numerical results obtained for rate-independent plasticity was discussed (for example, see [64, 75, 50]). Obviously, the numerical formulation and calculations always lead, in view of the algebraic character of the finite element technique, to approximate solutions. But for well–posed BVP the results obtained by FEM should converge to the real analytical solutions. Of course, this is true under the condition that the problems are mathematically well–posed. For ill–posed problems, one can expect extremely strong mesh sensitivity in computations and the results become meaningless.

For the purpose of this presentation let us accept now the following two definitions:

3.3.1 Definition 1

Mesh dependence of the first order (primary mesh dependence – PMD) is the one which follows directly from the mathematical ill–posedness of the BVP.

For ill-posed problems the uniqueness and the stability of the solution can not be proved. As a simple consequence, when the algebraic solution based on ill-posed formulation is constructed, a serious mesh sensitivity can be easily observed in computations. The results which are obtained in these cases exhibit different responses (particularly in the post-critical range), when the analysis is started from changed mesh. If this kind of mesh dependence appears (PMD), it should be recognized as an effect of ill-posedness of the original mathematical formulation. For ill–posed problems the results of numerical calculations are sensitive to the finite element mesh in an unexpected manner, without the possibility of estimation of the errors that appear.

3.3.2 Definition 2

Mesh dependence of the second order (secondary mesh dependence SMD) reflects the well–known influence of spatial and time discretization.

For well–posed problems it approaches in the limit the analytical solution when finer meshes are used.

In this paper we are going to avoid only the *mesh dependence of the first order* PMD which, in our opinion, is the most important aim in case we are dealing with plastic localization and softening problems.

The *mesh dependence of the second order* SMD can be always avoided by applying different sophisticated numerical techniques (eg. adaptive remeshing) and can be taken into consideration only if the primary problems are satisfied. All the methods of regularization which are shown in Box 1, introduce, in fact, in a different way a length scale parameter. It is achieved for example explicitly in embedded elements and in Cosserat formulation, or implicitly for nonlocal theories. The length scale parameter which is introduced in the formulation assures the well–posedness and the possibility of continuation of the analysis after localization criterion is satisfied and it influences the width of the localization zones as well. For the methods mentioned above which introduce the length scale parameter, one can observe that the value is chosen almost arbitrarily (it can depend on the grain size).

It seems to be natural that, basing on the experimental observations which show that all the materials have some viscous properties, we choose PERZYNA's viscoplasticity [72, 73, 74] as the tool of regularization.

3.4 Rate dependent – viscoplastic regularization

The rate-dependent material model which is expressed in a form transformed with respect to the rotation group $\mathbf{Q}$ can be performed in the following manner

$$\dot{\bar{\mathbf{T}}} = \mathbf{C} : \left(\bar{\mathbf{D}} - \bar{\mathbf{D}}^{vp}\right), \tag{26}$$

where $\bar{\mathbf{D}}^{vp}$ is the viscoplastic rate of deformation tensor and is assumed as [72]

$$\bar{\mathbf{D}}^{vp} = \varphi \langle \phi(S, p, \eta^p) \rangle \frac{\partial \phi}{\partial \bar{\mathbf{T}}} \tag{27}$$

In the above formula φ denotes the viscosity and the associative type of plasticity is taken into account, ϕ is empirical overstress function and $\langle \cdot \rangle$ denotes the Macauley bracket which is understood as

$$\langle \phi(F) \rangle = \begin{cases} \Phi(F) & \text{if } F > 0, \\ 0 & \text{if } F \leq 0. \end{cases} \tag{28}$$

The viscosity φ is sometimes denoted by $\varphi = \frac{1}{T_m}$ where T_m is relaxation time for mechanical disturbances. The relaxation time which can be used for soil-like materials is of the order of $10^{-3}s$. The evolution equation for the irreversible part of the porosity changes is

$$\dot{\eta}^{vp} = -\eta tr(\bar{\mathbf{D}}^{vp}). \tag{29}$$

The important feature derived from the viscoplastic formulation of the problem for numerical calculation is the existence, uniqueness and stability of the abstract Cauchy problem. The discussion of this problem, after the condition had been specified by KATO [39] and HUGHES, KATO & MARSDEN [37], was presented by PERZYNA in [75, 76].

Since in numerical calculations we restrict our attention to 2–D plane strain problems, let us now specify the necessary expressions which describe the constitutive model for this case. The matrix **C** takes now the form

$$\mathbf{C} = 2G\mathbf{I} + (k - G)\mathbf{1} \otimes \mathbf{1}, \tag{30}$$

where $k = K + \frac{1}{3}G$, and the function ϕ is expressed similarly to (Eq. 59) as

$$\phi = (p - c)^2 + \frac{1}{2}(\frac{S}{d})^2 - (\frac{\mu_c c}{d})^2, \tag{31}$$

with the material function c given by (59). The viscoplastic part of the rate of deformation is now represented by

$$\bar{\mathbf{D}}^{vp} = \varphi \langle \phi(S, p, \eta^p) \rangle [\frac{S}{d^2}\bar{\mathbf{n}} - (p - c)\mathbf{1}], \tag{32}$$

where $\bar{\mathbf{n}} = \bar{\mathbf{S}}/S$.

3.5 Numerical solution

3.5.1 Numerical integration of constitutive relations (Gauss point level)

To approximate the value of any function y in the vicinity of certain point where it is equal to y_0, one can adopt the general form of the well-known trapezoidal operator such as the relation

$$y = y_0 + t[(1 - \Theta)f(y_0, 0) + \Theta f(y, t)], \tag{33}$$

for $0 \leq \Theta \leq 1$, where t represents the time interval and f time derivative of y. Using the limit values of Θ we arrive at the so-called fully implicit (for $\Theta = 1$) or fully explicit (for $\Theta = 0$) estimates, respectively. Particular application of this trapezoidal operator produces the following numerical approximation of Kirchhoff stresses

$$\bar{\mathbf{T}}_{n+1} = \bar{\mathbf{T}}_n + t[(1 - \Theta)\dot{\bar{\mathbf{T}}}_n + \Theta\dot{\bar{\mathbf{T}}}_{n+1}], \tag{34}$$

where subscripts n and $n + 1$ denote the two neighbouring states. Assuming the evolution of stresses as in (64) and viscoplastic rate of deformation as (68) in (34) for 3–D case, one can arrive at

$$\bar{\mathbf{T}}_{n+1} = \overbrace{\bar{\mathbf{T}}_n + 2G\Delta\bar{\mathbf{e}} + k\Delta e \cdot \mathbf{1}}^{\bar{\mathbf{T}}^{pre}_{n+1}} - 2G \cdot t(1 - \Theta)\dot{\gamma}^{vp}_n \bar{\mathbf{n}}_n - 2Gt\Theta\dot{\gamma}^{vp}_{n+1} \cdot \bar{\mathbf{n}}_{n+1}, \quad 0 \leq \Theta \leq 1 \tag{35}$$

where $\Delta\bar{\epsilon} = \bar{D}\cdot t$, $tr(\bar{D})\cdot t = tr(\Delta\bar{\epsilon}) = \Delta e$ and $\dot{\gamma}_n^{vp} = \varphi\langle\phi_n(\cdot)\rangle$. If we apply the full backward operator ($\Theta = 1$), and after the decomposition of Kirchhoff trial (elastic predictor) stresses into deviatoric and mean pressure components we obtain

$$\bar{S}_{n+1} = \bar{S}_{n+1}^{pre} - 2Gt\dot{\gamma}_{n+1}^{vp}\cdot\bar{n}_{n+1},$$
$$\bar{p}_{n+1} = \bar{p}_{n+1}^{pre} + k\Delta e^{vp}. \tag{36}$$

The first trial (elastic predictor) step of numerical algorithm gives

$$\bar{T}_{n+1}^{pre} = \bar{T}_n + 2G\Delta\bar{\epsilon} + k\Delta e\mathbf{1},$$
$$\bar{S}_{n+1}^{pre} = \bar{S}_n + 2G\Delta\bar{\epsilon},$$
$$p_{n+1}^{pre} = p_n - k\Delta e. \tag{37}$$

Evaluating the estimation of porosity and using the same way as before (backward operator), we arrive at

$$\eta_{n+1}^p = \eta_n^p - \eta\cdot t\cdot tr(\bar{D}) = \eta_n^p - \eta\Delta e, \tag{38}$$

where η could be obtained directly from the evolution law $\dot{\eta} = -\eta tr\bar{D}$ by exact integration. So we have

$$\eta = \eta_0 e^{-\Delta e}. \tag{39}$$

It is more convenient to present the last result in terms of p_{n+1}^{pre}, then in the short form Δe using the last equation of (38). Then the evolution of porosity η has the form as follows

$$\eta_{n+1} = \eta_n \cdot e^{\frac{p_{n+1}^{pre} - p_n}{k}}. \tag{40}$$

Instead of cumbersome integration, the problem reduces now to the solution of the following nonlinear system of equations

$$\bar{S}_{n+1} = \bar{S}_{n+1}^{pre} - 2Gt\varphi\langle\phi_{n+1}(\cdot)\rangle\bar{n}_{n+1},$$
$$p_{n+1} = p_{n+1}^{pre} + k\Delta e^{vp},$$
$$\eta_{n+1} = \eta_n e^{\frac{p_{n+1}^{pre} - p_n}{k}}. \tag{41}$$

Let us observe that after multiplication of the first equation of the above system by $\bar{n}_{n+1}$ we arrive at the purely algebraic following system of equations

$$\bar{S}_{n+1} = \bar{S}_{n+1}^{pre} - 2Gt\varphi\langle\phi_{n+1}(\cdot)\rangle,$$
$$\bar{p}_{n+1} = \bar{p}_{n+1}^{pre} + k\Delta e_{n+1}^{vp},$$
$$\eta_{n+1} = \eta_n e^{\frac{p_{n+1}^{pre} - p_n}{k}}. \tag{42}$$

The last equation of (42) can be directly used in the equations (42.1) and (42.2) because ϕ_{n+1} is the function of η_{n+1} and also $\Delta e_{n+1}^{vp} = tr(\Delta\bar{\epsilon}^{vp}) = t\cdot tr(\bar{D}^{vp}) = t\cdot tr(\varphi\langle\phi(\cdot)\rangle\frac{\partial\phi}{\partial\mathbf{T}})$, so finally we will use the system of only two algebraic equations.

The system of equations, which has to be solved at each integration point at all iterations, can be highly nonlinear, so we have to be sure that we are able to obtain a good convergence of results for a wide class of nonlinear expressions. For this purpose the BROWN solver [24] of nonlinear system of algebraic equations was tested for a lot of sophisticated functions and finally applied to the general elastic predictor – plastic corrector algorithm. It was numerically proven that even for very difficult and almost singular nonlinear functions starting from the arbitrary point the solver achieved the solution with good accuracy after only a few iterations. One can imagine, that the efficiency of this solver, which has to be used several times at each integration point and at each increment, is crucial for the finite element calculations. According to the numerical experience collected with this solver, we have never observed any difficulties with its convergence.

This integration procedure which is performed on the level of the Gauss point was included into the definition of the user subroutine UMAT that includes users own constitutive law in the commercial program ABAQUS [1]. In this case ABAQUS serves as the finite element framework to the solution of nonlinear mechanical problem. This attractive possibility which allows to include our own material law through subroutine UMAT requires the definition of the constitutive Jacobian matrix. It can influence the speed of convergence, for example the quadratic one, if a consistent linearization is used. In the other formulations of constitutive Jacobian matrix one can expect a slower convergence, but it should not influence the lack of generally convergent properties of the algorithm.

The summary of the main steps of constitutive algorithm is shown in BOX 2.

BOX 2

> **Summary of the constitutive algorithm**
>
> 1. Calculate the trial stress and normal mean pressure
>
> $$\bar{\mathbf{T}}^{pre}_{n+1} = \mathbf{T}_n + \mathcal{C}[\Delta E_{n+1}], \quad p^{pre}_{n+1} = -\frac{1}{3}tr(\bar{\mathbf{T}}^{pre}_{n+1})$$
>
> 2. Deviatoric trial
>
> $$\bar{\mathbf{S}}^{pre}_{n+1} = \bar{\mathbf{T}}^{pre}_{n+1} + p^{pre}_{n+1}\mathbf{I},$$
>
> 3. If $\Phi(\cdot) \leq 0$ then the deformation is elastic
> the constitutive algorithm is complete
> else
> continue
> 4. Solve the system of algebraic equations (eqn. 36)
> 5. Calculate the radial return factor λ_{n+1}
> 6. Update stresses
>
> $$\bar{\mathbf{T}}_{n+1} = \lambda_{n+1}\bar{\mathbf{S}}^{pre}_{n+1} + p^{pre}_{n+1}\mathbf{I}$$
>
> $$\mathbf{T}_{n+1} = \mathbf{Q}_{n+1}\bar{\mathbf{T}}_{n+1}\mathbf{Q}^T_{n+1}$$

3.5.2 Numerical integration of the equations. Level of BVP

The method of regularization which is used in the paper, namely the viscoplastic one, requires fully dynamical formulation. ABAQUS offers the dynamic analyses both the explicit and implicit operators to solve the incremental problem. Explicit schemes determine the values of quantities at time t_{n+1} based on the available values at time t_n, but the procedure is only conditionally stable and the length of time increment that defines the stability limit is approximately equal to the time for an elastic wave to cross the smallest element dimension in the model. Implicit schemes remove this bound on the time step size but nonlinear system of equations has to be solved. Usually in structural problems implicit integration schemes give good solutions with the time steps by one or two orders of magnitude larger than the stability limits of the explicit version. In the finite element calculations of the dynamic boundary value problems we have used the wide spectrum of ABAQUS possibilities.

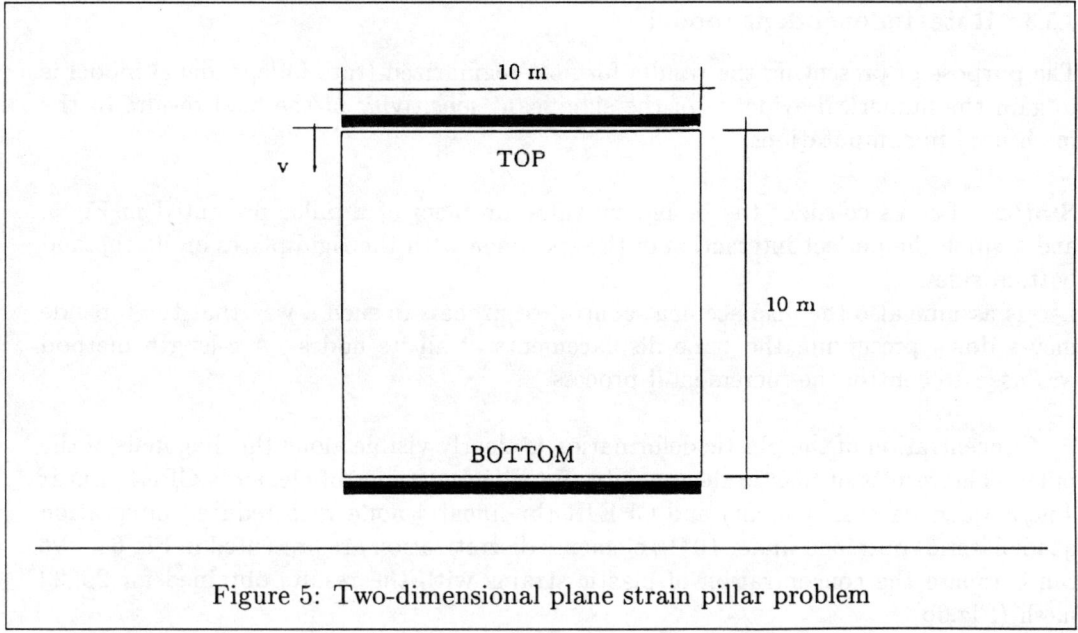

Figure 5: Two-dimensional plane strain pillar problem

Test problem Let us examine a set of examples concerning a 10m by 10m pillar problem treated as a 2–D plane strain case under different mesh density; Fig.5. Basically we will use three meshes 10*10, 20*20 and 40*40 of rectangular elements of a linear shape functions.

The results for both the rate-independent (without regularization) and rate-dependent models are presented and discussed. We will show the critical mesh dependence (sensitivity) of the first order for a rate-independent model in static or dynamic cases, and significant improvement of the results by using regularized viscoplastic dynamic formulation.

The interpretation of the mesh dependence is according to ŻYCZKOWSKI's classification [99] presented on the level of structure (Force–displacement space) and on the level of point (distribution of the equivalent plastic strains $\bar{\varepsilon}^{pl} = \int \sqrt{\frac{2}{3}\mathbf{D}^p : \mathbf{D}^p}\, dt$).

The data accepted according to DRESCHER [21] in the finite element calculations reflect the parameters for medium granular sand: E = 29.4 MPa, ν = 0.3, α = 4.905, n = 0.58, c_0 = 7.8, $\eta_0 = 1.64 kg/m^3$.

The results presented here concern mainly the 4 node bi–linear reduced–integration elements with hourglass control.

3.5.3 Rate–independent model

The purpose of presenting the results for non–regularized (rate-independent) model is to gain the numerical evidence of the significant sensitivity of the final results to the mesh used in computations.

Statics Let us consider the boundary value problem of a pillar presented in Fig.5, and assume the perfect interaction of the specimen with the rigid plates on its top and bottom sides.
Let us assume also the displacement-controlled process in such a way that the top side moves down preserving the same displacements of all its nodes. Arc-length method was used to control the incremental process.

Concentration of the plastic deformation is clearly visible along the diagonals of the pillar. The results of plastic shear strains for different types of elements CPE4 (linear 4-node quadrilateral element) and CPE4R (bi–linear 4-node with reduced integration quadrilateral) obtained under 10*10 elements discretization are presented in Fig.6a. We can compare the concentration of plastic strains with the results obtained for 20*20 mesh (Fig.6b).

It is natural that for this rate-independent formulation the width of localization should tend to zero. The tendency observed in numerical experiments confirms this expectation and shows more concentrated strains around the diagonals for finer meshes. The confirmation of a strong mesh sensitivity on the global level is shown in Fig.7 where the diagram of the total forces versus the displacement of the top side of the pillar is plotted. For the static case, the different meshes used in the calculations lead to different behaviour in the post–critical states (after the peak load has been reached), and also predict different maximal values of forces that the structure can carry.

Dynamics The same specimen was loaded in a dynamical way. The top side moves with the constant velocity v = 3.125 m/s so that after t = 0.16 s the displacement of this side is 0.5 m. In this case, contrary to the static one, the localization does not propagate symmetrically but it starts first from the top side. One can also observe the influence of the elastic wave propagation which in the whole process runs over the pillar, reflects from the bottom side and comes back to the top. This wave plays the role of imperfection, so no other artificial disturbances are necessary to enforce the place of localization.

In Fig.8 we can observe the development of equivalent plastic shear strain zones obtained for the 40*40 mesh calculations. Comparing the results of distribution of the plastic equivalent shear strains for two meshes at the same time $t = 0.16s$, one can easily recognize the significant difference (see Fig.8d and Fig.9) which is the result of PMD.

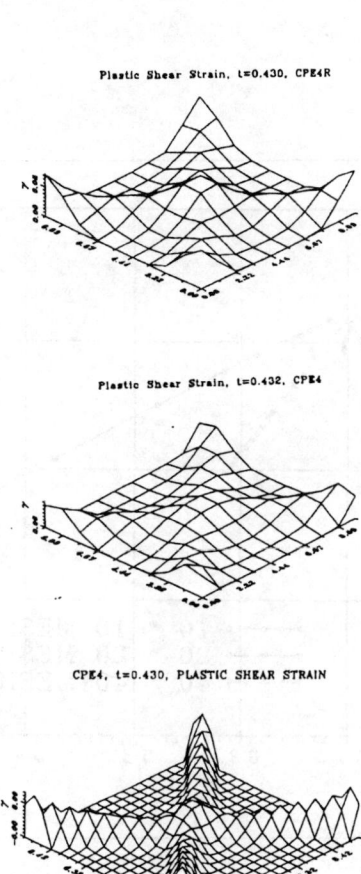

Figure 6: Distribution of plastic shear strain under static loading for 10*10 meshes and 20*20 mesh; a) CPE4 (bi-linear elements) and CPE4R (bi-linear elements with reduced integration) 10*10 meshs, b) CPE4 (bi-linear elements 20*20 mesh

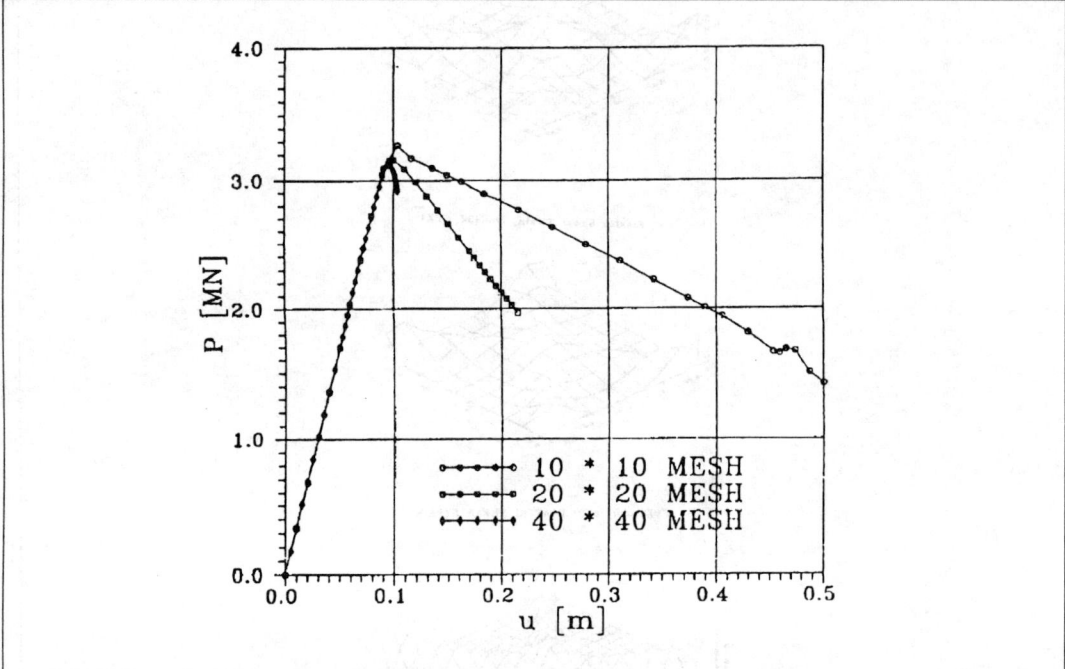

Figure 7: Mesh dependence for static case on the global level. Total force vs. displacements

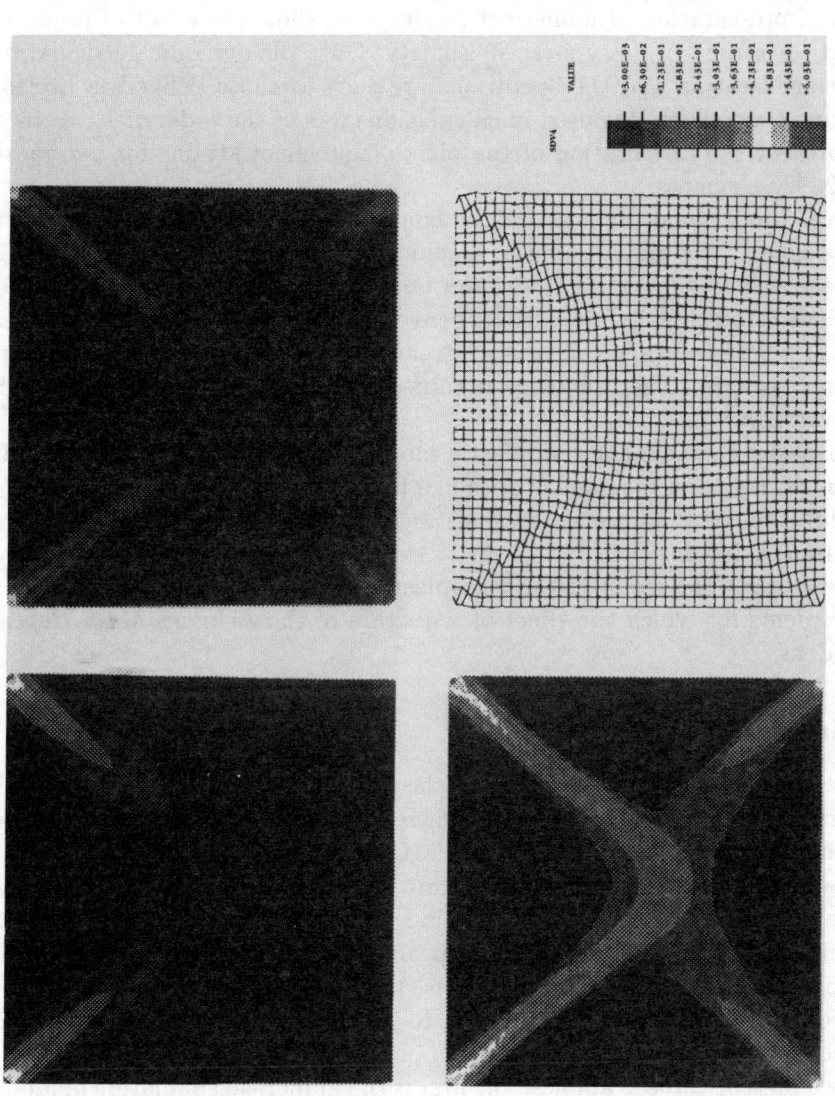

Figure 8: Distribution of the equivalent plastic strain for dynamic case without regularization for 40*40 mesh. Development of localized strain zones in time: a) $t = 0.08s$, b) $t = 0.12s$, c) $t = 0.16s$, d) deformed mesh. The assumed velocity of top side movement $v = 3.125 m/s$.

3.5.4 Avoiding of spurious mesh sensitivity in test problem

Let us now examine our pillar problem using a regularized model. The main goal of this part of presentation of numerical results is to show the effect of minimization and to avoid the strong primary mesh sensitivity. Only the dynamical calculations are taken into consideration and the conditions are such as those defined in Section 3.4. The relaxation time which was used in calculations was of the order of $T_m = 10^{-3}s$. In Fig.10 we present the distribution of the plastic equivalent strains for two meshes of 10*10 and 20*20 elements.

The three-dimensional plots as well as contour plots of these strains confirm the good agreement of both results for the specific time step t = 0.16 s. The results for the plastic equivalent strains are close from both the qualitative and the quantitative points of view. Additionally, very good convergence of the obtained results can be seen in Fig.11 where the values of plastic strains for 3 meshes are compared along the cross-section $x = 8m$. The difference obtained for 20*20 and 40*40 meshes do not exceed the value of 5%.

One can observe that for a viscoplastic model the localization zones are diffused. Their widths do not tend to singular lines like those for rate-independent models. The degree of diffusion of those zones depends significantly on the viscosity (relaxation time) used in calculations.

In Fig.12 we present the result of the viscoplastic regularization in the space of forces and displacements for which the effect of reduction of the *Primary Mesh Dependence* PMD is evident.

3.6 Engineering applications

The set of engineering examples, including classical stability of slopes and a halfspace loaded by rigid block, were studied to confirm numerically the well posedness of the IBVP via the disappearing of the effect of PMD. It should stress here that for complicated engineering problems that include strain softening behaviour the computations can be time consuming (expensive) even using very powerful computers. In our cases the the computation time for some of the problems was of order hours using Fujitsu or Cray supercomputers. In these conditions, the decision on recalculating the IBVP under consideration with different FE mesh, to prove the well posedness, can be driven by economic constrains.

Below, we present three examples: the first is the numerical simulation of laboratory test and the second and the third are the dynamic analyses of the slope stability and a caisson type breakwater founded on the softening halfspace.

3.6.1 Simulation of laboratory tests

In serious of works DESRUES [19, 20] studied experimentally the classical biaxial soil mechanics tests. In the first step a constant pressure acts on the vertical boundaries

Initial Boundary Value Problems for Metals and Soils

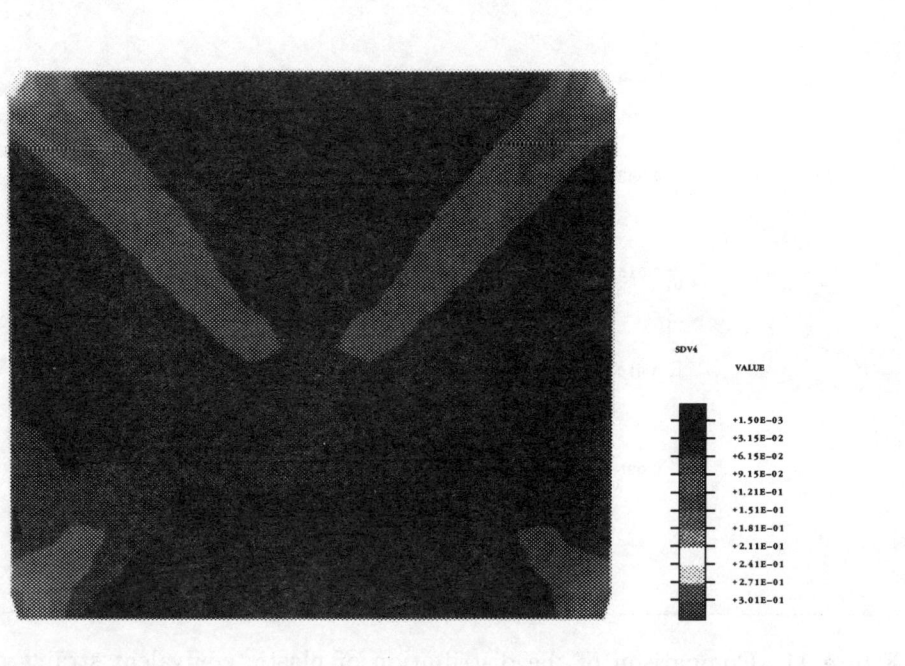

Figure 9: Distribution of the equivalent plastic strain for dynamic case without regularization for 20*20 mesh and $t = 0.16s$. (For comparison with Fig.8c).

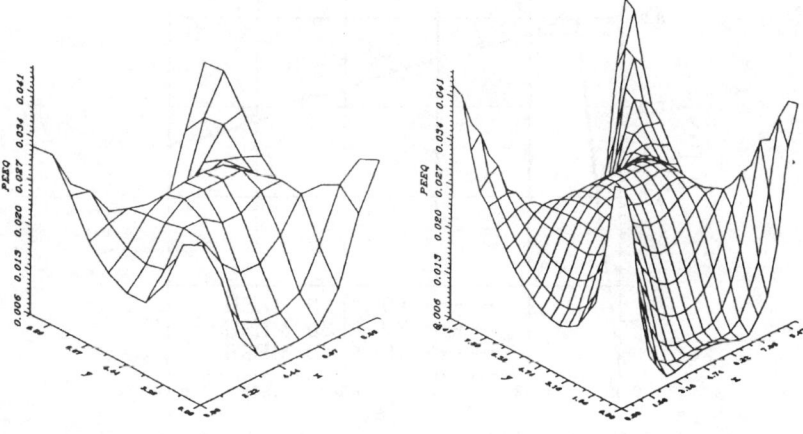

Figure 10: Distribution of the equivalent plastic strain PEEQ for two meshes 10*10 and 20*20 for viscoplastic regularized cases.

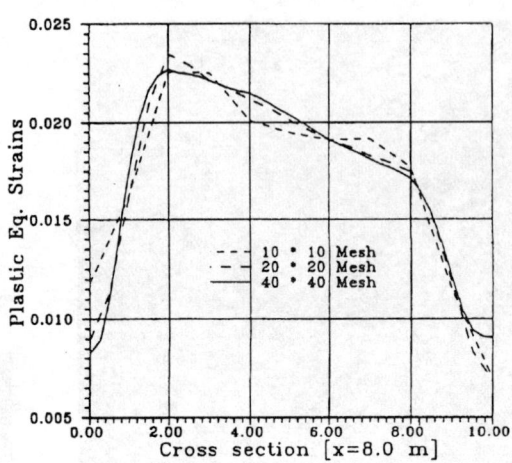

Figure 11: Comparison of the distribution of plastic equivalent strains along the pillar specimen for $x = 8.0m$ for meshes 10*10, 20*20 and 40*40 (regularized case).

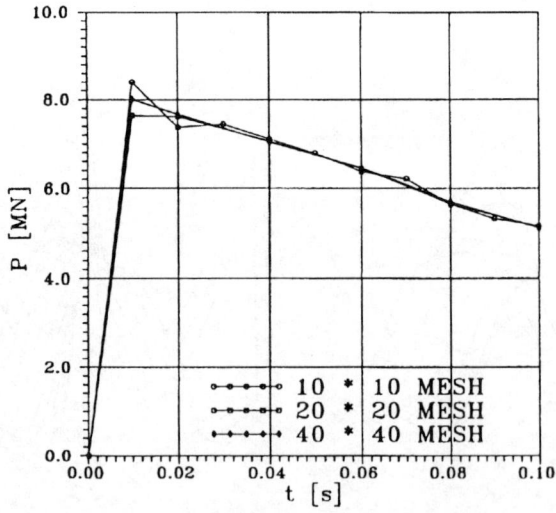

Figure 12: The effect of viscoplastic regularization for dynamic case total force vs. time (velocity driven problem).

of the specimen introducing the initial stress state. During the second step the upper plate of the device compresses the specimen with constant velocity. The plane strain state is analyzed for the specimen of 175 mm width and 350 mm depth.

The results which show the distribution of relative densities η (Fig.13), plastic equivalent strains and irreversible part of density changes η^p qualitatively confirm the experimental tests of Desures.

Both Figs.13 and 14 clearly show the places of strain localization where the material softens (places of decreasing relative densities).

The example was used to check the usefulness of the formulation of the constitutive model in numerical simulation of laboratory tests, and to show the possible foreseeing of the results of real physical experiments.

For different meshes the results obtained on both levels: global in $P - \delta$ space and local strains distribution confirm the convergence and mesh insensitivity in the meaning of PMD. The results obtained for coarse mesh (12*24) and for fine mesh (48*96) are close one to the other.

3.6.2 Breakwater example

Caisson-type breakwaters are traditionally designed by using rigorous simplifications concerning loading conditions, material behaviour of the foundation and a caisson structure, contact conditions between the breakwater and the foundation, stress distribution and fluid flow through the rubble mound. Usually the quasistatic loads are accepted to model the waves for examination of the stability of the structure against sliding, overturning and slip failure. This conventional design leads to uneconomical structures and does not predict their safety. This was experimentally confirmed and reported by OUMERACI [67] particularly for vertical breakwaters. Moreover, this simplified design is unable to explain the variety of failure mechanisms that appear in the structure under the critical load conditions. Therefore, a more sophisticated method is needed which will be able properly account for at least the most important possible failure modes e.g. sliding and tilting of the caisson and geotechnical failure of a half-space.

The observations of the existing structures and experimental measurements at Large Wave Flume confirm that the mechanisms responsible for shear failure of the foundations of caisson breakwaters are complex and in general there is not possible to extract the single effect. Various influences are linked together. For this purpose the numerical model should include the capability for modelling the majority of physical phenomena established in experiments as responsible for failure of the breakwater structures. The important aspect of the numerical model is how effective will be the model when starting large scale problems.

The loading conditions were accepted from the experimental results done at Large Wave Flume in Hannover and they simulate how the waves act on walls and the foundation of the breakwater.

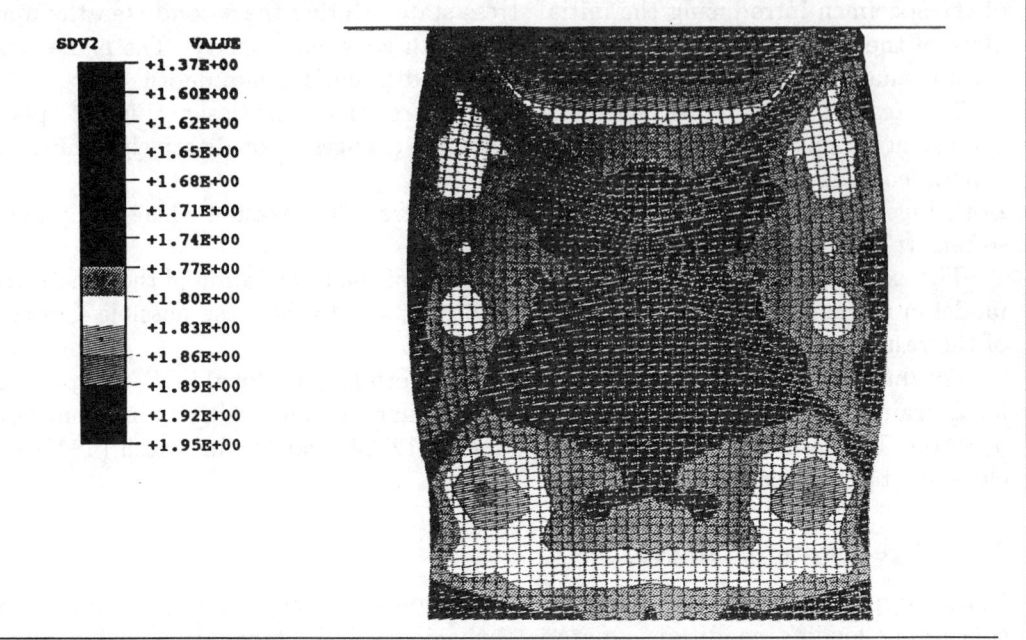

Figure 13: The distribution of relative densities in Desrues' specimen.

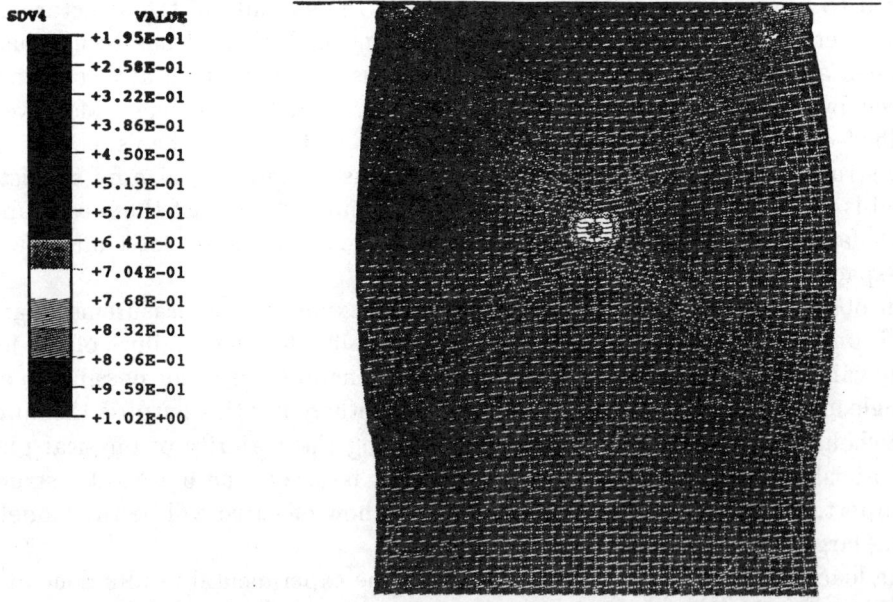

Figure 14: The distribution of plastic equivalent strains in Desrues' specimen.

Table: Loading Conditions

	t_0	t_1	t_2	t_3	t_4	t_5	a	b	Δt	p_{max}	p_{min}
	s	s	s	s	s	s	s	s	s	kN/m^2	kN/m^2
v	0.0	0.45	0.475	0.525	0.55	1.00	0.005	0.1	0.0	23.3	14.4
h	0.0	0.45	0.475	0.525	0.55	1.00	0.005	0.1		6.4	2.3

The changes of horizontal and vertical pressure in time are shown in Fig.15 and the corresponding numerical values accepted for computations in Table.

The duration of the whole simulation process equals $1.0 sec$. In this time the pressure increases up to its maximum level and decreases back to the minimum level.

The simulation of the creation and the failure process of the caisson-type breakwater consists of several steps: in-situ stress state due to e.g. self-weight, creation of the rubble mound as a foundation of the caisson, placement of the caisson e.g. from a ship on top of the rubble mound and the wave loading that acts on the breakwater surface. The results presented in the following Figs 17, 18 and 19 show for selected time (close to the maximum of pressure) the distribution of relative densities, plastic equivalent strains and irreversible part of relative densities.

In the simulated process of the breakwater behaviour one can observe the tilting of the caisson landwards and after the peak load retilting. In this case the gap occurs on the landside. This phenomenon, however often discussed in literature was not till now simulated numerically. In the next Figs 20 and 21 the nodal velocity plots illustrate the tilting and retilting of the caisson what takes place just before and after the peak load.

The other aspects of the numerical simulation of the behaviour of caisson type breakwater, in particular, the stage of dynamic placement of the caisson on a halfspace and the consolidation of the material of the halfspace were also numerically studied and have been reported by LODYGOWSKI et al. in [55].

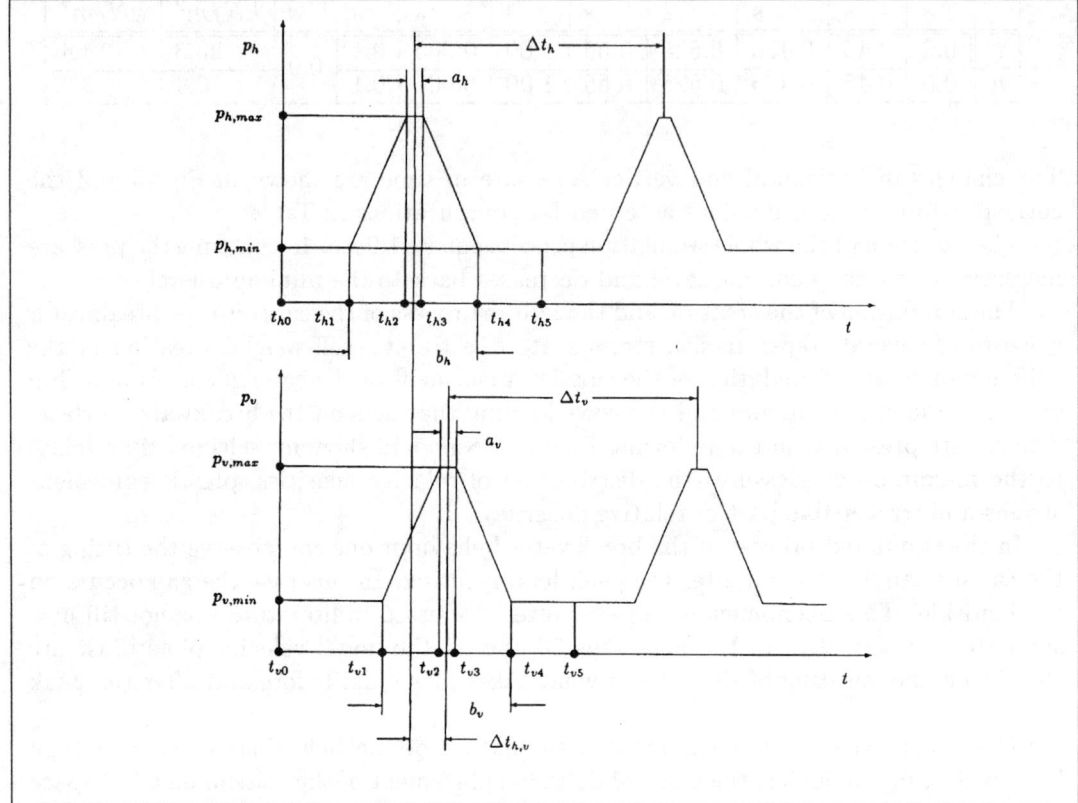

Figure 15: The distribution of horizontal and vertical pressure which act on a breakwater in time

Initial Boundary Value Problems for Metals and Soils

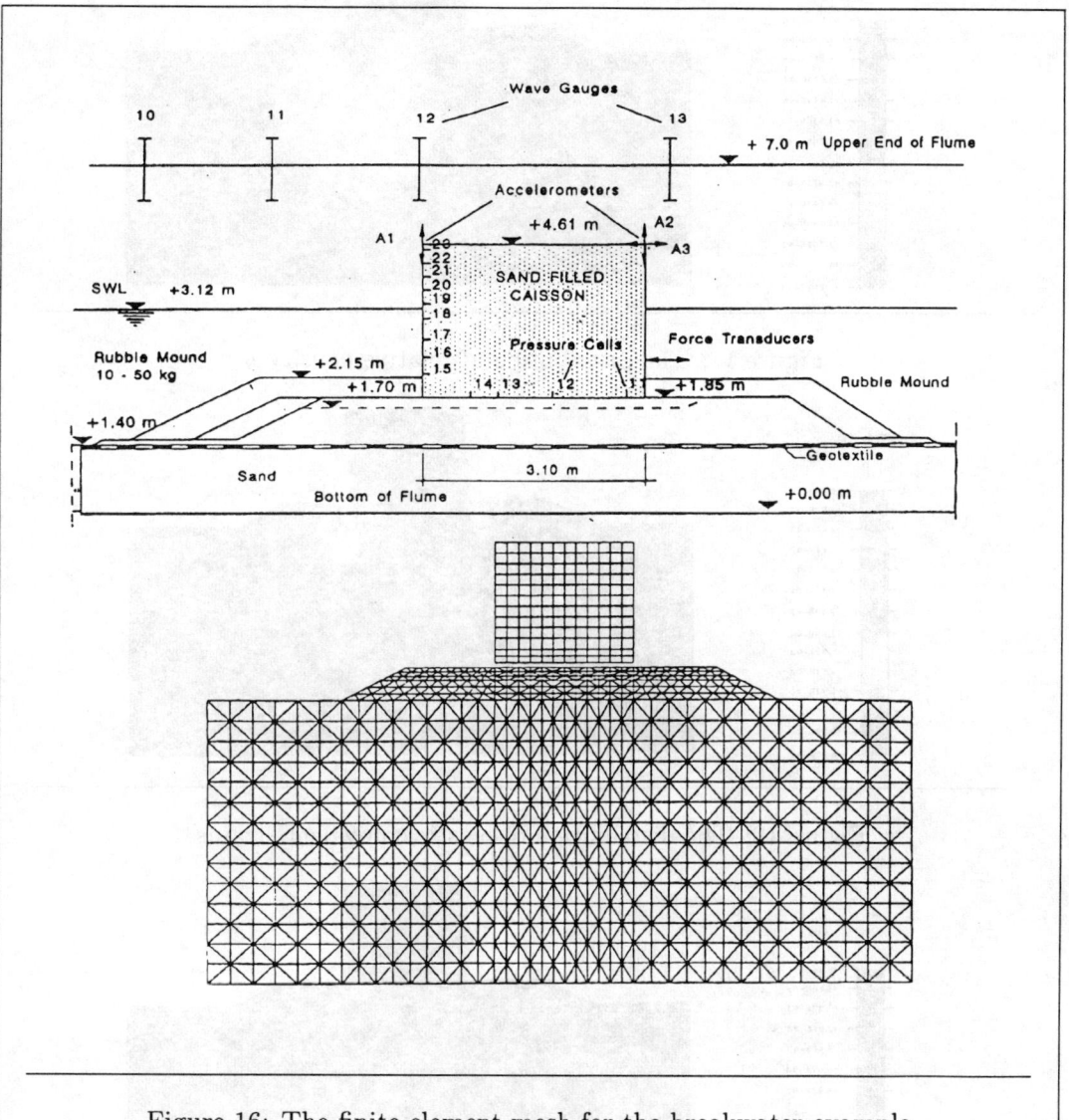

Figure 16: The finite element mesh for the breakwater example

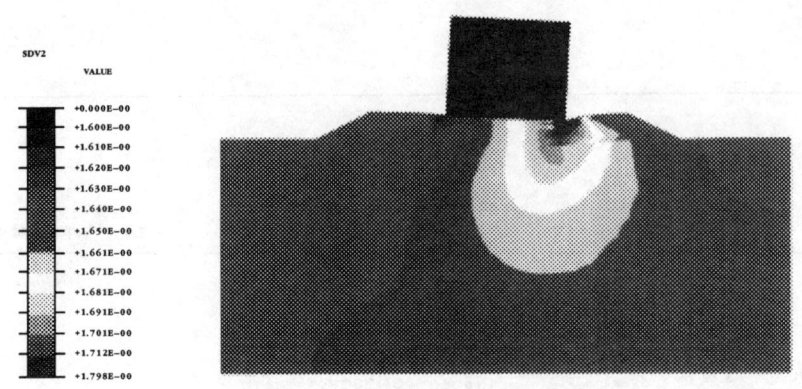

Figure 17: The distribution of relative density η

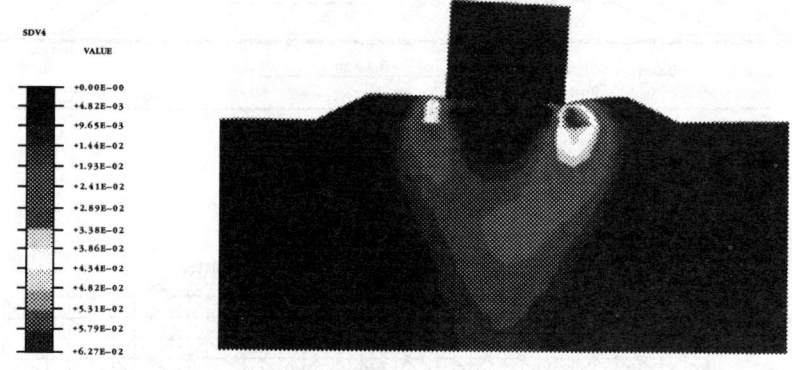

Figure 18: The distribution of plastic equivalent strains

Figure 19: The distribution of plastic part of relative density

Figure 20: Nodal velocity plot before the peak load

Figure 21: Nodal velocity plot after the peak load

4. Rate dependent models for metals

4.1 Formulation

The examples which summarize this Section refer to the problem of biaxial impact loading. The attention is paid to the mesh insensitivity in the meaning of avoiding of PMD. The influence of the different initial conditions on the placement of localization domains in documented. The good agreement of the results for different finite element meshes confirm efficiency of the rate dependent regularization. A strong influence of initial boundary conditions on the final results (the width and direction of localization domains) is easily visible.

The main ideas of this Section were reported by the author in [51] and with coworkers [54].

4.1.1 Kinematics

Following the notation of GURTIN [32] we assume two configurations: $\mathcal{B}_\tau$ at time $t = \tau$, for which the state is in equilibrium, and $\mathcal{B}_t$ at time $t = \tau + \Delta t$ which the continuum is going to occupy in the current configuration. Then the problem is to determine the equilibrium state of the body at each material point in $\mathcal{B}_t$ that based on the given constitutive law which is discussed in the next section. From the viewpoint of numerical formulation and then computations it is crucial to assure the consistency with the set of constitutive equations, numerical stability and objectivity what means that the proposed algorithm should be invariant with respect to the rigid body motions. The summary of incrementally objective integration schemes was recently presented by ZABARAS & ARIF [97].

The requirements of objectivity starts from the definition of two motions $\mathbf{\Phi}(\mathbf{X}, t)$ and $\mathbf{\Phi}^*(\mathbf{X}, t)$. It can be stated that those motions are objectively equivalent if they fulfill

$$\mathbf{\Phi}^*(\mathbf{X}, t) = \mathbf{Z}(t)(\mathbf{\Phi}^*(\mathbf{X}, t) - \mathbf{o}) + \mathbf{c}(t), \tag{43}$$

where the vector $\mathbf{o}$ collects coordinates of a fixed point in space; $\mathbf{Z}(t)$ and $\mathbf{c}(t)$ are the rotation and the translation time dependent functions, respectively. Moreover the rotation $\mathbf{Z}(t)$ satisfies

$$\mathbf{Z}^T \mathbf{Z} = \mathbf{I} \quad and \quad det\mathbf{Z} = 1. \tag{44}$$

As a consequence of the above assumption the tensorial quantity is frame independent if for any equivalent motions (43) we obtain the following tensor law transformations for scalar s and tensors $\mathbf{a}, \mathbf{B}, \mathcal{C}$ of first, second and fourth order, respectively;

$$\begin{aligned} s^* &= s, \\ \mathbf{a}^* &= \mathbf{Z}\mathbf{a}, \\ \mathbf{B}^* &= \mathbf{Z}\mathbf{B}\mathbf{Z}^T, \\ \mathcal{C}^* &= \mathbf{Z}[\mathbf{Z}\mathcal{C}\mathbf{Z}^T]\mathbf{Z}^T. \end{aligned} \tag{45}$$

4.1.2 Constitutive model

Assuming after LEE [45] the multiplicativity of the deformation gradient in the form $\mathbf{F} = \mathbf{F}^e\mathbf{F}^p$ it follows that the total deformation rate $\mathbf{D}$ is simply a sum of its elastic $\mathbf{D}^e$ and plastic $\mathbf{D}^p$ parts. In GURTIN's [32] notation, one can write the evolution of the Cauchy stress tensor in the form

$$\overset{\triangledown}{\mathbf{T}} = \mathcal{C}^e[\mathbf{D} - \mathbf{D}^p], \qquad (46)$$

where $\overset{\triangledown}{\mathbf{T}}$ is Jaumann rate of Cauchy stress and $\mathcal{C}^e = 2G\mathbf{I} + (K - \frac{2}{3}G)\mathbf{1} \otimes \mathbf{1}$ is an elastic isotropic modulus. In the last G and K are the known shear and bulk moduli, respectively, and $\mathbf{I}$ and $\mathbf{1}$ denote fourth rank and second rank unit tensors.

To specify the essence of the plastic approach one has to contain the fundamental definition of inelastic part of deformation rate, generally in the form

$$\mathbf{D}^p = F(\mathbf{S}, \Theta, \boldsymbol{\mu}), \qquad (47)$$

where $\mathbf{S}$ is the deviatoric stress tensor, Θ is the temperature and $\boldsymbol{\mu}$ represents the vector of internal state variables which can consist of scalar, vector or tensor components. The variety of models can be analyzed and/or constructed using in general the following basic relations:

- a flow rule of tensorial character,

- the necessary evolution equations which describe the evolution of internal variables $\boldsymbol{\mu}$,

- a kinetic equation (eg. balance of energy) of scalar type that relates stresses, inelastic parts of strain rate and temperatures.

We will restrict our attention to the isothermal processes and assume the *flow rule* in the form

$$\mathbf{D}^p = \Lambda \tilde{\mathbf{n}}, \qquad (48)$$

where Λ denotes a scalar valued function, and $\tilde{\mathbf{n}}$ represents a tensor of second rank, respectively.

Rate independent plasticity If we assume the flow rule in the form (48) and the definition of scalar function Λ as follows:

$$\Lambda = \begin{cases} \dot{\lambda} & : \quad \text{if} \quad \phi = 0 \quad \text{and} \quad \tilde{\mathbf{n}} : \mathcal{C} : \bar{\mathbf{D}} > 0, \\ 0 & : \quad \text{if} \quad \phi \leq 0 \quad \text{or} \quad \phi = 0 \quad \text{and} \quad \tilde{\mathbf{n}} : \mathcal{C} : \bar{\mathbf{D}} \leq 0, \end{cases} \qquad (49)$$

where ϕ represents the yield condition, we arrive at the definition of rate independent plasticity. For $\tilde{\mathbf{n}} = \frac{\partial g}{\partial \mathbf{T}}$ we specify the associative plasticity if $g = f$, or nonassociative one if $g \neq f$.

The parameter $\dot{\lambda}$ is derived from the consistency conditions.

Rate dependent plasticity If we assume that Λ is an isothermal function of the type $\Lambda(\tilde{\mathbf{S}}, \boldsymbol{\mu})$ we are starting to declare the rate dependent law which has to be supplemented by the next two relations which were pointed out above. It means, that for rate dependent flow law we arrive at

$$\mathbf{D}^p = \Lambda(\tilde{\mathbf{S}}, \boldsymbol{\mu})\tilde{\mathbf{n}}. \tag{50}$$

Additionally, we have to define the evolution equations for the internal state variables. Among the variety of internal parameters $\boldsymbol{\mu}$ let us now restrict our attention to only two, namely: the scalar value which describes the yield limit s, which is responsible for isotropic hardening/softening and tensorial value $\mathbf{B}$ which is called back stress tensor (symmetric, traceless tensor) that defines the kinematic hardening effect, both of stress units.

The *evolution of internal state variables* can be proposed, for example in the following form:

$$\dot{s} = r\,\Lambda, \tag{51}$$
$$\dot{\mathbf{B}} = H\,\mathbf{D}^p - \Lambda\,C\,\mathbf{B}, \tag{52}$$

where r is a hardening/softening parameter, and also H and C are parameters which are assumed to be functions of the list of variables $\boldsymbol{\mu}$.

The different function can be adopted to specify Λ what defines the type of rate dependency.

Let us for example assume

$$\mathbf{D}^p = \sqrt{\frac{3}{2}}\dot{\varepsilon}^p \tilde{\mathbf{n}} \tag{53}$$

and

$$\tilde{\mathbf{n}} = \sqrt{\frac{3}{2}}\frac{\mathbf{S}}{\tilde{S}}, \tag{54}$$

where $\mathbf{S} = \mathbf{T}' = \mathbf{T} - \frac{1}{3}tr(\mathbf{T})\mathbf{1}$ is the deviatoric part of the Cauchy stress tensor, and

$$\tilde{S} = \sqrt{\frac{3}{2}\mathbf{S} : \mathbf{S}} \tag{55}$$

is the equivalent stress.

Additionally, the plastic equivalent strain rate $\dot{\varepsilon}^p = (\frac{2}{3}\mathbf{D}^p : \mathbf{D}^p)^{\frac{1}{2}}$ is prescribed as a function of current equivalent stress $\tilde{S}$ and state variables $\boldsymbol{\mu}$

$$\dot{\varepsilon}^p = f(\tilde{S}, \boldsymbol{\mu}). \tag{56}$$

To complete the system of equations it is necessary to add the evolution equations for the state variables

$$\dot{\boldsymbol{\mu}} = m(\tilde{S}, \boldsymbol{\mu}), \tag{57}$$

where
$$m(\tilde{S},\boldsymbol{\mu}) = h(\boldsymbol{\mu})\,\dot{\varepsilon}^p, \tag{58}$$

and $h(\boldsymbol{\mu})$ denotes the hardening/softening function.

The selection of functions $f(\tilde{S},\boldsymbol{\mu})$ and $m(\tilde{S},\boldsymbol{\mu})$ base on phenomenological theories and should be strongly related to micromechanical observations and the experimental results obtained in physics of solids.

For the further numerical consideration, now let us restrict the class of functions $f(\tilde{S},\boldsymbol{\mu})$ in such a way that $\boldsymbol{\mu}$ is represented only by a scalar value s. If we accept then the yield function in the form

$$f(\tilde{S},s) = \begin{cases} \phi(\frac{\tilde{S}}{s} - 1.0)^n & : \quad \text{if } \tilde{S} \geq s, \\ 0 & : \quad \text{if } \tilde{S} < s, \end{cases} \tag{59}$$

we arrive at the restricted version of viscoplasticity, originally introduced by PERZYNA [72, 73, 74], where ϕ is the viscosity which is the reciprocal of the relaxation time of mechanical disturbances T_m, ($\phi = \frac{1}{T_m}$).

If we accept for example the function f the form

$$f(\tilde{S},s) = \dot{\varepsilon}_0 \left(\frac{\tilde{S}}{s}\right)^{\frac{1}{m}}, \tag{60}$$

we define the known viscoplastic power law. There are also the other possibilities of choosing $f(\tilde{S},s)$ functions which successfully serve in variety of particular cases.

Rotation neutralized description Following the discussion in [94, 95] for convenience, we introduce rotational – neutralized form of our constitutive model. The so called bar form of Cauchy stress tensor $\bar{\mathbf{T}}$ is now expressed as

$$\bar{\mathbf{T}} = \mathbf{Q}^T \mathbf{T} \mathbf{Q}, \tag{61}$$

where the rotation tensor $\mathbf{Q}(t)$ was introduced here as the solution of the following initial value problem

$$\begin{aligned}\dot{\mathbf{Q}}(t)\mathbf{Q}^T(t) &= \mathbf{W}(t) \text{ for } t_n \leq t \leq t_{n+1}, \\ \mathbf{Q}(t_n) &= \mathbf{I}. \end{aligned} \tag{62}$$

The spin $\mathbf{W}(t)$ is defined as an nonsymmetric part of the velocity gradient $\mathbf{L}$. This significantly simplifies the equation (46), also see NAGTEGAAL [63], to the form

$$\dot{\bar{\mathbf{T}}} = \mathbf{Q}^T \overset{\triangledown}{\mathbf{T}} \mathbf{Q} = \mathcal{C}[\bar{\mathbf{D}} - \bar{\mathbf{D}}^p]. \tag{63}$$

Now using the bar formulation we obtain the following system of equations that together with eqn. (63) describes our rate–dependent model

$$\bar{\mathbf{D}}^p = \sqrt{\frac{3}{2}}\dot{\varepsilon}^p \bar{\mathbf{n}}(\bar{\mathbf{S}}, \tilde{S}), \tag{64}$$

$$\bar{\mathbf{n}}(\bar{\mathbf{S}}, \tilde{S}) = \sqrt{\frac{3}{2}}\frac{\bar{\mathbf{S}}}{\tilde{S}}, \tag{65}$$

$$\tilde{S} = \sqrt{\frac{3}{2}\bar{\mathbf{S}} : \bar{\mathbf{S}}}, \tag{66}$$

$$\dot{\varepsilon}^p = f(\tilde{S}, s). \tag{67}$$

The evolution equation for only one scalar value s is

$$\dot{s} = m(\tilde{S}, s) = h(s)\,\dot{\varepsilon}^p. \tag{68}$$

Then for different choice of functions $f(\tilde{S}, s)$ we can define the Perzyna's type viscoplasticity (59), or creep model (60). The integration of a such prepared system (bar formulation) is computationally more efficient.

The above system of equations describes the active process of plastification. In computations the elastic unloading is treated classically as for passive processes; see e.g. WASZCZYSZYN [93].

4.2 Numerical aspects

4.2.1 Time integration procedure

The goal of this integration is to find the state represented by Cauchy stress tensor $\mathbf{T}$ and the scalar independent variable s at time $t = \tau + \Delta t = t_{n+1}$ knowing the state $(\mathbf{T}_n, s_n)$ at a time $t = \tau = t_n$. The values of unknown variables can be obtained as

$$\mathbf{T}_{n+1} = \mathbf{Q}_{n+1}(\mathbf{T}_n + \int_{t_n}^{t_{n+1}} \mathcal{C}[\mathbf{D} - \sqrt{\frac{3}{2}}\dot{\varepsilon}^p\,\bar{\mathbf{n}}]\,dt)\mathbf{Q}_{n+1}^T, \tag{69}$$

$$s_{n+1} = s_n + \int_{t_n}^{t_{n+1}} \dot{s}\,dt, \tag{70}$$

where $\mathbf{Q}_{n+1}$ is the rotation tensor at time $t+1$ relative to the configuration at t_n.

Using the following classical approximation

$$\begin{aligned} s_{n+\beta} &= s_n + \beta \dot{s}_{n+\beta}\Delta t, \\ \dot{s}_{n+\beta} &= m(\tilde{S}_{n+\beta}, s_{n+\beta}), \\ \bar{\mathbf{T}}_{n+\beta} &= (1-\beta)\bar{\mathbf{T}}_n + \beta\bar{\mathbf{T}}_{n+1}, \end{aligned} \tag{71}$$

one can generate the different integration schemes. We will use here for $\beta = 1$ so called full backward integration method. Then we arrive at

$$\bar{\mathbf{T}}_{n+1} = \mathbf{T}^{pre}_{n+1} - \sqrt{6}G\Delta t f(\tilde{S}_{n+1}, s_{n+1})\bar{\mathbf{n}}_{n+1}, \tag{72}$$

$$s_{n+1} = s_n + \Delta t m(\tilde{S}_{n+1}, s_{n+1}), \tag{73}$$

where

$$\mathbf{T}^{pre}_{n+1} = \mathbf{Q}_{n+1}\bar{\mathbf{T}}_n\mathbf{Q}^T_{n+1} + \mathcal{C}[\Delta\mathbf{E}_{n+1}], \tag{74}$$

and the strain increment $\Delta\mathbf{E}_{n+1}$ is

$$\Delta\mathbf{E} = \mathbf{Q}_{n+1}(\int_{t_n}^{t_{n+1}} \bar{\mathbf{D}}dt)\mathbf{Q}^T_{n+1}. \tag{75}$$

Finally, the problem is reduced to determining the scalar values s_{n+1} and $\tilde{S}_{n+1}$ from the pair of algebraic equations

$$s_{n+1} - s_n - \Delta t\, m(\tilde{S}_{n+1}, s_{n+1}) = 0,$$
$$\tilde{S}_{n+1} - \tilde{S}^{pre}_{n+1} + 3G\Delta t\, f(\tilde{S}_{n+1}, s_{n+1}) = 0. \tag{76}$$

Locally this system of equations has to be solved for each point of integration at each increment (iteration). So the efficiency of the solver is crucial for the good convergence of the results and the power of the realized algorithm. The subroutine BROWN that was adopted for this purpose appears to be unfailing in the variety of tests which were performed. After the solution of the system (76) one can easily predict the corrector step and update the stresses.

Constitutive algorithm The scheme of integration of the global system of equations (for dynamics) implicit or explicit is realized by the environment of the ABAQUS [1] program.

The summary of necessary steps to integrate the system of rate dependent equations that consists of two classical steps (elastic predictor and plastic corrector) is finally presented in BOX 3.

BOX 3

Summary of the constitutive algorithm

1. Calculate the trial stress and normal mean pressure

$$\bar{\mathbf{T}}^{pre}_{n+1} = \mathbf{T}_n + \mathcal{C}[\Delta \mathbf{E}_{n+1}], \quad p^{pre}_{n+1} = \frac{1}{3}tr(\bar{\mathbf{T}}^{pre}_{n+1})$$

2. Deviatoric trial and equivalent tensile stresses

$$\bar{\mathbf{S}}^{pre}_{n+1} = \bar{\mathbf{T}}^{pre}_{n+1} + p^{pre}_{n+1}\mathbf{I}, \quad \tilde{S}^{pre}_{n+1} = \sqrt{\frac{3}{2}\bar{\mathbf{S}}^{pre}_{n+1} : \bar{\mathbf{S}}^{pre}_{n+1}}$$

3. If $\tilde{S}^{pre}_{n+1} \leq s_n$ then only elastic deformations occur

$$s_{n+1} = s_n, \quad \mathbf{T}_{n+1} = \bar{\mathbf{T}}^{pre}_{n+1} = \mathbf{Q}_{n+1}\bar{\mathbf{T}}^{pre}_{n+1}\mathbf{Q}^T_{n+1}$$

the constitutive algorithm is complete
else $\longmapsto$ continue

4. Solve the system of algebraic equations

$$s_{n+1} - s_n - \Delta t g(\tilde{S}_{n+1}, s_{n+1}) = 0$$

$$\tilde{S}_{n+1} - \tilde{S}^{pre}_{n+1} + 3G\Delta t f(\tilde{S}_{n+1}, s_{n+1}) = 0$$

5. Calculate the radial return factor

$$\eta_{n+1} = \frac{\tilde{S}_{n+1}}{\tilde{S}^{pre}_{n+1}}$$

6. Update stresses

$$\mathbf{T}_{n+1} = \eta_{n+1}\bar{\mathbf{S}}^{pre}_{n+1} + p^{pre}_{n+1}\mathbf{I}$$

$$\mathbf{T}_{n+1} = \mathbf{Q}_{n+1}\bar{\mathbf{T}}_{n+1}\mathbf{Q}^T_{n+1}$$

Test problems The numerical calculations were supported by using a general purpose finite element program ABAQUS. Using the open possibility of the code, which allows the user to create his own constitutive relation, the all necessary parts of the program were introduced by a procedure called UMAT.

This time we will omit the examples which show for both static and dynamic cases the significant mesh sensitivity when dealing with rate independent material resulting from the change of governing operator during the process of deformation.

The UMAT procedure is called for each Gauss point at every iteration. Among other information that are introduced by this procedure the most important are the definition of the actual jacobian stiffness matrix of the material and the way of integration of the constitutive relation on the local level. Because the integration of the constitutive law at the local level basically is reduced to the solving of nonlinear algebraic system of equations one can imagine how important is to use practically reliable solver. This role is played by the adapted procedure BROWN which, due to sophisticated tests which were chosen to check its validity, proved its high efficiency and very fast convergence even if the starting points for the iterations were picked up arbitrary. In the practical computations this important local part of the code never failed and together with such a formulation of integration the constitutive relation confirmed its high quality usefulness.

The hardening/softening function $h(s)$ is assumed to be constant in the presented examples.

4.2.2 Biaxial impact loading – force controlled process

The specimen of dimensions $60mm \times 120mm$ treated as a plain strain shown in Fig. 22a was loaded dynamically at the top side in longitudinal direction by the force $F(t)$, and to brake the symmetric behaviour by the horizontal force $0.1F(t)$.

The force $F(t)$ changes in time as it is depicted in Fig.22b so that after the time period $t_d = 15 \times 10^{-6} s$ reaches its maximal value and keeps it constant for the rest of the process. The whole duration of the physical process takes $t_m = 1.65 \times 10^{-4} s$. The material data used in the computations were as follow:
$E = 11920.0 N/mm^2$, $\nu = 0.49$, $\sigma_0 = 100.0 N/mm^2$, $n = 1.0$, $\varrho = 5.0 g/mm^3$. However the author has the experience with different spatial discretization (4–Node linear and bilinear and 8–Node quadratic elements) the results presented herein restrict themselfs to the using of 4–Node bilinear reduced integration elements. To study numerically the mesh sensitivity four meshes 3×6, 6×12, 12×24 and 48×96 elements were used in the calculations.

The boundary conditions are formulated as follows:

- The bottom side of the specimen is pinned (the displacements of all the nodes that lie on this edge in both directions are zero),

- The top side can rigiditly rotate but the nodes remain on the same straight line (MPC – multi point constrain option was used to declare this behaviour).

On the global level of the integration of nonlinear dynamical system of equations the explicit method was used with the time step $\Delta t = 1.5 \times 10^{-6} s$ which assures the stability of the integration for the defined above set of parameters.

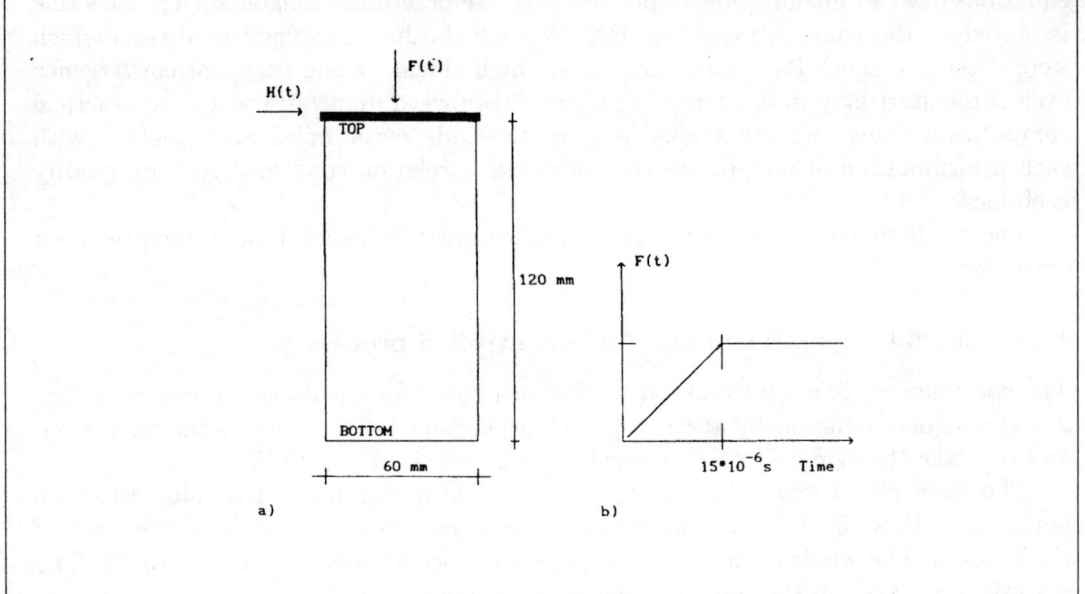

Figure 22: The definition of the IBVP; a) geometry of the specimen, b) loading history

The whole process is realized in one step that declares simultaneous vertical and horizontal increment of loadings.

The process of creation of the zones of localized plastic strains is shown qualitatively in Figs.23 and 24.

Let us notice that the scales of indicated plastic equivalent strains which accompanies the results in Fig.23 are different for different time steps, whereas in Fig.24 remain constant. These results were obtained for the mesh $12*24$ elements. The stress relaxation time was of the order of $T_m = 10^{-4}s$.

The final state of the localization of plastic strain zones is shown in Fig.25 for different meshes.

To compare the influence of mesh discretization on the results obtained, we propose also the discussion on the integrated level of information that means by presenting the result in $P-\delta$ space and on the local level showing, for example, by the distribution of plastic equivalent strains. In Fig.26 the rigid rotation of the top side against the time is depicted. One can observe very good agreement of the curves nevertheless the mesh $3*6$ is very coarse one. For this coarse mesh the important differences could be visible on the local level (distribution of plastic strains).

The contour plots of plastic equivalent strains are presented in Fig.27a-c for the meshes under consideration.

Of course, in this case the poor approximation obtained from the $3*6$ mesh is evident, but the remaining meshes show fast convergence of the results and the reduced effect of SMD.

Even better the convergence of the results is visible in Fig.28, which presents for the chosen meshes the distribution of the plastic equivalent strains along the line $x = 40.0mm$ in the specimen for a specific time $t = t_m$.

The results presented in the last two figures computationally confirm that the rate dependent formulation drives to the solutions which are free of spurious mesh dependency observed when the formulations of problems become ill–posed.

In the viscoplastic formulations the internal length scale is introduced implicitly by the relaxation time for stresses (the relaxation time plays the role of regularization parameter) and the velocity of elastic wave propagation. Generally, it drives to the stronger localization for the shorter relaxation times. It is interesting to notice that for the cases under consideration the different, but reasonable for the used material parameters, relaxation times that changes between $T_m = 10^{-6} \div 10^{-4}s$ slightly influence the distribution and the values of plastic equivalent strains, see Fig.29.

The results obtained here confirm the physical bases of the viscoplastic formulation and the possibility of using the material parameter (T_m) for the regularization of mathematical model.

4.2.3 Biaxial impact loading – velocity controlled process

Now, for the specimen of the same dimmensions as shown in Fig.22, the loading process is realized into two steps. The first one is static and kinematically controlled. The

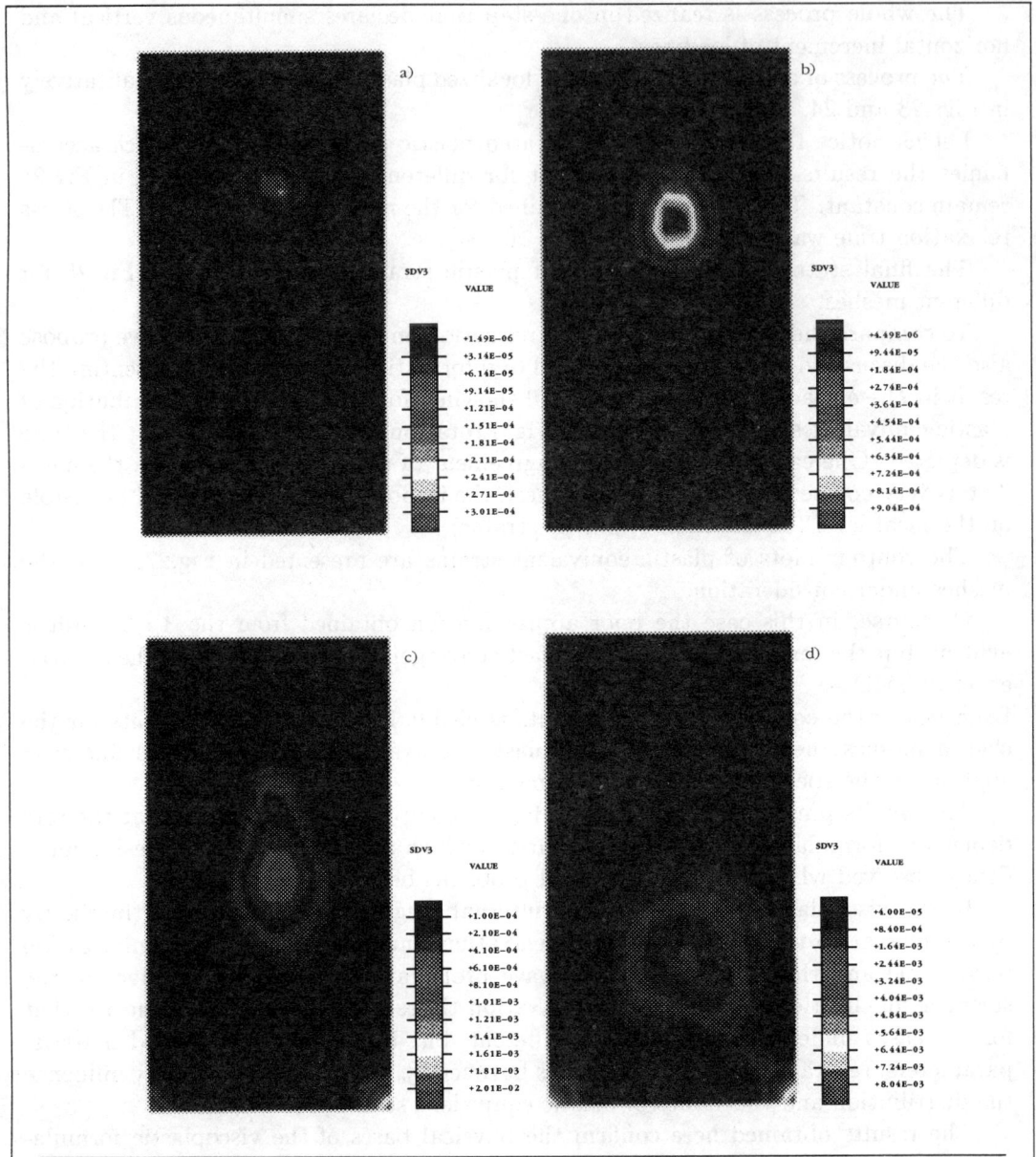

Figure 23: The development of zones of plastic deformation PEEQ for load-controlled dynamic analysis: a) $6.0 \times 10^{-5}s$, b) $7.5 \times 10^{-5}s$, c) $9.0 \times 10^{-5}s$, d) $1.05 \times 10^{-5}s$

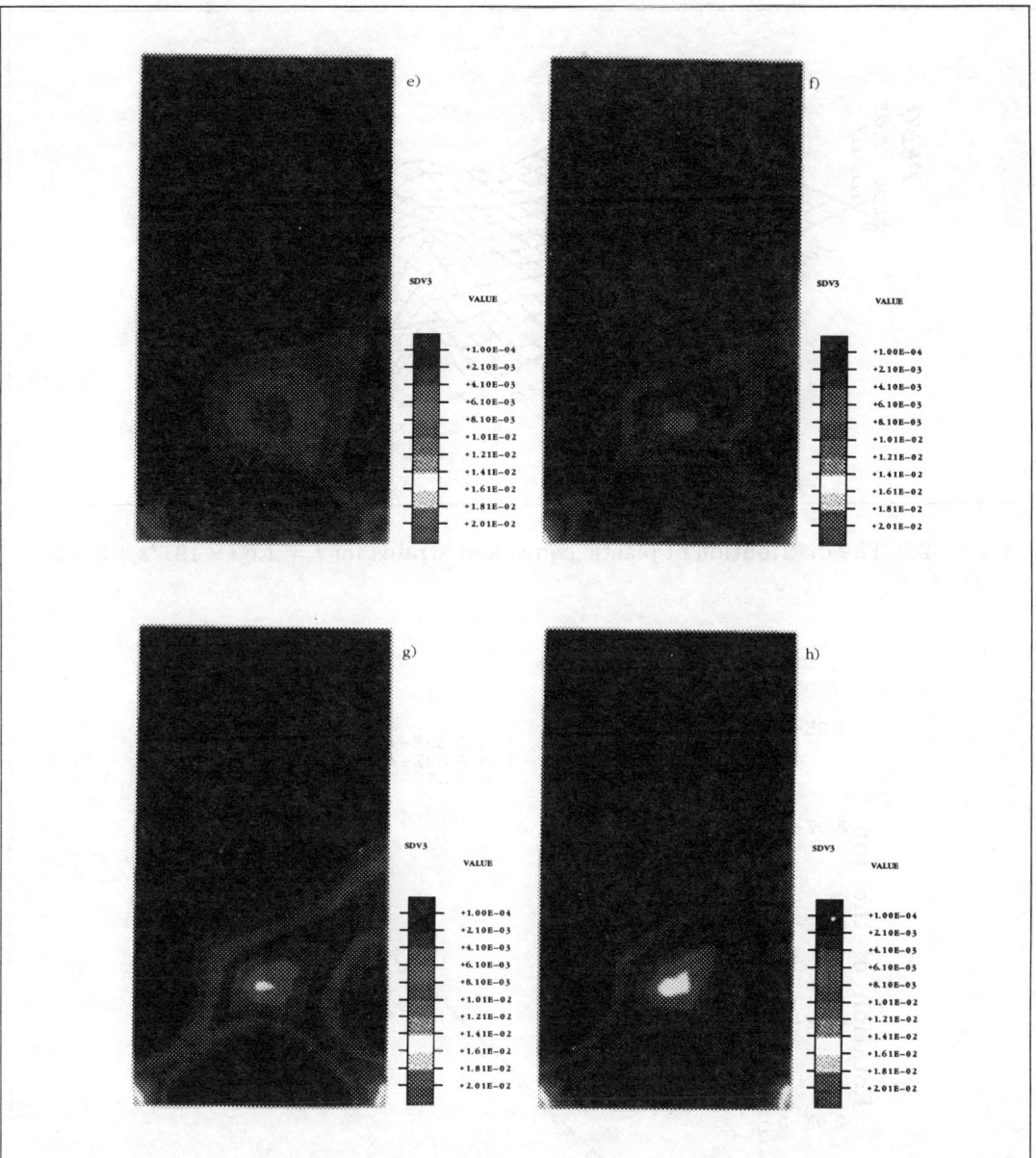

Figure 24: The development of zones of plastic deformation PEEQ for load-controlled dynamic analysis [cont.]: e) $1.2 \times 10^{-5} s$, f) $1.35 \times 10^{-5} s$, g) $1.50 \times 10^{-5} s$, h) $1.65 \times 10^{-5} s$

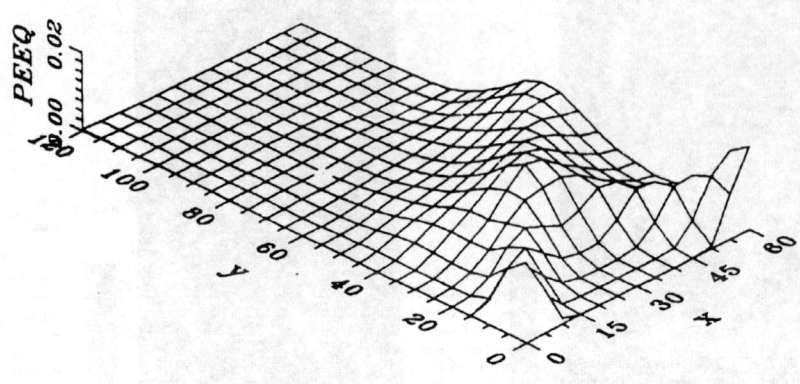

Figure 25: The distribution of plastic equivalent strains for $t = 1.65 \times 10^{-4}s$, 12×24 mesh

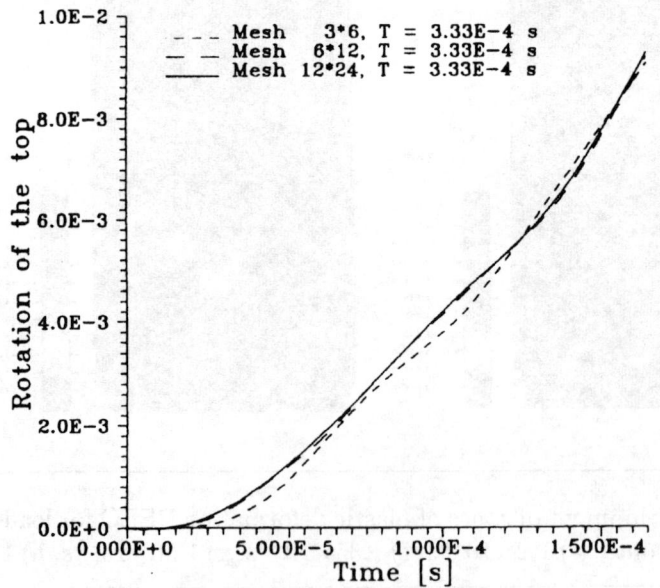

Figure 26: Rotation of top side vs. time

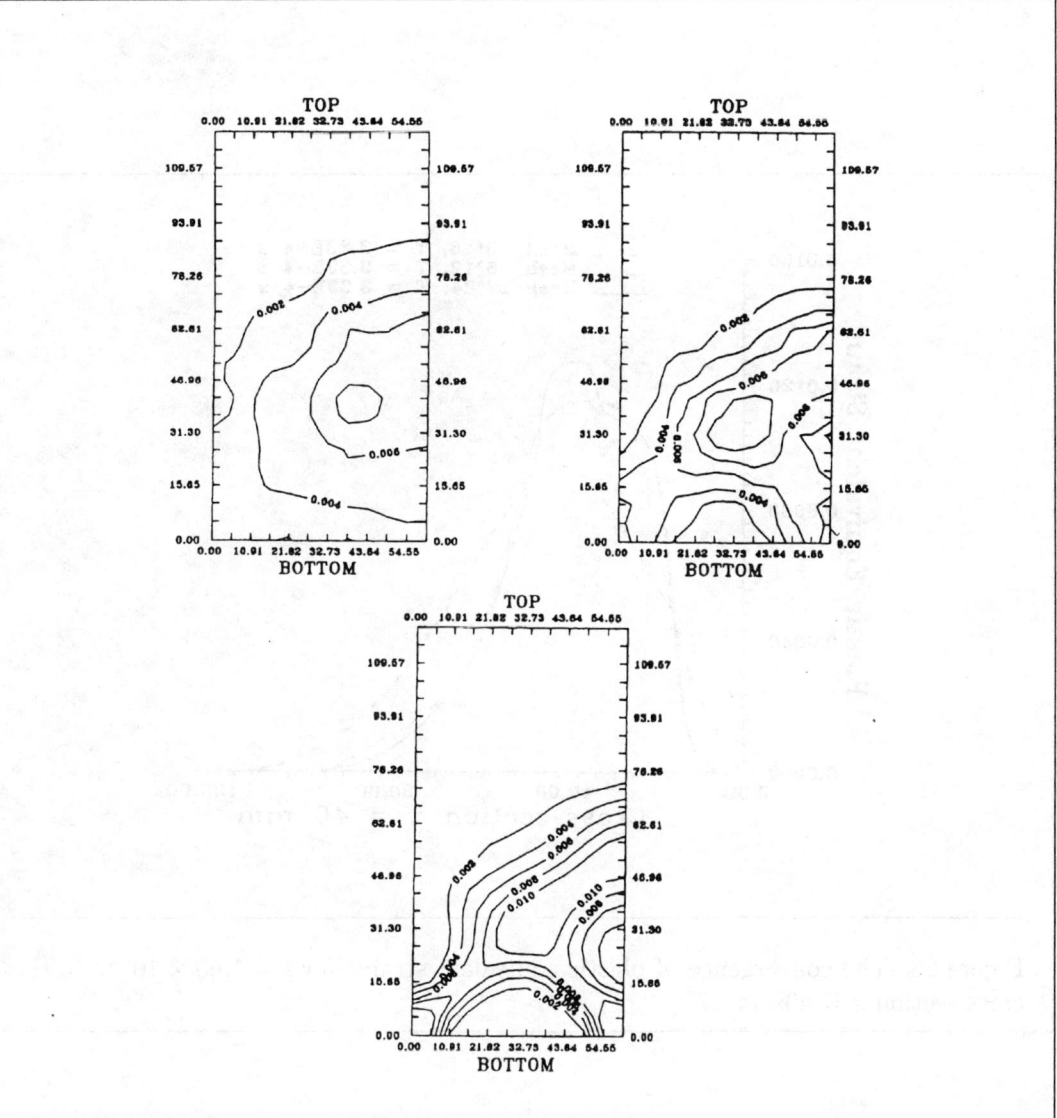

Figure 27: Comparison of plastic equivalent strains for $t = 1.65 \times 10^{-4}s$ for different meshes including very coarse (3*6)

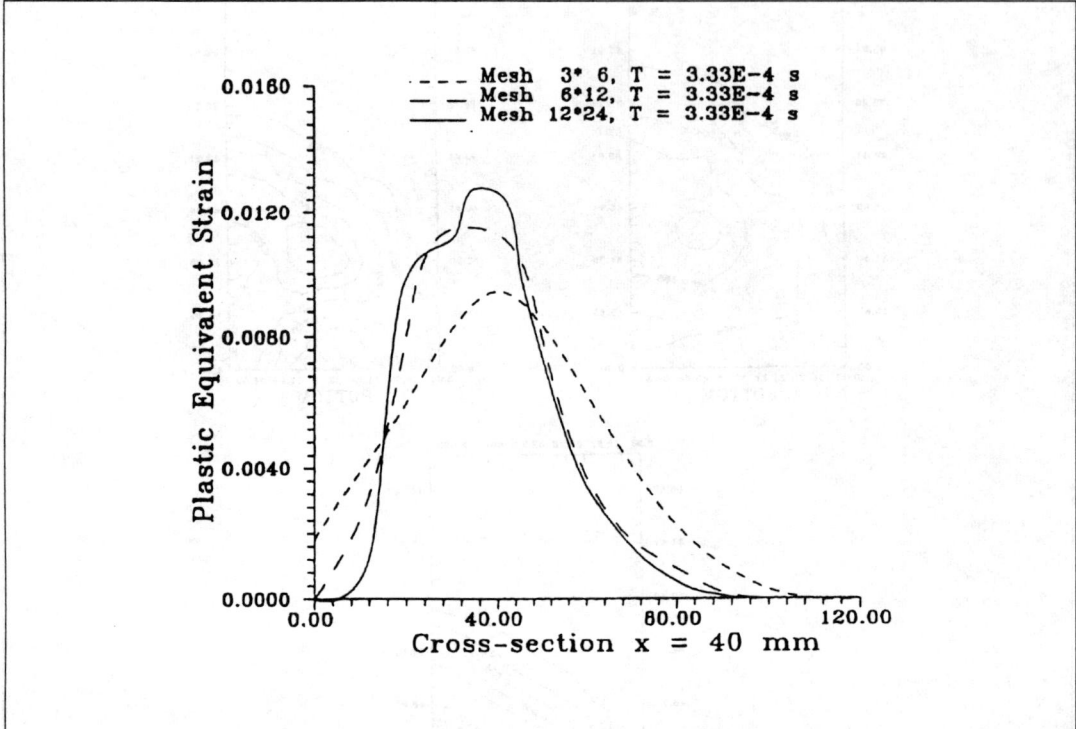

Figure 28: The convergence of plastic equivalent strains for $t = 1.65 \times 10^{-4}s$ in the cross section $x = 40mm$

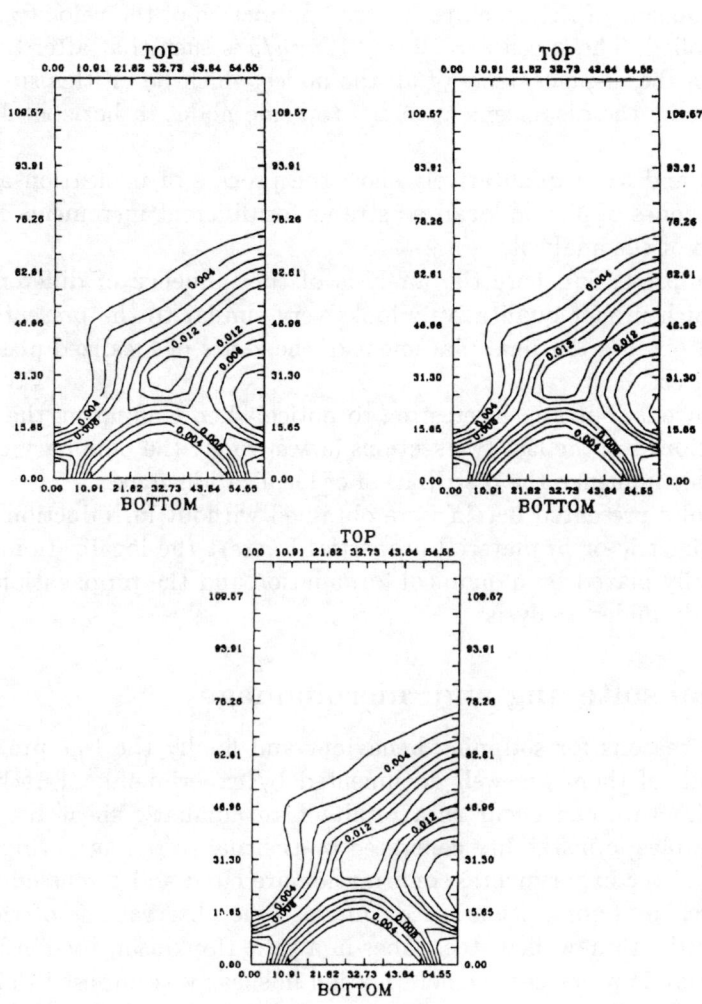

Figure 29: Comparison of the distribution of plastic equivalent strains for different relaxation times: a) $T_m = 10^{-4}s$, b) $T_m = 10^{-5}s$, c) $T_m = 10^{-6}s$

rigid top edge of the specimen is horizontally moved (without rotation) by the value of $2mm$. This introduces the initial state of stress for the second dynamical step. Then, on the top side of the specimen the ramp function of the velocity in the vertical direction is applied. The speed $v = 12121.212 mm/s$ is such that after the time period $t = 3.3 \times 10^{-4} s$ the displacements of all the nodes which lie on this side are equal to $4mm$. In this case the displacement of the top side nodes in horizontal direction are not allowed.

The Fig.30 and 31a-f qualitatively show the process of nucleation and the development of the zones of plastic localized strains for different increments for the second dynamical step of the analysis.

We are not presenting here the analysis of the influence of different spatial discretization, which in fact qualitatively looks very similar to the presented results. In Fig.31 one can observe the final placement of the zones of localized plastic strains on the specimen area.

What is obvious, but still interesting to notice when comparing the results of the examples mentioned in the last two sections how strongly the boundary conditions and initial conditions influence the distribution of localization zones.

All the results presented herein were obtained without introduction any local imperfections geometric or in material properties to start the localization process. This effect is naturally played by dynamical formulation and the propagation of the waves in the solid body under analysis.

4.3 Thermal softening and microdamage

There are two reasons for softening behaviour and finally the fracture of the loaded specimen. Both of them are well documented by experiments. Firstly, in dynamical processes fracture can occur as a result of an adiabatic shear band localization attributed to a plastic instability generated by thermal softening during plastic deformation. Some of the experimental confirmation are cited and discussed in [58].

The researchers among others made microscopic observations of the shear band localization on the thin-walled steel tubes in a split Hopkinson torsion bar. Secondly, in static or dynamic processes the deformation in shear was imposed to produce shear bands. It was found whenever the shear band led to fracture of the specimen, the fracture occurred by a process of void nucleation, growth and coalescence.

The investigation of localized shear band phenomenon, when softening occurs, requires the using of any regularization method which assures the well-posedness of the formulation. This in turn leads to avoiding of so called spurious mesh sensitivity in numerical calculations; see [52]. In particular, for dynamic processes in ductile materials the rate dependent formulation has a good physical background and successfully serves as a regularization method, cf. [30],[77]. In this case the relaxation time plays the role of regularization parameter.

In recent years ZBIB and JURBAN [98] have investigated a 3-D problem involving

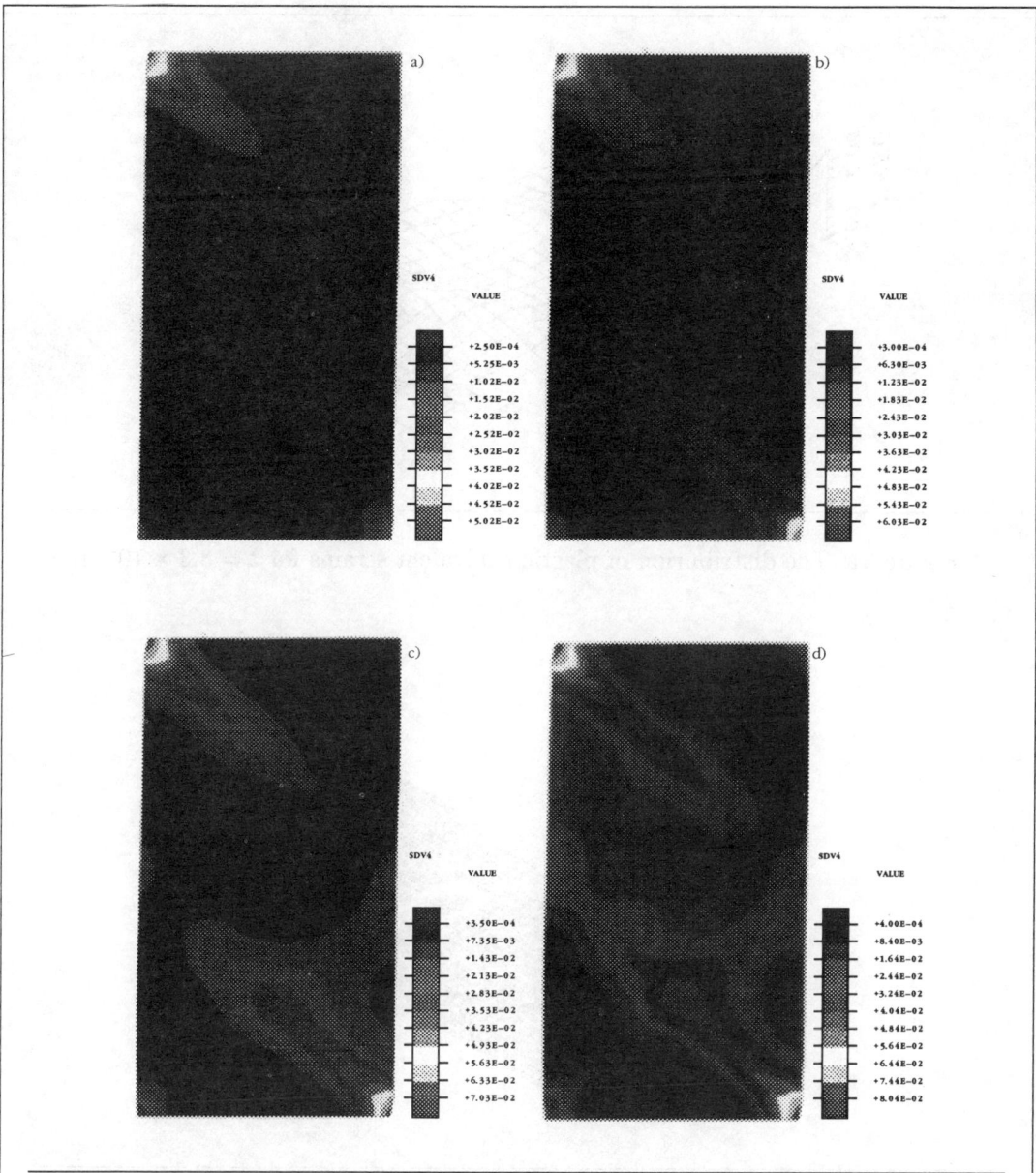

Figure 30: The development of zones of plastic deformation PEEQ in kinematically controlled process: first step static, second step dynamic. Plots of the dynamic step only: a) $6.0 \times 10^{-5}s$, b) $1.2 \times 10^{-4}s$, c) $1.8 \times 10^{-4}s$, d) $3.3 \times 10^{-4}s$.

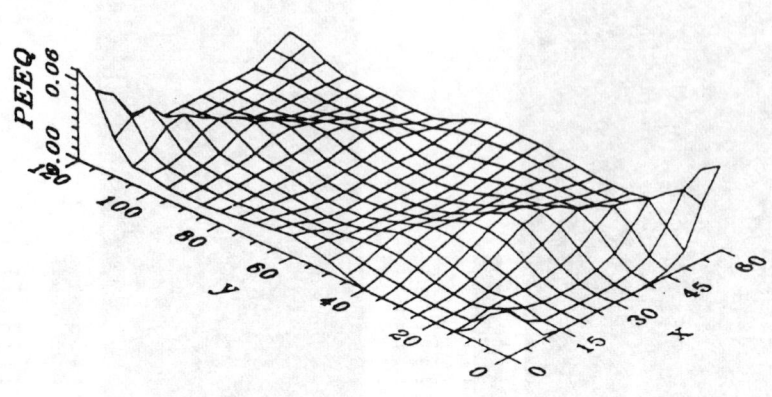

Figure 31: The distribution of plastic equivalent strains for $t = 3.3 \times 10^{-4}$s

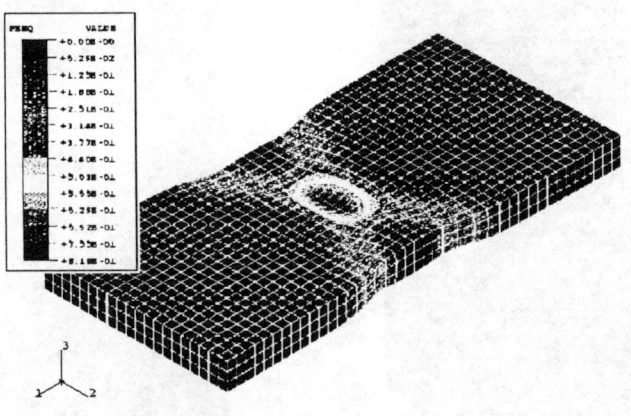

Figure 32: Localization of PEEQ in a specimen

the development of shear bands in a steel bar pulled in tension and BATRA and ZHANG [5], the 3-D dynamic thermomechanical deformations of a steel thin tube twisted in a split Hopkinson bar. The authors presented their numerical achievements in the analysis of plastic strain localization for 2-D and 3-D cases with the discussion of the finite element modelling problems in [30, 57, 53].

Selected elements of the constitutive formulation are recalled in this paper. The objective of this presentation is the application of a recently developed viscoplastic--damage type constitutive theory, see Perzyna [77], for both, high strain rate flow process and ductile fracture to the problem of shear band localization and fracture of dynamically and quasi statically loaded tests specimens.

The examples show separately the influence of the softening by the rise of temperature in high speed dynamic adiabatic processes and evolution of voids in quasi static loading. The variation of boundary and initial conditions and the material parameters influence the maximum load capacity, the position and the width of localized zones.

Two types of specimens are investigated in this presentation using in modeling 2-D and 3-D elements.

The rate type constitutive structure for an elastic–viscoplastic material in which the effects of the micro–damage mechanism and thermomechanical coupling are taken into consideration.

There are introduced the following axioms in the formulation: 1) axiom of the existence of the free energy function which depends on the deformation gradient $\mathbf{F}$, a temperature field ϑ and the internal state variable vector $\boldsymbol{\mu}$; 2) axiom of objectivity which states that the constitutive structure is invariant with respect to any rotation and stretching; and 3) the axiom of entropy production.

It is also postulated that the vector of internal state variables $\boldsymbol{\mu} = (\boldsymbol{\zeta}, \xi)$ is the function of $\boldsymbol{\zeta}$ the new internal state vector which describes the dissipation effects generated by viscoplastic flow phenomena and ξ the volume fraction porosity parameter which takes account for micro–damage mechanism.

The plastic potential function for damaged material is postulated in the form

$$f = J_2 + n\xi J_1^2, \qquad (77)$$

where $n = n(\vartheta)$ is the temperature dependent material function, J_1 and J_2 are the first invariant of Kirchhoff stress tensor and the second invariant of deviatoric state, respectively.

In particular the evolution equation for plastic part of rate of total deformation $\mathbf{d}^p$ is proposed as follows

$$\mathbf{d}^p = \Lambda \mathbf{P}, \qquad (78)$$

where the elastic–viscoplastic Perzyna's type model of a material is assumed

$$\Lambda = \frac{1}{T_m} \langle \Phi(f - \kappa) \rangle, \qquad (79)$$

T_m denotes the relaxation time for mechanical disturbances and κ is the isotropic work-hardening parameter, Φ is the empirical overstress function and the bracket $\langle \cdot \rangle$ defines the ramp function. For associative plasticity it is assumed: $\mathbf{P} = \frac{1}{2\sqrt{J_2}} \frac{\partial f}{\partial \boldsymbol{\tau}}$.

The isotropic hardening–softening material function κ can be assumed in two alternative forms. The first after Perzyna

$$\kappa = \kappa_0^2 \{q + (1-q) \exp[-h(\vartheta) \, \in^p]\}^2 \left[1 - \left(\frac{\xi}{\xi^F}\right)^{\frac{1}{2}}\right], \tag{80}$$

and the second following Gurson model with the modifications proposed by Tvergaard and Needleman

$$\kappa = 1 + q_3 \xi^2, \tag{81}$$

where $q = \frac{\kappa_1}{\kappa_0}$, κ_0 and κ_1 denote the yield and saturation stress of the matrix material (both can be temperature dependent functions), respectively, $h = h(\vartheta)$ is the temperature dependent strain hardening function for the matrix material, $\in^p = \int_0^t (\frac{2}{3}\mathbf{d}^p : \mathbf{d}^p)^{\frac{1}{2}} dt$ is the equivalent plastic deformation, ξ^F denotes the value of porosity at which the incipient fracture occurs; the overstress viscoplastic function Φ is postulated in the form

$$\Phi(f - \kappa) = \left(\frac{f}{\kappa} - 1\right)^m, \quad \text{where } m = 1, 3, 5, \ldots \tag{82}$$

The detailed discussion on the set of evolution equations one can find in [58, 57].

The experimental observation have shown that the micro–damage process consists of nucleation, growth and coalescence of microvoids and that coalescence mechanism can be treated as nucleation and growth process on a smaller scale. Then the porosity or the void volume fraction parameter ξ can be determined by $\dot{\xi} = (\dot{\xi})_{nucl} + (\dot{\xi})_{grow}$.

Based on a heuristic suggestion that the nucleation of microvoids in dynamic loading processes which are characterized by very short time duration is governed by the thermally–activated mechanism, we postulate the evolution law for porosity nucleation in the form

$$(\dot{\xi})_{nucl} = \frac{1}{T_m} h^*(\xi, \vartheta) \left[\exp \frac{m^*(\vartheta) \, | \sigma - \sigma_N(\xi, \vartheta, \in^p) |}{k\vartheta} - 1\right], \tag{83}$$

where k denotes the Boltzmann constant, $h^*(\xi, \vartheta)$ represents a void nucleation material function which is introduced to take account of the effect of microvoid interaction, $m^*(\vartheta)$ is a temperature dependent coefficient, $\sigma = (1//3)J_1$ is the mean stress and $\sigma_N(\xi, \vartheta, \in^p)$ is the porosity, temperature and equivalent plastic strain dependent threshold stress for microvoid nucleation.

For the growth mechanism it is postulated

$$(\dot{\xi})_{grow} = \frac{1}{T_m} \frac{g^*(\xi, \vartheta)}{\sqrt{\kappa}} [\sigma - \sigma_{eq}(\xi, \vartheta, \in^p)], \tag{84}$$

where $T_m\sqrt{\kappa}$ denotes the dynamic viscosity of a material, $g^*(\xi,\vartheta)$ represents a void growth material function and takes account for void interaction and $\sigma_{eq}(\xi,\vartheta,\epsilon^p)$ is the porosity, temperature and equivalent plastic strain dependent void growth threshold mean stress.

Previous Eqns determine the evolution function for porosity. For a certain set of parameters also the Gurson's criteria could be met.

The fracture criterion on the evolution of the porosity internal state variable assumes that for $\xi = \xi^F$ catastrophe takes place, that is

$$\kappa = \hat{\kappa}(\epsilon^p, \vartheta, \xi)\,|_{\xi=\xi^F} = 0. \tag{85}$$

This condition describes the main feature observed experimentally that the load tends to zero at the fracture point. It is noteworthy that the isotropic hardening–softening material function $\hat{\kappa}$ proposed in particular form (80) satisfies the fracture criterion (85).

4.3.1 Adiabatic problems

Without going into details let us recall that an adiabatic inelastic flow process is defined as follows. Find velocity vector v, mass density ρ_M, Kirchhoff stress τ, porosity ξ and temperature ϑ as functions of time t and spatial variable $\mathbf{x}$ such that

(i) the field equations

$$\dot{v} = \frac{1}{\rho_M^0(1-\xi_0)}\left(\frac{\tau}{\rho_M}\mathrm{grad}\rho_M + \mathrm{div}\tau - \frac{\tau}{1-\xi}\mathrm{grad}\xi\right),$$

$$\dot{\rho}_M = \frac{\rho_M}{1-\xi}\Xi - \rho_M \mathrm{div}v,$$

$$\dot{\tau} = \left[\mathcal{L}^e - \frac{1}{c_p \rho_{Ref}}\vartheta\mathcal{L}^{th}\frac{\partial\tau}{\partial\vartheta}\right]:\mathrm{sym}Dv + 2\mathrm{sym}\left(\tau:\frac{\partial v}{\partial x}\right)$$

$$-\left[\left(\frac{\chi^*}{\rho_M(1-\xi)c_p}\mathcal{L}^{th}\tau + \mathcal{L}^e + \mathbf{g}\tau + \tau\mathbf{g}\right):\mathbf{P}\right]\frac{1}{T_m}\langle\left(\frac{f}{\kappa}-1\right)^m\rangle$$

$$-\frac{\chi^{**}\mathcal{L}^{th}}{\rho_M(1-\xi)c_p}\Xi, \tag{86}$$

$$\dot{\xi} = \Xi,$$

$$\dot{\vartheta} = \frac{\vartheta}{c_p \rho_{Ref}}\frac{\partial\tau}{\partial\vartheta}:\mathrm{sym}Dv + \frac{\chi^*}{\rho_M(1-\xi)c_p}\tau:\mathbf{P}\frac{1}{T_m}\langle\left(\frac{f}{\kappa}-1\right)^m\rangle$$

$$+\frac{\chi^{**}}{\rho_M(1-\xi)c_p}\Xi;$$

(ii) the boundary conditions

(a) displacement ϕ is prescribed on a part ∂_ϕ of $\partial\phi(\mathcal{B})$ and tractions $(\tau \cdot \mathbf{n})^a$ are prescribed on part ∂_τ of $\partial\phi(\mathcal{B})$, where $\partial_\phi \cap \partial_\tau = 0$ and $\overline{\partial_\phi \cup \partial_\tau} = \partial\phi(\mathcal{B})$;

(b) heat flux $\mathbf{q} \cdot \mathbf{n} = 0$ is prescribed on $\partial\phi(\mathcal{B})$;

(iii) the initial conditions

ϕ, $\boldsymbol{v}$, σ_M, ϑ, ξ and τ are given at each particle $X \in \mathcal{B}$ at $t = 0$;

are satisfied.

In the field equations (86) ρ_M and ρ_M^0 denote the actual and reference mass density of the matrix material, respectively, ξ_0 is the initial porosity of a material and $D\boldsymbol{v}$ denotes the spatial velocity gradient. The above system of governing equations is viewed in the form of so called nonhomogeneous abstract Cauchy problem which allows the discussion of well–posedness.

Using the above formulation confirms that the localization of plastic deformations in an elastic–viscoplastic solid body arises only as the result of the reflection and interaction of waves. In computations any other artificial imperfections are not necessary to start the process of localized deformations.

4.3.2 Numerical examples

Evolution of porosity for quasi–static case There was tested a rectangular strip of the length $l = 25.4mm$, the width $w = 12.7mm$ and the thickness $t = 1.95mm$. The specimen was divided into 2400 8-node 3-D elements ($40 * 20 * 3$). The following data were accepted as: Young modulus $E = 200,000MPa$, yield limit $\sigma_P = 1634MPa$, mass density $\rho = 7850kg/m^3$, initial porosity 0.04 and critical porosity 0.3.

The incremental process was controlled by the velocity that acts at the shorter edges $v = 1mm/s$ in the tension direction. In Fig.32 we present the deformation of the plate (total elongation is about 25%) and the distribution of plastic equivalent strains (PEEQ) which clearly shows the localized zones of plastic deformations.

In Figs.33 and 34 the evolution of PEEQ, the changes of relative density (RD) and void volume fraction (VVF) along the longer edge of the specimen are presented. The localization appears symmetrically in the middle of the plate. Locally the PEEQ reaches the value of 80%. The detailed study was presented in [31].

Dynamic tension of thin plate The adiabatic case of a thin plate ($t = 0.33mm$) was modeled by using 4-node shell elements. The constitutive parameters were the same as before and additionally the relaxation time of the material $T_m = 2.5 * 10^{-6}s$, specific heat $460J/kg^0K$ and heat fraction 0.9. The computed set of examples included both: double–sided symmetrically loaded specimens and one–sided ones. For slow (quasistatic) processes the results for both cases approach the same responds, however for the processes in which the inertial effects and the effects of wave propagation were strong the results depend significantly on the boundary and initial conditions.

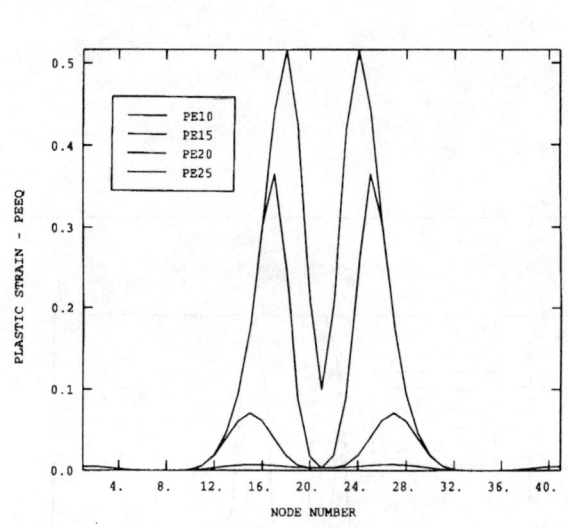

Figure 33: The in the distribution of PEEQ in time

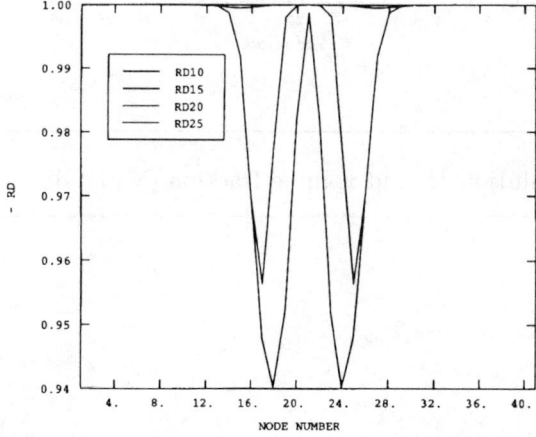

Figure 34: The changes in relative density (RD) in time

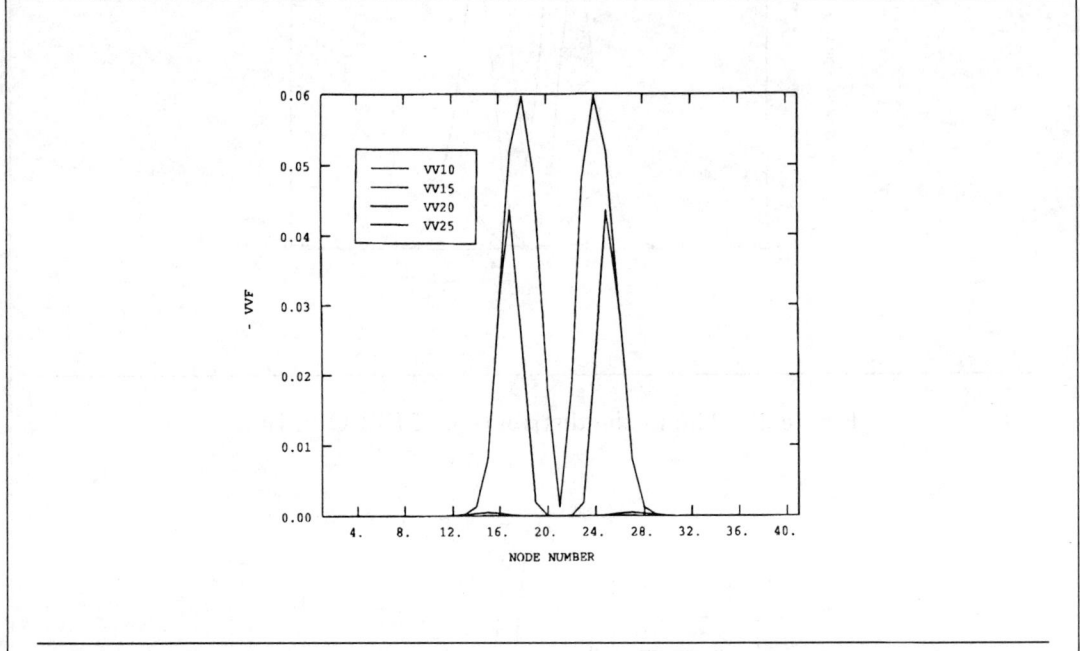

Figure 35: The evolution of void volume fraction (VVF) distribution in time

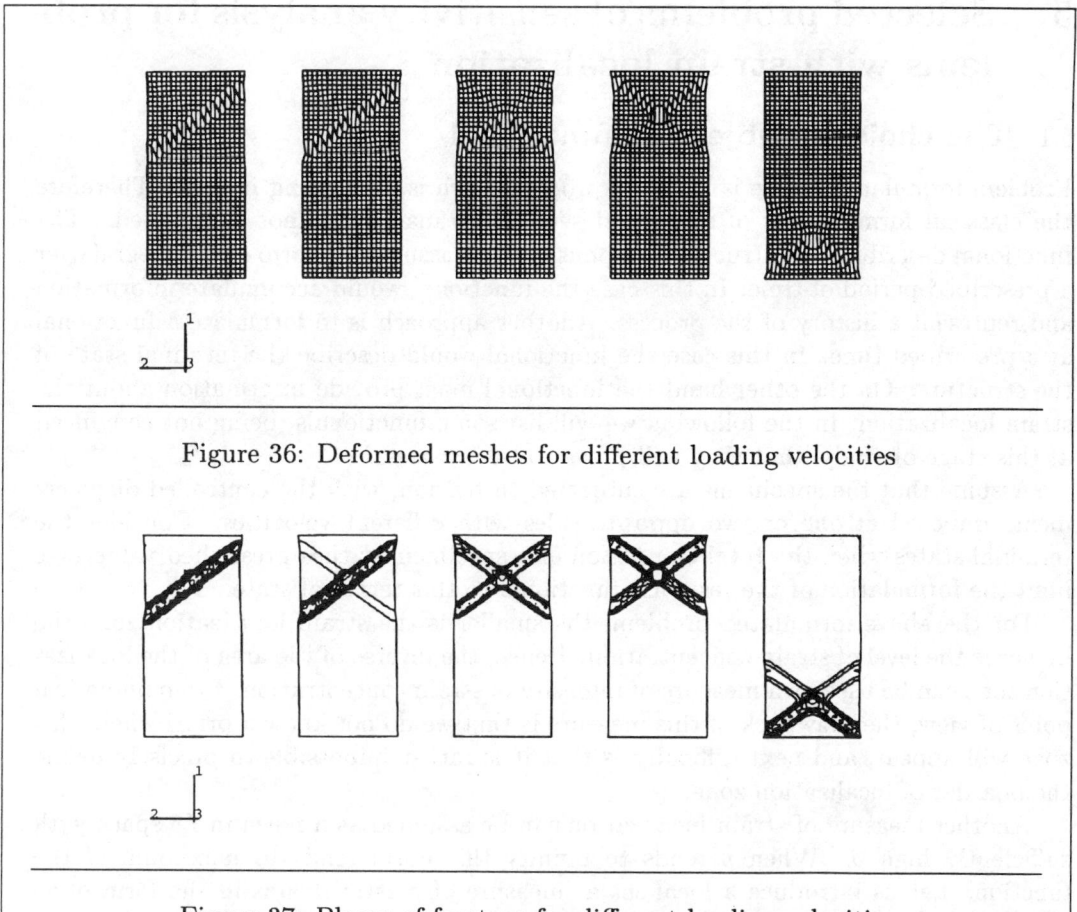

Figure 36: Deformed meshes for different loading velocities

Figure 37: Places of fracture for different loading velocities

The velocities of displacements acting on the edge change from $1m/s$ to $100m/s$. The fracture criterion for metal which was used in computations was reached when the value of PEEQ approached 100The places of localization different velocities of loading $(2m/s, 5m/s, 10m/s, 20m/s, 50m/s)$ are visualized the deformed meshes in Fig.36.

The form of fracture are confirmed in Fig.37 by the distribution of PEEQ. The results are monitored for the same elongation of the specimens.

As it is expected the values of the forces which accompany these states differ reaching the higher limit values for faster processes.

The aspects of numerical modelling, in particular the choise of the FE mesh, the method of time integration for the problems under consideration were discussed in A.GLEMA et al. [29].

5. Selected problems of sensitivity analysis for problems with strain localization

5.1 The choice of objective functional

Problem formulated before is in fact a process which is developing in time. Therefore the classical formulations of structural sensitivity analysis cannot be applied. The functional describing the structural response can be assumed in form of an integral over a prescribed period of time. In this case the functional would accumulate information and represent a history of the process. Another approach is to formulate a functional at a prescribed time. In this case the functional would describe the terminal state of the structure. On the other hand the functional must provide information about the strain localization. In the following we will list some functionals, being not convinced at this stage of study, that they will prove to be a success.

Assume that the specimens are subjected to tension, with the controlled displacements imposed at one or two opposite sides with different velocities. Consider the terminal states when the total elongation of a specimen reaches prescribed value. We limit the formulation of the response functional to this terminal state.

For the above formulated problem, the smaller is the strain localization zone the higher is the level of strain concentration. Hence, the inverse of the area of the localization zone can be used as a measure of intensity of strain concentration. From numerical point of view, the drawback of this measure is that we do not know a priori where this zone will appear, and next difficulty is that it is rather impossible to precisely define the boarder of localization zone.

Another measure of strain localization can be assumed as a norm in L^p space with sufficiently high p. When p tends to infinity this norm tends to maximum of the function. Let us introduce a local scalar measure of plastic strains in the form of so called plastic equivalent strain

$$\varepsilon_{PEEQ} = \int_t (\frac{3}{2}\mathbf{d}^p : \mathbf{d}^p)^{\frac{1}{2}} dt \tag{87}$$

Consider a functional which is the global measure of strain localization intensity in the form of norm L^p

$$\| \varepsilon_{pl} \| = [\int_V (\varepsilon_{PEEQ})^p dV]^{\frac{1}{p}} \tag{88}$$

where the integration is carried our over the whole volume of the specimen. Energetic norms can be assumed, too. Next problem is what we define as design variables. In other words, with respect to which parameters we should carry out the sensitivity analysis to obtain useful results. Interesting can be the sensitivity with respect to the velocity of imposed displacements, distribution of the imposed displacements, boundary of the specimen or with respect to parameters of the physical model.

Numerical examples and the parametric study of sensitivity will be presented to provide the general information about the problem under discussion. The study will be further developed aiming at the theoretical and numerical verification of the applicability of the above mentioned functionals and design variables to such a complex nonlinear problem. A mathematical background of the sensitivity analysis allowing for the physical and/or geometrical nonlinearity was presented in [41], whereas the initial distortions were discussed in [26].

5.2 Influence of the relaxation time, initial and boundary conditions

In the numerical examples presented in the paper we studied the sensitivity of location and the width of the localized plastic zones with respect to initial and boundary conditions for dynamic tension of axisymmetric specimen; see DUSZEK-PERZYNA et al. [22]. Basically, two general cases of boundary and initial conditions were considered: (1) symmetric (duble side) tension of the specimen which results in symmetric pattern of deformations, (2) asymmetric (single side) tension of the specimen with the opposite side fixed, which leads to non-symmetric deformation. The axisymmetric specimen under analysis is 19.05 mm long and has the radius of 3.175 mm. The other properties were assumed as follows: Young modulus E = 200 GPa, Poisson ratio = 0.3, initial mass density = 7850, inelastic heat fraction 0.9, specific heat $460 J/kg^0C$, the yield stress = 1634 MPa for initial temperature which changes (softens) nonlinearly to 1006 MPa for the rise of the temperature up to 610^0C. The viscosity $D = 1/T_m$ was assumed also as a function of temperature and varies between 40 1/s (for initial temperature) and 100 1/s for the temperature 200^0C. For the set of symmetric examples the velocities changed between = 0.1 m/s and = 20 m/s. In Fig.38 the evolution of localization zone of plastic deformations for the velocity = 5 m/s is presented. The other symmetric cases of loading with different velocities also result in symmetric pattern of plastic deformation however for higher speed of deformation two zones of shear bands appear close to the loaded edges. The distributions of plastic equivalent strains (PEEQ) for different velocities are shown in Figs.39 and 40. In the study there was also used the failure criteria which assumed that the element brakes down if the value of PEEQ reaches 100%.

The interesting response confirming the strong interaction of longitudinal waves is presented in Fig.41. In this figure the changes of forces which act on the edges versus total elongation of the specimen are presented. For slower processes ("quasi static") the post-critical behaviour is close one to the other while for the cases where the inertial effects are stronger the significant oscillations are observed.

For non-symmetric (single side) loading the place of plastic strain localization varies depending on the assumed initial velocity that acts on the upper edge of the specimen. For quasi static (slow) cases zones of plastic concentrations tend to the middle of a specimen, Fig.42. For higher velocities this position changes and finally appears close

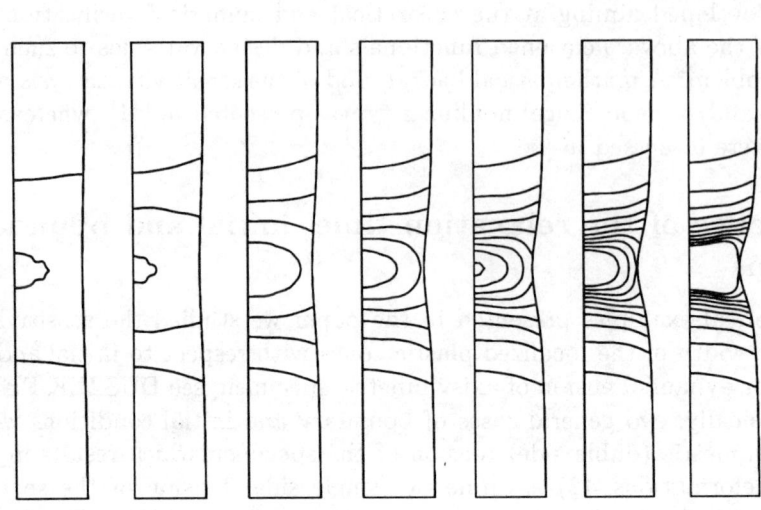

Figure 38: Deformation history for duble-side velocity $v = 5m/s$

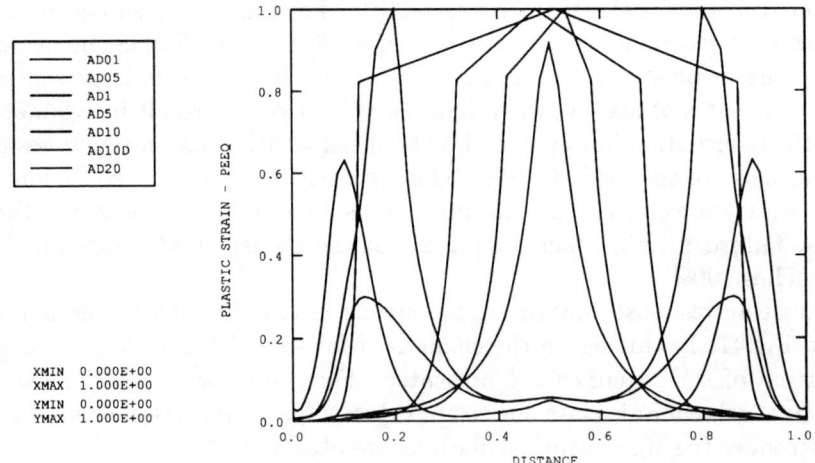

Figure 39: PEEQ strains along deformed specimen axis for different double-side velocities

Figure 40: PEEQ strain contours (5-50%) for different double-side velocities

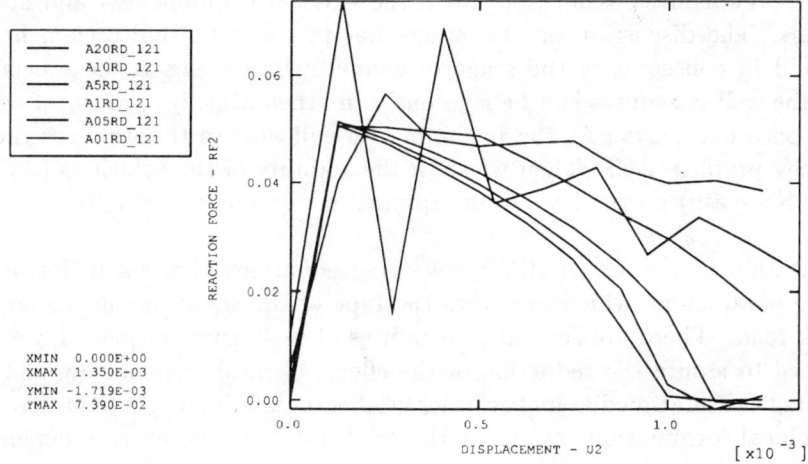

Figure 41: Reaction force versus displacement for different double-side velocities

to the loaded edge. Similarly to the previous case (symmetric) the sensitivity of loads acting on the fixed edge to the velocity of the process is observed in Fig.43.

In the processes under consideration for which the total elongation of the specimen was of order 25% the changes of temperature locally reached several hundreds degrees C.

Beside of the results presented in this paper also the evident sensitivity of the width of zones of localized deformations to variations of constitutive parameters (viscosity) was observed for other types of the specimens (e.g. thin plates). Basically, for shorter relaxation times (higher viscosity) the shear bands were of the narrower width.

Next numerical examples will be demonstrated during the presentation.

6. Final remarks and conclusions

The numerical modelling of the failure in softening materials requires a proper material modelling of the problem. Trying to stay in the convention of phenomenological approach, one has to remember, that simple mapping of load-displacement data obtained in experiments with a descending behaviour in post-critical states provides negative constitutive stiffnesses and in the classical, rate-independent continuum result in loss of well-posedness of IBVP.

For typical set of equations, which we work with, when modelling any IBVP it is very difficult, or even not possible, to prove the existence, uniqueness and stability of the solutions. The discussion on this stage that precedes the finite element weak formulation and in consequence the stage of computations seems to be crucial. To some extend the well posedness can be rigorously (mathematically) proven, if we split the attention onto two parts. At the beginning we will discuss the well posedness of abstract Cauchy problem ACP. Then we show the stability of the solutions (its insensitivity to the FE mesh) on the level of interpretation of numerical results.

A regularization of the set of IBVP governing equations consists in introducing of length scale parameters, which preserves the type of the equations unchanged also after the peak load. There are several possibilities of regularization which are intensively developed to assure the reduction of the effects of mesh dependency and mesh alignment (using Cosserat media, higher order gradients theories or gradient dependent plasticity, nonlocal formulations etc.). In the work we have chosen rate-dependency (viscoplasticity) which is physically well motivated, particularly for the cases under dynamic loading, and simultaneously plays the role of a mathematical tool that secures the well-posedness of the problems.

One of the aims of this presentation was the numerical investigation of the effects of introducing the viscoplastic regularization procedure for plastic flow localization phe-

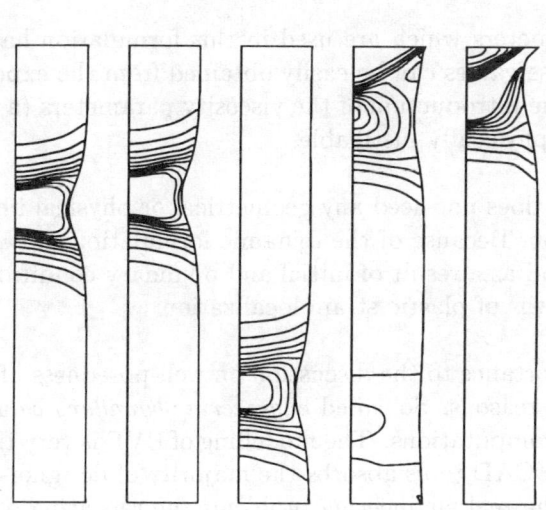

Figure 42: PEEQ strain contours (5-50%) for different single-side velocities

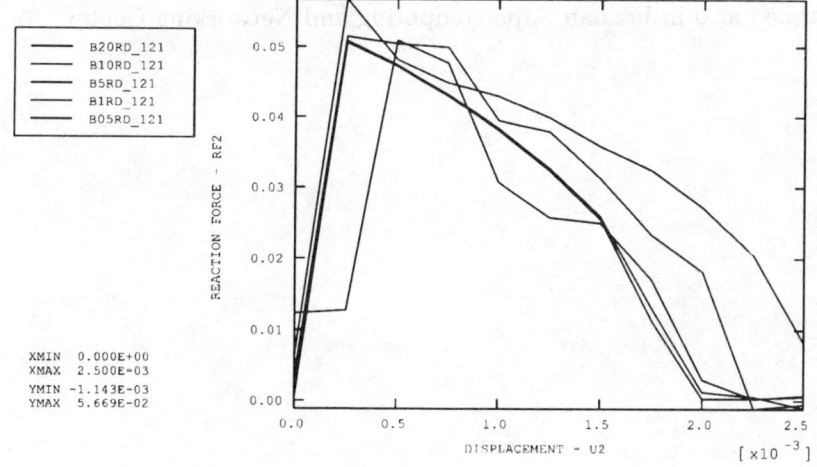

Figure 43: Reaction force versus displacement for different single-side velocities

nomena in a ductile and soil like materials.

The material parameters which are used in this formulation have a good physical interpretation and these values can be easily obtained from the experiments. Specially for ductile materials the introduction of the viscosity parameters (a relaxation time for stresses T_m) is deeply physically justifiable.

The viscoplasticity does not need any geometrical or physical imperfection to start the place of localization. Because of the dynamic formulation the waves that travel in the domain interact and as a result of initial and boundary conditions they choose the directions and the widths of plastic strain localization.

We attach the importance to the discussion on well-posedness of the IBVP also because of very practical reasons. So called *engineering boundary value problems* usually lead to the large scale computations. The modelling of BVP is very time consuming and in the using of modern CAD tools absorbs the majority of designers' activity. Because of the dimensions of the real engineering problems the repeating of the computations for different meshes can be questionable and sometimes economically not motivated. Therefore, the using the softening constitutive laws should be seriously discussed on the level of mathematical formulation before we start any computations.

Acknowledgment: The paper has been prepared within the research programme sponsored by the Committee of Scientific Research under Grant 7 T07 A013 10 and the local grant DS-11-555/98 of Poznan University of Technology. The computations has been performed also in Poznan Supercomputing and Networking Center.

References

[1] ABAQUS. Manuals for v. 5.6. Reports, Hibbitt, Karlsson & Sorensen, Inc., 1997.

[2] T. ADACHI and F. OKA. Constitutive equations for normally consolidated clay based on elasto-viscoplasticity. *Solids and Foundations*, 22(4):57–70, 1982.

[3] T. ADACHI, F. OKA, and M. MIMURA. Mathematical structure of an overstress elasto-viscoplastic model for clay. *Soils and Foundations*, 27(4):31–42, 1987.

[4] R. C. BATRA. Numerical solutions of initial-boundary-value problems with shear band localization. *Proceedinngs of Advanced School on Localization and fracture phenomena in inelastic solids, Udine (in this volume)*, 1997.

[5] R. C. BATRA and X. ZHANG. On the propagation of shear band in a steel tube. *Int.Jour. of Plasticity*, 1993.

[6] Z. P. BAZANT, J. PAN, and G. PIJAUDIER-CABOT. Softening in reinforced concrete beams and frames.

[7] T. BELYTSCHKO and J. FISH. Spectral superposition on finite elements for shear banding problems. In *V Int. Symp. on Numerical Methods in Engineering, Lausanne, 11-15 Sept. 1989, Part I*.

[8] T. BELYTSCHKO and J. FISH. Embedded hinge lines for plate elements. *Comp. Meth. in Appl. Mech. Eng.*, 76:67–86, 1989.

[9] T. BELYTSCHKO, J. FISH, and B. E. ENGELMANN. A finite element with embedded localization zones. *Comp. Meth. in Appl. Mech. Eng.*, 70:59–89, September 1988.

[10] T. BELYTSCHKO and D. LASRY. A study of localization limiters for strain-softening in statics and in dynamics. *Comp. Struc.*, 33(3):707–715, 1989.

[11] T. BELYTSCHKO, X.-J. WANG, Z. P. BAZANT, and Y. HYUN. Transient solutions for one-dimensional problems with strain softening. *Jour. Appl. Mech.*, 54:513–518, September 1987.

[12] D. BIGONI and T. HUECKEL. Uniqueness and localization–I. associative and non-associative elastoplasticity. *Int. Jour. of Plasticity*, 28(2):197–213, 1991.

[13] R. de BORST. Simulation of strain localization: A reappraisal of the Cosserat continuum. *Eng. Comp.*, 8:317–332, 1991.

[14] R. de BORST. Gradient-dependent plasticity: Formulation and algorithmic aspects. *Int. J. Num. Meth. in Eng.*, 35:521–539, 1992.

[15] R. de BORST. Fundamental issues in finite element analyses of localization of deformation. *Eng. Comp.*, 10:99–121, 1993.

[16] R. de BORST, H.-B. MUEHLHAUS, J. PAMIN, and L. J. SLUYS. Computational modelling of localisation of deformation. In D. R. J. Owen, E. Onate, and E. Hinton, editors, *Proc. Third Intern. Conference on Computational Plasticity, Fundamentals and Applications, Barcelona, 1992*, pages 483–508, Swansea, April 4-9 1992. Pineridge Press.

[17] A. K. CHAKRABARTI and J. W. SPRETNAK. Instability of plastic flow in the directions of pure shear: I. theory. *Metallurgica Transactions*, 6A:733–747, April 1975.

[18] D. R. CURRAN, L. SEAMAN, and D.A. SHOCKEY. Dynamic failure in solids. *Physics Today*, pages 46–55, 1977.

[19] J. DESRUES. *La localisation de la deformation dans les materiaux granulaires*. PhD thesis, L'Universite Scientifique et Medical et L'Institut National Polytechnique de Grenoble, 1984.

[20] J. DESRUES and R. CHAMBON. Shear band analysis for granular materials: The question of incremental non-linearity. *Ingenieur-Archiv*, 59:187–196, 1989.

[21] A. DRESCHER. Zagadnienia doświadczalnej weryfikacji modelu ciała o wzmocnieniu gestościowym. *Rozp. Inż.*, 3:351–387, 1972.

[22] M. DUSZEK-PERZYNA, A. GARSTECKI, A. GLEMA, and T. ŁODYGOWSKI. Sensitivity of strain localization - parametric study. In A.Gastecki and J.Rakowski, editors, *Proceedings of Polish Conference on Computer Methods in Mechanics, PCCMM'97 - Poznan*, volume 1, pages 345–352, 1997.

[23] G. DUVAUT and J. L. LIONS. *Inequalities in Mechanics and Physics*. Springer, Berlin [u.a.], 1976.

[24] G. ENGELN-MÜLLGES and F. REUTER. *Formelsammlung zur Numerischen Mathematik mit Standard-FORTRAN 77-Programmen*. BI Wissenschaftsverlag, 1988.

[25] J. FISH and T. BELYTSCHKO. A general finite element procedure for problems with high gradients. *Comp. Struc.*, 35(4):309–319, 1990.

[26] A. GARSTECKI and A. GLEMA. Sensitivity analysis and optimal redesign of columns in the state of initial distortions and prestress. *Structural Optimization*, 3, 1991.

[27] A. GAWĘCKI. Sprężysto–plastyczne konstrukcje prętowe z luzami. *Rozprawy Politechniki Poznańskiej*, (185), 1987.

[28] A. GAWĘCKI and B. JANIŃSKA. Problemy analizy i identyfikacji w procesach nieustalonego przeplywu ciepla. *ZNPP*, 39:103–131, 1995.

[29] A. GLEMA, W. KĄKOL, and T. ŁODYGOWSKI. Numerical modelling in adiabatic shear band formation in a twisting test. *Eng. Trans.*, 3, 1997.

[30] A. GLEMA and T. ŁODYGOWSKI. Plastic strain localization and failure of a ductile specimen. In *ZAMM'96*, volume 77, pages S97–98, 1997.

[31] A. GLEMA, T. ŁODYGOWSKI, and P. PERZYNA. Effects of microdamage in plastic strain localization. In A.Gastecki and J.Rakowski, editors, *Proceedings of Polish Conference on Computer Methods in Mechanics, PCCMM'97 - Poznan*, volume 2, pages 451–458, 1997.

[32] M. E. GURTIN. *An Introduction to Continuum Mechanics*. Academic Press, 1981.

[33] J. HADAMARD. *Lesons sur la Propagation des Ondes et les Equations de L'Hydrodynamique*. Paris, 1903.

[34] R. HILL. A general theory of uniqueness and stability in elastic–plastic solids. *J. Mech. Phys. Solids*, 6:236–249, 1958.

[35] R. HILL. Acceleration waves in solids. *J. Mech. Phys. of Solids*, 10:1–16, 1962.

[36] R. HILL and J. W. HUTCHINSON. Bifurcation phenomena in the plane tension test. *J. Mech. Phys. Solids*, 23, 1975.

[37] T. J. R. HUGHES, T. KATO, and J. E. MARSDEN. Well-posed quasi-linear second-order hyperbolic systems with applications to nonlinear elastodynamics and general relativity. *Arch. Rat. Mech. Anal.*, 63:273–294, 1977.

[38] IONESCU I.R. and M. SOFONEA. *Functional and numerical methods in viscoplasticity*. Oxford University Press, Oxford, New York, Tokyo, 1993.

[39] T. KATO. The Cauchy problem for quasi–linear symmetric hyperbolic systems. *Arch. Rational Mech. Anal.*, 58:181–205, 1975.

[40] K. KIBLER, M. LENGNICK, and T. ŁODYGOWSKI. Selected aspects of the well–posedness of the localized plastic flow processes. *CAM & ES (submitted for publication)*, 1994.

[41] M. KLEIBER. Shape and non-shape structural sensitivity analysis for problems with any material and kinematic non-linearity. *Comp. Meth. in Appl. Mech. Eng.*, 108:73–97, 1993.

[42] M. KLISIŃSKI, K. RUNESSON, and S. STURE. Finite element with inner softening band. *Jour. of Eng. Mech., ASCE*, 17(3):575–587, 1991.

[43] A. KORBEL. Structural and mechanical aspects of homogeneous and heterogeneous deformation of solids. *Proceedinngs of Advanced School on Localization and fracture phenomena in inelastic solids, Udine (this volume)*, 1997.

[44] S. KURCYUSZ. *Matematyczne podstawy teorii optymalizacji*. PWN - Warszawa, 1982.

[45] E.H. LEE. Elastic–plastic deformations at finite strains. *J. Appl. Mech.*, 36, 1969.

[46] J. LEMAITRE and J.-L. CHABOCHE. *Mechanics of solid materials*. Cambridge University Press, 1985.

[47] M. LENGNICK, T. ŁODYGOWSKI, P. PERZYNA, and E. STEIN. On regularization of plastic flow localization in a soil material. *Eng. Trans.*, 44(3), 1996.

[48] S. LEROUEIL, M. KOBBAJ, F. TAVENAS, and R. BOUCHARD. Stress-strain-strain rate relation for the compressibility of sensitive natural clays. *Geotechnique*, 35:159–180, 1985.

[49] T. ŁODYGOWSKI. Numerical analysis of softening structures. In *V Int. Symp. on Numerical Methods in Engineering, Lausanne, 11-15 Sept. 1989, Part I*, pages 371–376, 1989.

[50] T. ŁODYGOWSKI. Mesh independent beam elements for strain localization. *Comp. Meth. in Civil Eng.*, 3(3):9–24, 1993.

[51] T. ŁODYGOWSKI. On avoiding of spurious mesh sensitivity in numerical analysis of plastic strain localization. *CAM & ES*, 2(3):231–248, 1995.

[52] T. ŁODYGOWSKI. Theoretical and numerical aspects of plastic strain localization. *Wyd. politechniki Poznańskiej*, 1996.

[53] T. ŁODYGOWSKI and A. GLEMA. Numerial modelling of plastic strain localization in an adiabatic twisting test. In Cz. Rymarz, editor, *XII Conf. CMM, Warsaw-Zegrze, May 9-13, 1995*, 1995.

[54] T. ŁODYGOWSKI, M. LENGNICK, P. PERZYNA, and E. STEIN. Viscoplastic numerical analysis of dynamic plastic strain localization for a ductile material. *Arch. Mech.*, 46(4):541–557, 1994.

[55] T. ŁODYGOWSKI, M. LENGNICK, and E. STEIN. Numerical analysis of localization phenomena in ductile and brittle materials. In B.H.V. Topping, editor, *Second International Conference on Computational Structures Technology, Athena, Greece, August 30 - September 1, 1994*, 1994.

[56] T. ŁODYGOWSKI, M. LENGNICK, and E. STEIN. On rate dependent regularization of localization phenomena in a brittle degradating material. In R. de Borst N. Bicanic, H. Mang, editor, *EURO-C'94 Conference on Numerical Modelling of Concrete Structures, Innsbruck, March 22-25, 1994*, volume 1, pages 333-342, 1994.

[57] T. ŁODYGOWSKI and P. PERZYNA. Localized fracture in inelastic polycrystalline solids under dynamic loading processes. *Int. Journ. Damage Mech.*, 6(4):364, 1997.

[58] T. ŁODYGOWSKI and P. PERZYNA. Numerical modelling of localized fracture of inelastic solids in dynamic loading processes. *Int. Journ. Num. Meth. Eng. Sci.*, 40, 1997.

[59] B. LORET. An introduction to classical theory of elastoplasticity. In F.Darve, editor, *Geomaterials: constitutive equations and modelling*, pages 149-186. Elsevier Applied Science, London and New York, 1990.

[60] A.M. LYAPUNOV. The general problem of the stability of motion. *Int. Journ. Control*, 55, 1992.

[61] J. MANDEL. Conditions de stabilite et postulat de Drucker. In J. Kravtchenko and P.M. Sirieys, editors, *Rheology and soil mechanics*, pages 58-68. Springer, Berlin, 1966.

[62] A. MARCHAND and J. DUFFY. An experimental study of the formation process of adiabatic shear bands in a structural steel. *J. Mech. Phys. of Solids*, 1987.

[63] J. C. NAGTEGAAL. On the implementation of inelastic constitutive equations with special reference to large deformation problems. *Comp. Meth. Appl. Mech. Eng.*, 33:469-486, 1982.

[64] A. NEEDLEMAN. Material rate dependence and mesh sensitivity in localization problems. *Comp. Meth. in Appl. Mech. Eng.*, 67:69-85, 1988.

[65] M. K. NEILSEN and H. L. SCHREYER. Bifurcation in elastic-plastic materials. *Int. J. Solids Struc.*, 30(4):521-544, 1993.

[66] M. ORTIZ, Y. LEROY, and A. NEEDLEMAN. A finite element method for localized failure analysis. *Comp. Meth. in Appl. Mech. Eng.*, 61, 1987.

[67] H. OUMERACI. Review and Analysis of Vertical Breakwater Failures. MAST G6-S/Project 2, wave impact loading on vertical structures, Franzius-Institute, 1992.

[68] J. PAMIN. *Gradient-dependent plasticity in numerical simulation of localization phenomena*. Delft University Press, 1994.

[69] C. V. PAO. The existence and stability of the solutions of nonlinear operator differential equations. *Arch. Rat. Mech. Analysis*, 35:16–29, 1969.

[70] C. V. PAO and W. G. VOGT. On the stability of nonlinear operator differential equations, and applications. *Arch. Rat. Mech. Analysis*, 35:30–46, 1969.

[71] A. PAZY. *Semigroups of linear operators and applications to partial differential equations*. Springer-Verlag, 1983.

[72] P. PERZYNA. The constitutive equations for rate sensitive plastic materials. *Quart. Appl. Math.*, 20:321–332, 1963.

[73] P. PERZYNA. Fundamental problems in viscoplasticity. In C.-S. Yih, editor, *Advances in Applied Mechanics*, volume 9, pages 243–377. Academic Press, 1966.

[74] P. PERZYNA. Thermodynamic theory of viscoplasticity. In *Advances in Applied Mechanics*, volume 11, pages 313–354. Academic Press, 1971.

[75] P. PERZYNA. Constitutive equations of dynamic plasticity. In D. R. J. Owen, E. O nate, and E. Hinton, editors, *Computational Plasticity, Fundamentals and Applications*, pages 483–508, Swansea, 1992. *Barcelona, April 6-10*, Pineridge Press.

[76] P. PERZYNA. Analysis of the fundamental equations describing thermoplastic flow process in solid body. *Arch. Mech.*, 43:287–296, 1993.

[77] P. PERZYNA. Instability phenomena and adiabatic shear band localization in thermoplastic flow processes. *Acta. Mech.*, 94:1–31, 1994.

[78] P. PERZYNA. Constitutive modelling of dissipative solids for localization and fracture (single cristals and polycristalline solids). *Proceedinngs of Advanced School on Localization and fracture phenomena in inelastic solids, Udine (this volume)*, 1997.

[79] S. PIETRUSZCZAK and Z. MRÓZ. Numerical analysis of elastic–plastic compression of pillars accounting for material hardening and softening. *Int. J. Rock Mech. Min. Sci. Geomech.*, 17:199–207, 1980.

[80] S. PIETRUSZCZAK and Z. MRÓZ. Finite element analysis of deformation of strain–softening materials. *Int. J. Num. Meth. in Eng.*, 17:327–334, 1981.

[81] M. RENARDY and R.C. ROGERS. *An Introduction to Partial Differential Equations*. Springer-Verlag, New York, 1992.

[82] J. R. RICE. The localization of plastic deformation. In W. T. Koiter, editor, *Theoretical and Applied Mechanics*, pages 207–220. North-Holland Publishing Company, 1976.

[83] J. W. RUDNICKI and J. R. RICE. Conditions for the localization of deformations in pressure–sensitive dilatant meterials. *J. Mech. Phys. of Solids*, 23:371–394, 1975.

[84] L. J. SLUYS. *Wave propagation, localisation and dispersion in softening solids.* Dissertation, Delft University of Technology, Department of Civil Engineering, 1992.

[85] L. J. SLUYS. Different regularization methods for solution of the initial boundary value problems (geotechnical materials). *Proceedinngs of Advanced School on Localization and fracture phenomena in inelastic solids, Udine (this volume)*, 1997.

[86] L. J. SLUYS, J. BLOCK, and R. de BORST. Wave propagation and localization in viscoplastic media. In E. Hinton D. Owen, E. Oñate, editor, *III Int. Conf. on Comp. Plasticity, Fundamentals and Applications, COMPLAS III, Barcelona, Spain, April 4-9, 1992*, pages 539–550, 1992.

[87] P. STEINMANN and K. WILLAM. Localization within the framework of micropolar elasto–plasticity. In *Advances in Continuum Mechanics.* Springer, 1991.

[88] H. P. STÜWE. Experimental aspects of crystal plasticity. *Proceedinngs of Advanced School on Localization and fracture phenomena in inelastic solids, Udine (this volume)*, 1997.

[89] F. TAVENAS and S. LEROUEIL. The behaviour of embankments on clay foundations. *Canadian Geotechnical Journal*, 17:236–260, 1980.

[90] F. TAVENAS, S. LEROUEIL, P. La ROCHELLE, and M. ROY. Creep behaviour of an undisturbed lightly overconsolidated clay. *Canadian Geotechnical Juornal*, 15:402–423, 1978.

[91] A. TIKHONOV and V. ARSENIN. *Methods for the solution of incorrect problems (in Russian).* Nauka, Moscow, 1979.

[92] C. TRUESDELL and W. NOLL. *The nonlinear field theories.* In Handbuch der Physik, Band III/3. Springer, Berlin, Heidelberg, New York, 1965.

[93] Z. WASZCZYSZYN. Computational methods and plasticity. Report lr-583, TU Delft, 1989.

[94] G. WEBER and L. ANAND. Finite deformation constitutive equations and a time integration procedure for isotropic, hyperelastic–viscoplastic solids. *Comp. Meth. in Appl. Mech. Eng.*, 79:173–202, 1990.

[95] G. WEBER, A. M. LUSH, A. ZAVALANGOS, and L. ANAND. An objective time-integration procedure for isotropic rate–independent and rate–dependent elastic-plastic constitutive equations. *Int. J. of Plasticity*, 6:701–744, 1990.

[96] D. M. WOOD. *Soil Behaviour and Critical State Soil Mechanics*. Technical report, Cambridge University Press, 1990.

[97] N. ZABARAS and A. F. M. ARIF. A family of integration algorithms for constitutive equations in in finite deformation elasto-viscoplasticity. *Int. J. Num. Meth. in Eng.*, 33:59–84, 1992.

[98] H. M. ZBIB and J. S. JUBRAN. Dynamic sheat banding: A three–dimensional analysis. *Int. Jour. of Plasticity*, 8:619–641, 1992.

[99] M. ŻYCZKOWSKI. *Combined Loadings in the Theory of Plasticity*. PWN-Polish Scientific Publishers, 1981.